32nd Ethylene Producers Conference 2020

AA009296

Topical Conference at the 2020 AIChE Spring Meeting and 16th Global Congress on Process Safety

Online
17-21 August 2020

ISBN: 978-1-7138-2337-7

Printed from e-media with permission by:

Curran Associates, Inc.
57 Morehouse Lane
Red Hook, NY 12571

Some format issues inherent in the e-media version may also appear in this print version.

Copyright© (2020) by AIChE
All rights reserved.

Printed with permission by Curran Associates, Inc. (2021)

For permission requests, please contact AIChE
at the address below.

AIChE
120 Wall Street, FL 23
New York, NY 10005-4020

Phone: (800) 242-4363
Fax: (203) 775-5177

www.aiche.org

Additional copies of this publication are available from:

Curran Associates, Inc.
57 Morehouse Lane
Red Hook, NY 12571 USA
Phone: 845-758-0400
Fax: 845-758-2633
Email: curran@proceedings.com
Web: www.proceedings.com

TABLE OF CONTENTS

591157 Securing Enterprises via Cybersecurity Understanding, Resiliency, and Education

Abstract:

Securing Enterprises via Cybersecurity Understanding, Resiliency, and Education

Industry 4.0 initiatives are enabled by leveraging the deployment of information technologies into the operational technology environment. While a logical approach in defining the future of smart manufacturing, it remains fret with challenges and risks that must be mitigated assure enterprise resilience to cyber and cyber-physical attacks. As economies recalibrate in the aftermath of the current public health emergency, definitions of what constitutes "critical infrastructure" are broadening while, at the same time, attack surfaces and the threat landscape are expanding. This means companies must place a greater emphasis on taking the steps and making the investments necessary to overcome the challenges and navigate these challenges and risks to maximize their resilience and secure and protect their enterprises. This presentation will explore the link between cybersecurity and process safety in the context of a safety culture. It will also identify cybersecurity challenges in industry 4.0 initiatives, the integration of information technologies (IT) and operational technologies (IT), the delivery of training, and the need for credentialing and ongoing professional development.

Copyright © American Institute of Chemical Engineers. All rights reserved.

590220 Global Ethylene/Propylene: Refinery Scale Petrochemical Investments and the Impact on Global Markets

Abstract:

Previous announcements have become a reality concerning new refineries built on a chemicals platform with fuel as the byproduct. Long term these type of investments represent a trend that could truly change the market balance. In this presentation we will examine the differing motivations and technologies behind refinery scale olefin investments, how does North America fit into the supply/demand picture, and other long term factors that will play into petrochemical investment decisions.

Copyright © American Institute of Chemical Engineers. All rights reserved.

590229 Disruptive Trends and their Implications

Abstract:

Braskem, known as a leading innovator in petrochemical sustainability, also has a robust transformation agenda preparing for megatrends that could reshape the industry as we know it. In his talk, Kevin will share some real-world examples of exponential change and how disruptive it can be to the most well-established of industries. Exponential change is all around us, but we as humans are often terrible at perceiving and forecasting that change until large disruptions occur. Kevin will talk through four key megatrends impacting the petrochemical industry – Mobility, Energy, Digitization, and Sustainability – and give some examples of how these areas could be disrupted and how Braskem is preparing and embracing these changes.

Copyright © American Institute of Chemical Engineers. All rights reserved.

OPTIMIZING TURBOMACHINERY FOR MAXIMUM PROCESS PERFORMANCE: METHODOLOGY AND CASE STUDY FOR TURBOEXPANDER CONTROL IN ETHYLENE PLANT

Dr. Stefano Belloni, Compressor Controls Corporation, Des Moines, IA

Abstract

Turbomachinery is highly integrated into industrial plants, acting as the primary mover of key process units. Turbomachinery control directly impacts Key Performance Indicators (KPIs) including safety, process, reliability, production, yield, energy consumption, environmental emissions and maintenance costs. However, in existing assets, controls are often based on antiquated hydro-mechanical methods or simplistic implementations that significantly limit production rates, process availability, energy optimization and emission reduction efforts. Understanding and quantifying these limitations is key to justifying improvement projects for existing assets.

This paper presents a new methodology for executing turbomachinery optimization studies. Leveraging knowledge, advanced analytics tools and collaborative engineering processes, this method has proven effective in multiple installations, including the control optimization study of an ethylene plant's cryogenic expander-recompressors discussed in this paper. By analyzing the correlation between production yield and expander control design, CCC determined the most effective primary and constraint control objectives. Compressor dimensional or non-dimensional analysis then identified shortfalls in control system design and an optimal operating point. The use of advanced data analytics software in combination with thermo-properties database provides effective tools to study turbomachinery operation through extensive time periods.

The paper further presents CCC proven strategy for optimal control of cryogenic expanders in ethylene plants. Implemented with CCC's patented design, this coordinated control of guide vanes, the JT valve and recompressor antisurge valve maximizes production and economic return.

Introduction

Turbomachinery is highly integrated into industrial plants, acting as the primary mover of key process units in oil and gas applications ranging from upstream and midstream to refining and petrochemicals. In most of these industrial applications, turbomachinery is closely interconnected with process equipment that requires specific operating conditions, either in the form of operating set points or operating limits. Due to batch or changeover operations, process equipment can generate discrete instabilities in load, production rate, composition and state variables.

Turbomachinery control is critical to ensuring the safe, reliable and optimized operation of both the machine itself and, most importantly, the connected process. Turbomachinery control directly impacts plant KPIs including safety, process reliability, production, yield, energy consumption, environmental emissions and maintenance costs.

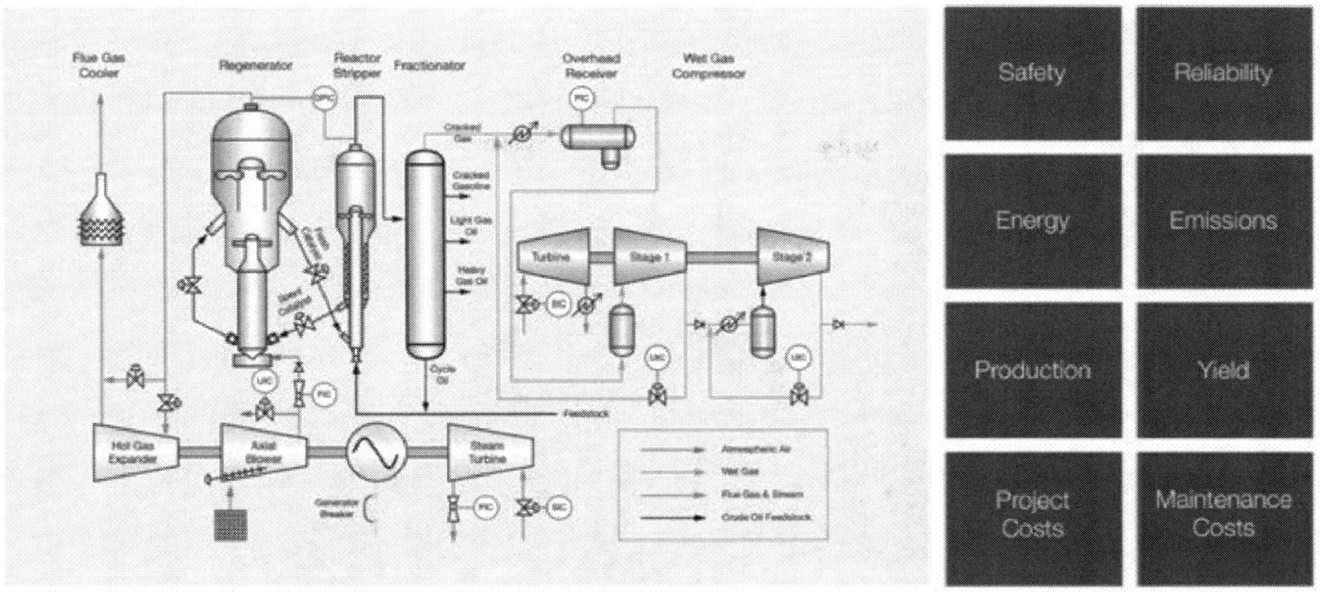

Figure 1: Values of turbomachinery optimization for the process unit

Effective turbomachinery control has the potential to optimize plant operations. However, common control methods—often in the form of antiquated hydro-mechanical strategies or simplistic digital implementations—can cause significant deficiencies in existing assets. In these situations, an operations team often controls the machine manually. While safety and uptime can still be achieved through manual control, this method can significantly affect production, energy consumption and other KPIs critical to plant profitability. Common signs of poor turbomachinery control design and commissioning include:

- Manual operation of valves, speed set point, inlet guide vanes and other control elements
- Unstable control of primary control variables
- Constant opening of antisurge recycle valves
- Vibrations during defined operating conditions
- High safety control margins due to trip conditions
- Unplanned trips
- Excessive wear on mechanical components
- Antisurge recycle valves plugged with gas condensates
- Complex manual operations for machine startup
- Coast-down surging
- Instabilities or surging due to variations in gas composition
- Driver overload and trips

Beyond these fully visible issues, a more detailed analysis typically uncovers additional deficiencies that are less evident but no less critical to turbomachinery optimization. Understanding and quantifying these limitations is key to justifying improvement projects for existing assets. Turbomachinery optimization studies are a valuable service for site personnel or headquarter management seeking reliable assessments to guide investment discussions.

Methodology

This paper proposes a systematic methodology for turbomachinery optimization studies combining turbomachinery control knowledge, specialized engineering tools and collaborative processes. CCC's methodology is composed of multiple sequential steps that, taken as a whole, have proven effective in multiple installations.

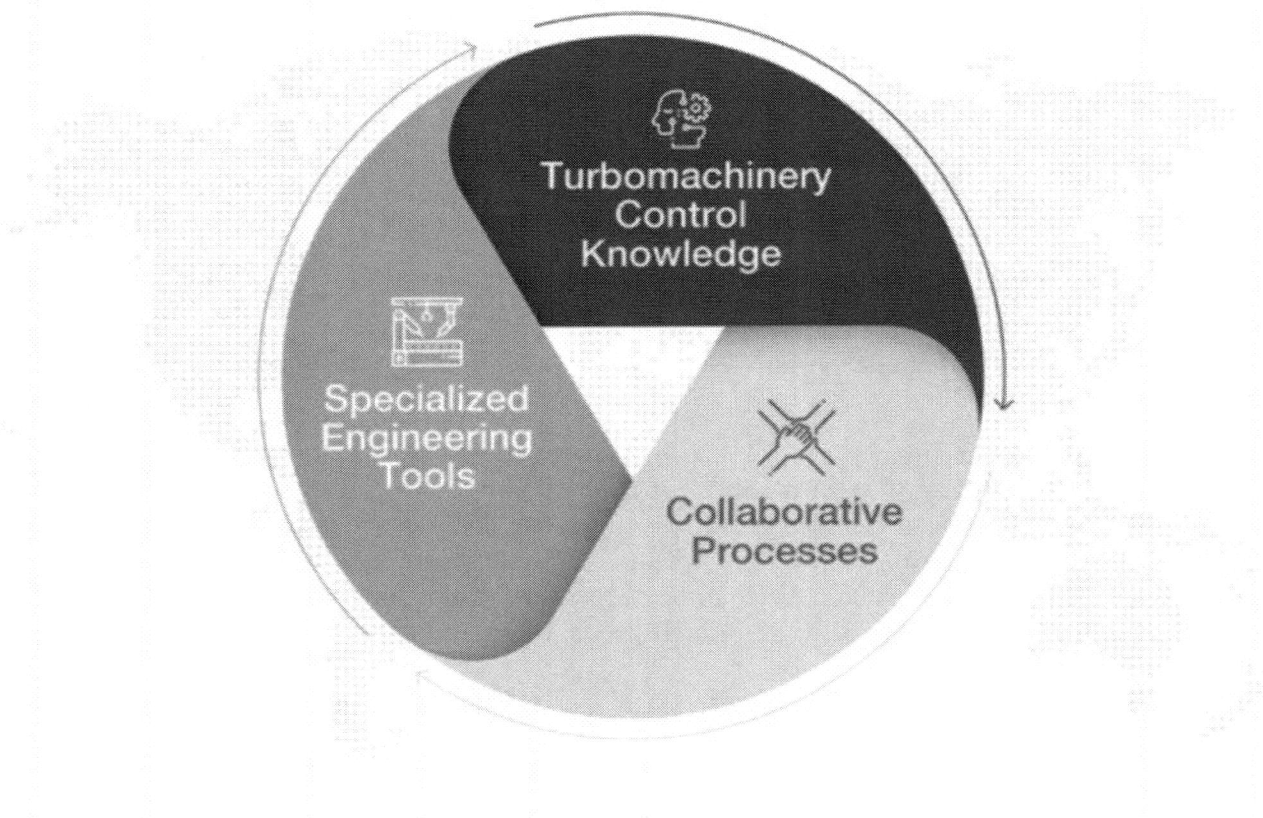

Figure 2: Turbomachinery optimization study methodology

1. **Data Collection.** Collect a variety of data before and during the site visit. This includes plant and equipment design data; plant production and economic data; and operational data for turbomachinery and connected processes, extracted from the plant historian and typically spanning 12 months of operation.
2. **Site Visit and Stakeholder Engagement.** Conduct a site visit to discuss the plant's overall process control, production and economics, operational practices, shutdown history and maintenance records.

   This activity is conducted in close collaboration with a variety of disciplines on-site, including operation, process and technology, rotating equipment, control and instrumentation, and maintenance teams. This site visit is a crucial opportunity to collect information and viewpoints from all key stakeholders. Often, it also serves as a discovery and learning process for site personnel. During the site visit, stakeholders share and cultivate a clear understanding of the

KPIs and limitations of the process unit, and turbomachinery optimization teams collect any missing data required for their analysis.

3. **Engineering Analysis.** Conduct a thorough engineering analysis of operational data extracted from the historian over a 12-month period. It is critical to use operational data to confirm the stories, suggestions and knowledge of typical optimization strategies collected during site discussions.

   Traditional turbomachinery analysis involves determining the operating point position on dimensional or, preferably, adimensional coordinate sets. Adimensional coordinate sets are utilized for proper antisurge control, and proven engineering practices are established to calculate these variables. An example of an adimensional coordinate set for the plot of the compressor's surge limit line is presented in *Fig. 3*, based on reduced head and reduced flow. This analysis is beneficial for gauging the impact of changing gas compositions or suction conditions, and for correctly identifying the location of the compressor operating point. It also has proven effective in identifying any existing inadequacy of controls based on dimensional variables.

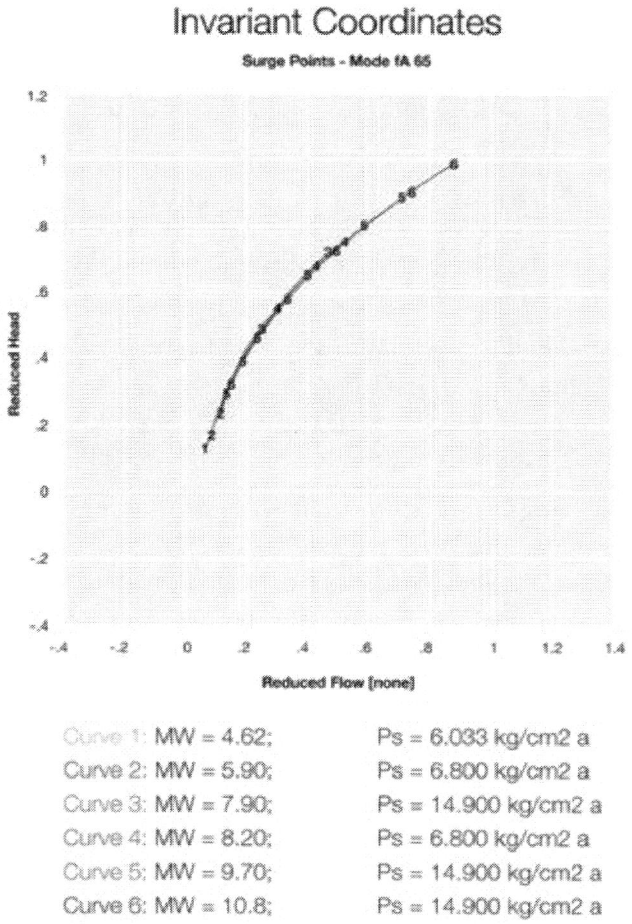

Figure 3: Adimensional coordinate set for plot of compressor surge limit line with varying mol weight

Traditional engineering tools have posed a challenge for teams conducting adimensional coordinate calculations on data sets spanning 12 months of operation. However, data analytics software has developed significantly over the past decade and, subsequently, its use is becoming more prominent in connection with plant historians. The functionalities developed for the analysis of time-series data are beneficial not only for real-time data but also for data extracted from the plant's historian. CCC has adopted Seeq advanced analytics software and collaborated with the Seeq engineering team to adapt it to turbomachinery analysis. By combining advanced analytics functions available in software, Equation of State packages and turbomachinery calculation formulas, CCC teams leverage established knowledge as they adopt new software tools.

Integrated with the Equation of State package, Seeq (or advanced analytics software) may be used to conduct time-series calculations of thermo-properties based on constant or time-variable gas composition. This enables access to built-in features for data cleansing, condition searching, profile searching, boundary condition definition, prediction, regression, statistical coefficient calculation and beyond. These tools enable the easier, more effective analysis of time-series data and support the identification of anomalies, off-design conditions, correlation between process variables and other key findings. During engineering analysis, optimized operating conditions can be estimated and quantified in terms of energy savings, production increases, process reliability increase, maintenance cost reductions and emission reductions. These improvements can then be translated into corresponding economic benefits on an annual basis. In certain cases, multiple improvement options are available for discussion with the end user.

4. **Report Delivery and Discussion.** Deliver the report and hold an on-site meeting to discuss the results with relevant personnel. This step is critical to verifying that the findings—which may or may not be expected—are reasonable and agreeable to the site team, which has daily experience with the design, operation and maintenance of the process unit.

   This on-site meeting is also an opportunity to select the modification options that deliver the most value to the plant while aligning with the end user's strategy for that machine and process unit. Once these options are selected, a site survey is performed to collect the information needed for the modifications and other elements of the full scope project.

5. **Scope of Modifications and ROI Calculations.** Define and document the scope of work, divided into work packages, and the budgetary estimate of the full project costs. Key project financial indicators are calculated based on the projected economic benefits and project costs. These can include metrics such as payback period, net present value or others required by the end user's project evaluation practice.

6. **Final Report Delivery.** The final report is delivered and presented to end user. The report is a valuable tool for presenting the investment opportunity to management teams.

Case Study

Cryogenic expanders play a key role in increasing the overall efficiency of an ethylene plant. Increased ethylene recovery is achieved through the deeper refrigeration of tail gas, while increased energy recovery is achieved through subsequent fuel gas recompression. The expander-recompressor is tightly integrated with the charge gas refrigeration train, absorber, demethanizer and fuel system.

The charge gas from the "hot" section is dehydrated and then cooled down, condensed and separated through a series of heat exchangers, coolers and separators. The stream is then fed into the absorber tower; here, methane and hydrogen at the top are separated from ethylene and higher components at the bottom. Through a lever control, liquid from the bottom of the absorber cascades into the demethanizer, where the final separation of methane and hydrogen from ethylene occurs. This tail gas is then expanded and cooled down to about -140 °C. This cryogenic stream provides refrigeration duty to the absorber tower and to heat exchangers of the incoming charge gas. The refrigeration duty provided by the expander diminishes the duty required from the propylene and ethylene compressors.

After exchanging heat, the tail gas from the expander is returned to the recompressor. The energy recovered by the expander drives the recompressor, allowing it to deliver the tail gas at the desired pressure needed for the fuel gas system.

In the case of an expander trip or when an expander reaches maximum capacity, JT valves are controlled with a slightly different set point. In a less efficient expansion process, gas from the JT valves is cooled only to about -100 °C, bypassing some of the cooling duty users.

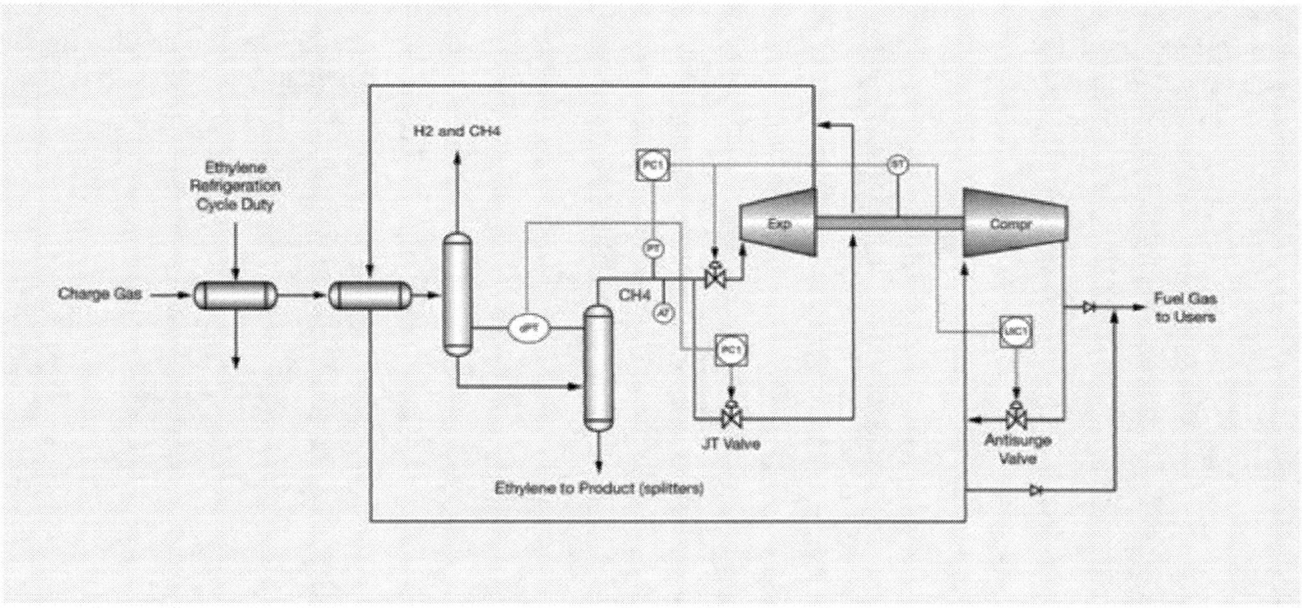

Figure 4: Cryogenic expander in ethylene process

In this case study, the expander and JT valves were controlled based on differential pressure between the absorber and demethanizer, with a slightly lower set point for the second PID control loop. The antisurge control was implemented as a simple flow controller operated in manual override mode. The system was equipped with a gas analyzer that continuously measured gas composition at key process locations.

Results

CCC analyzed data from the plant historian using a correlation matrix that identified whether there was positive, negative or no correlation between key process variables. By rigorously analyzing the input variables affecting the ethylene volume flow in the tail gas, CCC uncovered critical insights.

- The differential pressure between the absorber and demethanizer exhibited weak-to-no correlation.
- Plant production exhibited a positive correlation, indicating a higher increase of ethylene volume fraction at higher production rates.
- Absolute pressure at the demethanizer exhibited a negative correlation, indicating a lower ethylene volume fraction at a higher operating pressure.

To further analyze the relationship between ethylene volume fraction and absolute pressure at the demethanizer, CCC employed scatter plot and prediction functions to create a polynomial interpolation curve. This identified an optimized operating condition for minimizing the ethylene volume fraction in the tail gas, which when improperly controlled represents lost ethylene product. CCC proposed selecting this optimum pressure as the primary control variable for the expander and JT valve controls rather than the differential pressure between absorber and demethanizer, which the plant previously employed. Through further analysis, CCC identified that the latter would be better configured as a constraint control variable to allow a minimum differential pressure that, in turn, guaranteed optimal liquid flow from the bottom of the absorber to the demethanizer. This optimization is estimated to deliver 100,000 USD per year in improved ethylene production yield.

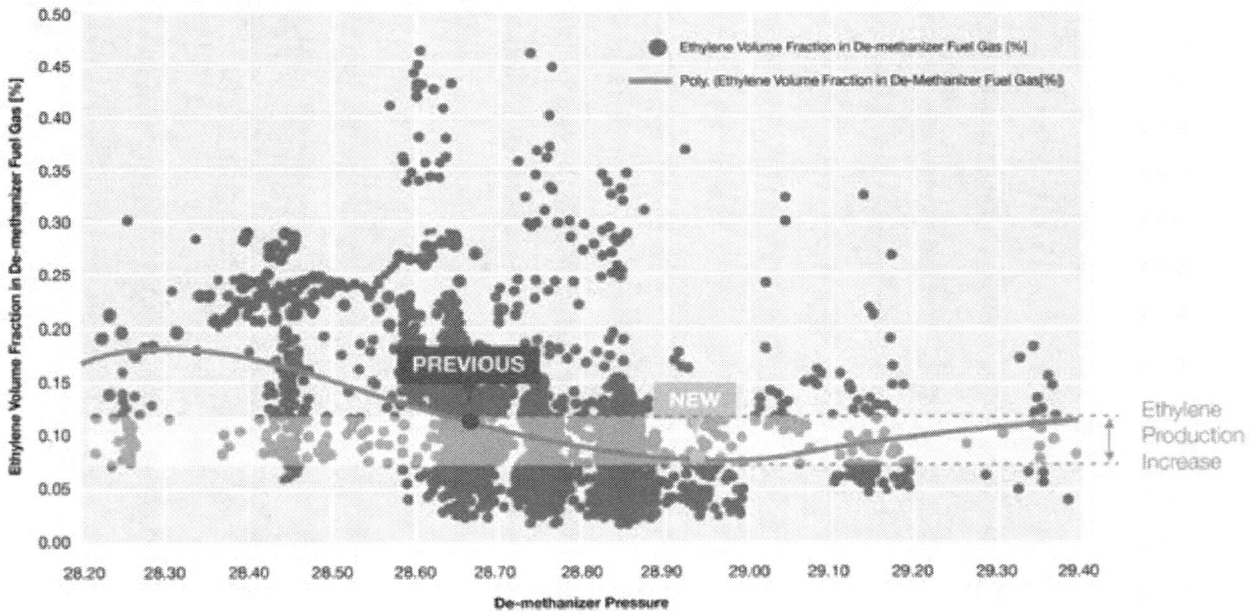

Figure 5: Improved ethylene recovery scatter plot

As a source of refrigeration duty, the turboexpander is significantly more efficient than the JT valve in lowering gas pressure and temperature, which it achieves through isentropic expansion. In this specific case study, the expander was subject to vibrations when moving above a certain speed. An Advanced Process Control (APC) was implemented to open JT valves when that speed was reached in order to avoid exceeding it. By calculating the head and flow through the machine during the 12 months of operation, the optimization study identified that the machine was often operating close to or into surge when in the lower speed range. The compression ratio on these machines ranges from 1.2 at minimum speed to 1.8 at maximum, and flow instabilities at low speeds are likely to come in the form of a rotating stall rather than a full surge. This was identified as a likely source of the vibrations, which would increase at a higher speed due to the rotating stall turning into a classic surge. A vibration analysis was not conducted at this time.

Drawing from this analysis, CCC proposed an optimized solution that implemented proper antisurge control, kept the compressor away from the surge area and removed the speed limitation implemented in the APC. By removing the speed limitation, flow capacity is estimated to increase by an average of 5%, corresponding to a 75kW increase in refrigeration duty. This, in turn, leads to reduced duty request at the ethylene refrigeration compressor for the cooling of the charge gas flow from the hot section. The annual savings in steam consumption at the ethylene compressor was valued at approximately 80,000 USD. These improvements are achievable by applying a specialized control solution for the stable, secure and efficient control of turboexpanders.

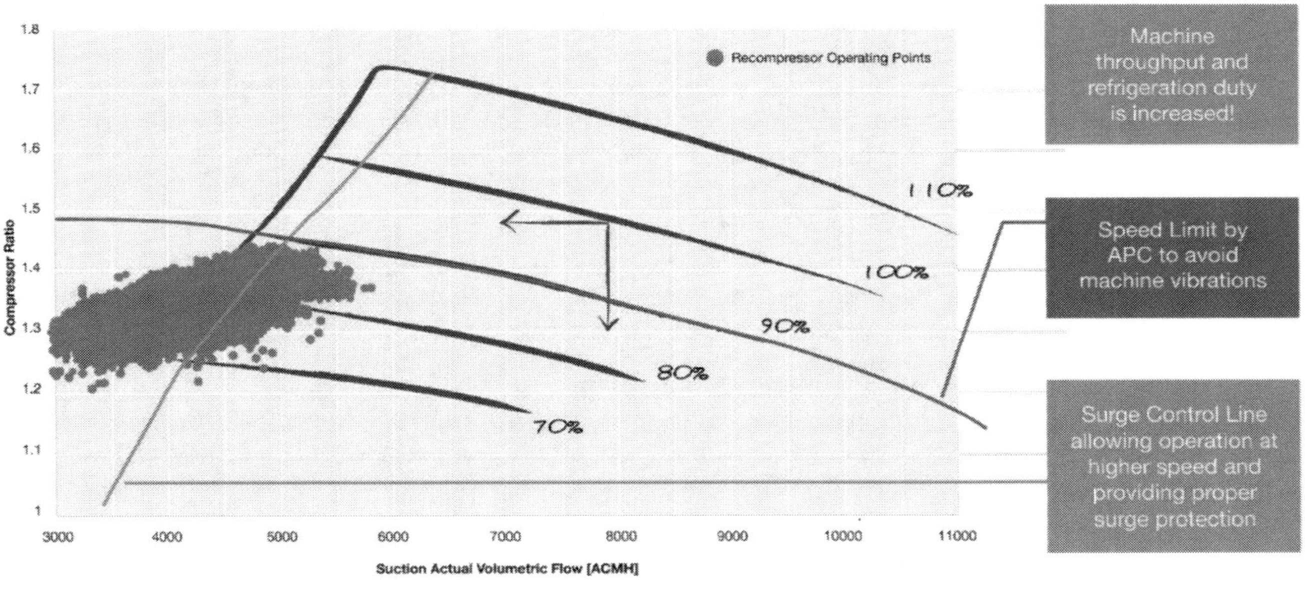

Figure 6: Increased recovery of refrigeration duty from the expander

In this situation, given the low compression ratio and the stable gas composition, it was possible to use dimensional compressor maps for proper visualization of the operating conditions. However, in other cases it could be necessary to use adimensional coordinates in accordance with the ASME PTC-10 performance test code or a company's standard practices for antisurge control system engineering.

Optimization Strategies

Cryogenic turboexpanders in ethylene and NGL plants have many similarities. In both cases, the objective is to separate heavier hydrocarbons from natural gas. Conventional methodologies for turboexpander control include the use of a simple PID loop with a split range for control of the IGV and JT valve; the limitation of maximum differential pressure across the expander rather than maximum speed; and the use of minimum flow control or maximum pressure differential for surge control.

Through engineering and commissioning hundreds of turboexpanders, CCC has developed and patented a number of best practices for control responses. Applicable to both NGL and ethylene plants, these strategies allow engineering teams to optimize control of the machine and the larger connected process—maximizing condensate production and economic return.

- **Separate control loops for IGV and JT valve**. Separate control loops allow for the dedicated tuning of each loop, which is necessary to account for the different response in each. It also allows the immediate and easy use of the JT valve opening to control rapid disturbances.
- **Pre-positioning JT valve upon expander trip.** JT valve pre-positioning is based on the equivalent capacity of the expander trip to the trip itself. It allows the JT valve opening to step to a particular position and, from there, begin closed-loop control. This minimizes the process disturbance and quickly transitions the flow from the expander to the JT valve.
- **Speed limitation by antisurge valve opening**. A turboexpander's speed is a result of the balance on the shaft between the power of the expander and the compressor load. Typically, the method for limiting expander speeds involves reducing the IGV opening and, as a result, reducing the flow through the expander. A more efficient solution is to first limit the speed of the expander through the opening of the antisurge valve, which in turn increases the load on the shaft. This enables the expander to continue operating at a higher IGV opening, which maximizes the flow through the expander and, therefore, condensate production and refrigeration duty recovery.
- **Over-speed avoidance by antisurge valve opening**. Over-speed is a very dangerous event for turboexpanders. Although over-speed trip protection is mandated to avoid this issue, open-loop control responses can be utilized to avoid reaching the speed that triggers the over-speed protection and trips the machine. By coordinating an open-loop response with a simultaneous step-close of the IGV and step-open of the antisurge valve, the control system can impact the power balance on both the driver and driven side to avoid reaching the over-speed trip. This over-speed prevention strategy leverages a combined, coordinated response to speed and antisurge control.
- **Coordinated sequencing during startup**. Some expander manufacturers require a startup sequencing of the expander from "0 Speed" to "Rated Speed," which involves following a specified sequence with ramp rates, idle speeds and critical speed band avoidance. The coordinated control of IGV, antisurge valve and JT valve is critical to achieving this sequencing while still maintaining the primary control variable. Much like an opera performance in which one character is the lead, one control loop takes control while the other is exiting from the closed-loop control.
- **Surge control based on adimensional coordinate sets**. The use of adimensional coordinate sets for antisurge control provides a more reliable control characterized by less variation to

changes in suction conditions or gas composition. The implementation of reliable and stable control avoids manual override of the antisurge control, which affects production and energy consumption.

Conclusion

CCC's turbomachinery optimization methodology includes specialized control responses cultivated over 45 years of experience and thousands of installations. Together, these control responses enable the optimization of production and refrigeration duty recovery for cryogenic expanders. By analyzing an existing installation using a proven methodology and advanced computational tools, stakeholders can identify and quantify the improvements needed to optimize their turbomachinery control and, in turn, justify improvement projects that maximize production and profitability.

References

1. Jacobson, Wayne, and Mirsky, Saul. "Control for cryogenic turboexpanders." Hydrocarbon Engineering. February 2014.
2. Mirsky, Saul. "Modern compressor and turboexpander controls maximize production of condensate at NGL facilities." Hydrocarbon Engineering. September 2000.

589074 Low Temperature Considerations for Centrifugal Compressor Casings in Ethylene Service

Abstract:

Centrifugal compressor casings with Ethylene process gas often have a Minimum Design Metal Temperature (MDMT) between -29°C and -115°C, and the design temperature may be even lower under certain conditions. Compressor casings are fabricated from various grades of carbon steel, low-alloy steel or stainless steel and follow the guidelines set forth by API 617. The material selection and the fabrication procedures utilized for is heavily dependent upon the Minimum Design Metal Temperature as this is the most critical and demanding design parameter of an ethylene compressor.

This presentation will review the low temperature limitations of steel alloys used for ethylene compressor casings, the testing which is used to qualify these materials for low temperature service, and the relative commercial impact of the alloy selection

Copyright © American Institute of Chemical Engineers. All rights reserved.

AIChE Paper Number: XXXX

User Friendly Steam Turbine Design to Ensure Easy Operability

Takuro Koda takuro_koda@compressor.mhi.co.jp
Team Leader, Turbine Engineering & Design Section, Mitsubishi Heavy Industries Compressor Corporation, Hiroshima, Japan. He has 14 years of experience as a turbine engineer. Mr. Koda graduated from Kyoto Institute of Technology.

Akito Kunimoto akito_kunimoto@compressor.mhi.co.jp
Engineer, Turbine Engineering & Design Section, Mitsubishi Heavy Industries Compressor Corporation, Hiroshima, Japan. He has 11 years of experience as a turbine engineer. Mr. Kunimoto graduated from Miyajima Technical High School.

Yusuke Oishi yusuke_oishi@compressor.mhi.co.jp
Engineer, Turbine Engineering & Design Section, Mitsubishi Heavy Industries Compressor Corporation, Hiroshima, Japan. He has 11 years of experience as a turbine engineer. Mr. Oishi graduated from Tokuyama College of Technology (Mechanical & Electrical Engineering).

Prepared for Presentation at the 2020 Spring National Meeting
Houston, Texas, March 30th, 2020

AIChE and EPC shall not be responsible for statements or opinions contained in papers or printed in its publications

User Friendly Steam Turbine Design to Ensure Easy Operability

Takuro Koda takuro_koda@compressor.mhi.co.jp
Team Leader, Turbine Engineering & Design Section, Mitsubishi Heavy Industries Compressor Corporation, Hiroshima, Japan. He has 14 years of experience as a turbine engineer. Mr. Koda graduated from Kyoto Institute of Technology.

Akito Kunimoto akito_kunimoto@compressor.mhi.co.jp
Engineer, Turbine Engineering & Design Section, Mitsubishi Heavy Industries Compressor Corporation, Hiroshima, Japan. He has 11 years of experience as a turbine engineer. Mr. Kunimoto graduated from Miyajima Technical High School.

Yusuke Oishi yusuke_oishi@compressor.mhi.co.jp
Engineer, Turbine Engineering & Design Section, Mitsubishi Heavy Industries Compressor Corporation, Hiroshima, Japan. He has 11 years of experience as a turbine engineer. Mr. Oishi graduated from Tokuyama College of Technology (Mechanical & Electrical Engineering).

ABSTRACT

The operability and working environment are critical items for steam turbine design when evaluating the turbine life cycle in petrochemical plants. These critical items are just as important, if not more important, than the efficiency, reliability, and cost of the turbomachinery. This paper will introduce key technology features on steam turbines to ensure "User Friendly" experience.

Compact design and easy maintenance for high pressure components, suitable work space around the steam turbine from the point of view of HFE (human factor engineering), and HSSE (Health, safety, security and environment) are all key points when reviewing steam turbine designs. TV-GV combined type (Steam Valves) design is shown as a compact high pressure component solution to obtain larger work space while maintaining safe operations in this presentation. Installation area can be reduced 40% from the conventional type. Also the number of parts can be reduced 50% when compared to the conventional type which allows for easier maintenance and spare parts management.

Automatic operation of steam turbine systems (incl. condensate system, auxiliary system) is required in order to minimize or eliminate field operations. Examples such as Speed controlled by governor from 0 rpm, automated engage/disengage turning device, automated drain valve, automated gland condenser and automated air evacuation system will be explained throughout the paper.

Noise attenuation technology is important when considering the field working environment and the overall surroundings of petrochemical plants. Special insulation for high pressure casings and special acoustic technology for the low pressure casing are shown as the preferred technology to achieve low noise levels while keeping high reliability. The results of these technologies will be shown through verification tests within this paper.

INTRODUCTION

During the past decade the capacity of ethylene plants has continuously increased. To meet this capacity increase it has been necessary to apply larger sub-components including large flow governing valves, longer blades, and longer bearing span rotors.

In these larger plants customers are requiring compact designs and easy maintenance from the point of view of HFE (human factor engineering) and HSSE (Health, safety, security and environment). In addition, customers are requiring automatic operation in order to minimize or eliminate field operations as well as minimizing steam turbine noise when considering field working environment and the overall surroundings of petrochemical plants.
The authors introduce the technology development for the following items in this paper.

(1) Compact design and easy maintenance for the high pressure components
(2) Automatic operation system
(3) Noise attenuation structure

Ethylene Plant Trend

The capacities of ethylene plants have increased continuously over the last decade as shown in Figure-1. Consequently, flow rates in the high-pressure and low-pressure stages of the steam turbine driver have also increased. To meet this capacity increase it is necessary to apply larger sub-components including large flow governing valves, longer blades, and longer bearing span rotors. Compressor and turbine trains with capacities in excess of 1.8 MMTA have already been designed, manufactured, and are successfully operating in existing plants.

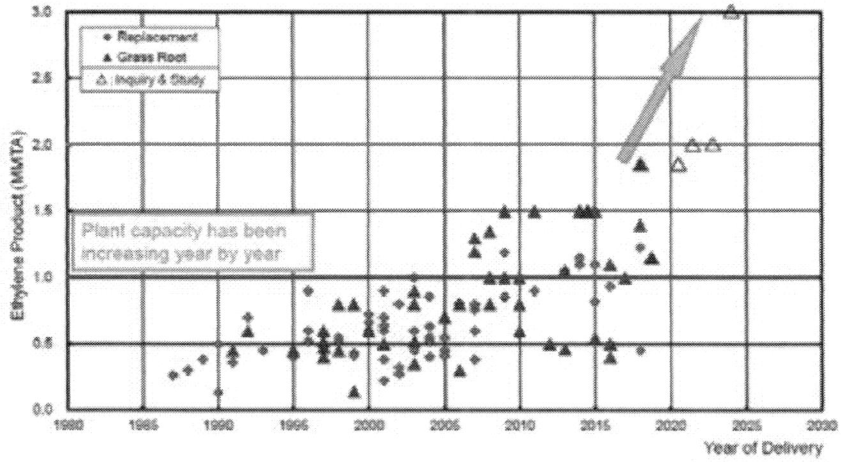

Figure.1 Ethylene Plant Capacity Increase Supported by Upgraded Equipment and Technology

Figure 2 shows a recent Cracked Gas compressor drive steam turbine for Mega ethylene crackers. This steam turbine has more than 70MW output power. As you can see there are dual trip throttle valves for large HP steam flow.

Figure.2 Steam Turbine for Mega Crackers (1.3MMTA)

Every compressor's performance, including rotor dynamics, is measured and verified as a part of the manufacturing process. In addition, the steam turbine and compressors can be coupled and the complete train is tested to confirm combined mechanical performance, such as vibration and bearing condition, on a large test stand prior to shipment. Figure 3 shows a 1.5MMTA class mega ethylene CGC train in the same configuration condition as the site. Through these processes the OEM can guarantee high quality machines will be provided to the clients.

Figure.3 Steam Turbine for Mega Crackers (1.5MMTA)

Challenging Technologies for Critical Components

To produce the increase power the required inlet steam flow (HP section) and exhaust steam flow (LP section) become larger and a larger component size for the flow path is necessary. Figure.4 shows key technologies in the steam turbine design in order to meet the requirement for large plants. Optimized impulse blades are applied for the inlet and middle stage flow paths in order to increase efficiency and minimize the thrust load due to differential pressure across each blade inlet and outlet.

The last 3 stages in the LP section utilize a z-lock integral shroud and high rigidity blade profiles in which reduce resonance vibration stresses in high variable speed and to increase stage efficiencies.

A critical design concept in modern rotor design is to increase rigidity as much as possible by optimizing blade stage number, bearing span, and HP/LP section efficiency at the required operating speed as well as applying high dynamic and static performance of larger sized journal bearings. A high rigidity rotor can reduce vibration response due to unbalance and increase vibration stability. Also, for large size journal bearings, self-aligning bearing pad to support rotor is applied to get the high dynamic and static performance.

Recently operating intervals of over 8 to 10 years have been required to meet customer demand. For this requirement, several new technologies have been developed and applied to existing machines.

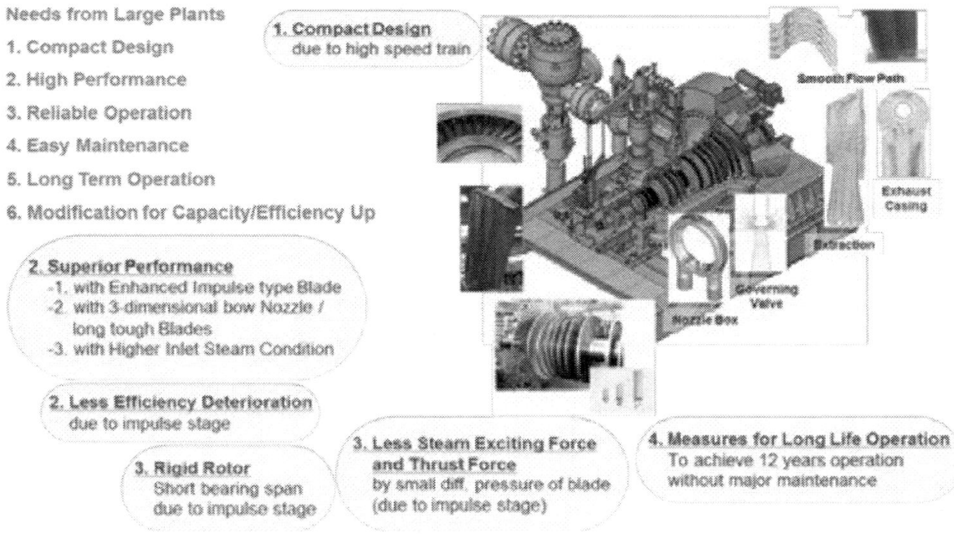

Figure.4 Steam Turbine Key Technologies

Compact Design and Easy Maintenance

Regarding compact design and easy maintenance for high pressure components a suitable work space is required around the steam turbine from the point of view of HFE (human factor engineering) and HSSE (Health, safety, security and environment). Therefore a compact design is one very important factor. Figure 5 shows TV-GV combined type (Steam Valves) is an example where compact high pressure components are utilized to obtain the larger work space. Installation area can be reduced approximately 40 % from the conventional type. In addition total weight is also an important factor. For TV-GV combined type (Steam Valves), the total weight can be reduced approximately 20% versus the conventional type.

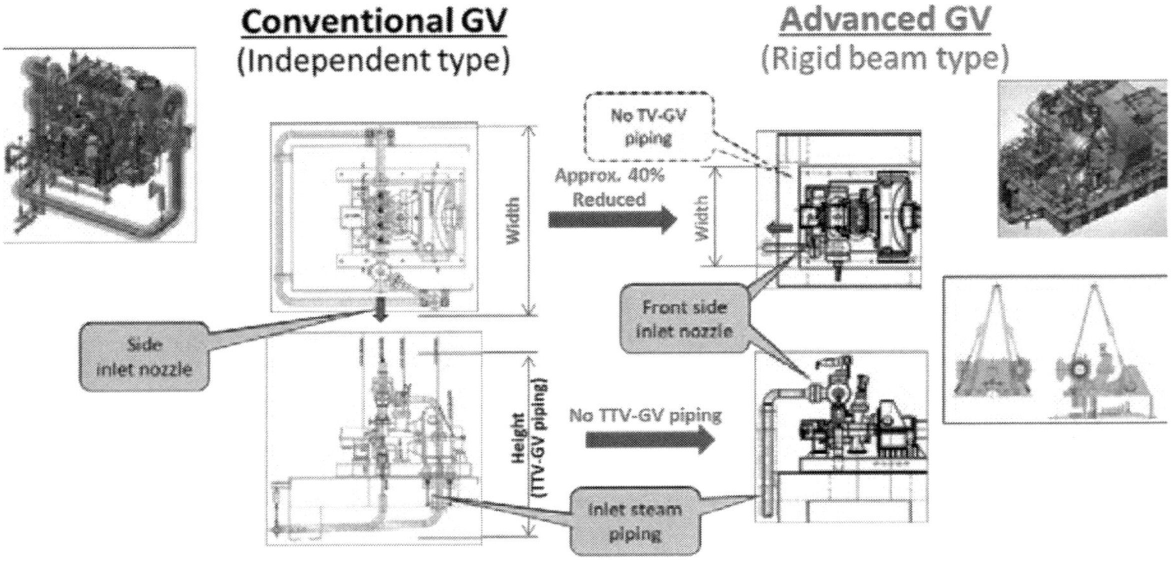

Turbine size [Width]	-	Approx. 40% reduced
Total weight	-	Approx. 20% Reduced

Figure.5 Comparison of Advanced and Conventional types

Shown in Figure 6 the number of parts can be reduced 50% when compared to the conventional type which allows for easier maintenance and spare parts management.

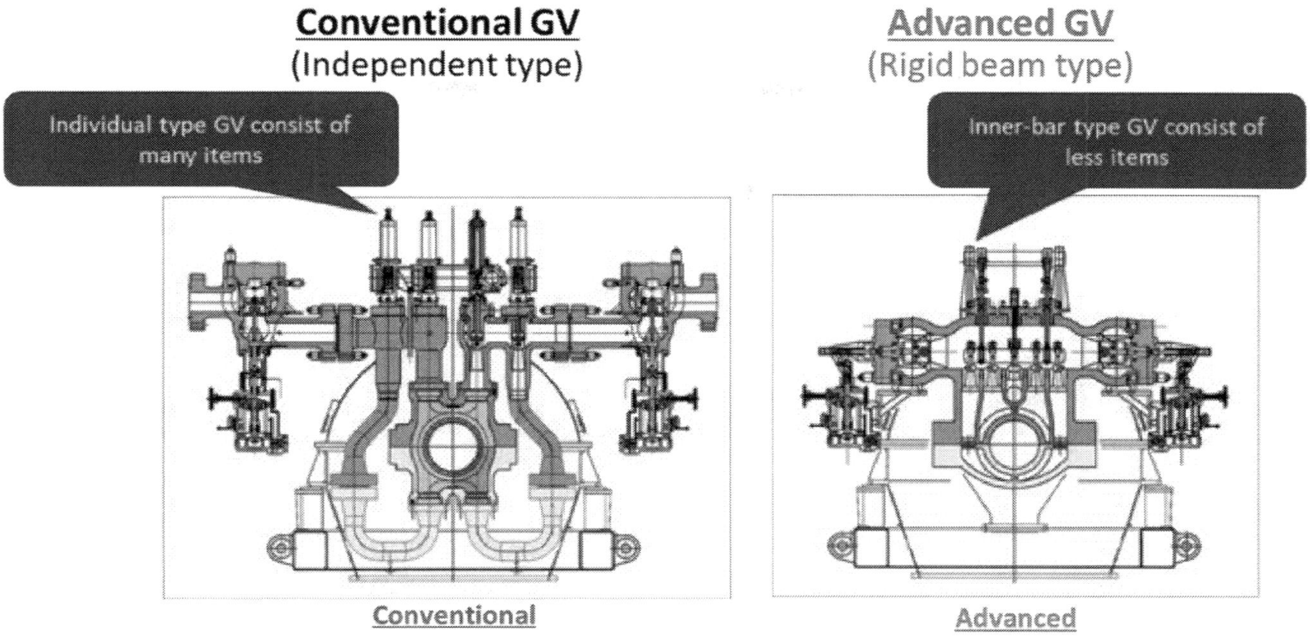

Parts Number	-	**Approx. 50% reduced**

Figure.6 Reduction of steam turbine part number for easy maintenance

Automatic Operation

Recently customers are requesting automatic operation for steam turbine systems due shortage of manpower and reduction skilled operations personnel. Automatic operation of steam turbine systems (incl. condensate system, auxiliary system) is required in order to minimize or eliminate field operations. In steam turbine operations there is speed control by governor from 0 rpm, automated engage/disengage turning device, automated drain valve, automated gland condenser and automated air evacuation system. Figure 7 shows full automatic operation of steam turbine including auxiliary system, semi-automatic operation and manual operation. Each operation mode can be customized according to customer requirement.

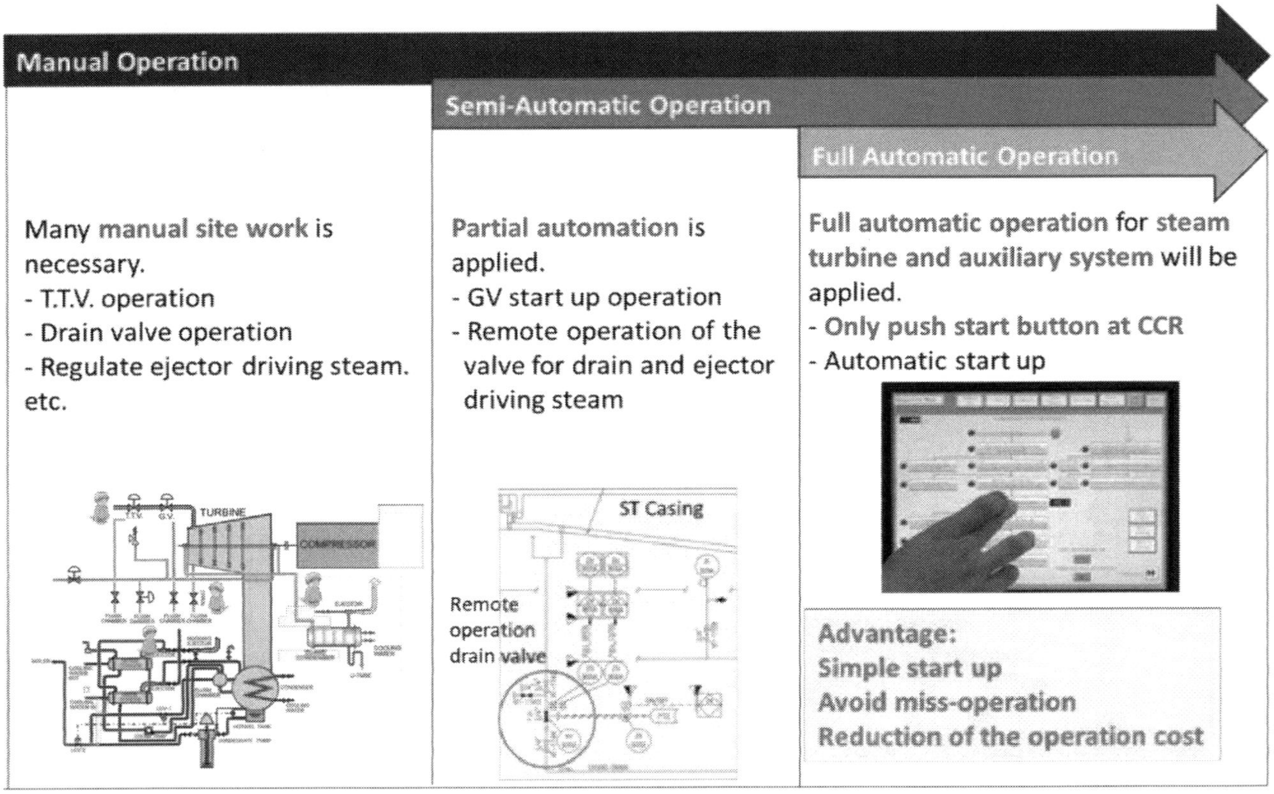

Figure.7 Full automatic operation of steam turbine including auxiliary system

Figure 8 shows 3D overview and outline of full automatic operation. One of the biggest advantages is to avoid improper operation due to human error. Also, in Figure 9, you'll find a standard sequence for full automatic operations.

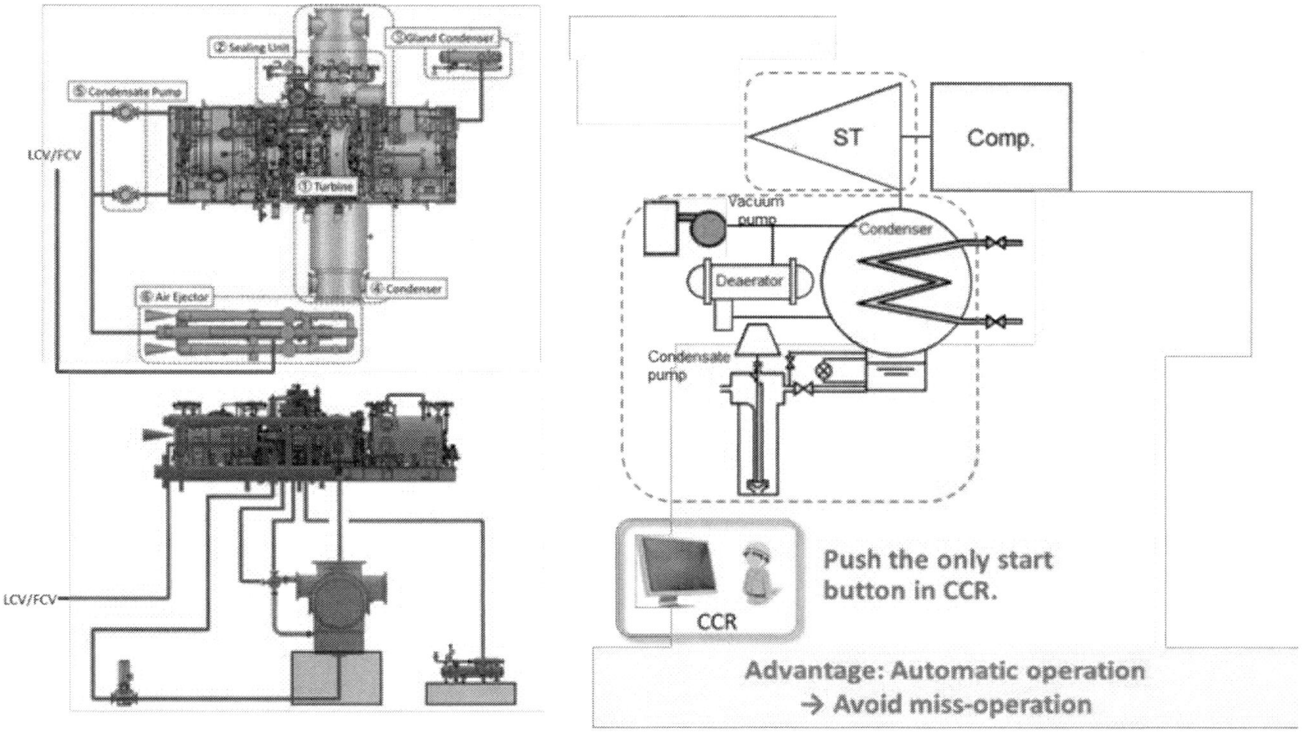

Figure.8 3D overview of full automatic operation

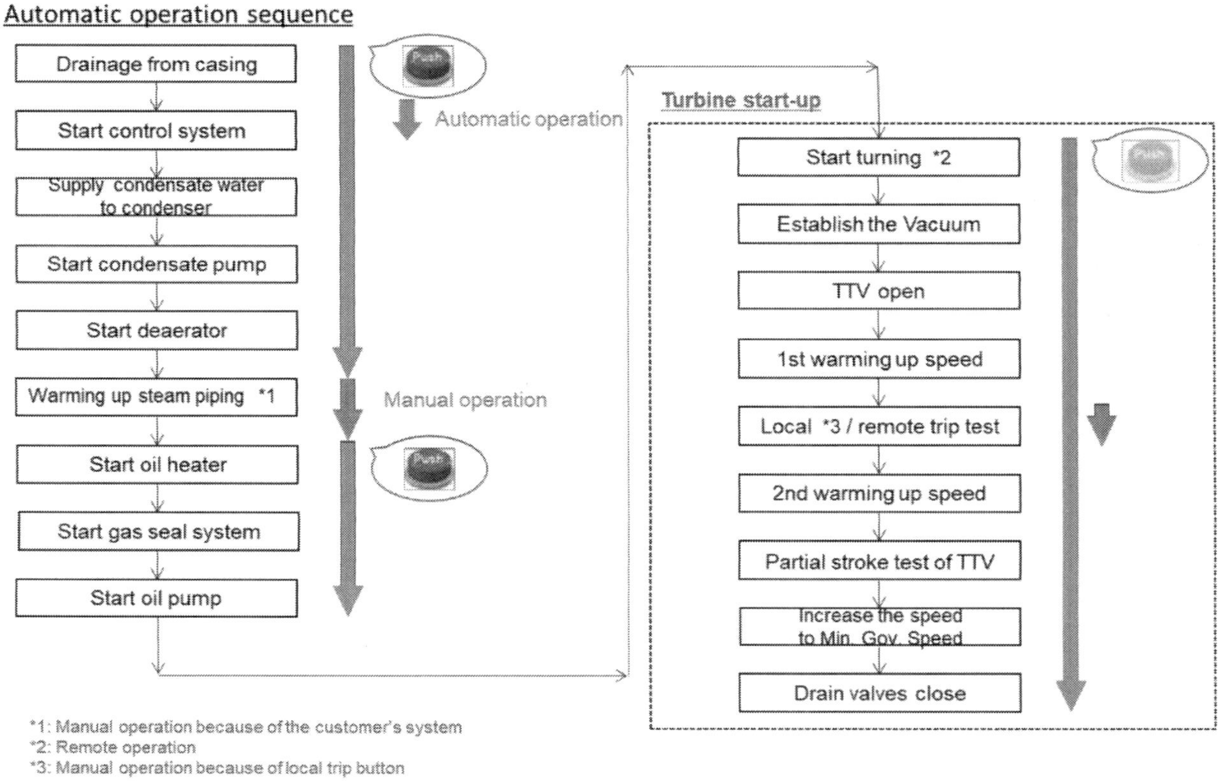

Figure.9 Full automatic operation sequence

Figure 10 shows the verification test in full automatic operation. These automatic steam turbine systems (incl. condensate system, auxiliary system) were successfully verified without any technical or safety issues.

Schematic of the verification test

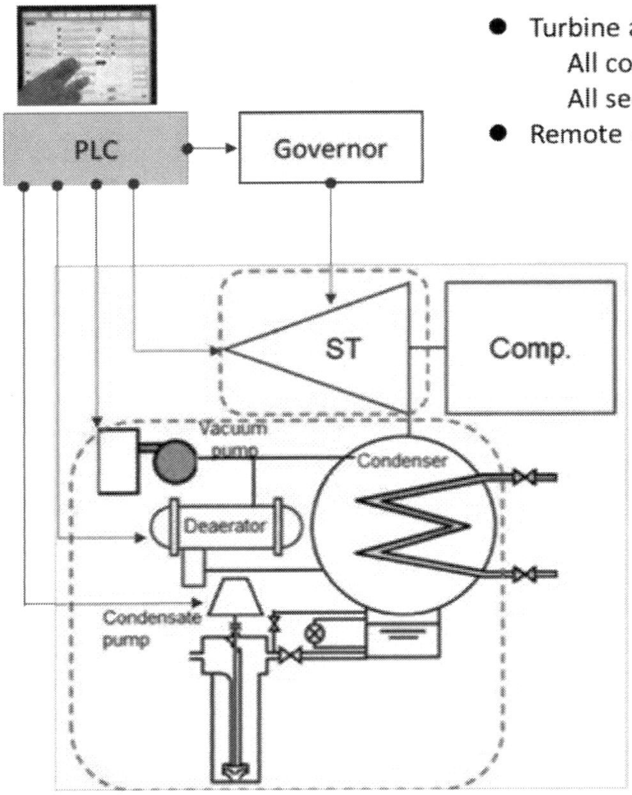

Function of the full automatic / remote operation

- Turbine and condensate system can be operated by PLC.
  All components are connected to PLC.
  All sequence from start up to shut down is controlled by PLC.
- Remote or Automatic operation can be selected.

#8 shop test turbine

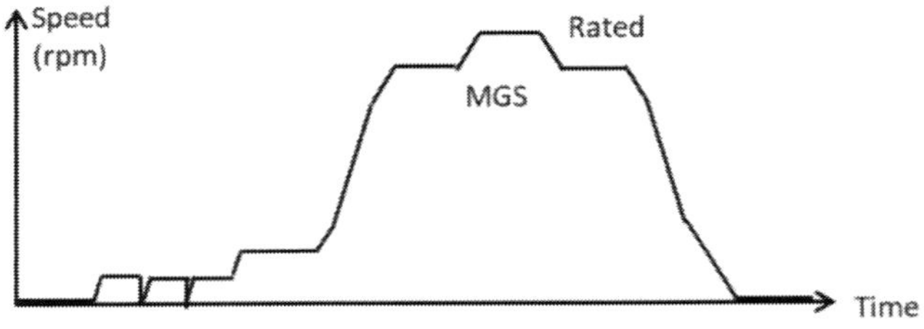

Verification test results for full remote operation

<Judgement>
Accepted.
The followings were confirmed.

- **Turbine and condensate system could be operated** by PLC (remote operation, no site operator).
- All sequence from start up to shut down could be controlled by PLC.

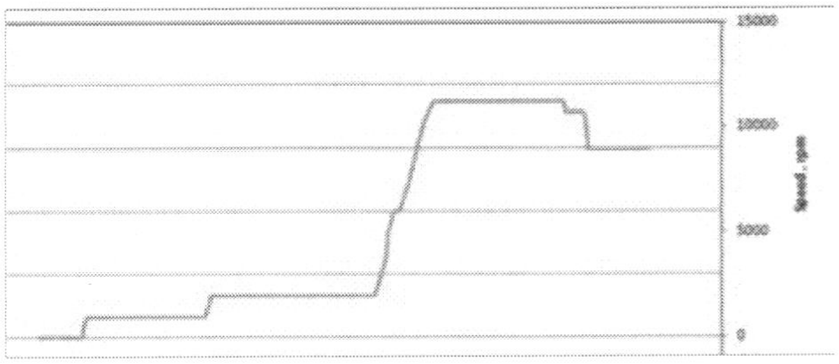

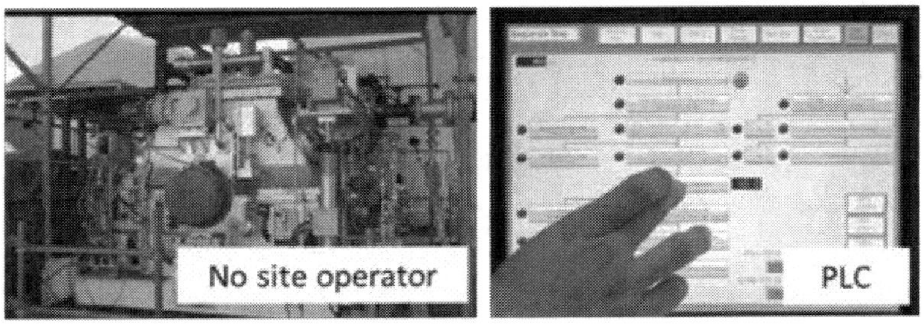

Figure.10 Verification test result in full automatic operation

Noise attenuation structure

Noise attenuation technology is important when considering field working environment and the overall surroundings of petrochemical plants. Special insulation for high pressure casings and special noise attenuation structure for low pressure casings are shown in Figure 11 and Figure 12. The key technologies in both applications assist in achieving lower noise levels while keeping high reliability.

Figure.11 Advanced noise insulation for high pressure portion

Figure.12 Noise attenuation structure for low pressure portion

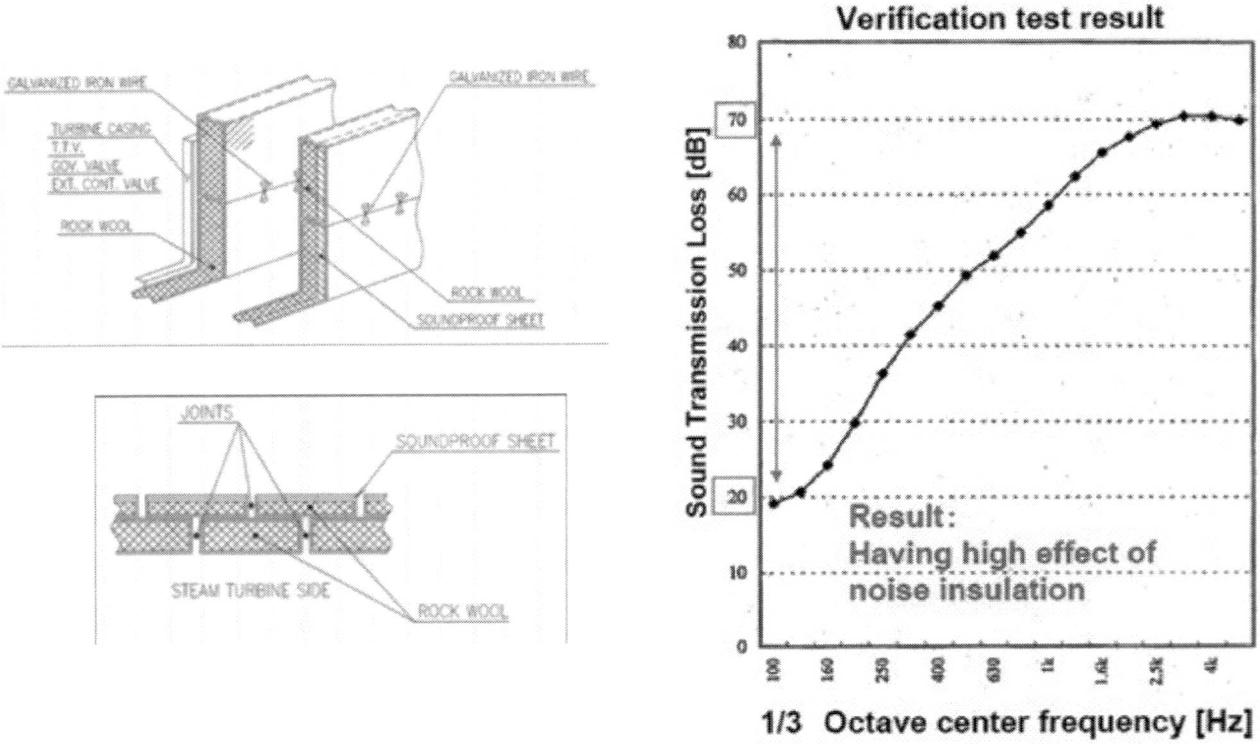

b) Performance of advanced type for high pressure portion

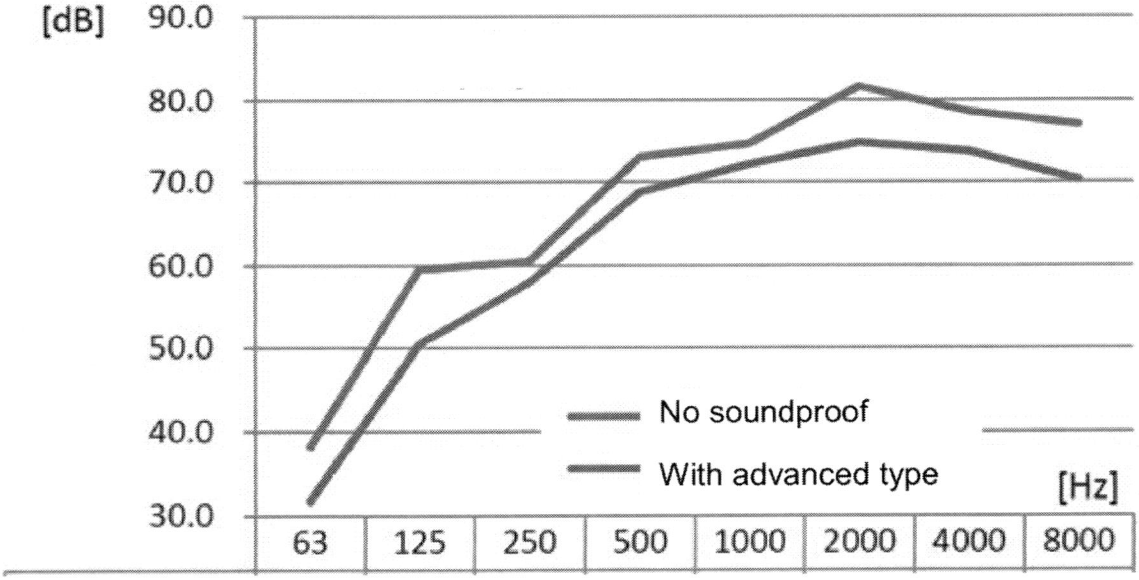

a) Performance of advanced type for low pressure portion

Figure.13 Verification test to noise reduction

Conclusion

1. From the point of view of HFE (human factor engineering) and HSSE (Health, safety, security and environment) a compact design with easy maintenance is one of the most important factors. Advanced TV-GV combined type (Steam Valves) has been successfully designed to obtain the larger work space as well as reduce the number of parts by up to 50% which can allow for easier maintenance and spare parts management.
2. Full automatic operation of steam turbine systems (incl. condensate system, auxiliary system) has been verified through actual products/projects. This solutions assists in minimizing field operations.
3. Noise attenuation technology is important when considering the field working environment and the overall surroundings of petrochemical plants. Each advanced type noise attenuation structure for both high and low pressure portion helped achieve low noise levels while keeping high reliability.

Reference

Satoshi Hata, et al, 2016, ATPS-Tutorial 9 _" Technical Challenges for Compressor and Steam Turbine for Efficient and Sustainable Operation in Mega Ethylene Plants"

589775 Data Infrastructure As Foundation to Operational Excellence and Digital Transformation

Abstract:

Is technology a solution looking for a problem? A successful digital transformation builds on the foundations of Operational Excellence. Innovations in the technology we use further these initiatives. The building blocks of these improvements are rooted in the data we collect, compile, and contextualize into operational intelligence. Successful implementations identify desired outcomes and build the data and technical infrastructures to achieve them.

A structured Operational Excellence program begins with clear definitions for:

- What's important?
- Who's responsible for action, who's accountable for success?
- How is success measured?
- How are excursions identified, monitored, and corrected?
- What communications are necessary to ensure improvement?
- What skills and tools are necessary to continuously improve?

Compelling events for transformation may include regulatory compliance, hazard recognition, sustainability, cost controls, quality improvements, and asset reliability as examples. The data required covers the spectrum of real-time operational data, environmental monitoring, quality data, and asset information. A connected infrastructure runs from edge connected IoT devices to the cloud, sharing information across individual assets, plants, enterprises, and communities of connected services.

However, the final transformation does not have to be the first implementation. Like any solution, digital tools evolve as our needs and understanding of the problem change. Our use of the data infrastructure will necessarily evolve by:

- Collecting data into the data historian
- Contextualizing an operational "chart of accounts"
- Integrating with external data sources such as LIMS, ERP, CMMS, etc.
- Formulating streaming real-time analytics, first law calculations, and KPI's
- Detecting and notifying on errors and events
- Leveraging complex analytic engines to develop predictive and prescriptive models

Data should be collected once, as close to the source as possible. The data infrastructure will interface with these source systems. As part of the transformation effort, organizations evaluate missing data for the potential value of the information against the expense of collection. Where possible, organizations can leverage IoT technologies or collect manually.

Demonstrated use cases include monitoring site heat index to prevent heat stress, extrapolating corrosion curves to facilitate mechanical integrity programs, performing both golden batch and multivariate analyses, and collecting asset utilization data for reporting among others. Successful digital transformations provide competitive advantages for these enterprises. We believe people with data can transform our world.

Copyright © American Institute of Chemical Engineers. All rights reserved.

590961 How to Realize Your Digital Transformation

Abstract:

Today, few companies doubt the benefits of Industry 4.0. It is relatively easy to talk about the need for a digital transformation from a high level, but when it comes to planning and executing a specific digital transformation roadmap, it tends to be more challenging. But it doesn't have to be that way. Industry 4.0 has now been around for a few years, which has allowed tried and tested approaches and methodologies to emerge that can be applied to pretty much any size company in any industry.

We will look at the value chain of chemical processing plants identifying application areas for Industry 4.0 technologies such as digital twins. The focus of this industry specific approach is on structured and open data models that break down stand-alone and isolated knowledge silos - in effect integrating the industry relevant value chains. Actionable items include, but is not limited to: the use of multiple digital twins as controlled copies handling business critical information for data-based business management; the continuous assessment of the correct technologies and technology levels to be implemented for each business relevant area; business and site level consulting on industry 4.0 until maturity is reached and digitalization is business as usual in the chemical processing industries

In addition, we will gain insights into how to shape your own future by combining novel and mature technologies in new and exciting ways, which includes project prioritization. It is a critical exercise for companies in formulating a digital transformation because it enables them to identify areas that will generate the greatest value and evaluate each of them based on specific criteria. These can then be prioritized, which will help decision-making and effective resource allocation. This may sound like a simple step, but it requires dialog, involvement from several stakeholders and a framework to put it all together. Industry 4.0 is *not* a one-size-fits-all approach.

By attending this presentation, you will gain the following insights:

- Gain high-level insight into the concept of 4.0 technologies applied in chemical processing industries.
- Gain site-level insight into digitalization use cases transforming our industry now!
- Basic understanding of how to identify Industry 4.0 projects.

View of how to best prioritize them in terms of resources, ROI, digital maturity level and more.

Copyright © American Institute of Chemical Engineers. All rights reserved.

589365 Piloting Digitalization at Dupont

Abstract:

Okay, you've decided that it's critically important to begin going down the digitalization path. Now what? This presentation will cover DuPont's efforts to date to implement pilot digitalization projects making use of data from existing assets and provide some learnings that you may benefit from in your own piloting efforts.

Specific questions that will be answered and topics that will be covered include...

- Implementation in a production facility making use of real world data
- Selection of digitalization options to be piloted
- Who was involved in the implementation process?
- How long did it take to complete the piloting?
- What support did you receive during the piloting?
- What specific digitalization tools/apps were piloted?
- What was the outcome of the pilot testing?
- What would DuPont have done differently?
- What are the next steps?

Copyright © American Institute of Chemical Engineers. All rights reserved.

590251 Panel Discussions - Future of Olefins

Abstract:

Olefins industry has been historically used to cyclic nature of business following the economic cycles. The regional growth has been driven by the regions with demand side or by advantaged feedstock. There are many historical and unique factors at play that will influence the future of olefins industry. In this session, we have invited industry leaders to share their ideas and thoughts about the factors that can significantly impact the industry in mid to long term. Some of the factors that may drive the future include:

- Historical factors like GDP growth, population growth, urbanization trends, quality of life improvements etc.
- Geo-political factors like trade barriers, tensions in Middle East, sanctions etc.
- Environmental pressures like recycling/recovery of plastics, carbon taxation/greenhouse gas emissions, ban of single use plastics etc.
- Changing energy dynamics driven by focus of oil majors away from fuels to petrochemicals
- Large and complex facilities with high capital intensity resulting in challenges to develop, build and operate them safely and achieving desired performance and financial outcomes
- Disruptive technologies like shale developments, transportation systems, sharing platforms, alternate energy sources etc.

Copyright © American Institute of Chemical Engineers. All rights reserved.

AIChE Paper Number: 124a

Mixed Plastic Waste To Olefins: From Sorting, Over Impurities To Product Yields And Coke Formation

Marvin Kusenberg
ir.
Ghent University, Ghent, Belgium

Azd Zayoud
Dr. ir.
Ghent University, Ghent, Belgium

Thi Hang Dao
Dr.
Ghent University, Ghent, Belgium

Marko R. Djokic
Dr. ir.
Ghent University, Ghent, Belgium

Nicolas Mys
Dr. ir.
Ghent University, Ghent, Belgium

Uros Kresovic
ir.
Indaver N.V. Belgium, Mechelen, Belgium

Nick Alderweireldt
ir.
Indaver N.V. Belgium, Mechelen, Belgium

Steven De Meester
Prof. Dr. ir.
Ghent University, Ghent, Belgium

Kevin M. Van Geem
Prof. Dr. ir.
Ghent University, Ghent, Belgium

Prepared for presentation at the EPC Technology session: Ethylene Plant Technology Fundamentals, at the 2020 Spring National meeting

Houston, Texas, August 16 – 20, 2020

AIChE and EPC shall not be responsible for statements or opinions contained in papers or printed in its publications.

Mixed Plastic Waste To Olefins: From Sorting, Over Impurities To Product Yields And Coke Formation

Marvin Kusenberg
ir.
Ghent University, Ghent, Belgium

Azd Zayoud
Dr. ir.
Ghent University, Ghent, Belgium

Thi Hang Dao
Dr.
Ghent University, Ghent, Belgium

Marko R. Djokic
Dr. ir.
Ghent University, Ghent, Belgium

Nicolas Mys
Dr. ir.
Ghent University, Ghent, Belgium

Uros Kresovic
ir.
Indaver N.V. Belgium, Mechelen, Belgium

Nick Alderweireldt
ir.
Indaver N.V. Belgium, Mechelen, Belgium

Steven De Meester
Prof. Dr. ir.
Ghent University, Ghent, Belgium

Kevin M. Van Geem
Prof. Dr. ir.
Ghent University, Ghent, Belgium

Abstract:
Plastic waste recycling plays a major role in tackling global challenges such as climate change and an ever growing environmental waste crisis. Mechanical recycling, often claimed to be the most efficient way of utilizing plastic waste, poses limitations for highly mixed plastic waste

streams such as plastic packaging. These obstacles may lead to low quality of recycled products or restrictions for the re-use in, for instance, food packaging. Chemical recycling of plastic waste to base chemicals via pyrolysis and subsequent steam cracking of pyrolysis oils shows great potential to overcome the limitations in present means of plastic waste recycling. In this scenario, the largest concern is the feasibility. Are plastic waste pyrolysis products acceptable steam cracking feedstocks in terms of composition, product yields and radiant coke formation? In this work, steam cracking experiments of pure pyrolysis oil from mixed polyolefins (MPO) as well as of pyrolysis oil from polyethylene (PE)-foils blended with fossil naphtha were performed without prior treatment. Product yields, coke formation and process related issues were benchmarked to fossil naphtha as an industrial feedstock. It was found that plastic waste pyrolysis oils show reasonable product yields in the range of fossil naphtha with roughly twofold coking tendencies. The coke formation could be explained with high olefin contents as well as contaminants in the pyrolysis oils. Possible pre-treatment processes have been suggested to reach feedstock qualities comparable with fossil naphtha. Finally, it has been demonstrated that steam cracking of untreated plastic waste pyrolysis oils shows highly promising results in terms of product yields. It has been concluded that pyrolysis oils of mixed polyolefin waste have the potential to be used as steam cracking feedstock after treatment.

Introduction

During recent years, energy and carbon efficiency became increasingly important in the transition from a linear to a circular economy, which aims to decouple economic growth from the consumption of finite resources. According to the Ellen MacArthur Foundation [1], greenhouse gas emissions could be reduced by 22-44 % in 2050 if the current linear economic path is transitioned to a circular economy.

In this scenario, recycling of plastics plays a key role due to their energy and material intensive production, their abundant use and their long lifetime as a pollutant. Global plastics production exceeded 350 million tons in 2018 and is expected to double within the coming two decades [2]. Roughly one quarter of the produced plastics is used as packaging material, from which only about 5 % is recycled while 95 % is practically single-use material which is either disposed of in landfills or lost to the environment [3]. The enormous volume of plastic waste produced every year creates an increasingly severe waste-handling problem with an estimated 4800-12700 kt of plastics ending up in marine environments [4, 5].

Post-consumer plastic waste is a mixture of several different polymers such as polyethylene (PE), polypropylene (PP), polystyrene (PS), polyvinylchloride (PVC) and polyethylene terephthalate (PET), which further contains a number of additives and auxiliary materials such as stabilizers, flame retardants, plasticizers and colorants. Polyolefins (PE and PP) are the most prominent components found in plastic waste since they are the main ingredients in plastic packaging which represents by far the largest part of the waste streams. [5, 6]

The two main recycling options for plastic waste are mechanical recycling and chemical recycling. Mechanical recycling aims for the conversion of sorted plastic waste streams into products of similar quality and application. For this, single-polymer waste streams are needed which are rarely available due to the highly mixed waste streams, large amounts of additives and the substantial sorting effort required for high purity streams [7, 8]. The large complexity of post-consumer plastic waste makes mechanical recycling of plastic packaging waste challenging due to certain limitations regarding separation and re-use of materials (i.e. food-contact applications or value-added materials) [9]. Additionally, sophisticated multilayer packaging poses considerable challenges to waste sorting facilities. Therefore, mechanically recycled products are often of lower value than the original products. [3]

The highly diverse nature of mixed plastic waste requires a recycling technology which is able to produce high-quality virgin feedstock applicable for any kind of purpose. Therefore, thermochemical decomposition into base chemicals represents a promising recycling technology. Base chemicals such as ethylene and propylene are used to produce new virgin plastics thus closing the material loop [10]. For this purpose, a two-step process can be applied, namely pyrolysis of solid plastic waste and subsequent steam cracking of the pyrolysis oils.

In a pyrolysis reactor, solid plastics are thermally decomposed in an inert atmosphere yielding a distribution of gaseous, liquid and solid products depending on process conditions such as temperature, pressure and residence time [11, 12]. Depending on the polymer, different decomposition mechanisms occur. For instance, polystyrene decomposes in a depolymerization reaction yielding mainly monomers. PE and PP, however, decompose in a random fragmentation reaction yielding a wide range of products. [8, 9, 13] The solid pyrolysis product typically consists of carbon and the inorganic content of the original feedstock [11]. Liquid products of mixed plastic waste pyrolysis contain a complex mixture of hydrocarbons with a wide carbon number distribution. These may be utilized as petrochemical feedstock after upgrading with processes such as hydrotreatment and distillation. Impurities in the pyrolysis oil such as oxygenates, sulfur or nitrogen containing compounds are known to diminish the quality of the liquid product making further treatment necessary in order to obtain petrochemical feedstocks of high quality such as naphtha. [8]

Base chemicals such as ethylene and propylene are produced via steam cracking which is a thermal process utilizing substantial amounts of energy in order to decompose large hydrocarbon molecules into light olefins [6]. Steam cracking of plastic waste pyrolysis oils requires a liquid feedstock that is compatible with industrial steam cracking units in terms of physical properties (e.g. viscosity, boiling point), molecular composition (e.g. olefinic, paraffinic, naphthenic and aromatic contents) and potential contaminant limits (e.g. oxygenates, sulfur, nitrogen, chlorine, metals) [14]. Furthermore, sufficient amounts of high quality feedstock are needed to meet the huge scale of operation of industrial steam crackers. Therefore, it is likely that in an industrial scenario, plastic waste pyrolysis products will be blended with fossil naphtha in order to reach sufficient production scales rather than entirely replacing fossil feedstock [15]. Next to the light olefin yields (i.e. ethylene, propylene, 1,3-butadiene), coke formation is a crucial factor which will determine the economic success of plastic waste pyrolysis oils as steam cracker feedstock, since it is a major factor for the time-on-stream of industrial plants [16].

In this work, waste plastic feedstocks have been pyrolyzed and the liquid products have been steam cracked in a bench-scale steam cracking unit in order to study the feasibility of plastic waste as petrochemical feedstock. As a reference feedstock, fossil naphtha has been steam cracked in a separate experiment to compare product yields as well as coke formation to an industrially accepted feedstock. Plastic pyrolysis oils, as well as steam cracking effluents, have been analyzed using comprehensive two-dimensional gas chromatography coupled to a flame ionization detector (GC × GC-FID). Coke formation was measured using an infrared meter in a controlled decoking procedure after completion of the respective steam cracking experiments.

Materials and methods

Plastic waste samples: Two separate post-consumer plastic waste feedstocks, provided by Indaver N.V. (Belgium), have been used, namely a mixed polyolefin (MPO) sample, mainly consisting of PE and PP and a PE-film sample consisting primarily of PE-foils with a random contamination of other polymers such as PP and PET. Samples provided have been manually sorted, hence no PVC was present. A typical industrial light naphtha was used as a steam cracking reference feedstock.

Analytical setups: Comprehensive two-dimensional gas chromatography coupled to a flame ionization detector (GC × GC-FID) was used to determine the detailed composition of the different feedstocks. The feedstock analyses (off-line), as well as the reactor effluent analyses (on-line), were performed using Thermo Scientific TRACE GC × GC setups (Interscience, Belgium) equipped with a two-stage cryogenic liquid CO_2 modulator [17] and coupled to a flame ionization detector (FID) [18, 19]. As an internal standard, 3-chlorothiophene was used.

Experimental setups and methods: Plastic waste samples have been pyrolyzed in a pyrolysis pilot setup at the Laboratory for Chemical Technology (LCT) at Ghent University, Ghent, Belgium (shown in Figure 1).

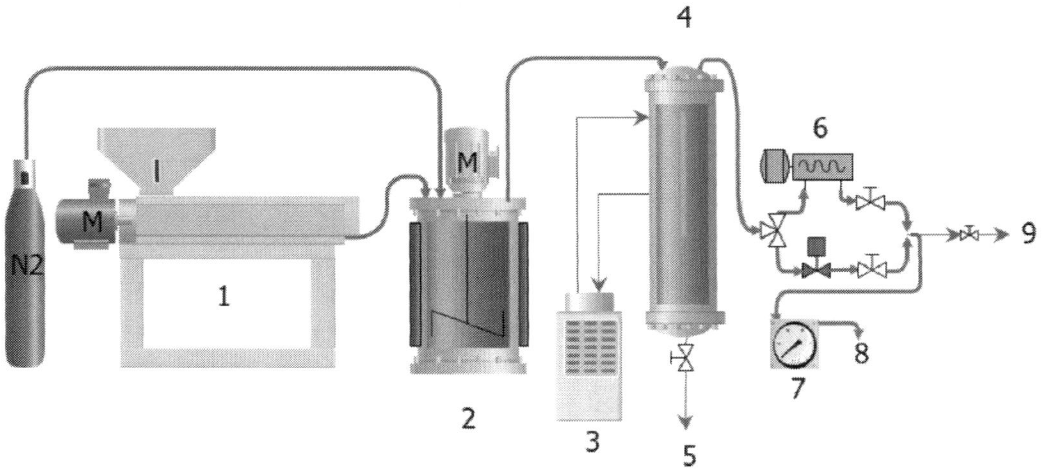

Figure 1: Pyrolysis pilot setup installed at the Laboratory for Chemical Technology (LCT) at Ghent University, Ghent, Belgium: (1) extruder, (2) stirred-tank reactor, (3) cooler, (4) condenser, (5) liquid product sampling, (6) vacuum pump, (7) volumetric gas flow-meter, (8) vent, (9) gaseous product sampling.

The setup consists of an extruder connected to an externally heated stirred-tank reactor vessel. The plastic feedstock is fed in pellet form to the extruder which is used to melt and feed the plastic to the pyrolysis reactor. Plastics are thermally cracked and leave the reactor in vapor form. In the downstream section of the reactor, liquid products are condensed and collected.

The flow of non-condensable gaseous products is measured using a volumetric gas flow-meter. Solid products remain in the reactor and are weighted separately after completion of each experiment in order to close the mass balance. The pyrolysis oils which were steam cracked in this study were produced at 500 °C and 1 bar for PE-film and at 450 °C and 5 bar for MPO. Both experiments were performed with feeding rates of 1 kg/h.

Subsequent steam cracking experiments with pyrolysis oils produced in the plastic waste pyrolysis setup have been carried out at a bench-scale steam cracking setup (BSSC) at the LCT (Ghent University, Belgium). A schematic representation of the bench scale unit is shown in Figure 2.

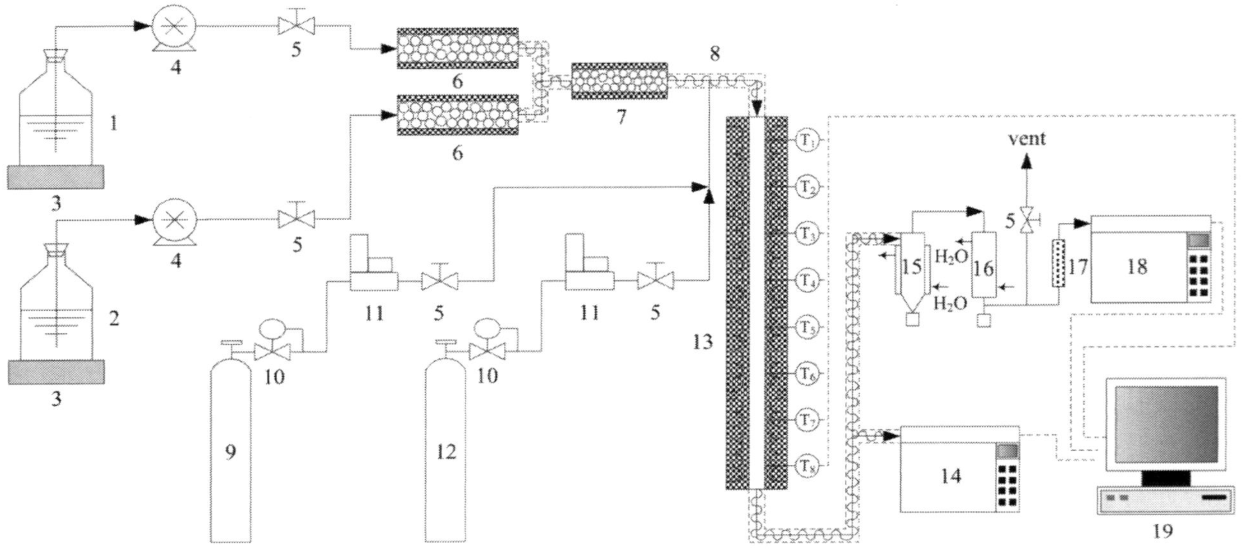

Figure 2: Schematic representation of the bench scale unit: (1) water; (2) hydrocarbon; (3) electronic balance; (4) pump; (5) valve; (6) evaporator; (7) mixer; (8) heater; (9) nitrogen; (10) pressure regulator; (11) mass flow controller; (12) air; (13) reactor; (14) GC × GC; (15,16) condensers; (17) dryrites; (18) RGA; (19) data acquisition. [20]

The setup consists of three sections, namely the feeding section, the furnace/reactor section, as well as the analysis section. Since a detailed description of the setup has already been published elsewhere [20, 21], only a brief description of the setup will be given. The feeding section controls the supply of the feedstock and the water to the reactor coil. The flow of liquids is regulated by Coriolis flow-meter (CORI-FLOW) controlled pumps (Bronkhorst, The Netherlands). Liquid feedstock bottles may be heated using electrical heating mantles in case of more viscous feedstocks to avoid solidification and to guarantee a stable operation of the feeding section. Steam and hydrocarbon feedstock are pre-heated and mixed in electrically heated ovens, filled with quartz beads, enabling an efficient and uniform mixing of feedstock and diluent prior to entering the reactor. The mixture enters the 1.475 m long reactor which is placed vertically in a rectangular electrically heated furnace. There are eight thermocouples positioned along the axial reactor coordinate, measuring the process gas temperature at different positions. The furnace is divided into eight separate sections which can be controlled independently to set a specific temperature profile. The pressure in the reactor is controlled by a back pressure regulator positioned downstream from the outlet of the reactor. Two pressure transducers, placed at the inlet and outlet of the reactor, indicate the coil inlet (CIP) and the coil outlet pressure (COP), respectively. At the reactor outlet, the injection of nitrogen provides an internal standard for analytical measurements and contributes to a certain extent to the quenching.

The reactor effluent is sampled online, i.e. during operation, at high temperature (300 °C). Via a valve-based sampling system and uniformly heated transfer lines, a gaseous sample of the reactor effluent is injected onto the comprehensive two-dimensional gas chromatograph (GC × GC) coupled to a flame ionization detector (FID). The high-temperature 6-port-2-way sampling valve is placed in an oven in which the temperature is kept above the dew point of the effluent sample [18]. Further downstream, the reactor effluent is cooled, water and condensed heavier products (pyrolysis gasoline and pyrolysis fuel oil) are removed, while the remainder of the gaseous effluent stream is sent to the vent. Before reaching the vent, a fraction of the effluent is withdrawn for analysis on the so-called Refinery Gas Analyzer (RGA) using a built-in gas sampling valve system (80 °C).

The reactor is made of Incoloy 800HT (Ni: 30-35, Cr: 19-23, Fe: > 39.5 wt.%). The length of the reactor where the reaction occurs is approximately 1.475 m long, with 0.006 m internal diameter.

The C_{4-} analysis of the quenched effluent gases is performed on the refinery gas analyzer (RGA). Hydrogen, carbon dioxide, carbon monoxide, nitrogen, methane, ethane, ethylene and acetylene are all detected by a thermal conductivity detector (TCD). While the remaining C_1 to C_4 compounds are analyzed by a flame ionization detector (FID) leading to an internal validation. Each detector is mounted on the Interscience Trace GC Ultra. Comprehensive two-dimensional GC × GC-FID is used to detect the C_{5+} compounds. A detailed description of the on-line effluent analysis procedure performed at LCT, Ghent, Belgium has been published elsewhere [18, 22].

Two different coil outlet temperatures (COT) have been tested in the experiments, namely 820 °C and 850 °C. The coil inlet temperature (CIT) was set to 500 °C for all experiments. For both COTs, temperature profiles were set using 8 separate heating zones in the reactor. The coil outlet pressure (COP) was set to 1.7 bar for all experiments using a back pressure regulator downstream of the reactor. A steam dilution of 0.5 was chosen for all experiments. To mimic industrial conditions, 250 ppm of sulfur was continuously added to the steam flow in form of dimethyl disulfide (DMDS), which is known to reduce the CO formation and coking tendency [23]. An overview of the process conditions is given in Table 1.

Table 1: Overview of the performed experiments at the BSSC setup at LCT, Ghent, Belgium.

Temperature profile 1 [°C]	500-550-590-630-670-700-740-780-820
Temperature profile 2 [°C]	500-550-590-630-680-720-770-810-850
Coil outlet pressure [bar]	1.7 bar
Liquid hydrocarbon flow [g/h]	150 g/h
Steam flow [g/h]	75 g/h
Continuous sulfur addition	250 ppm from DMDS

Steam cracking of plastic waste pyrolysis oils was performed without any treatment prior to the experiment. Both pyrolysis oils had a high visible viscosity as well as solid impurities suspended in the liquid. MPO pyrolysis oil, which was in liquid state at ambient conditions, was steam cracked without further mixing with another hydrocarbon. However, due to the viscosity, the feed bottle was heated during operation. PE-film pyrolysis oil had a waxy constitution and was in a solid state at ambient conditions with a melting point around 80 °C. For the steam cracking experiments, it was blended with naphtha for two reasons. First, due to its high melting point it was not practical to feed it pure to the steam cracking unit. Second, plastic waste pyrolysis oils are unlikely to entirely replace fossil feedstock in the future but rather complement it. Therefore, PE-film pyrolysis oil was blended with naphtha in a ratio of 25 wt.% of plastic pyrolysis oil and 75 wt.% of naphtha. The chosen ratio exceeds the expected mixing ratio in industrial steam crackers which is more likely to range around 10 % due to different orders of magnitude in scale between industrial naphtha crackers and plastic waste sorting plants.

Next to steam cracking, controlled decoking experiments were performed after each experiment. For this, the amount of coke which was deposited during each respective steam cracking run was quantified. The quantification was performed using an infrared meter which measures the volumetric concentration of CO and CO_2 in the reactor effluent produced while decoking the reactor coil. For the decoking experiment, air was fed to the reactor which was heated up to 850 °C in all zones. The decoking experiment was finished once CO and CO_2 were no longer detected in the reactor effluent. The coke was quantified using a volumetric drum-type gas meter and referred to the total steam cracking duration per feedstock.

Results and Discussion

Feedstock characterization: Two plastic waste pyrolysis oils (MPO and PE-film) were steam cracked in this study. Prior to the experiments the samples have been analyzed using GC × GC-FID. The molecular composition of both plastic waste pyrolysis oils is shown in Table 2. As a comparison, the molecular composition of fossil naphtha as a benchmark industrial steam cracking feedstock is shown as well. It is to be noted that the composition of plastic waste pyrolysis oils differs significantly depending on the process conditions during pyrolysis as well as on the exact composition of the plastic waste. This has been shown in several studies [9, 11, 13].

Table 2: Molecular composition of the plastic waste pyrolysis oils (MPO, PE-film) and fossil naphtha reference analyzed with GC × GC-FID.

Feedstock	MPO	PE-film	Fossil naphtha
Carbon number range	$C_5 - C_{51}$	$C_5 - C_{47}$	Up to C_9
Paraffins [wt.%]	41	37	79
Olefins [wt.%]	40	55	-
Naphthenes [wt.%]	11	7	17
Aromatics [wt.%]	8	1	4

As an example, Figure 3 provides the GC × GC-FID color plot of MPO pyrolysis oil which underlines the multitude of compounds present. The respective compound families are marked as well.

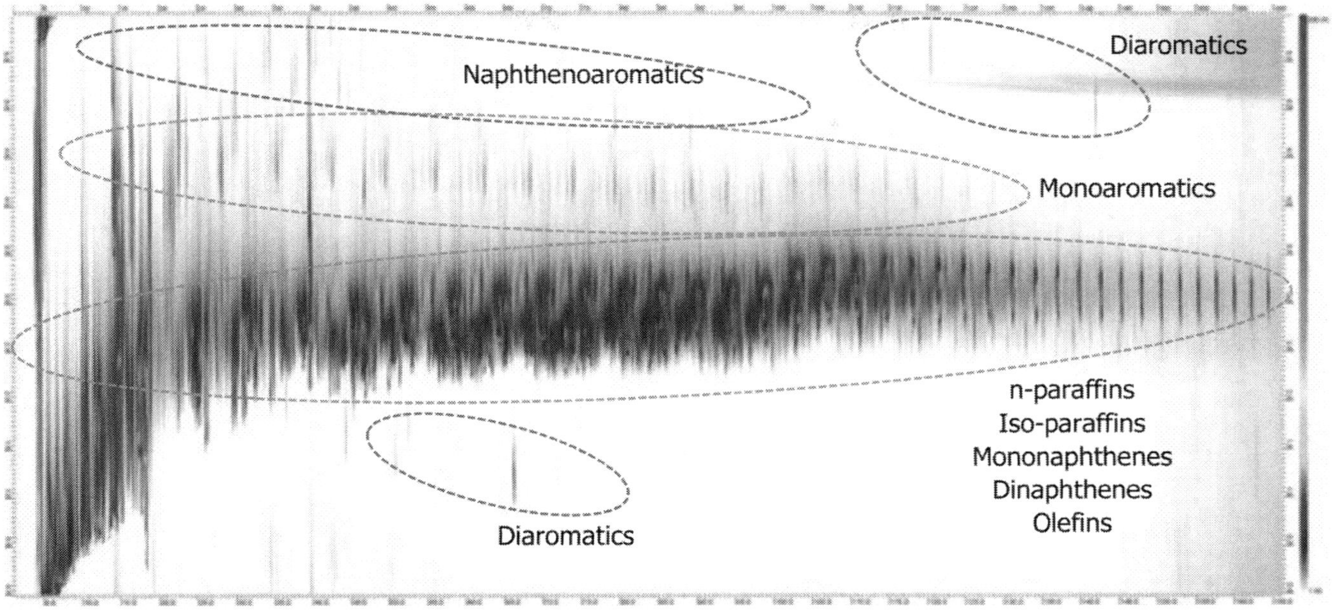

Figure 3: GC × GC-FID color plot of MPO pyrolysis oil. Compound families are marked individually.

It can be seen from the feedstock analysis that the plastic waste pyrolysis oils have a wide carbon range distribution from C_5 up to C_{51} which is in contrast to the carbon range of the light naphtha reference feedstock. This large discrepancy makes the pyrolysis oil problematic for an industrial naphtha cracker without prior fractionation steps. Since the presented bench-scale steam cracking setup used in this study is able to handle feedstocks with boiling points up to 650 °C, no further fractionation steps were performed. Considering the ongoing transition in industry towards heavier petrochemical feedstocks for steam crackers such as crude oil, fractionation into lighter fractions might not be necessary for future use [24, 25].

Regarding the molecular composition, there are large differences between the pyrolysis oils and fossil naphtha especially in the olefin content which would make the pyrolysis oils unattractive for industrial steam crackers due to issues such as high coking and fouling potential as well as the inhibition of paraffin cracking reactions [26, 27]. Processes such as hydrotreatment could be performed to adjust the olefin content as well as to remove nitrogen, sulfur and oxygen-containing compounds which may lead to coke formation, fouling and catalyst poisoning in later upgrading stages [28]. However, the aim of this work is to proof the concept of steam cracking raw plastic pyrolysis oils without further treatment in terms of light olefin yields, coke formation as well as processability in a bench-scale steam cracking unit. It has to be noted that due to the blending of PE-film pyrolysis oil with fossil naphtha, the olefin content is naturally reduced to a far lower content (to ~14 wt.%) by virtue of the absence of olefins in fossil naphtha.

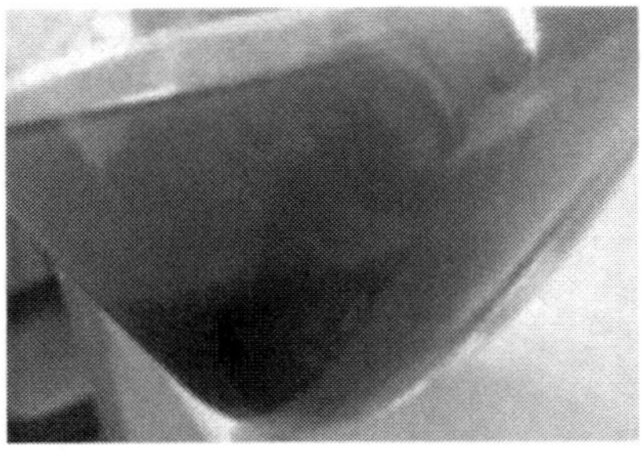

Figure 4: Residue in the feed bottle of PE-film pyrolysis oil.

In the steam cracking experiments observations could be made regarding the processability of the plastic feedstocks. First, due to the very high viscosity of the feedstocks, additional measures were needed such as heating of the feed bottle and tracing of the feeding lines. Second, as already mentioned, both plastic pyrolysis oils contained solid impurities which ultimately caused issues with the feeding pumps since particles in the feedstock eventually precipitated from the liquid (see Figure 4). To overcome this problem, hot filtration for the removal of solids is advisable.

Cracking product yields: Figure 5 shows the product yields of methane and the most important light olefins ethylene, propylene and 1,3-butadiene in wt.% for the two respective COTs 820 °C and 850 °C.

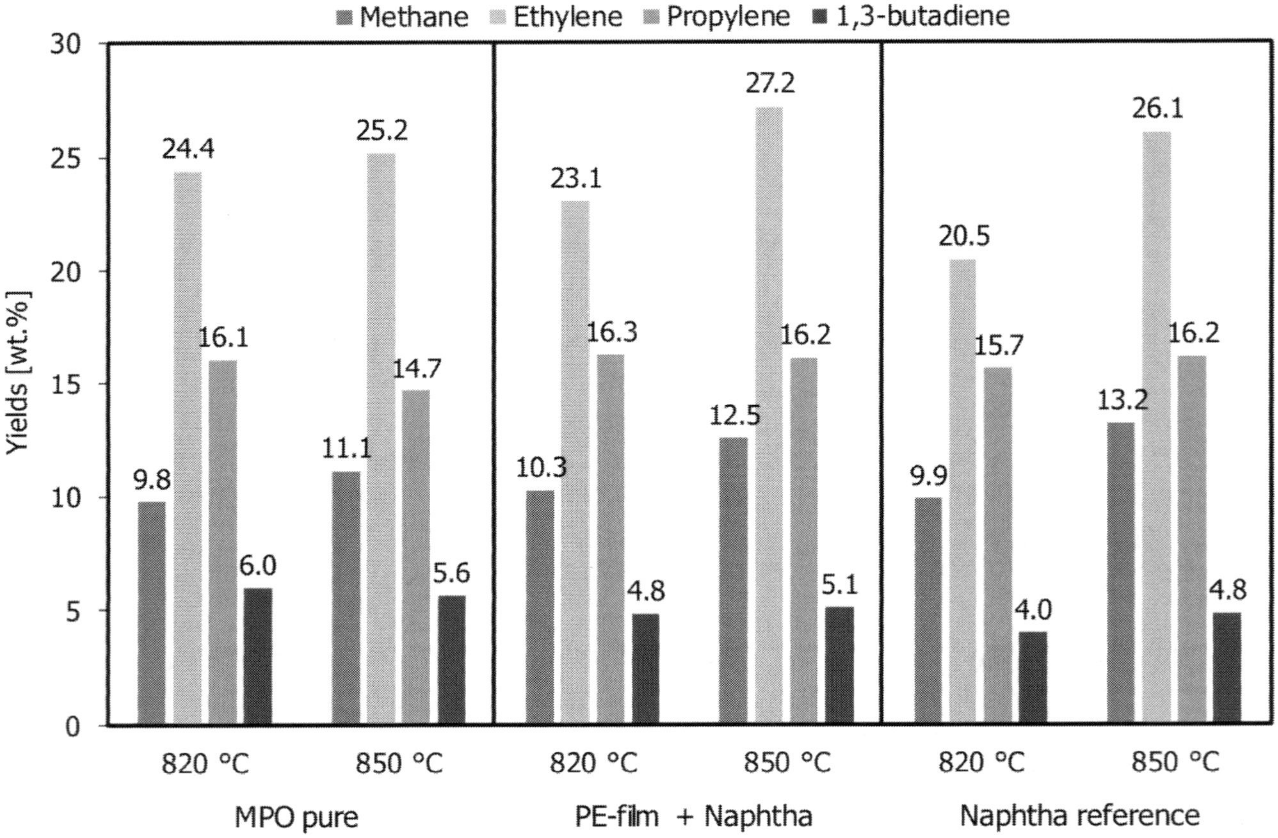

Figure 5: Light olefin yields of steam cracking pure MPO pyrolysis oil, PE-film pyrolysis oil blended with naphtha and naphtha reference in wt.%.

From Figure 5 it can be seen that light olefin yields for all three feedstocks are within the same range with a clearly visible dependency on the temperature. For all three feedstocks, higher cracking severity led to an increase in ethylene yield with a corresponding decrease in propylene yield. This trend is most prominent for fossil naphtha with a sharp increase from 20.5 wt.%

ethylene at a COT of 820 °C to 26.1 wt.% at a COT of 850 °C. The temperature dependency of the yields of pure MPO pyrolysis oil are rather low indicating the influence of the high olefin content of more than 40 wt.% which inhibits paraffin cracking reactions. At a COT of 820 °C, however, the lights' yields of the two pyrolysis oils exceed these of fossil naphtha which indicates that the chosen lower temperature profile is, of course, not optimal for steam cracking of fossil naphtha. Detailed yields of methane, ethylene, propylene and 1,3-butadiene next to the total C_{4-} yields for all experiments are listed in Table 3. The presented values are averaged from at least three stable measurements for each condition, respectively.

Table 3: Result summary of the product yields of the most important light products during steam cracking of the three feedstocks at the chosen process conditions. Numbers given are averaged mass yields of at least three stable measurements for each condition, respectively.

Feedstock	**MPO pure**		**PE-film + Naphtha**		**Naphtha reference**	
COT [°C]	820	850	820	850	820	850
Total C_1-C_4 Species [wt.%]	69.51	68.94	68.02	72.25	68.33	77.45
			Yields [wt.%]			
Methane	9.78	11.14	10.27	12.54	9.89	13.19
Ethylene	24.37	25.17	23.06	27.17	20.48	26.07
Ethane	3.52	3.41	3.17	3.18	2.74	3.05
Propylene	16.10	14.66	16.28	16.15	15.66	16.21
1-butene	2.73	1.76	2.82	2.14	2.82	2.19
Iso-butene	3.30	2.96	3.06	2.75	3.54	3.26
1,3-butadiene	6.00	5.64	4.82	5.09	3.99	4.83

With respect to Table 3, it can be seen that although the plastic waste pyrolysis oils show high ethylene, propylene and 1,3-butadiene yields, the total light olefin yields (C_{4-}) yield lie well below the one of fossil naphtha for a COT of 850 °C. For the lower COT of 820 °C, the values are similar which is confirmed by the fairly low ethylene and propylene yields of fossil naphtha at this cracking severity. The lower total C_1-C_4 species yields of plastic waste pyrolysis oils at 850 °C compared to fossil naphtha can be explained by the respective feedstock molecular composition. Very heavy feedstocks yield more heavy products which are in the range of pyrolysis gasoline (C_5-C_{10}) and pyrolysis fuel oil (C_{10+}) which may induce process fouling. This can be the consequence of high olefin contents which are more prone to secondary reactions yielding heavier products. Additionally, more stable compounds in the feedstock may be unconverted in the cracking experiments [27]. As already mentioned, the olefin content of pure MPO pyrolysis oil is significantly higher than the one of the PE-film – naphtha blend (~40 wt.% vs. ~14 wt.%, respectively) which explains the low total C_{4-} values at a COT of 850 °C compared to the PE-film – naphtha blend and the naphtha reference.

Nevertheless, the presented results underline the feasibility of steam cracking plastic waste pyrolysis oils in terms of product yields. It was shown that steam cracking of untreated plastic waste pyrolysis oils is feasible with light olefin product yields in the range of an industrial naphtha feedstock.

Coking data: A highly important aspect in steam cracking is the formation of coke. Coke depositions on the inner reactor wall act as heat transfer inhibitor increasing the heat demand. Furthermore, the pressure drop along the reactor increases, eventually causing a costly shutdown in order to remove the deposits via a controlled decoking process. [16] In general, coke formation is a direct consequence of feedstock impurities such as heteroatoms but also high olefinic and aromatic contents [26, 29]. Decoking experiments were performed after cracking each respective feedstock according to the described method. Results of the decoking experiments are presented in Figure 6. The total mass of coke was determined using a volumetric drum-type gas meter and put in relation to the total duration of the respective steam cracking runs.

It can be seen from Figure 6 that the plastic waste pyrolysis oils show twofold coke formation compared to fossil naphtha as industrial benchmark. It is further visible that both plastic waste pyrolysis oils show almost equal tendencies (33.4 mg/h for pure MPO pyrolysis oil and 32.8 mg/h for PE-film pyrolysis oil blended with naphtha), which is interesting since pure MPO pyrolysis oil was cracked while PE-film pyrolysis oil was blended with naphtha. This observation implies that the coking behavior of pure PE-film pyrolysis oil would have been substantially higher than the one of MPO pyrolysis oil which indicates that the PE-film pyrolysis oil contains further impurities next to olefins and aromatics which promote coke formation. Considering the mixing ratio of 1:3 of the PE-film + naphtha blend, the theoretical estimated coke formation of pure PE-film pyrolysis oil would be higher than 130 mg/h.

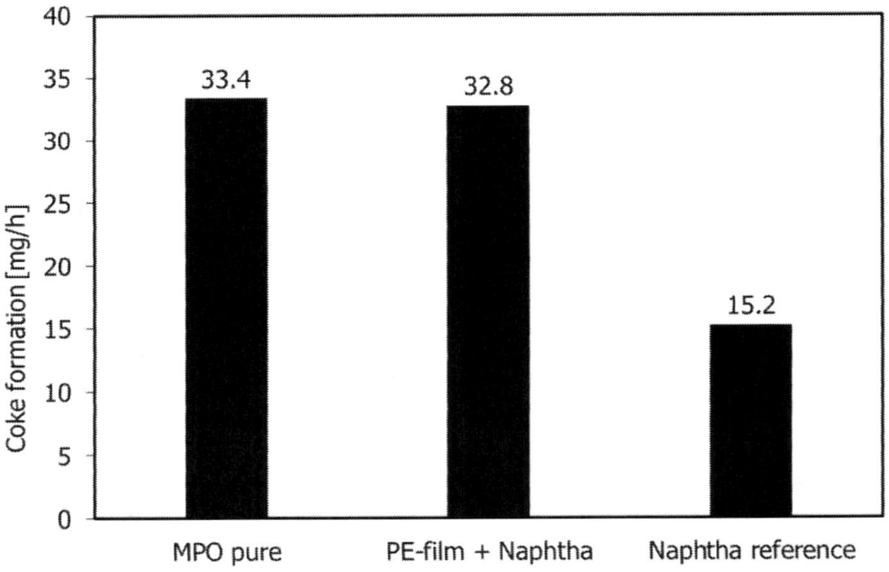

Figure 6: Coke formation in mg/h for MPO pure, PE-film + naphtha as well as fossil naphtha. Data was obtained in a controlled decoking experiment after a total steam cracking duration of at least 4 hours.

Although plastic waste pyrolysis oils show higher coking tendencies than fossil naphtha, the results are still in a reasonable and well explainable range which indicate that steam cracking of these oils is feasible from a coking point of view. The results further suggest that certain treatment steps prior to steam cracking such as hydrotreatment should rather be taken in order to reduce olefin contents, as well as contaminants.

Regarding the PE-film pyrolysis oil – naphtha blend which combined has a reasonably low olefin content of ~14 wt.%, the high coking tendency indicates further impurities which promote coke formation. Further analyses such as elemental analysis (CHNS/O), ICP-AES/OES/MS for metals or ion chromatography for halogens are advised in order to comprehensively understand the composition of post-consumer plastic waste pyrolysis oils. This is especially valid for the plastic foil fractions since these fractions are considered to contain higher amounts of dirt and food residuals as well as metal containing inks [30]. Since contaminants are and will be the determining factor for the industrial use of these products, further research is of paramount importance.

Next to coke formation in the reactor coil, issues such as transfer line exchanger fouling might occur. These issues can be observed by monitoring the pressure difference between reactor inlet and outlet as a function of the reaction time since heavy products condense and deposit in the colder downstream section of the reactor causing a pressure build-up. Fouling issues can furthermore be caused by the presence of sulfur and nitrogen compounds in the feedstock which may react with diolefins in the downstream section. The Δp between the reactor inlet and the reactor outlet it shown in Figure 7 measured during steam cracking pure MPO pyrolysis oil as well as fossil naphtha reference. It can be seen that the pressure difference for fossil naphtha stays constant for a straight cracking duration of 5 hours. MPO, however, shows a steady increase with a sharp rise after 3.5 hours due to a blockage in the downstream section of the reactor. The curve for PE-film pyrolysis oil – naphtha is not shown since the experiment was stopped due to a blockage in the convection section after roughly 4 hours presumably caused by not evaporated feedstock or solid impurities in the liquid feedstock as shown in Figure 4.

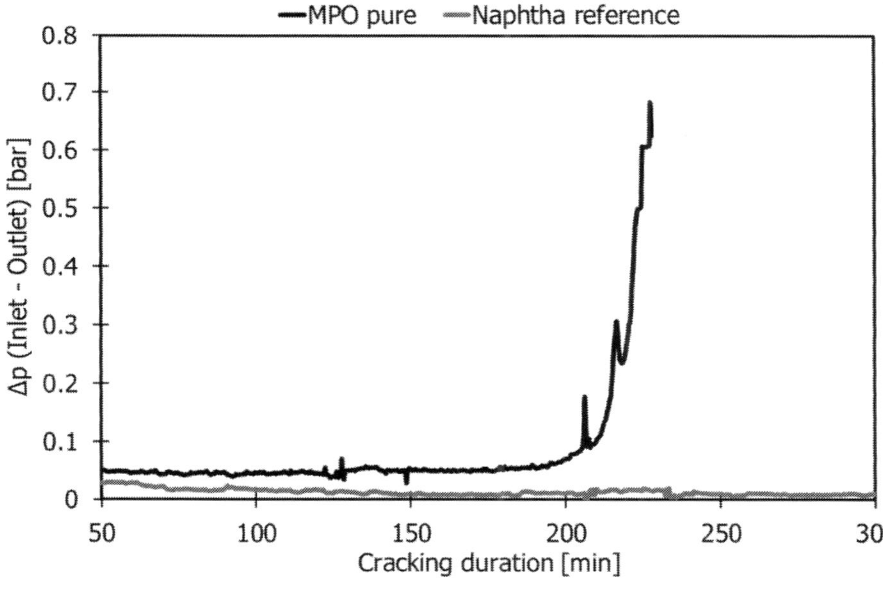

Figure 7: Pressure difference between reactor inlet and reactor outlet as a function of cracking duration for raw MPO pyrolysis oil and fossil naphtha.

The process-related observations suggest that the plastic waste pyrolysis oils require pre-treatment steps such as hot-filtration to remove suspended solids as well as fractionation steps removing heavy residual parts of the oil.

Conclusions and Outlook

Our research clearly shows that steam cracking of post-consumer plastic waste pyrolysis oils is feasible with reasonable product yields and coking tendencies. The pyrolysis oils show potential as steam cracking feedstocks after treatment steps such as hydrotreatment and fractionation via distillation. This has been demonstrated on the basis of steam cracking experiments of raw, untreated plastic pyrolysis oils. Fouling potential leading to blockages in the convection section or the transfer line exchanger has been observed with untreated feedstocks underlining the necessity of pre-treatment steps. These "proof-of-concept" experiments were not intended to ultimately evaluate plastic waste pyrolysis oils as steam cracking feedstocks but rather to demonstrate the feasibility of the streams without prior treatment. The results underline the potential of the feedstocks when pre-treatment steps are applied. Distillation steps for the production of a naphtha range product fraction might reduce the reported fouling and coking issues substantially enabling an even better benchmarking to an industrial naphtha feedstock. In further treatment steps such as hydrotreatment the molecular composition can be tuned to a more naphtha-like composition in respect of olefin and aromatic contents. Additionally, analyses such as elemental analysis (CHNS/O), ICP-AES/OES/MS or ion chromatography will shed light on the concentration of contaminants such as metals and heteroatoms (N, O, S, Cl) which are a further source for coke formation. With these steps further knowledge will be obtained on the influence of the plastic pyrolysis oil composition on steam cracking yields and coke formation as well as process related issues such as transfer line exchanger fouling.

Moreover, thermal pre-treatment of plastic waste could diminish obstacles such as fouling or low quality pyrolysis oils. Post-consumer plastic waste contains a broad range of additives such as stabilizers, colorants and plasticizers as well as chlorinated compounds especially from small amounts of PVC. Due to their high corrosiveness, chlorinated compounds are to be removed as early in the process as possible. By thermally removing volatiles prior to pyrolysis, it is expected to increase the pyrolysis oil quality substantially. Future work, therefore, aims to combine the feeding of the plastic waste to the pyrolysis reactor with a thermal pre-treatment within the extruder by variation of temperature, application of vacuum and screw speed. This way, volatile organic compounds, halogenic contaminants and a range of added compounds may be removed from the plastic melt prior to pyrolysis.

Acknowledgment

The authors acknowledge financial support by the Catalisti-ICON project (HBC.2018.0262) MATTER (Mechanical and Thermochemical Recycling of mixed plastic waste) funded by Flanders Innovation & Entrepreneurship (VLAIO).

References

[1] Ellen MacArthur Foundation, "Completing the picture how the circular economy tackles climate change," 2019, vol. V.3.

[2] Plastics Europe - Association of Plastics Manufacturers, "Plastics - The Facts 2019," 2019.

[3] Ellen MacArthur Foundation, "The New Plastics Economy: Rethinking the future of plastics," 2016,

[4] Jambeck, J. R., Geyer, R., Wilcox, C., Siegler, T. R., Perryman, M., Andrady, A., Narayan, R., and Law, K. L., "Plastic waste inputs from land into the ocean," *Science,* vol. 347, no. 6223, pp. 768-771, 2015, doi: 10.1126/science.1260352.

[5] Geyer, R., Jambeck, J. R., and Law, K. L., "Production, use, and fate of all plastics ever made," *Science Advances,* vol. 3, no. 7, p. e1700782, 2017, doi: 10.1126/sciadv.1700782.

[6] Thunman, H., Berdugo Vilches, T., Seemann, M., Maric, J., Vela, I. C., Pissot, S., and Nguyen, H. N. T., "Circular use of plastics-transformation of existing petrochemical clusters into thermochemical recycling plants with 100% plastics recovery," *Sustainable Materials and Technologies,* vol. 22, p. e00124, 2019, doi: 10.1016/j.susmat.2019.e00124.

[7] Brems, A., Baeyens, J., and Dewil, R., "Recycling and recovery of post-consumer plastic solid waste in a European context," *J Thermal Science,* vol. 16, no. 3, 2012, doi: 10.2298/TSCI120111121B.

[8] Al-Salem, S. M., Lettieri, P., and Baeyens, J., "The valorization of plastic solid waste (PSW) by primary to quaternary routes: From re-use to energy and chemicals," *Progress in Energy and Combustion Science,* vol. 36, no. 1, pp. 103-129, 2010, doi: 10.1016/j.pecs.2009.09.001.

[9] Ragaert, K., Delva, L., and Van Geem, K., "Mechanical and chemical recycling of solid plastic waste," *Waste Management,* vol. 69, pp. 24-58, 2017, doi: 10.1016/j.wasman.2017.07.044.

[10] Garforth, A. A., Ali, S., Hernández-Martínez, J., and Akah, A., "Feedstock recycling of polymer wastes," *Current Opinion in Solid State and Materials Science,* vol. 8, no. 6, pp. 419-425, 2004, doi: 10.1016/j.cossms.2005.04.003.

[11] Buekens, A. G., "Introduction to Feedstock Recycling of Plastics," in *Feedstock Recycling and Pyrolysis of Waste Plastics*, J. Scheirs and W. Kaminsky, Eds.: John Wiley & Sons, Ltd, 2006, pp. 1-41.

[12] Lopez, G., Artetxe, M., Amutio, M., Bilbao, J., and Olazar, M., "Thermochemical routes for the valorization of waste polyolefinic plastics to produce fuels and chemicals. A review," *Renewable and Sustainable Energy Reviews,* vol. 73, pp. 346-368, 2017, doi: 10.1016/j.rser.2017.01.142.

[13] Angyal, A., Miskolczi, N., and Bartha, L., "Petrochemical feedstock by thermal cracking of plastic waste," *Journal of Analytical and Applied Pyrolysis,* vol. 79, no. 1, pp. 409-414, 2007, doi: 10.1016/j.jaap.2006.12.031.

[14] Sundaram, K. M. and Stancato, B., "How much is too much? Feed contaminants and their consequences," presented at the AIChE 2018 Spring National Meeting, Orlando, Florida, 22-26 April 2018.

[15] Amghizar, I., Vandewalle, L. A., Van Geem, K. M., and Marin, G. B., "New Trends in Olefin Production," *Engineering,* vol. 3, no. 2, pp. 171-178, 2017, doi: 10.1016/J.ENG.2017.02.006.

[16] Patil, M., Djokic, M. R., Marin, G. B., Reyniers, M.-F., and Van Geem, K. M., "Effect of Contaminants on Coke Formation during Steam Cracking of Propane," in *2019 Spring Meeting and 15th Global Congress on Process Safety*, 2019: AIChE.

[17] Adahchour, M., Beens, J., and Brinkman, U. A. T., "Recent developments in the application of comprehensive two-dimensional gas chromatography," *Journal of Chromatography A,* vol. 1186, no. 1, pp. 67-108, 2008, doi: 10.1016/j.chroma.2008.01.002.
[18] Van Geem, K. M., Pyl, S. P., Reyniers, M.-F., Vercammen, J., Beens, J., and Marin, G. B., "On-line analysis of complex hydrocarbon mixtures using comprehensive two-dimensional gas chromatography," *Journal of Chromatography A,* vol. 1217, no. 43, pp. 6623-6633, 2010, doi: 10.1016/j.chroma.2010.04.006.
[19] Pyl, S. P., Schietekat, C. M., Van Geem, K. M., Reyniers, M.-F., Vercammen, J., Beens, J., and Marin, G. B., "Rapeseed oil methyl ester pyrolysis: On-line product analysis using comprehensive two-dimensional gas chromatography," *Journal of Chromatography A,* vol. 1218, no. 21, pp. 3217-3223, 2011, doi: 10.1016/j.chroma.2010.12.109.
[20] De Bruycker, R., Anthonykutty, J. M., Linnekoski, J., Harlin, A., Lehtonen, J., Van Geem, K. M., Räsänen, J., and Marin, G. B., "Assessing the Potential of Crude Tall Oil for the Production of Green-Base Chemicals: An Experimental and Kinetic Modeling Study," *Industrial & Engineering Chemistry Research,* vol. 53, no. 48, pp. 18430-18442, 2014, doi: 10.1021/ie503505f.
[21] Harper, M. R., Van Geem, K. M., Pyl, S. P., Marin, G. B., and Green, W. H., "Comprehensive reaction mechanism for n-butanol pyrolysis and combustion," *Combustion and Flame,* vol. 158, no. 1, pp. 16-41, 2011, doi: 10.1016/j.combustflame.2010.06.002.
[22] Pyl, S. P., Schietekat, C. M., Reyniers, M.-F., Abhari, R., Marin, G. B., and Van Geem, K. M., "Biomass to olefins: Cracking of renewable naphtha," *Chemical Engineering Journal,* vol. 176-177, pp. 178-187, 2011, doi: 10.1016/j.cej.2011.04.062.
[23] Dhuyvetter, I., Reyniers, M.-F., Froment, G. F., Marin, G. B., and Viennet, D., "The Influence of Dimethyl Disulfide on Naphtha Steam Cracking," *Industrial & Engineering Chemistry Research,* vol. 40, no. 20, pp. 4353-4362, 2001, doi: 10.1021/ie001131b.
[24] Corma, A., Sauvanaud, L., Mathieu, Y., Al-Bogami, S., Bourane, A., and Al-Ghrami, M., "Direct crude oil cracking for producing chemicals: Thermal cracking modeling," *Fuel,* vol. 211, pp. 726-736, 2018, doi: 10.1016/j.fuel.2017.09.099.
[25] IHS Markit. (2019, 19/02/2020). *Crude Oil-to-Chemicals (COTC).* Available: https://cdn.ihs.com/www/pdf/0519/COTCBrochurePEP.pdf
[26] Kopinke, F. D., Zimmermann, G., Reyniers, G. C., and Froment, G. F., "Relative rates of coke formation from hydrocarbons in steam cracking of naphtha. 2. Paraffins, naphthenes, mono-, di-, and cycloolefins, and acetylenes," *Industrial & Engineering Chemistry Research,* vol. 32, no. 1, pp. 56-61, 1993, doi: 10.1021/ie00013a009.
[27] Albright, L. F., Crynes, B. L., and Corcoran, W. H., *Pyrolysis: Theory and Industrial Practice.* New York: Academic Press Inc, 1983.
[28] Toraman, H. E., Dijkmans, T., Djokic, M. R., Van Geem, K. M., and Marin, G. B., "Detailed compositional characterization of plastic waste pyrolysis oil by comprehensive two-dimensional gas-chromatography coupled to multiple detectors," *Journal of Chromatography A,* vol. 1359, pp. 237-246, 2014, doi: 10.1016/j.chroma.2014.07.017.
[29] Kopinke, F. D., Zimmermann, G., Reyniers, G. C., and Froment, G. F., "Relative rates of coke formation from hydrocarbons in steam cracking of naphtha. 3. Aromatic

hydrocarbons," *Industrial & Engineering Chemistry Research,* vol. 32, no. 11, pp. 2620-2625, 1993, doi: 10.1021/ie00023a027.

[30] Brouwer, M., Picuno, C., Thoden van Velzen, E. U., Kuchta, K., De Meester, S., and Ragaert, K., "The impact of collection portfolio expansion on key performance indicators of the Dutch recycling system for Post-Consumer Plastic Packaging Waste, a comparison between 2014 and 2017," *Waste Management,* vol. 100, pp. 112-121, 2019, doi: 10.1016/j.wasman.2019.09.012.

AIChE Paper Number: 124b

Oxidative Dehydrogenation of Ethane

*Gunther Kracker**
Linde Engineering

Prepared for Presentation at the 2020 AIChE Spring National Meeting
EPC Technology Session: Ethylene Plant Technology Fundamentals

Houston, TX, March 30 – April 1, 2020

AIChE and the EPC shall not be responsible for statements or opinions contained in papers or printed in its publications

**Presenter*
Co Authors: Rainer Kemper, Desislava Tota, Mathieu Zellhuber, Martin Schubert

Oxidative Dehydrogenation of Ethane

Gunther Kracker
Linde Engineering

Abstract:

Currently steam cracking represents the state-of-the-art technology for olefin production, especially for ethylene and propylene. These base petrochemicals are used within a wide range of downstream processes, and as such continuing growth is expected. The steam cracking process is a mature and well-established technology.

The steam cracking feedstock shift to lighter hydrocarbons, primarily ethane, in the last decade has prompted research and development of alternative on-purpose production route to ethylene. In contrast to propylene (e.g. PDH technology), no relevant on-purpose production route has been developed to date for ethylene at commercial scale.

As a promising on-purpose approach the Oxidative Dehydrogenation of Ethane (ODH-E) as a new process technology is introduced in this paper. Linde, partnering with Clariant for the catalyst supply, developed an ODH-E process demonstrating a high selectivity to valuable products with increased profitability compared to steam cracker technology and with a significant reduction potential of CO2 footprint. This paper discusses the development of this promising process technology, its current status, and relevant performance information.

1 Introduction

The majority share of the global olefin production is still today provided by steam cracking which is a mature and well-established technology, but which shows limited development potential.

Now the oxidative dehydrogenation of ethane (ODH-E) offers an interesting alternative to the conventional steam cracking. By catalytic selective conversion of ethane and oxygen, ethylene and acetic acid are formed as commercial high value products under mild conditions at low pressure and temperatures of max. 400°C.

Principle of the ODH-E Process (main reactions to ethylene and acetic acid):

$C_2H_6 + 0.5\ O_2 \rightarrow C_2H_4 + H_2O$ $\qquad \Delta H_R^0 = -105$ kJ/mol

$C_2H_6 + 1.5\ O_2 \rightarrow CH_3COOH + H_2O$ $\qquad \Delta H_R^0 = -642$ kJ/mol

Significant steps have been achieved towards the first-of-its-kind commercial realization of this innovative technology. Presently, no comparable commercial process is available for licensing. The competitive advantage compared to steam cracking mainly comes from the reaction section, which can be operated continuously without decoking or regeneration due to the advantage of oxidative process.

The paper will discuss the ongoing development of this promising ODH-E process technology, its current status, and relevant performance information.

2 Technology Development – Review and Status

Linde´s ODH-E technology is being developed in close cooperation with partner Clariant as catalyst supplier. After successful laboratory and pilot plant tests, a Linde demonstration plant was built up and intensively operated under commercial conditions. A schematic overview of the demo plant arrangement is given in Figure 1. The reactor in the demo plant has been realized as a molten salt cooled, single-tube, fixed-bed reactor with commercial size tube geometry (w.r.t. length and inner diameter).

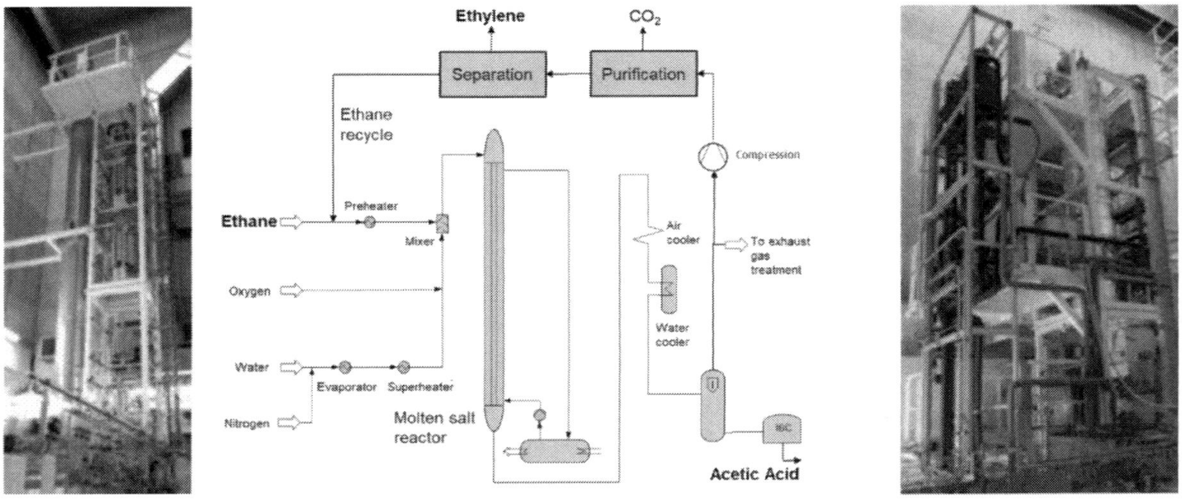

Figure 1: Linde ODH-E Demo Plant

The main focus of the first demonstration phase was on the reactor. Tests have been conducted with various ethane/oxygen/steam feed composition under variation of reaction conditions. The results confirmed laboratory findings regarding high selectivity to valuable products ethylene and acetic acid, with a wide and adjustable range of acetic acid/ethylene production ratios (see Figure 2).

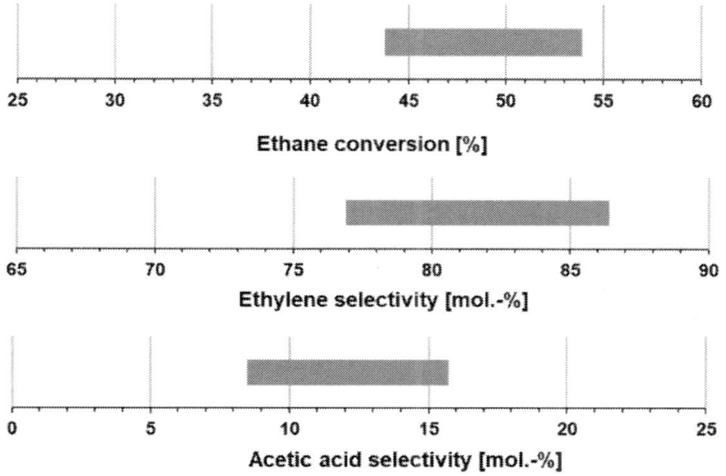

Figure 2: Demo Plant Performance Figures - Ethane Conversion and Product Selectivity

A significant contribution to the demonstration of the new technology has been achieved by operating the demo plant with recycled unconverted ethane. Extending the demo plant with the recycle, the performance data have been validated in a closed-loop plant setup including gas compression, CO2 removal, drying, C2-splitting and closed ethane recycle.

Considering the measured ethane conversion and the product selectivity, overall process performance was confirmed at commercial conditions. At a target ethane conversion at 50% and above, a total selectivity to commercial products (sum of selectivity to ethylene and acetic acid) above 95% was achieved. By choosing different process parameter sets (temperature, pressure, ethane/steam ratio), the desired product ratio of ethylene to acetic acid can be adjusted in a wide range, which is a very attractive feature for potential applications of the ODH-E technology.

As an example shown in Figure 3, a stable plant operation over 1000 hrs accumulated time was realized under commercial conditions. During the long operation campaign with and without closed loop operation, a nearly constant catalyst performance has been observed for all key reaction parameters.

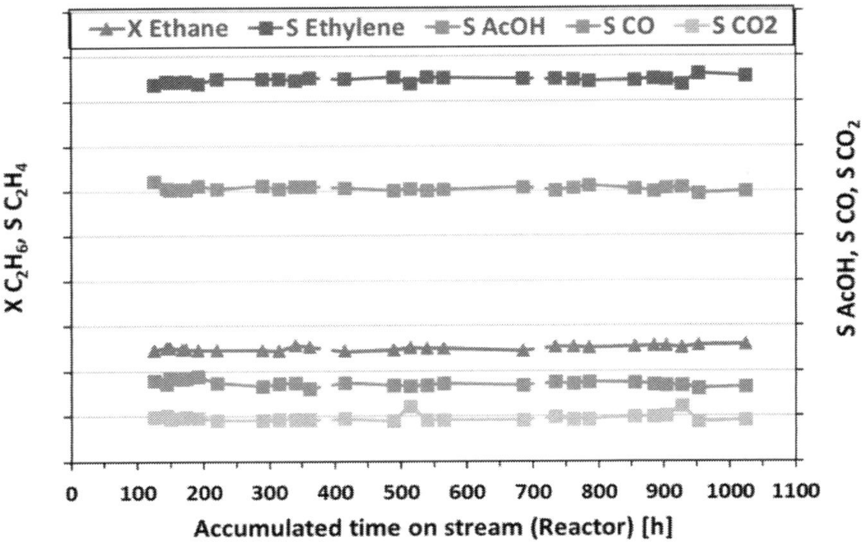

Figure 3: Stable plant operation over 1000 hours (accumulated time)

3 Commercial Technology Layout

The ODH-E technology can be applied as stand-alone facility and takes advantage of integration with a steam cracker. The stand-alone plant concept combines the new developed catalytic process with common well-known process units for purification and gas separation. The process design including a process gas dilution with steam and a cryogenic gas separation train was confirmed to be the most advantageous concept during a systematic evaluation study based on process simulation data. A conceptual design was executed for this preferred process option, resulting in a co-production of ethylene and acetic acid as value products.

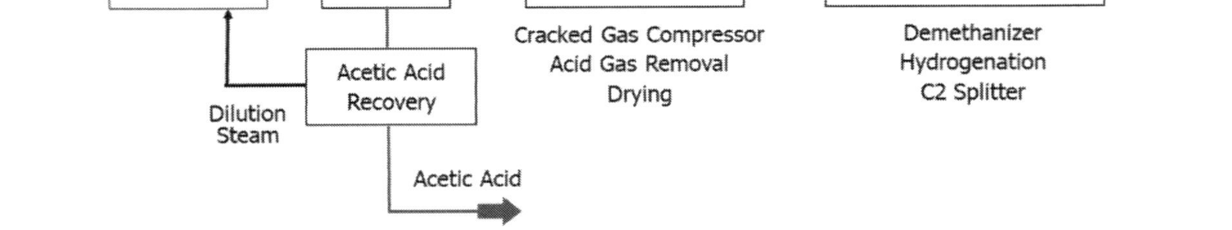

Figure 4: Process Flow Scheme of the Linde ODH-E Technology

Figure 4 provides a process flow overview for a stand-alone ODH-E plant. The fresh feedstock and the recycle streams are converted in the reaction section after mixing with dilution steam and oxygen. The process can take commercially available ethane qualities. An air separation unit upstream of the reactor is foreseen for providing the purified oxygen feed. By catalytic selective conversion of ethane and oxygen under mild conditions, ethylene and acetic acid are formed as main products. Due to parallel and series reactions, the reactor effluent also contains small amounts of by-products CO and CO2, which are removed from the process gas. The reactor effluent is cooled and fed to the water quench column to remove process water and acetic acid. The condensate is routed to a state-of-the-art acetic acid extraction unit to produce high purity acetic acid and recover water for dilution steam.

After separation of water and acetic acid, the process gas leaves the top section of the quench tower to be compressed in a multistage compressor. The compressed gas is then routed for CO2 removal in a purification system comprising an amine wash unit. Here the CO2 product is recovered with high purity for further utilization or storage. Downstream of the amine wash unit, remaining CO2 traces are removed in the caustic wash system.

After CO2 removal the process gas is dried to adsorb water traces before gas processing in the cold separation section. The dried process gas is fed to the cold separation train. After cooling and partial condensation of the C2 products in the low temperature section, the process gas and the condensate stream rich in C2 components are routed to the demethanizer column to separate C2 product from CO and any residual traces of Oxygen. The bottom product is processed in the C2 splitter to recover the ethylene product and recycle the unconverted ethane back to the ODH-E reactor section.

The ODH-E technology can also be integrated with a steam cracker, as shown in Figure 5, which gives an additional flexibility for the application of the ODH-E process. Various combinations, also with existing facilities are possible.

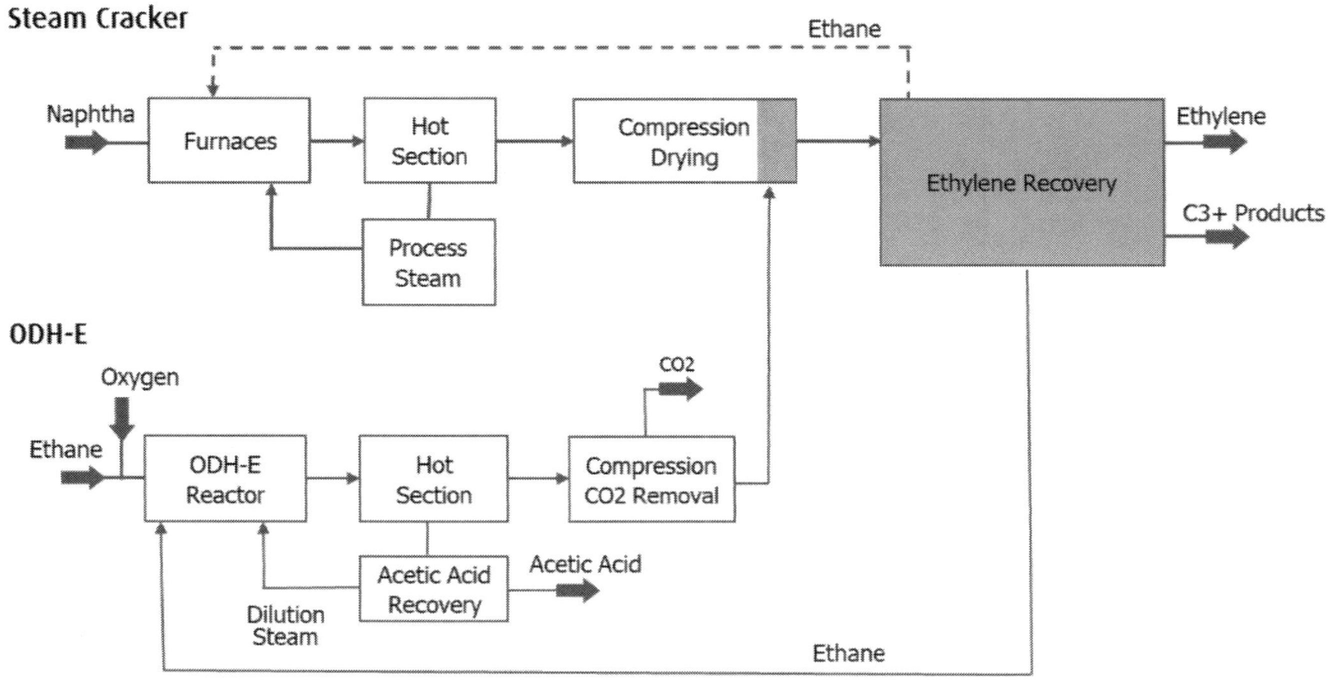

Figure 5: Linde ODH-E Technology Integration with a Steam Cracker

Add-on capacity of ODH-E reactor in existing steam cracker facilities is considered as an attractive business case which can provide additional ethylene production with low investment costs and potential for reducing the CO2 footprint of the combined facility.

4 Reactor Concept of the ODH-E Process

Using the performance data successfully demonstrated by the Linde demo plant and by detailed reactor simulations, a reactor design has been performed for a commercial ODH-E plant with a capacity of 250 kta ethylene.

The new exothermic ODH-E process utilizes an industrially known and well-proven multi-tubular fixed-bed reactor design with molten salt cooling. Figure 6 shows a process design of salt-cooled, fixed-bed reactor. Examples for other industrial applications of the ODH-E reactor system are the synthesis of acrolein/acrylic acid, phthalic and maleic anhydride.

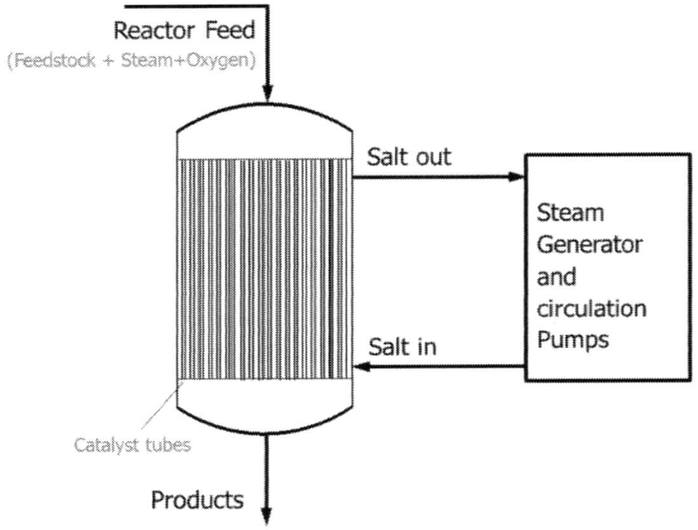

Figure 6: Multitube fixed bed reactor with liquid cooling

The reactor contains a tube bundles filled with the catalyst particles. Within the surrounding reactor jacket flows molten salt as heat transfer medium. The liquid heat carrier is circulated by pumps around the tube bundle to remove the reaction heat of the exothermic oxidative dehydrogenation reaction and to ensure near isothermal conditions. Due to the effective heat transfer a favorable temperature profile in the reactor can be maintained. The salt circulation system includes pumps and a steam generator to use the reaction heat for HP Steam production used for the compressor turbines. The reactor concept ensures highest process safety due to the direct cooling of the tubes.

5 Process Performance Figures

The techno-economic optimization of the commercial scale plant was performed based on rigorous process simulations. Applying the performance figures for a full scale commercial plant layout, an overall material balance and utility consumption have been calculated for a typical stand-alone ODH-E plant with a capacity of 250 kta ethylene.

For the calculated case, the ODH-E plant delivers approximately 0.36 tons of acetic acid per ton of ethylene product, requiring merely 15% higher ethane consumption compared to a steam cracker with the same ethylene capacity. Considering the carbon efficiency[1], the ODH-E process offers a 5% advantage compared to the steam cracking at the same capacity of 250 kta ethylene. This principal advantage of the exothermic ODH-E technology compared to the endothermic steam cracker technology is mostly due to the strongly reduced carbon losses in firing exhaust streams.

Further, the ODH-E technology includes already the carbon capture process step for delivering the CO2 generated in the ODH-E reactor as a by-product in the piping to battery limits for potential utilization in CO2 conversion technologies.

Table 1: Specific CO2 Emissions of the ODH-E Plant (ton CO2 per ton produced ethylene and acetic acid)

	ODH-E Plant including ASU (compressor with E-motor and green power) or ODH-E Plant with import of oxygen		ODH-E Plant including ASU (compressor with E-motor and conventional generated power)		Steam Cracker
	CO2 as carbon captured product	CO2 as emission	CO2 as carbon captured product	CO2 as emission	
Specific CO2 emissions, %	37	46	68	77	100

For the sustainability analysis of the new technology, characteristic values for the greenhouse impact have been calculated to compare the ODH-E process with the steam cracker process. In terms of CO2 emission, even in case of applying conventionally generated electrical power, the ODH-E process offers a considerable improvement compared to the steam cracker processes. For an ODH-E plant including the oxygen production in an Air Separation Unit (ASU) within the battery limits of the ODH-E unit, the specific CO2 emission is reduced by approx. 32% compared to conventional steam cracker in case the captured CO2 is utilized in a downstream application. If no CO2 processing is feasible and CO2 has to be vented, still a 23% CO2 reduction compared to the steam cracker can be claimed.

A further large improvement potential appears considering ODH-E plant including an ASU and utilizing electrical power originating from CO2 neutral or renewable sources. A significant beneficial aspect of the ODH-E technology is a process integration using "green" electricity combined with a "storage" of electrical energy in the form of liquid oxygen generated in the

[1] carbon from feed and fuel gas to carbon from commercial products

ASU. Such integration together with a downstream processing of captured CO2 reduces the specific CO2 emissions by 63% as indicated in Table 1 compared to conventional production with steam crackers.

The application of such a concept would enable an industrial base-chemical production adjusting its energy demand to the availability of renewable electricity, resulting in minimized CO2 footprint. Unlike Power-to-Gas, the objective is not to convert electrical energy to chemical energy for pure energy storage, but to establish more flexible steady production processes and to enable to balance fluctuations in the electrical grid without additional process steps.

6 Benchmarking with Steam Cracking

For the economic analysis, Linde ODH-E technology was compared to state-of-the-art ethane steam cracking technologies. As a reference case, a plant capacity of 250 kta ethylene production was considered. The ODH-E setup considered maximum ethylene production, with corresponding low acetic acid co-production.

A Total Installed Cost (TIC) estimation has been performed for an ethane steam cracker and an ODH-E plant resulting in a verifiable advantage for the ODH-E plant. Feedstock and product prices for the region USA Gulf Coast served as basis for the economic evaluation. Using the Internal Rate of Return (IRR) as indicator for the economic performance, the calculation shows that the ODH-E process profitability increases by 25% compared to the steam cracker. Based on detail analysis, the Linde ODH-E Plant provides higher profitability compared with the steam cracker at same ethylene capacity (at 250 kta ethylene) and is a competitive solution particularly for ethylene and acetic acid demand, allowing for cost-effective and self-sufficient production.

7 Markets

Linde's ODH-E process is hence an on-purpose ethylene technology, which besides ethylene also co-produces a certain amount of acetic acid, with a product split depending on the process conditions. Within the operating window the ODH-E process offers the advantage of a high overall selectivity to valuable products (ethylene + acetic acid selectivity > 95%).

Ethylene consumers, as well as consumers of both, ethylene and acetic acid are the potential target markets. Analysts [1] forecast an increase in the global acetic acid market value to around US$ 16.4 billion by 2026. The market value is estimated to grow at CAGR above 6.2% over the forecast time frame.

As depicted in Figure 7, the acetic acid is mainly converted in synthesis processes to vinyl acetate monomer (VAM), purified terephthalic acid (PTA), acetic anhydride and ethyl acetate.

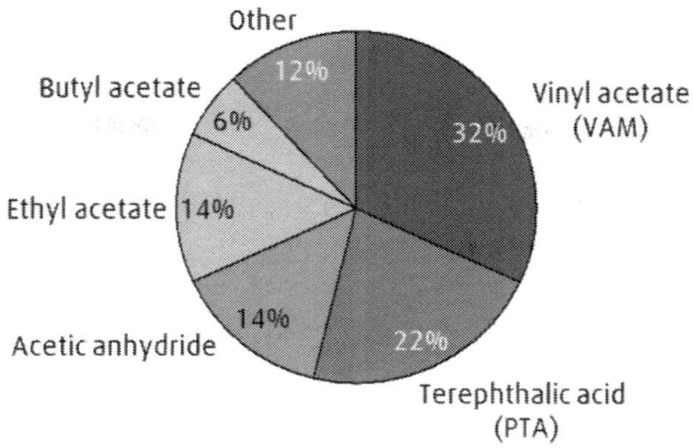

Figure 7: Acetic acid world consumption by end use [2]

VAM and ethyl acetate are products that consume also ethylene feedstock in their manufacturing process. These two products already account for about 46% of the world demand of acetic acid. Moreover, PTA is a product mainly used for the manufacture of polyethylene terephthalate (PET), which consumes ethylene in its value chain to polyethylene terephthalate (PET). The world consumption of acetic acid for PTA accounts for approx. 22%. These three products (VAM, ethyl acetate and PTA), which feature the consumption of ethylene and acetic acid in their production or value chain, account for approx. 60% of the world demand of acetic acid. Therefore, most of the acetic acid demand is linked to products consuming also ethylene. In Table 2 are shown some examples of the required ethylene to acetic acid ratio for these applications.

Table 2: Examples for acetic acid and ethylene consumption in downstream applications

	Acetic acid consumption (t/t product)	Ratio acetic acid/ ethylene (t/t)
Vinyl Acetate Monomer (VAM)	≈ 0.7	≈ 2
Terephthalic Acid (PTA)	≈ 0.05	≈ 0.17[1)]
Ethyl acetate	≈ 0.7	≈ 2
Ethyl Vinyl Acetate (EVA)	≈ 0.2	≈ 0.22

1) For PET production via PTA

On these grounds, the co-production of acetic acid with ethylene allows for a new potential promising market for applications simultaneously demanding ethylene and acetic acid.

8 Summary

A new ODH-E technology has been developed by Linde as an alternative to the conventional ethylene production by steam cracking. The performance of the new process has already been successfully demonstrated by Linde demo plant tests, confirming stable reaction conditions and very high selectivity to the valuable products ethylene and acetic acid, furthermore with an easily adjustable product ratio.

Conceptual design of a commercial ODH-E plant and an economic analysis have been performed for comparison with the conventional steam cracking technology. The Linde ODH-E plant provides higher profitability than steam cracker at same ethylene capacity (at 250 kta ethylene) and is a competitive solution for simultaneous ethylene and acetic acid demand, allowing for cost effective and self-sufficient production.

In terms of CO2 emission and greenhouse impact, the ODH-E process offers a strong improvement potential compared to the conventional steam cracker processes, as it delivers CO2 in the process gas carbon captured and it can directly and significantly benefit from the application of green electricity based on renewables. Therefore, the new ODH-E technology represents a significant step forward on the way to economically and environmentally sustainable solutions for base chemicals production.

References

1. Acumen Research and Consulting (2019), "Acetic Acid Market – Global Industry Size, Share, Trends and Forecast 2019-2026"

2. Chemical Economics Handbook (2013), IHS Chemical

Ab initio studies of the mechanism and kinetics of polymerization fouling in ethylene plants

Ying Yu, Dow, Freeport, TX
Mark Davis, Dow, Freeport, TX
Jorge H. Pazmino, Dow, Freeport, TX

Abstract

Polymerization fouling of butadiene and/or styrene can happen under favorable conditions in downstream separation sections of ethylene plants and is a major cause of shutdowns and safety issues. The general move of USGC crackers to lighter feedstocks and the corresponding longer residences times of many reactive chemicals in the plant can lead to higher fouling potential in different areas of the plant. In this work, the mechanism and kinetics of initiation and propagation reactions of the butadiene-styrene system were investigated using density functional theory (DFT) methods. The overwhelmingly dominant initiation mechanism under typical plant conditions was found to be peroxide decomposition over thermal initiation at ppm levels of peroxide impurities. The similarity between isoprene and butadiene in fouling chemistry was unveiled, such as monomer reactivity ratios in the copolymerization with styrene. The results suggest that isoprene is a good analogue of butadiene in lab studies for copolymerization fouling with the advantages of reducing the risk of chemical exposure and fire/explosion. Furthermore, DFT predictions of the instantaneous relative monomer consumption rates of isoprene-styrene system were validated by experimental results at different initial monomer compositions. It is predicted that in a butadiene-styrene system, butadiene will be consumed faster than styrene even at 70% of styrene in monomer composition, and the overall propagation rate constant will increase with higher styrene fractions. We recommend that anti-fouling treatments should focus on eliminating initiators such as peroxides, and aggressively tailoring additive programs based on radical-forming monomers such as styrene. Overall, this study built a basis for integrating computational study with experimental efforts in understanding the fundamentals of polymerization fouling, which could eventually help provide more effective anti-fouling strategies in the production of light hydrocarbons.

1. Introduction

Fouling is a major cause for shutdowns and safety issues in steam crackers and accounts for substantial economic losses. Polymerization fouling is considered to be the main mechanism of foulant formation in downstream separation sections, where favorable conditions for polymerization of butadiene and/or styrene are present. The shift to share gas has escalated the value of cracking lighter for USGC ethylene plants, leading to higher concentrations and/or longer residences times of many reactive chemicals in different areas of the plant that higher fouling potential. Currently, effective online monitoring or predicting tools for polymerization fouling are not yet available. Analytical studies of fouling face great challenges due to the high complexity of foulant composition. Consequently, rational design of anti-fouling treatments is limited by the lack of fundamental understanding of mechanism and kinetics. Given the variety of plant operating conditions, it is critical to obtain fundamental knowledge of polymerization fouling that could serve as a basis to assess and improve anti-fouling treatments and ultimately increase asset utilization. While experimental studies of the polymerization of hazardous monomers (e.g. butadiene) are presented with strict safety constraints, computational studies on such systems offer a safer and more efficient alternative to provide comprehensive mechanistic insights that help guide ongoing and future experiments, as well as the rational design of anti-fouling strategies.

Polymerization fouling falls in the category of free radical polymerization in which polymers form by the successive addition of free radical building blocks. This process is complex and typically consists of initiation, propagation, chain transfer, and termination reactions. Among them, initiation and propagation reactions are key in this study to both identify initiation mechanisms and elucidate co-propagation kinetics under typical operating conditions. With this information, optimum run conditions can be developed and effective inhibition strategies can be identified to reduce process fouling.

Initiation reactions are responsible for generating free radicals that kick off other reactions. In this study, two initiation mechanisms were proposed for the butadiene-styrene system under typical plant conditions with a temperature from 85 °C to 120 °C: thermal initiation and peroxide decomposition. Thermal initiation mechanism of styrene has been studied both experimentally and computationally, as summarized in Figure 1.[1-11] Two mechanisms were previously proposed for the formation of a key Diels-Alder dimer (**A_2H**) intermediate: either a concerted mechanism[2-3] through transition state **TS_{2-3}**, or a stepwise mechanism via a diradical intermediate (**DR_2**).[4] However, there is no consensus yet as to the actual **A_2H** formation mechanism. Therefore, both mechanisms were considered in this study. Molecule-assisted homolysis[10, 12] between **A_2H** and another styrene molecule results in monoradical species, **A_2•** and **HM_2•**, which initiate polymerization. Two main side reactions include: a **DR_2** ring-closure to form 1,2-diphenylcyclobutane (**DCB**); and an **A_2H** reaction with styrene (**M_2**) to generate an inactive trimer, **A_2M_2**. The thermal initiation mechanism of butadiene has not been reported in the literature. However, dimerization mechanisms of butadiene through Diels-Alder reactions have been widely studied,[13-19] with a simplified dimerization mechanism shown inside the box in Figure 2. The Diels-Alder dimer (**VCH**) is the main product, while a side product, 1,2-divinylcyclobutane (**DVCB**), can be formed by ring closure of the diradical intermediate (**DR_{1tt}**) and may undergo further rearrangement (not shown in Figure 2 as it is beyond the scope of this work).[14, 17-18, 20] In this study, the complete thermal initiation mechanism of butadiene beyond dimerization reactions was mapped as shown in Figure 2 by adapting the mechanism of styrene. **VCH** can react with a butadiene monomer and generate either two monoradicals (**A_1•** and **HM_1•**) that initiate polymerization, or an inactive trimer (**A_1M_1**). Considering that the conversion from *trans*-butadiene to *cis*-butadiene requires an enthalpy change of 3.14 kcal/mol and a barrier height of 7.43 kcal/mol,[21] we used the energy of *trans*-butadiene as the reference point.

The second proposed initiation mechanism is peroxide decomposition. As possible impurities in the cracked gas, organic peroxides can readily undergo decomposition reactions and form radicals to initiate polymerization. Cumene hydroperoxide (**CHP**) was chosen as a representative peroxide in this study. The mechanism and kinetics of **CHP** decomposition to generate two monoradicals has been reported, with a 0.5 reaction order on **CHP** concentration and rate coefficients listed in Table 1.[22] To determine the dominant initiation mechanisms under typical plant conditions, we reported, for the first time, the predicted thermal initiation energetics of styrene and butadiene systems from the same computational framework, and the comparison of initiation rates between thermal initiation and **CHP** decomposition. Plant conditions were approximated using the compositions of liquid samples from separation units, as listed in Table 2. These results led to quantitatively understanding initiation kinetics and to recommendations for rational design of anti-fouling treatments by inhibiting radical formation.

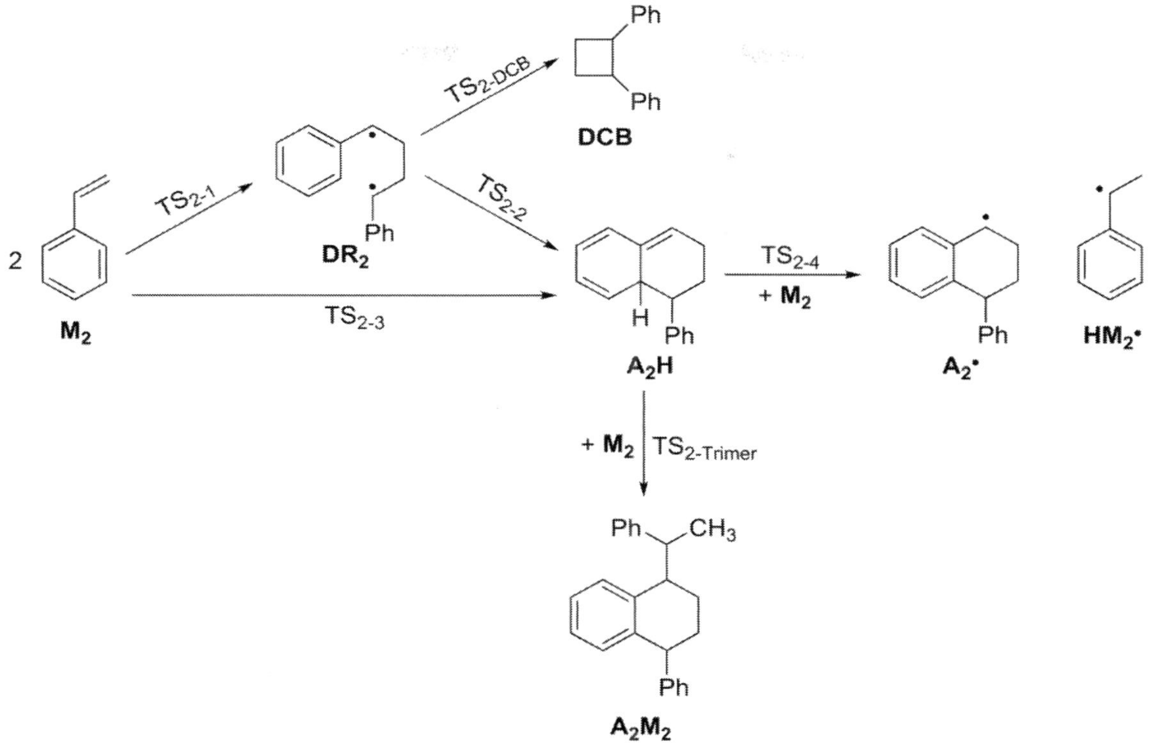

Figure 1. Thermal initiation mechanism of styrene (M_2)

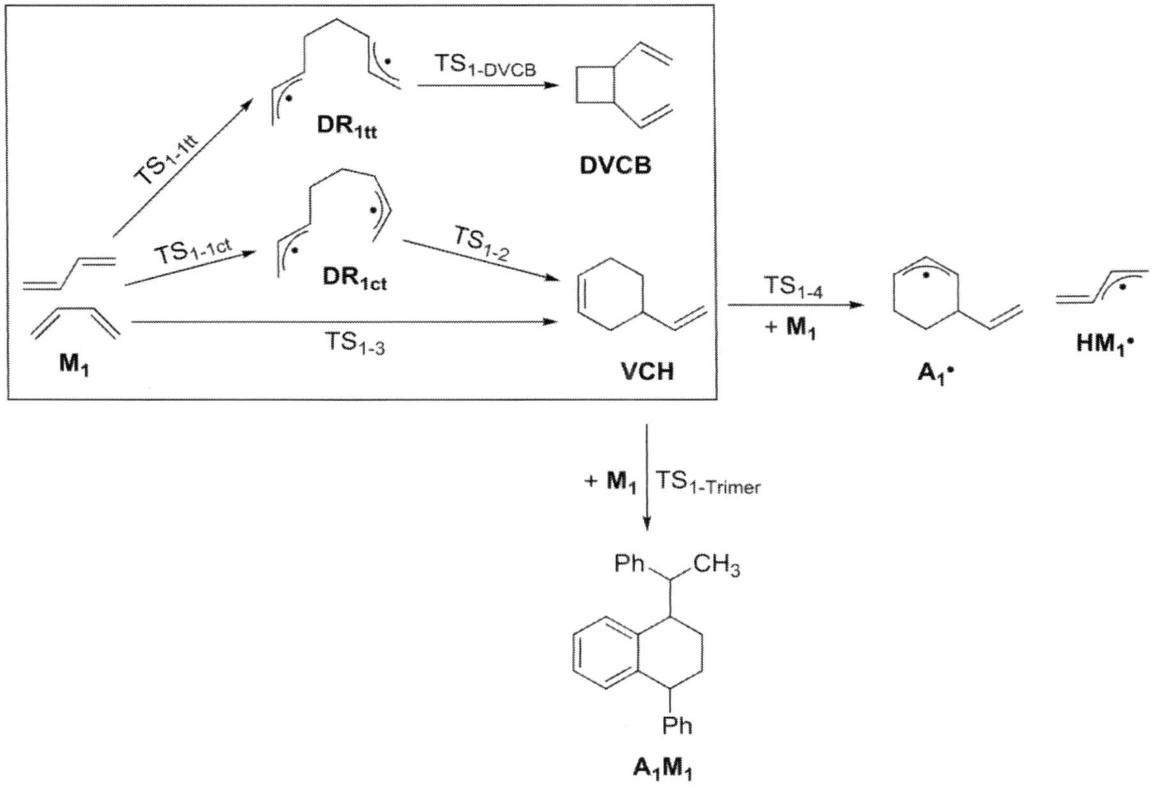

Figure 2. Proposed thermal initiation mechanism of butadiene (M_1)

Table 1. Rate coefficients of CHP decomposition

Overall reaction	E_a (kcal*mol^{-1})	A (mol$^{0.5}$*L$^{-0.5}$*s^{-1})
CHP → 2R•	29.16	1.78E+11

Table 2. Typical composition of process liquid.

Species	Weight Percentage (%)
Butadiene	1-10
Styrene	1-20

Propagation is the main reaction to increase chain length, in which long chain radicals add to unsaturated bonds of the monomer. The terminal model is the most commonly used and efficient model in the study of propagation kinetics in copolymerization. It assumes that chemical reactivity of the growing chain only depends on the unit at the chain end (the ultimate unit). The terminal model has been found to predict reactivity ratios, copolymer composition and sequence distribution accurately.[23] In this study, we adopted this model to describe co-propagation and listed the four propagation reactions of butadiene (**M_1**) and styrene (**M_2**) system in Scheme 1 with rate coefficients k_{ij}, where i and j denote the indexes of the ultimate unit and the monomer, respectively. Extensive experimental studies have been conducted on homo-propagation kinetics of styrene and butadiene, with the measured kinetic parameters listed in Table 3.[24-28] However, considerable variances exist in the reported values, possibly due to differences in reaction conditions. When cross-propagation is considered, monomer reactivity ratios quantify the relative rates of a radical chain reacting with its own type of monomer than with the other monomer, as defined in equations 1 and 2. Table 4 summarizes reported reactivity ratios of butadiene-styrene system from literature. Experimental results cover wide ranges: 1.40-2.20 for r_1, and 0.38-0.7 for r_2, presumably due to the difference in experimental methods and temperature range. Considering the large variances in the origins and values of reported kinetic parameters, a major goal of this study is to investigate co-propagation kinetics of butadiene-styrene system within the same computational framework to ensure consistency in results and enable rational comparison of propagation rates and reactivity ratios. In addition, when the same computational method is used for both initiation and propagation reactions, it will provide a basis for better understanding of the overall co-polymerization kinetics.

$$M_1 \bullet + M_1 \xrightarrow{k_{11}} M_1 \bullet$$

$$M_1 \bullet + M_2 \xrightarrow{k_{12}} M_2 \bullet$$

$$M_2 \bullet + M_1 \xrightarrow{k_{21}} M_1 \bullet$$

$$M_2 \bullet + M_2 \xrightarrow{k_{22}} M_2 \bullet$$

Scheme 1. Terminal model of free radical co-propagation of butadiene (M_1) and styrene (M_2). $M_1\bullet$ and $M_2\bullet$ represent radical chains with the ultimate unit of M_1 or M_2, respectively.

$$r_1 = \frac{k_{11}}{k_{12}} \tag{1}$$

$$r_2 = \frac{k_{22}}{k_{21}} \tag{2}$$

Table 3. Reported experimental kinetic parameters of homo-propagation of butadiene (M_1) and styrene (M_2) systems.

	E_a (kcal*mol^{-1})	**A** (L*mol^{-1}s^{-1})
	6.5±1	1.02E+06
$M_2\bullet + M_2$	5.75	9.72E+06
	7.77	4.27E+07
$M_1\bullet + M_1$	8.53	8.05E+07
	9.30	1.20E+08

Table 4. Reported monomer reactivity ratios of butadiene-styrene system

r_1	**r_2**
1.83[29]	0.65[29]
1.4-2.2[30]	0.5-0.7[30]
1.40[31-32]	0.38[31-32]

To further simplify the mechanistic model for butadiene-styrene co-polymerization, the following assumptions were applied.

1) Propagation rates are independent of radical chain length;

2) Steady state approximation is applied to radical concentration;
3) Chain transfer reactions are not included.

This co-polymerization kinetic model enables us to calculate key kinetic properties in addition to monomer reactivity ratios. Instantaneous relative monomer consumption rate according to Mayo-Lewis equation[33], $\frac{d[M_1]}{d[M_2]}$, and average co-propagation rate coefficient, $<k_p>$, can both be determined as in equations 3 and 4,

$$\frac{d[M_1]}{d[M_2]} = \frac{[M_1](r_1[M_1]+[M_2])}{[M_2]([M_1]+r_2[M_2])} = \frac{1+r_1\frac{f_1}{f_2}}{1+r_2\frac{f_2}{f_1}} \quad (3)$$

$$< k_p > = \frac{r_1 f_1^2+2f_1 f_2+r_2 f_2^2}{\frac{r_1 f_1}{k_{11}}+\frac{r_2 f_2}{k_{22}}} \quad (4)$$

where f_1 and f_2 are the molar fractions of monomer **M_1** and **M_2**, respectively. Predictions of these properties under diverse monomer fractions lead to suggestions of anti-fouling treatments. The fact that these properties can be monitored experimentally provides handles to valid computational predictions and can facilitate integrated computational and experimental studies in the future.

Another goal of this *ab initio* study is to evaluate the similarity of isoprene to butadiene in fouling chemistry for the design of safer lab experiments on polymerization fouling. While butadiene is highly volatile and hazardous, isoprene poses less danger to lab settings and can serve as a good analogue of butadiene. We used computational methods to quantify the similarity between butadiene and isoprene chemistry by comparing their initiation mechanisms and rationalizing their co-propagation kinetics with styrene. Furthermore, DFT predictions of co-polymerization kinetics of isoprene-styrene system was validated by in-house experiments. Ultimately, this study provided a basis for the integration of computational and experimental studies to better understand polymerization fouling and develop anti-fouling strategies.

2. Computational Methods

All DFT calculations were performed in the Gaussian 09 software package.[34] Geometries of all species were optimized in the gas phase using B3LYP[35-36] functional, and 6-31+G(d,p) basis set. This functional was chosen based on its successful application in predicting thermal initiation and propagation energetics of styrene free radical polymerization,[10, 37-39] and the dimerization energetics of butadiene.[16, 40] All TSs were verified by the presence of a single imaginary frequency and the intrinsic reaction coordinate (IRC) was followed in both directions.[41] The lowest free energy ground states and transition states (TS) were found through a combination of conventional geometry optimization and one-dimensional potential energy dihedral scans to improve accuracy. In the case of TS, the bond length corresponding to the vibration of imaginary frequency was kept frozen during the dihedral scans. Diradical species and transition states that lead to diradicals were treated with unrestricted B3LYP, which involves HOMO-LUMO mixing in the initial guess leading to unrestricted wave functions that had higher stability than the corresponding restricted wave functions. (Note: HOMO is short for highest occupied molecular orbital, LUMO is short for lowest unoccupied molecular orbital.) The temperature correction to the free energy was calculated using the frequencies from full optimization at B3LYP/6-31+G(d,p) level of theory with harmonic oscillator (HO) approximation based on standard formulae from statistical thermodynamics.[42] Free energy calculations also included the zero-point energy (ZPE) correction. To improve the accuracy of calculated electronic energy, single-point (SP) energy calculations were performed at the M06-2X/6-311+G(3df,2p)//B3LYP/6-31+G(d,p) level of theory, where M06-2X[43] is the hybrid metageneralized gradient approximation (hybrid-metaGGA) within DFT. Note that the same level of theory was used throughout this study to investigate thermal initiation and propagation reactions of styrene, butadiene and isoprene, which guaranteed the consistency in all DFT calculation results for further comparison and analysis.

Rate coefficients are calculated by combining electronic structure calculations with transition state theory (TST). Individual rate coefficients are calculated using transition state theory, via the expression shown below:

$$k(T) = \kappa(T)\sigma\frac{k_B T}{h}c^{(1-n)}e^{\frac{-\Delta G^{\neq}}{RT}} \tag{5}$$

where k_B is the Boltzmann's constant, T is the temperature at which the polymer synthesis occurs, h is Planck's constant, c is the standard state concentration, n is the molecularity of the reaction, R is the universal gas constant, and $\Delta G^{\neq}$ is the Gibbs free energy barrier of the reaction. The parameter σ represents the reaction path degeneracy, and $\kappa(T)$ is the tunneling correction factor which accounts for quantum mechanical tunneling. One approach to quantify this factor is given by the second order Wigner correction to the reaction rate,[44] in which the correction factor depends on the magnitude, $|v|$, of the imaginary frequency along the reaction coordinate:

$$\kappa(T) = 1 + \frac{1}{24}\left(\frac{h|v|}{k_B T}\right)^2 \tag{6}$$

Rate constants, k(T) were calculated under different temperatures from 25°C to 150°C, covering a wide range of possible operation conditions as well as experimental conditions. Linear regression was done to obtain Arrhenius rate parameters, activation energy E_a, and the pre-exponential factor, A.

3. Results and Discussion

The organization of this section is as follows. First, the thermal initiation mechanisms and energetics are described for styrene, butadiene and isoprene systems, respectively. The thermal initiation rate of butadiene-styrene system under plant conditions is estimated and compared with **CHP** decomposition rate to determine the dominant initiation mechanism. The similarity between isoprene and butadiene in thermal initiation chemistry is then concluded. Next, the co-propagation kinetics of the butadiene-styrene system is presented. Propagation rate parameters and monomer reactivity ratios are reported and validated against experimental results. Similarly, the isoprene-styrene system is examined to reveal its resemblance to the butadiene-styrene system. Finally, two key properties, instantaneous relative monomer consumption rate, $\frac{d[M_i]}{d[M_j]}$, and average propagation rate coefficient, $\langle k_p \rangle$, are predicted for both the butadiene-styrene and isoprene-styrene systems. Experimental validation of the instantaneous relative monomer consumption rate of the isoprene-styrene system is shown to prove the accuracy of DFT predictions. Recommendations for experimental design and anti-fouling treatments are proposed accordingly.

3.1. Initiation Mechanism and Kinetics

3.1.1. Styrene Thermal Initiation

DFT calculations were conducted for each elementary step with a comprehensive conformational search based on the reported mechanism of styrene thermal initiation. Free energy profiles are shown in Figure 3. The main path to produce monoradicals (in black line) and major side reactions (in colored lines) were captured.

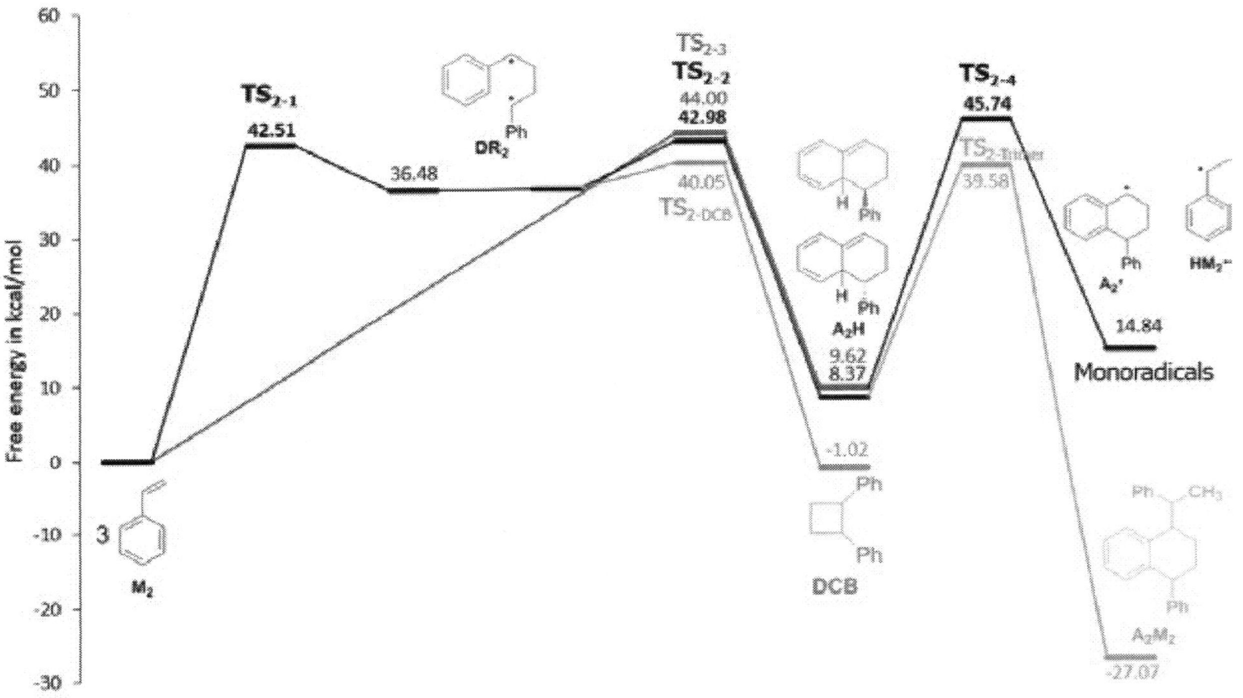

Figure 3. Free energy profiles of styrene thermal initiation at 100°C. Energies are calculated at M06-2X/6-311+G(3df,2p)//B3LYP/6-31+G(d,p) level of theory and are reported in kcal/mol.

Two competing reaction pathways for the formation of the Diels-Alder dimer, **A_2H**, show similar energetics in term of TS heights. The stepwise pathway (black line) is slightly more favorable than the concerted pathway (red line), which agrees with previous literature results.[10] The main side product, **DCB,** competes with **A_2H** in the later stage of stepwise pathway with a TS energy approximately 3 kcal/mol lower. This is in line with experimental evidence of **DCB** formation and the relative TS energies in literature, although a larger difference in TS energies (5.7 kcal/mol) was reported.[10] **A_2H** intermediates are unstable and can react with another styrene monomer to form two monoradicals (**A_2•** and **HM_2•**) or an inactive trimer (**A_2M_2**). The formation of **A_2M_2** is more favorable than monoradicals by a 6 kcal/mol difference in barrier heights. The rate-determining barrier in thermal initiation was found to be the barrier from **A_2H** to monoradicals (**$TS_{2\text{-}4}$**), which is approximately 3 kcal/mol higher than the barrier of **A_2H** formation (**$TS_{2\text{-}2}$**). Given the complexity of the mechanism with side reactions, it is not feasible to obtain a simple analytical rate expression for the rate of monoradical formation, i.e., the thermal initiation rate of styrene. On the other hand, the rate of $\mathbf{A_2H}$ formation reaction can be easily determined and can serve as an upper bound of thermal initiation rate. Considering the small difference in the free energies between $\mathbf{TS_{2\text{-}4}}$ and $\mathbf{TS_{2\text{-}2}}$, an approximation was made that the rate of thermal initiation can be represented by the rate of dimerization reaction as an upper limit. The approximated Arrhenius parameters of styrene thermal initiation were obtained: activation energy E_a=29.84 kcal/mol; and pre-exponential factor $A=9.94\times10^5$ $L{*}mol^{-1}s^{-1}$.

3.1.2. Butadiene Thermal Initiation

DFT calculations were conducted for each elementary step with a comprehensive conformational search based on the proposed mechanism of butadiene thermal initiation, and the free energy profiles are shown in Figure 4. The main path to produce monoradicals (in black line) and major side reactions (in colored lines) were captured. Free energies of all species are relative to that of *trans*-butadiene monomer reactants.

Unlike the thermal initiation mechanism of styrene, the formation of the Diels-Alder dimer (**VCH**) from two butadiene molecules is found to be dominated by the concerted reaction pathway via $\mathbf{TS_{1\text{-}3}}$. The stepwise pathway has a rate-determining TS ($\mathbf{TS_{1\text{-}1ct}}$) which is more than 4 kcal/mol higher in free energy than $\mathbf{TS_{1\text{-}3}}$. The formation of **DVCB** faces a higher barrier than **VCH**, making **VCH** the dominant dimerization product. In contrast to the low stability of $\mathbf{A_2H}$ intermediates in styrene thermal initiation, **VCH** dimers are highly stable with the free energies more than 20 kcal/mol lower compared with starting monomers. Although **VCH** can react with another butadiene monomer to form two monoradicals ($\mathbf{A_1{\bullet}}$ and $\mathbf{HM_1{\bullet}}$) via a transition state ($\mathbf{TS_{1\text{-}4}}$) lower in free energy than $\mathbf{TS_{1\text{-}3}}$, the extraordinary stability of VCH results in an actual free energy barrier of this elementary step as high as 52.34 kcal/mol which effectively shuts down the reaction. The formation of the inactive trimer, $\mathbf{A_1M_1}$, is also inhibited by high energy barriers, although less than the formation of monoradicals. Indeed, **VCH** serves as a “sink species” that traps most of the butadiene dimers in a deep well from further reactions. Considering the low temperature range in this study, it can be concluded that butadiene thermal initiation reactions make negligible contribution to radical formation compared with styrene thermal initiation. However, dimerization of butadiene proceeds much faster than that of styrene due to a free energy barrier approximately 7 kcal/mol lower.

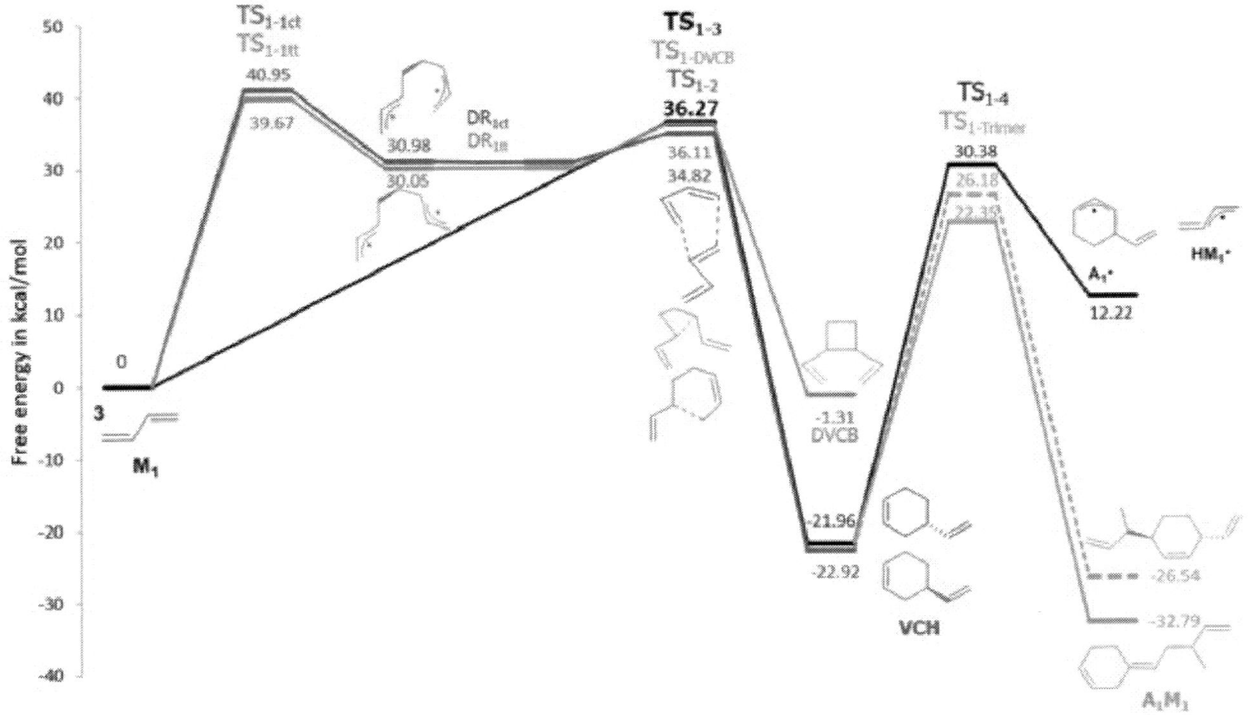

Figure 4. Free energy profiles of butadiene thermal initiation at 100°C. Energies are calculated at M06-2X/6-311+G(3df,2p)//B3LYP/6-31+G(d,p) level of theory and are reported in kcal/mol.

3.1.3. Isoprene Thermal Initiation

DFT calculations were conducted for isoprene thermal initiation following similar mechanism as in the case of butadiene. Free energy profiles are shown in Figure 5, including the main path towards monoradicals (in black line) and major side reactions (in colored lines). The free energies for all species are relative to that of *trans*-isoprene monomer reactants.

Thermal initiation energetics of isoprene show high similarity to those of butadiene. Concerted reaction pathway wins over stepwise reaction pathway towards the formation of Diels-Alder dimers from two isoprene molecules. Formation of Diels-Alder dimers is more favorable than that of the side product, 1,2-divinyl-1,2-dimethylcyclobutane (**DVMCB**). Like in the case of butadiene, Diels-Alder dimers of isoprene are highly stable with the free energies more than 20 kcal/mol lower compared with starting monomers, leading to a free energy barrier as high as 54.27 kcal/mol to form monoradicals. Consequently, reactions that generate monoradicals from Diels-Alder dimers are significantly inhibited, leading to negligible yield to radicals compared with styrene system. It can be concluded that isoprene resembled butadiene in terms of reaction mechanism and energetics of thermal initiation, and is a good analogue of butadiene in the regime of thermal initiation.

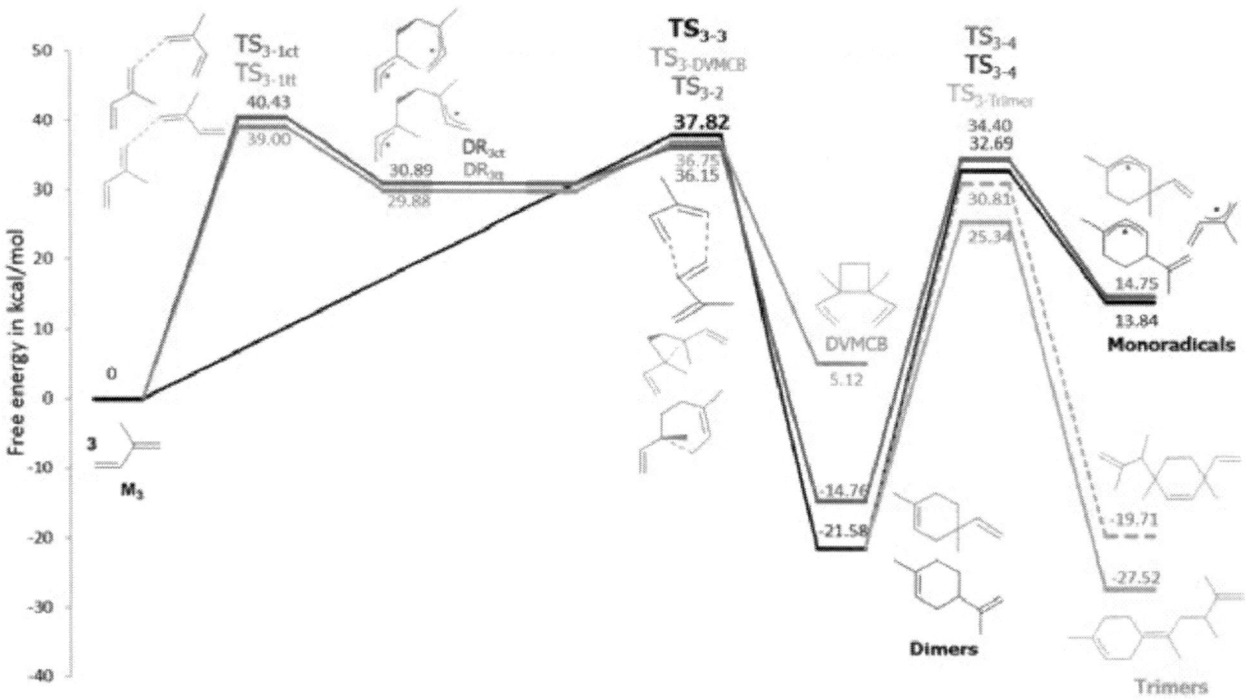

Figure 5. Free energy profiles of isoprene thermal initiation at 100°C. Energies are calculated at M06-2X/6-311+G(3df,2p)//B3LYP/6-31+G(d,p) level of theory and are reported in kcal/mol.

3.1.4. Determine the Dominant Initiation Mechanism

With the energetics of thermal initiation mapped and plant conditions determined (Table 2), a comparison was then made between thermal initiation rate and **CHP** decomposition rate at a representative temperature, 100 °C. The concentration of **CHP** is set to 1 ppm to represent low levels of peroxide within the detection limit. Based on previous results, the major pathway to generate radicals is styrene thermal initiation for both butadiene-styrene and isoprene-styrene systems. Therefore, the overall thermal initiation rate can be represented by the thermal initiation rate of styrene, which is approximated by the rate of stepwise dimerization reaction of styrene. Table 5 summarizes key kinetic parameters of the two initiation mechanisms and the corresponding initiation rates. At such a low **CHP** level, the initiation rate of **CHP** decomposition is three orders of magnitude faster than the thermal initiation rate. In fact, for **CHP** decomposition rate to match thermal initiation rate, only 1.92 ppt **CHP** is needed, which is beyond the limitations of current detecting methods. The stark contrast between initiation rates of the two mechanisms clearly points to peroxide decomposition as the dominant initiation mechanism. To inhibit radical formation from peroxide decomposition, two treatment methods can be proposed. First is to carefully dose proper free radical scavenger into the separation units; second is to lower the impurity level of peroxide or more generally, the level of oxygenates in the cracked gas to slow down the formation of free radicals.

Table 5. Comparison of thermal initiation kinetics between CHP decomposition (1 ppm) and styrene thermal initiation (10 wt% styrene in toluene) at 100 °C

Reaction	E_a (kcal*mol^{-1})	A (mol$^{0.5}$*L$^{-0.5}$*s^{-1})	k (100°C) (mol$^{0.5}$*L$^{-0.5}$*s^{-1})	r (100°C) (mol*L^{-1}s^{-1})
CHP Decomposition	29.16	1.78×10^{11}	1.48×10^{-6}	3.48×10^{-9}
	E_a (kcal*mol^{-1})	**A (L*mol^{-1}s^{-1})**	**k (100°C) (L*mol^{-1}s^{-1})**	**r (100°C) (mol*L^{-1}s^{-1})**
Styrene Thermal Initiation	29.84	9.94×10^{5}	3.30×10^{-12}	2.24×10^{-12}

3.2. Propagation Mechanism and Kinetics

3.2.1. Butadiene-styrene system

Upon the generation of radical species, propagation reactions start and monomers are being consumed at fast rates. Propagation kinetics of butadiene-styrene system (Scheme 1) were first investigated using the same level of theory as in the study of initiation kinetics.

DFT predicted kinetic parameters and rate constants are listed in Table 6. Styrene homo-propagation reaction ($M_2\bullet + M_2$) is almost ten times faster than butadiene homo-propagation reaction ($M_1\bullet + M_1$). However, in cross-propagation reactions ($M_2\bullet + M_1$ and $M_1\bullet + M_2$), butadiene monomer reacts more readily than styrene monomer. The monomer reactivity ratios are 2.04 for butadiene, and 0.41 for styrene at 100°C, which fall in the range of experimental results from literature as listed in Table 4, indicating good agreement with experimental measurements.

Table 6. Propagation kinetics of butadiene-styrene copolymerization

Reaction	E_a (kcal*mol^{-1})	A (L*mol^{-1}s^{-1})	k_p at 100°C (L*mol^{-1}s^{-1})
$M_2\bullet + M_2$	6.63	1.69×10^{6}	2.21×10^{2}
$M_1\bullet + M_1$	10.24	2.43×10^{7}	2.44×10^{1}
$M_2\bullet + M_1$	6.86	5.57×10^{6}	5.34×10^{2}
$M_1\bullet + M_2$	9.08	2.49×10^{6}	1.20×10^{1}

3.2.2. Isoprene-styrene system

Propagation kinetics of isoprene-styrene system were examined to evaluate how well isoprene resembles butadiene in polymerization chemistry. DFT-predicted kinetic parameters and rate constants are summarized in Table 7. Similar to the butadiene-styrene system, the styrene homo-propagation reaction ($M_2\bullet + M_2$) is more than one order of magnitude faster than isoprene homo-propagation ($M_3\bullet + M_3$). In cross-propagation reactions ($M_2\bullet + M_3$ and $M_3\bullet + M_2$), isoprene monomers react more readily than styrene monomers. The monomer reactivity ratios of 2.27 for isoprene and 0.48 for styrene at 100°C reveal the similarity of isoprene to butadiene in co-polymerization with styrene. Because isoprene and butadiene are similar in thermal initiation chemistry and in reactivity ratio, it can be concluded that isoprene is a good analogue of butadiene for the study of polymerization fouling.

Table 7. Propagation kinetics of isoprene-styrene copolymerization

Reaction	**E$_a$** (kcal*mol^{-1})	**A** (L*mol^{-1}s^{-1})	**k$_p$** at **100^0**C (L*mol^{-1}s^{-1})
$M_2 \bullet + M_2$	6.63	1.69×10^6	2.21×10^2
$M_3 \bullet + M_3$	10.17	7.97×10^6	8.80×10^0
$M_2 \bullet + M_3$	6.59	3.33×10^6	4.60×10^2
$M_3 \bullet + M_2$	10.07	3.06×10^6	3.87×10^0

3.3. Key Kinetic Properties of Co-polymerization Reactions

Based on the results of propagation kinetics, two additional key kinetic properties, instantaneous relative monomer consumption rate, $\frac{d[M_i]}{d[M_j]}$, and co-propagation rate coefficient, $\langle k_p \rangle$, can be determined for variant monomer fractions (equations 3 and 4).

Predictions of the instantaneous relative monomer consumption rate as a function of styrene molar fraction for butadiene-styrene and isoprene-styrene systems are shown in Figures 6 and 7, respectively. In both cases, the monomer consumption rate of butadiene or isoprene is higher than that of styrene up to approximately 70% styrene in monomer composition. For butadiene-styrene system, the ratio of monomer consumption rates, $\frac{d[M_1]}{d[M_2]}$, is approximately 4 at typical plant conditions (Table 2) where styrene fraction is 57%. For isoprene-styrene system, the ratio of monomer consumption rates, $\frac{d[M_3]}{d[M_2]}$, is 4.3 when the weight fraction is 5% for isoprene and 10% for styrene. Note that the instantaneous relative monomer consumption rate is an experimentally measurable property which can be used to validate computational predictions. Experiments were conducted for isoprene-styrene system at four different styrene fractions and were plotted in Figure 7 in comparison to DFT predictions. Experimental details are recorded in the supporting information. Across the styrene fraction range from 0.22 to 0.95, both DFT predictions and experimental measurements show decreasing trends of $\frac{d[M_3]}{d[M_2]}$ as a function of styrene fraction. In addition, the absolute values of $\frac{d[M_3]}{d[M_2]}$ from DFT predictions are in close proximity of experimental measurements, which validated the strong predictive power of DFT calculations on isoprene-styrene system to facilitate integrated computational and experimental studies in the future. Considering the similarity between isoprene and butadiene in polymerization fouling, computational results of butadiene-styrene system could be used with more confidence under different plant conditions.

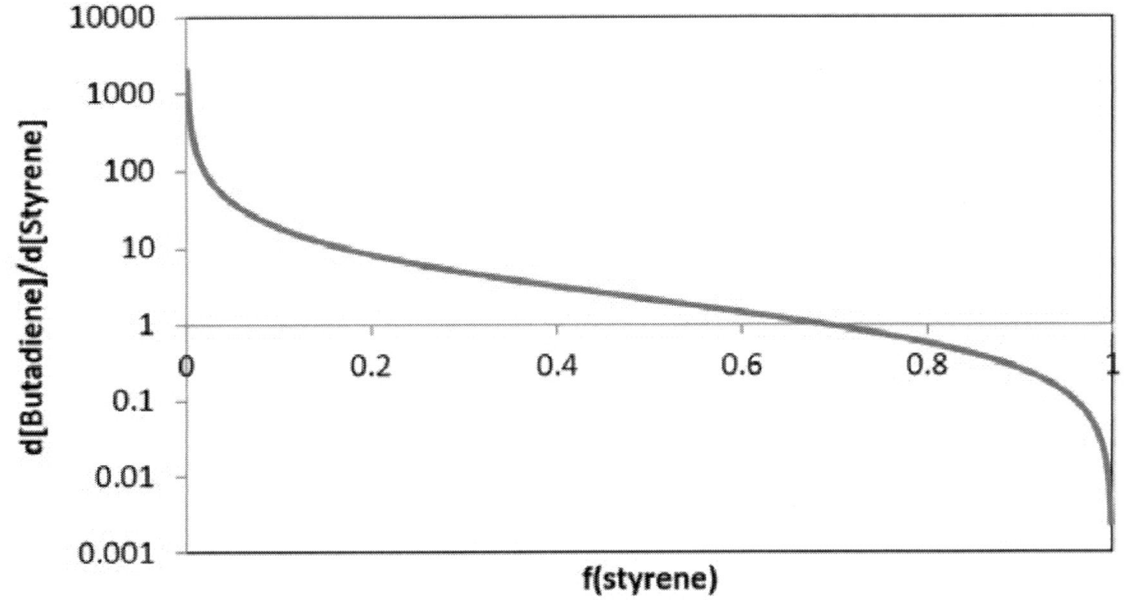

Figure 6. Relationship between styrene (M_2) fraction and the relative monomer consumption rate of butadiene (M_1)/styrene (M_2) system at 100°C.

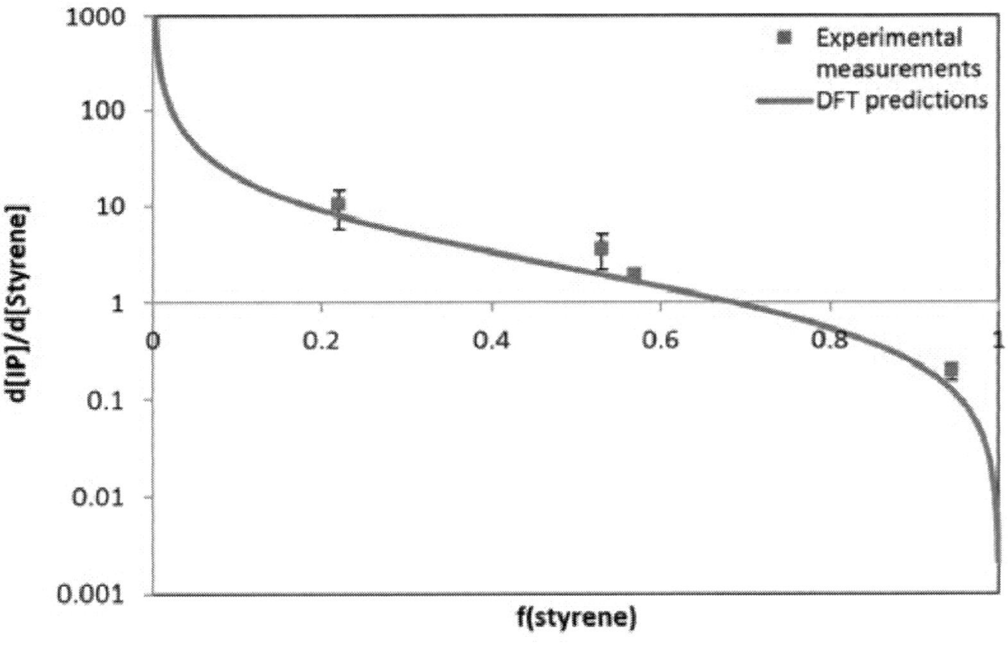

Figure 7. Relationship between styrene (M_2) fraction in monomer feed and the relative monomer consumption rate of isoprene (M_3)/styrene (M_2) system at 100°C. Experimental error bars show standard deviation.

The average co-propagation rate coefficient, $<k_p>$, as a function of styrene fraction were calculated for both butadiene-styrene and isoprene-styrene systems, as shown in Figure 8 and 9, respectively. In both cases, the co-propagation rate coefficient increases with styrene fraction, leading to faster formation of polymer foulant. The average co-propagation rate coefficient hits the highest value when the monomer composition is pure styrene, which is in line with the fact that styrene homo-propagation has the largest rate constants as in Tables 6 and 7. Correspondingly, a treatment method based on this relationship is to lower the fraction of styrene in the cracked gas. It may be achieved by lowering the temperature of the gas mixture and separating styrene at earlier stages.

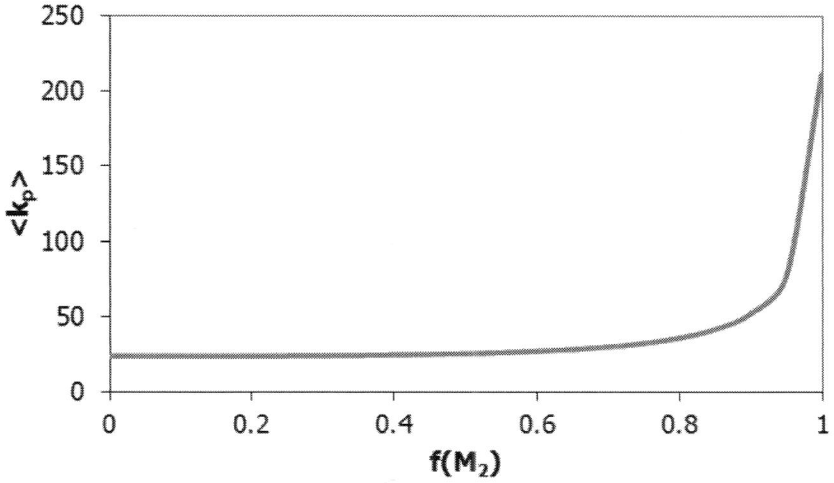

Figure 8. Relationship between styrene fraction and the overall propagation rate of butadiene-styrene system at 100 °C.

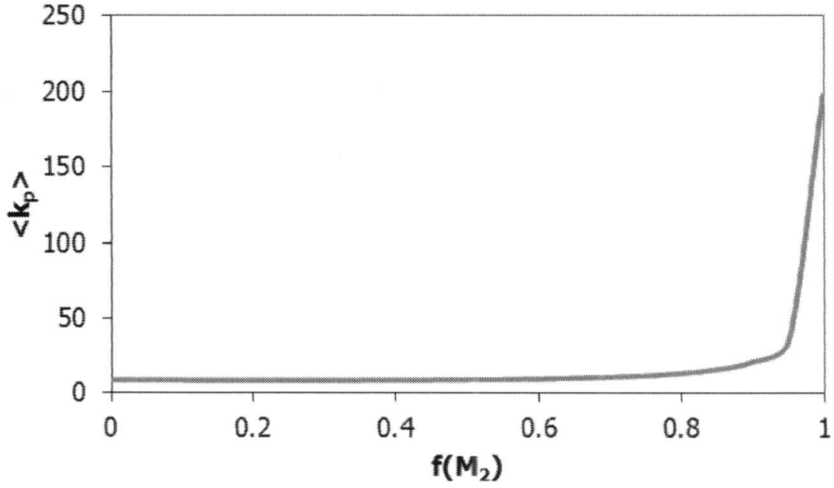

Figure 9. Relationship between styrene fraction and the overall propagation rate of isoprene-styrene system at 100 °C.

4. Conclusions

This study presents a broad yet in-depth investigation of the fundamentals of polymerization fouling using computational methods. Mechanism and kinetics of initiation and propagation reactions of the butadiene-styrene system were rationalized within the same computational framework. The dominant initiation mechanism at typical plant conditions was identified as peroxide decomposition over thermal initiation at ppm level of **CHP** impurity. Although both being extremely slow processes, thermal initiation of styrene is much more favorable than that of butadiene. This is due to the unusually high stability of butadiene Diels-Alder dimers that prevent them from further reactions. Results of co-propagation kinetics of butadiene-styrene system achieved good agreement with literature reports. The predicted reactivity ratios are 2.04 for butadiene and 0.41 for styrene, indicating a higher reactivity of butadiene monomers in propagation reactions. In addition, isoprene was validated as a good analogue to butadiene in polymerization fouling for lab experimental studies given safety considerations. Further analysis provided two more key kinetic properties that can be readily measured. One is the instantaneous relative monomer consumption rate, and the DFT predictions of this property were validated by experimental measurements for isoprene-styrene system; the other is the average co-propagation rate coefficient, which leads to a prediction that the overall polymerization would go faster with higher styrene fraction. Based on these results, we suggest that anti-fouling treatment should focus on lowering oxygenate levels, inhibiting free radical activity, and controlling styrene content in the cracked gas. This work also suggests future directions in elucidating the origins of trace peroxide species to better monitor and control polymerization fouling. Overall, this study provided a basis for integrating computational study with experimental efforts in understanding the fundamentals of polymerization fouling, which could eventually help provide more effective anti-fouling strategies for light hydrocarbon crackers.

References

1. Russell, K. E.; Tobolsky, A. V., Thermal Initiation of Styrene Polymerization. *J Am Chem Soc* **1953,** *75* (20), 5052-5054.
2. Mayo, F. R., Dimerization of Styrene. *J Am Chem Soc* **1968,** *90* (5), 1289-&.
3. Mayo, F. R., Chain Transfer in the Polymerization of Styrene .8. Chain Transfer with Bromobenzene and Mechanism of Thermal Initiation. *J Am Chem Soc* **1953,** *75* (24), 6133-6141.
4. Flory, P. J., The mechanism of vinyl polymerizations. *J Am Chem Soc* **1937,** *59*, 241-253.
5. Bengough, W. I.; Park, G. B., Thermal Polymerization of Styrene in Diethyladipate at Temperatures up to 160degrees. *Eur Polym J* **1976,** *12* (7), 431-433.
6. Barr, N. J.; Bengough, W. I.; Beveridge, G.; Park, G. B., Measurement of Rates of Initiation in Thermal Polymerization of Styrene. *Eur Polym J* **1978,** *14* (4), 245-250.
7. Chong, Y. K.; Rizzardo, E.; Solomon, D. H., Confirmation of the Mayo Mechanism for the Initiation of the Thermal Polymerization of Styrene. *J Am Chem Soc* **1983,** *105* (26), 7761-7762.
8. Buzanowski, W. C.; Graham, J. D.; Priddy, D. B.; Shero, E., Spontaneous Polymerization of Styrene in the Presence of Acid - Further Confirmation of the Mayo Mechanism. *Polymer* **1992,** *33* (14), 3055-3059.
9. Devonport, W.; Michalak, L.; Malmstrom, E.; Mate, M.; Kurdi, B.; Hawker, C. J.; Barclay, G. G.; Sinta, R., "Living" free-radical polymerizations in the absence of initiators: Controlled autopolymerization. *Macromolecules* **1997,** *30* (7), 1929-1934.
10. Khuong, K. S.; Jones, W. H.; Pryor, W. A.; Houk, K. N., The mechanism of the self-initiated thermal polymerization of styrene. Theoretical solution of a classic problem. *J Am Chem Soc* **2005,** *127* (4), 1265-1277.
11. Alsharaeh, E. H.; El-Shall, M. S., Ion mobility study of the mechanism of the gas phase thermal polymerization of styrene and the structures of the early oligomers. *Polymer* **2011,** *52* (24), 5551-5559.
12. Pryor, W. A.; Coco, J. H.; Daly, W. H.; Houk, K. N., Radical Generation from Polymolecular Reactions of Closed Shell Molecules - Molecule-Assisted Homolysis (Mah) - Hydrogen-Atom Transfer from a Diels-Alder Adduct to an Alkene. *J Am Chem Soc* **1974,** *96* (17), 5591-5593.
13. Caramella, P.; Quadrelli, P.; Toma, L.; Romano, S.; Khuong, K. S.; Northrop, B.; Houk, K. N., The three corrugated surfaces of 1,4-divinyltetramethylene diradical intermediates and their connections to 1,2-divinylcyclobutane, 4-vinylcyclohexene, 1,5-cyclooctadiene, and two butadienes. *J Org Chem* **2005,** *70* (8), 2994-3008.
14. Vaughan, W. E., The homogeneous thermal polymerization of 1,3-butadiene. *J Am Chem Soc* **1932,** *54*, 3863-3876.
15. Scarborough, D. L. A.; Kobayashi, R.; Thompson, C. D.; Izgorodina, E. I., Active space and basis set effects in CASPT2 models of the 1,3-butadiene-ethene cycloaddition and the 1,3-butadiene dimerization. *Int J Quantum Chem* **2015,** *115* (15), 989-1001.
16. Quadrelli, P.; Romano, S.; Toma, L.; Caramella, P., Merging and bifurcation of 4+2 and 2+4 cycloaddition modes in the archetypal dimerization of butadiene. A case of competing bispericyclic, pericyclic and diradical paths. *Tetrahedron Lett* **2002,** *43* (48), 8785-8789.
17. Aldeeb, A. A.; Rogers, W. J.; Mannan, M. S., Evaluation of 1,3-butadiene dimerization and secondary reactions in the presence and absence of oxygen. *J Hazard Mater* **2004,** *115* (1-3), 51-56.
18. Duncan, N. E.; Janz, G. J., The Thermal Dimerization of Butadiene, and the Equilibrium between Butadiene and Vinylcyclohexene. *J Chem Phys* **1952,** *20* (10), 1644-1645.
19. Stephenson, L. M.; Gemmer, R. V.; Current, S., Dimerization of Butadiene. *J Am Chem Soc* **1975,** *97* (20), 5909-5910.

20. Reed, H. W. B., The Formation of Trans-1-2-Divinylcyclobutane from Butadiene. *J Chem Soc* **1951,** (Mar), 685-687.
21. Durig, J. R.; Bucy, W. E.; Cole, A. R. H., Torsional Potential Function of Glyoxal. *Can J Phys* **1975,** *53* (19), 1832-1837.
22. Duh, Y. S.; Kao, C. S.; Hwang, H. H.; Lee, W. W. L., Thermal decomposition kinetics of cumene hydroperoxide. *Process Saf Environ* **1998,** *76* (B4), 271-276.
23. Odian, G. G., *Principles of polymerization*. 4th ed.; Wiley: Hoboken, N.J., 2004; p xxiv, 812 p.
24. Bamford, C. H.; Dewar, M. J. S., Studies in Polymerization .1. A Method for Determining the Velocity Constants Polymerization Reactions and Its Application to Styrene. *Proc R Soc Lon Ser-A* **1948,** *192* (1030), 309-328.
25. Deady, M.; Mau, A. W. H.; Moad, G.; Spurling, T. H., Evaluation of the Kinetic-Parameters for Styrene Polymerization and Their Chain-Length Dependence by Kinetic Simulation and Pulsed-Laser Photolysis. *Makromol Chem* **1993,** *194* (6), 1691-1705.
26. Buback, M.; Gilbert, R. G.; Hutchinson, R. A.; Klumperman, B.; Kuchta, F. D.; Manders, B. G.; Odriscoll, K. F.; Russell, G. T.; Schweer, J., Critically Evaluated Rate Coefficients for Free-Radical Polymerization .1. Propagation Rate Coefficient for Styrene. *Macromol Chem Physic* **1995,** *196* (10), 3267-3280.
27. Deibert, S.; Bandermann, F.; Schweer, J.; Sarnecki, J., Propagation Rate Coefficient of Free-Radical Polymerization of 1,3-Butadiene. *Makromol Chem-Rapid* **1992,** *13* (7), 351-355.
28. Morton, M.; Salatiello, P. P.; Landfield, H., Absolute Propagation Rates in Emulsion Polymerization .2. Butadiene in Hydroperoxide-Polyamine Systems. *J Polym Sci* **1952,** *8* (2), 215-224.
29. Gilbert, R. D.; Williams, H. L., Reactivity Ratios of Conjugated Dienes Copolymerized in Emulsion at 5-Degrees. *J Am Chem Soc* **1952,** *74* (16), 4114-4118.
30. Mitchell, J.; Williams, H. L., LABORATORY STUDY OF LOW RATIO BUTADIENE–STYRENE COPOLYMERS. *Canadian Journal of Research* **1949,** *27* (2), 35-46.
31. Orr, R. J., Thermochemical Aspects of Butadiene Styrene Copolymerization. *Polymer* **1961,** *2* (1), 74-82.
32. Kalfus, M.; Grzywa, E., Modeling and Simulation of Chemical Homogeneity of Butadiene-Styrene Copolymer in High-Conversion Emulsion Co-Polymerization. *Polym J* **1978,** *10* (6), 583-590.
33. Mayo, F. R.; Lewis, F. M., Copolymerization I A basis for comparing the behavior of monomers in copolymerization, the copolymerization of styrene and methyl methacrylate. *J Am Chem Soc* **1944,** *66*, 1594-1601.
34. Frisch, M. J.; Trucks, G. W.; Schlegel, H. B.; Scuseria, G. E.; Robb, M. A.; Cheeseman, J. R.; Scalmani, G.; Barone, V.; Mennucci, B.; Petersson, G. A.; Nakatsuji, H.; M., C.; Li, X.; Hratchian, H. P.; Izmaylov, A. F.; Bloino, J.; Zheng, G.; Sonnenberg, J. L.; Hada, M.; Ehara, M.; Toyota, K.; Fukuda, R.; Hasegawa, J.; Ishida, M.; Nakajima, T.; Honda, Y.; Kitao, O.; Nakai, H.; Vreven, T.; Montgomery, J., J. A.; E., P. J.; Ogliaro, F.; Bearpar, M.; Heyd, J. J.; Brother, E.; Kudin, K. N.; Staroverov, V. N.; Keith, T.; Kobayashi, R.; Normand, J.; Raghavachari, K.; Rendell, A.; Burant, J. C.; Iyengar, S. S.; Tomasi, J.; Cossi, M.; Rega, N.; Millam, J. M.; Klene, M.; Knox, J. E.; Cross, J. B.; Bakken, K.; Adamo, C.; Jaramillo, J.; Gomperts, R.; Stratmann, R. E.; Yazyev, O.; Austin, A. J.; Cammi, R.; Pomelli, C.; Ochterski, J. W.; Martin, R. L.; Morokuma, K.; Zakrzewski, V. G.; Voth, G. A.; Salvador, P.; Dannenberg, J. J.; Dapprich, S.; Daniels, A. D.; Farkas, O.; Foresman, J. B.; Ortiz, J. V.; Cioslowski, J.; Fox, D. J. *Gaussian 09*, Revision D.01; Gaussian, Inc.: 2013.
35. Becke, A. D., Density-Functional Thermochemistry .3. The Role of Exact Exchange. *J Chem Phys* **1993,** *98* (7), 5648-5652.
36. Lee, C. T.; Yang, W. T.; Parr, R. G., Development of the Colle-Salvetti Correlation-Energy Formula into a Functional of the Electron-Density. *Phys Rev B* **1988,** *37* (2), 785-789.

37. Coote, M. L., Quantum-Chemical Modeling of Free-Radical Polymerization. *Macromol Theor Simul* **2009,** *18* (7-8), 388-400.
38. Mavroudakis, E.; Liang, K.; Moscatelli, D.; Hutchinson, R. A., A Combined Computational and Experimental Study on the Free-Radical Copolymerization of Styrene and Hydroxyethyl Acrylate. *Macromol Chem Physic* **2012,** *213* (16), 1706-1716.
39. Dossi, M.; Storti, G.; Moscatelli, D., A Quantum Chemistry Study of the Free-Radical Copolymerization Propagation Kinetics of Styrene and 2-Hydroxyethyl Acrylate. *Polym Eng Sci* **2011,** *51* (10), 2109-2114.
40. Quadrelli, P.; Romano, S.; Toma, L.; Caramella, P., A bispericyclic transition structure allows for efficient relief of antiaromaticity enhancing reactivity and endo stereoselectivity in the dimerization of the fleeting cyclopentadienone. *J Org Chem* **2003,** *68* (15), 6035-6038.
41. Martinez, E. P.; Gonzalez, C. S.; Reyes, A. R., Variability and Increase in the Reaction-Time in the Early Diagnosis of Subclinical Hepatic-Encephalopathy. *Rev Esp Enferm Dig* **1990,** *77* (1), 29-32.
42. McQuarrie, D. A., *Statistical Mechanics*. University Science Books: 2000.
43. Zhao, Y.; Truhlar, D. G., The M06 suite of density functionals for main group thermochemistry, thermochemical kinetics, noncovalent interactions, excited states, and transition elements: two new functionals and systematic testing of four M06-class functionals and 12 other functionals. *Theor Chem Acc* **2008,** *120* (1-3), 215-241.
44. Wigner, E., The transition state method. *T Faraday Soc* **1938,** *34* (1), 0029-0040.

CFD FOR TURBOMACHINERY
WHAT TO EXPECT—AND WHAT NOT TO EXPECT

James R. Hardin, Elliott Group, Jeannette, PA, USA

Biography

James Hardin is a Senior Engineer in the Research and Development department at Elliott Group, where he performs computational fluid dynamics (CFD) and other aerodynamic analyses for turbines and compressors. Previous experience includes CFD and other analyses on shipboard propulsion and piping systems with Westinghouse Electric Corporation, and turbine design support and testing at Elliott Group. He has 38 years of engineering experience in aerodynamics and fluid systems. Mr. Hardin received a B.S. degree (Mechanical Engineering, 1981) from Carnegie-Mellon University and is a registered Professional Engineer in the State of Pennsylvania.

Introduction

It hasn't taken long for computational fluid dynamics (CFD) to move from a fringe academic pursuit to a routine part of every OEM's design process. Thirty years ago, an industrial user was thrilled to run a 10,000-node model on a single core of an expensive Unix workstation. Now we think nothing of running tens of millions of nodes on over a hundred cores spread among networked PCs, servers in a compute cluster, or machines out there in the "cloud." This expansion has been fueled by advances in CFD methods, incredible increases in computer capabilities, and the never-ending demand by users for more thorough and more accurate analyses.

But what do you get from all this computational firepower? Especially if you're not the one doing the CFD analyses, but need to work with the analysts who are?

CFD has some unique advantages:

- It shows more details of the flow field than would be practical to capture with a test. A test can accommodate only a limited number of instruments, especially when the test is being done on a unit in the field instead of a dedicated test rig. Seeing these flow details can show *why* performance is what it is and drive design changes or troubleshooting based on data.
- It enables predicting performance on machines that have not yet been tested, or perhaps even built. It is true that CFD is sometimes over-hyped as a "virtual test stand," and its predictions should be taken with several grains of salt. Nonetheless, properly done CFD results are of useful accuracy, even if not perfect accuracy.
- It enables evaluating designs quickly. CFD is slow compared to other analytical methods, but is fast compared to building and testing parts.

But CFD also has some important limitations:

- Like any analytical method, it will not predict reality exactly. CFD can predict differences between similar models better than it can predict real-world performance.

- It requires considerable expertise. There are many ways for a CFD analysis to provide unrealistic results even if it seems to run well, many of which are discussed in this paper. Knowledge of CFD and practical fluid mechanics is necessary to avoid these pitfalls.
- It requires considerable time. In some cases, especially field problems with limited equipment downtime available, decisions may have to be made sooner than the CFD analyses can be completed.
- It requires considerable computer power. For larger models with several operating points to analyze, cases are often run on hundreds of computer cores for several days.
- All this time and computer power can add up to considerable cost.

Fortunately, some upfront planning can go a long way toward maximizing those advantages while avoiding much of the potential harm caused by those limitations.

CFD is not an "all or nothing" proposition. It's not just deciding whether to perform CFD analyses, but rather, deciding how complex a CFD analysis is required. This is driven largely by what question needs to be answered, but can also be influenced by how much time, computational resources, and money is available. In practical terms, it usually comes down to deciding how little one can get away with and still get the needed answers. A "good enough" answer is far more useful than a needlessly precise answer that arrives too late to be used.

Any engineering job should be planned at the outset, but this is especially true of CFD analyses, since they have the potential to waste so much time and money. A fully transient run of an entire machine with a very fine mesh may give exceptionally accurate results, but that level of precision may not be needed to answer the question at hand.

Regardless of whether you're a CFD analyst or are working with others who are, it's helpful to be familiar with some of the different kinds of CFD analyses available and when they should be used. Some of the more common analysis types are discussed below. Other variations are possible, and aspects of these can be mixed and matched as needed for a particular problem.

Steady-State Single-Passage

In this approach, just one blade passage of each turbomachinery component is modeled, as shown in Figure 1. The model may comprise just one component, such as an impeller, or single passages for multiple components can be assembled end-to-end, making up a stage or multiple stages. The components don't have to have the same number of passages per circle; any CFD code intended for turbomachinery will correctly adjust the mass flow as the circumferential extent changes from component to component.

The important assumptions are that the flow is *steady-state*, not changing over time (no pulsations or vortex shedding or that kind of thing) and that it's *periodic*, that is, the flow in every passage of a component is the same. Despite how the model looks, flow does not have to be bounded within each passage; it's free to move over to the passage next door, as shown in Figure 2. But there must be an identical flow coming into the modeled passage from the passage next door on the other side. The boundaries in the circumferential direction are called, appropriately, periodic boundaries. They are not walls; they just enforce constraints that flows, pressures, temperatures, and everything are the same on both sides.

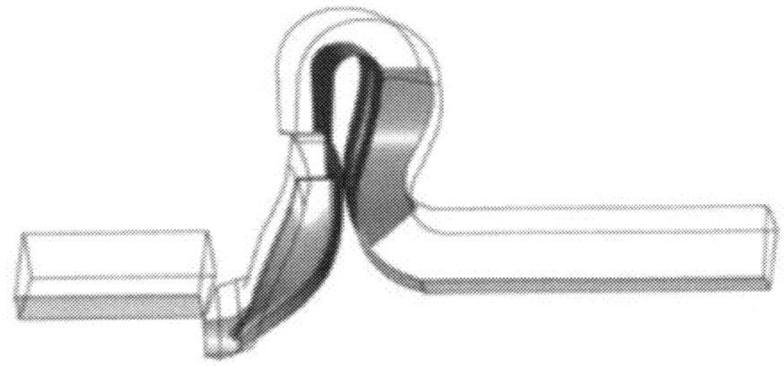

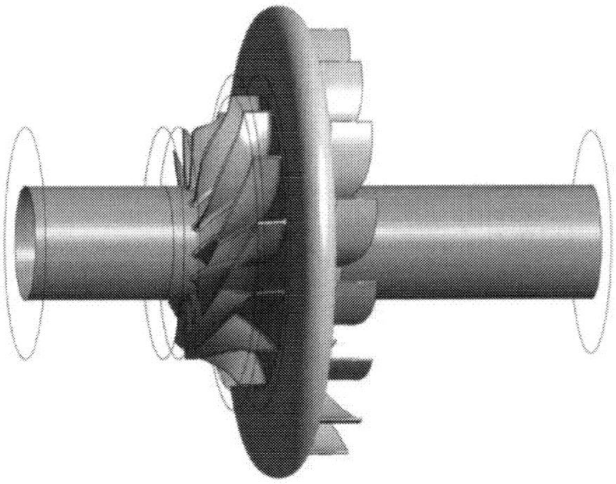

Figure 1. The single-passage model on the left represents the full-circle geometry on the right.

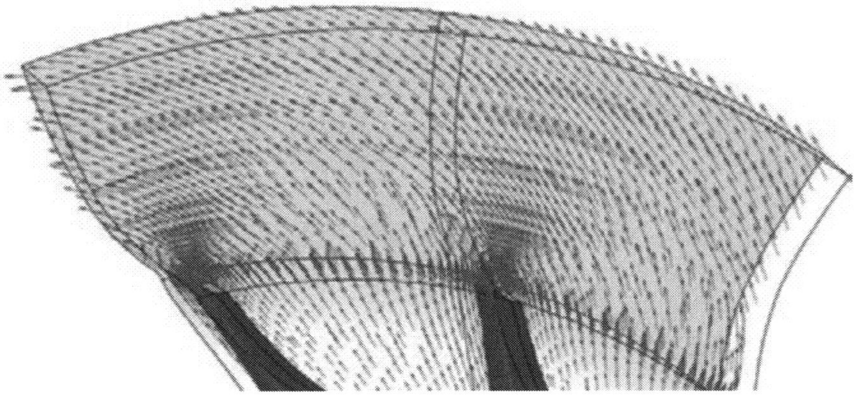

Figure 2. Flow moves freely from one passage to the next through periodic boundaries.

What happens when flow crosses between rotating and stationary components? Usually, one of two assumptions is made. Which is more appropriate depends on some difficult judgment calls rather than firm rules. It comes down to understanding what effects are important to answer a particular question.

In a *stage interface*, flow is averaged circumferentially as it crosses the interface, essentially smearing out to be uniform in the circumferential direction while keeping its spanwise variation, as shown in Figure 3. This approach is usually used for adjacent blade and vane rows in turbomachinery. It assumes that the circumferential position of each blade passage doesn't matter much. This works well if the vane and blade counts across the interface are similar, say within about 50 percent, and the rotational speed is high enough that wake effects and such tend to be spread across the passage instead of localized relative to the blades or vanes.

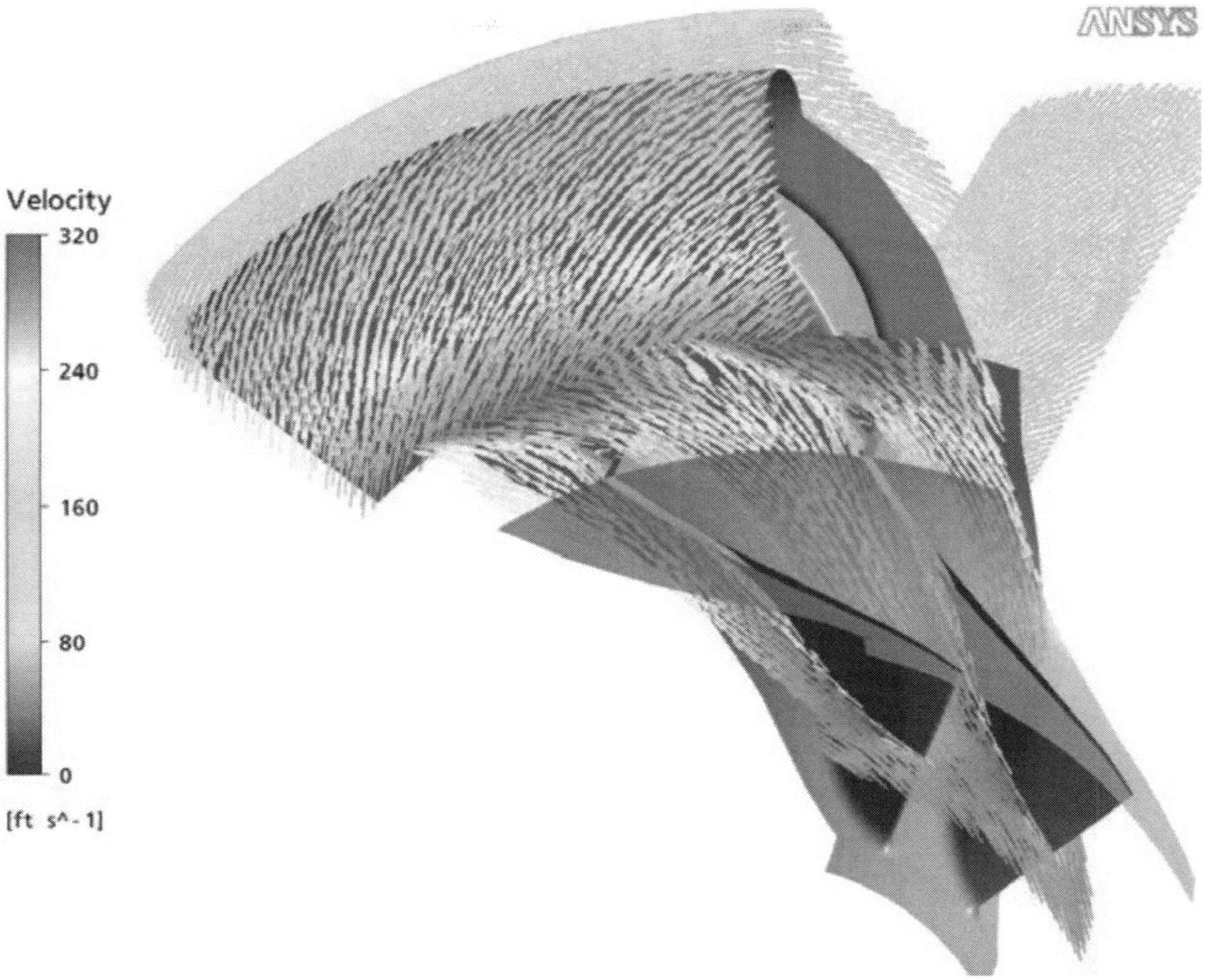

Figure 3. Flow crossing a stage interface is circumferentially averaged and changes its frame of reference.

In a *frozen rotor interface*, the flow field is mapped point by point across the interface, stretching or compressing in the circumferential direction as necessary, and the frame of reference is changed. Up- and downstream circumferential extents should be as similar as possible, say within about 20 percent, so the stretching or compressing of the flow profile doesn't affect results too much. Frozen rotor interfaces are needed when rotor position does matter, such as when there are strong blade/vane interaction effects and rotation is slow enough that these effects don't simply smear out. An important feature is that the rotor doesn't move (it's "frozen"), so a wake doesn't spread out with rotation, which focuses its effects on one point. Results depend on how the vanes and blades are aligned in the model, so it may be necessary to explore different relative positions if blade/vane interaction appears to be strong. Consider whether the flow field would be significantly different if the rotor were moved.

Neither of these approaches gives a true picture of the flow field at any moment. After all, the flow field between stationary and moving blades is changing through time, so no steady-state solution can really get it right. The error compared to the real world can vary widely depending on how important transient effects are and what's being measured, with performance parameters easily being 10 percent off or more in some cases. But the accuracy is good enough for making decisions in many cases, as discussed in the section on CFD strategy below.

Steady-state single-passage models are commonly used for design work, test comparisons, troubleshooting, quick checks of likely flow stability, and general once-over checks to make sure there are no surprises like separations or reverse flow that could cause unexpected losses. As CFD analyses go, models like this are easy to build and quick to run, sometimes taking less than an hour to run each operating point, depending on model complexity and computer power available.

Steady-State Full-Circle

When flow in every passage of a component is not identical, the periodic assumption breaks down and single-passage models can't be used. Examples include evaluating the effects of a side inlet, or a volute (as shown in Figure 4), or a partial admission turbine. For that matter, some components simply don't have individual passages, like a volute. In these cases, the full 360-degree model is required. Components can still be stacked end-to-end, and stage or frozen rotor interfaces can still be used.

Steady-state full-circle models are used for problems similar to those for steady-state single-passage models, but where circumferential nonuniformity is important for the question being investigated.

As a general rule, steady-state full-circle models require frozen rotor interfaces, so important circumferential variation isn't simply averaged away by a stage interface. But stage interfaces can be used to reduce node counts by switching between single-passage and full-circle components in the same CFD model—the stage interface doesn't care if the circumferential extent on one side happens to be 360 degrees. This approach wipes out the circumferential variation at the interface, so it must not be used at any location where the circumferential variation is important for answering the question at hand. For example, if you just want to study the flow in a compressor inlet and only need the impeller to draw flow through the inlet, a single-passage impeller model may be good enough, saving a lot of model size and run time. But if you want to see the effect of that nonuniform inlet on the impeller performance, then you'll need the full-circle impeller (and maybe a transient run).

Transient Full-Circle

Transient runs, also known as "unsteady" or "time-variant" runs, are used when the flow field is not constant with time. Realistically, this is true for most flows of interest, and for any flow involving stationary and moving blades in the same model. Consequently, transient CFD runs are considered the most accurate. But they require much more time to run than steady-state runs, typically on the order of ten times more, and are not necessary for many cases where the steady-state assumption is good enough.

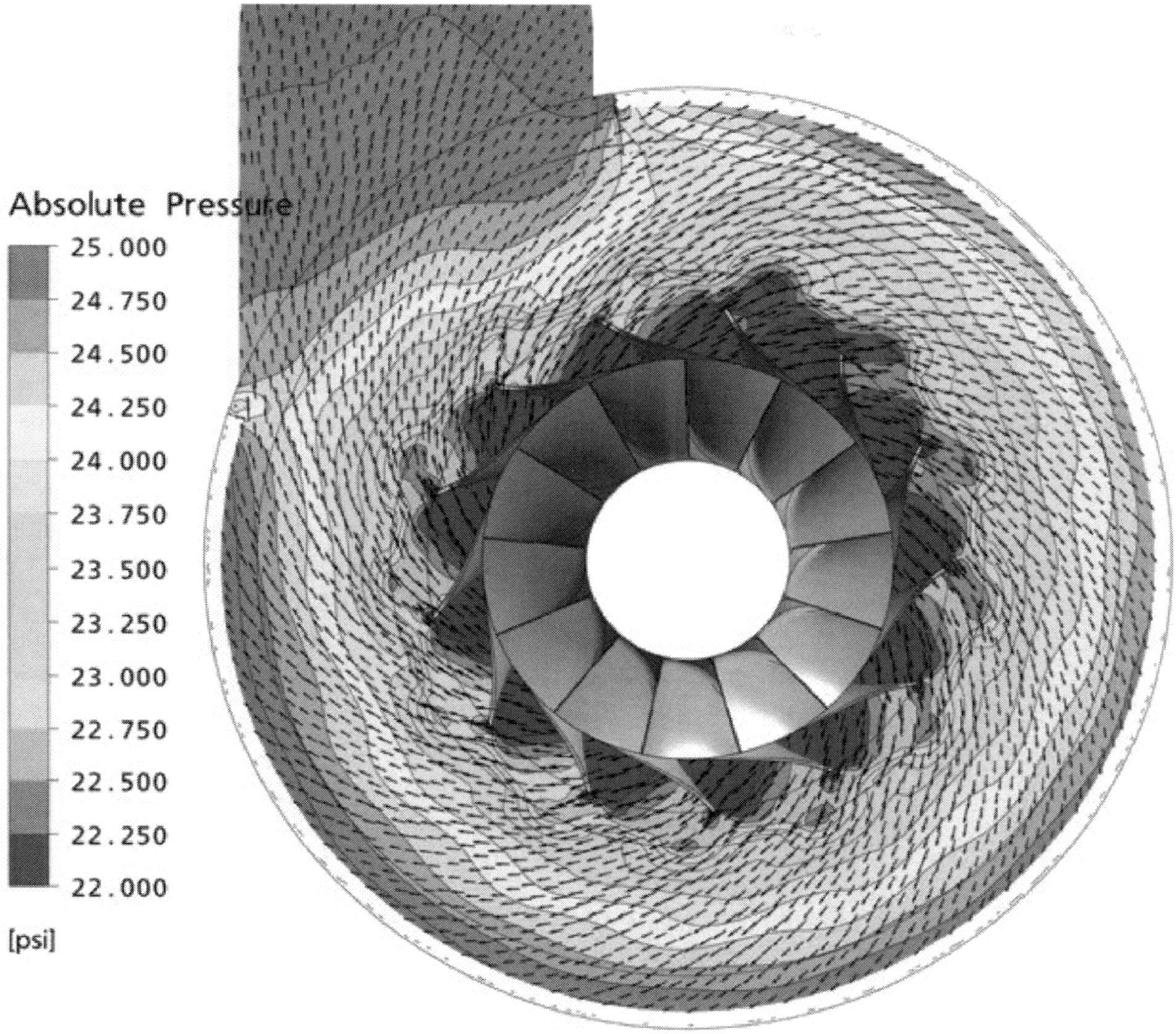

Figure 4. Pressure varies among blade passages because of the downstream volute.

Transient analyses solve the flow field at some point in time, then move time forward by a time step set by the user and solve the flow field again, and continue this process until they reach some specified total simulated time or someone stops them. The amount of simulated time required depends on the problem being solved, but is typically a few blade-passings for turbomachinery or the length of a transient for a sudden change, like opening or closing a valve. If the model geometry is moving, such as a rotating rotor, the appropriate motion is imposed for each time step. Choosing the right time step is a difficult balancing act. Smaller time steps usually run more stably and give more accurate answers, but require a lot of time to run. On the other hand, the time step must be small enough to capture the variations that are being searched for. Also, if you plan to do a Fast Fourier Transform analysis of the results to extract frequencies, you have to divide the total simulated time into enough steps to satisfy the

Nyquist criterion for the highest frequency of interest. You might also have to choose a time step that gives a power of 2 steps in a cycle of interest, depending on the analysis planned.

Since transient runs are solving for each moment, they don't need the assumptions in stage or frozen rotor interfaces. Flow simply advances across each interface, changing its frame of reference as it goes.

Transient runs are necessary to model startups and shutdowns, such as opening or closing a valve. In turbomachinery applications, they are often used to model periodic unsteady flow fields, like blade passing effects that repeat in time. ("Periodic" in this context refers to being periodic in time, different from the periodic geometry described for single-passage models.) In these cases, the model is run until the solution begins to repeat in the expected time period. That period is usually either a blade passing frequency or a full revolution, depending on what you're trying to find out. Average performance values can then be calculated across a full cycle, and individual moments within the cycle can be investigated further.

Transient analyses are needed when the assumption of steady-state flow is not good enough, which can be a difficult judgment call. Rotating stall, details of blade row interactions, and effects of strong circumferential nonuniformities like side inlets and discharges are a few phenomena that require transient analyses. Other unsteady flow effects might require transient analyses to resolve properly. Analysts still argue over whether compressor blade stall, for example, can be predicted adequately with steady-state analyses or transient runs are required.

It is possible to run transient single-passage analyses, with geometric periodic boundaries. It would be unusual to have an unsteady flow that was always unsteady in the same way at every moment in all vane or blade passages, so it would be unusual to build an entire model this way. The approach is useful, however, for a portion of a model whose transient effects are not of direct interest. In Figure 5, the transient interaction between the inlet and compressor impeller was being studied, and the rest of the stage was included simply for overall performance calculations. A stage interface was used to transition to a single-passage return channel model downstream of the impeller.

"Pseudo-Transient"

There are several mathematical schemes that approximate transient analyses but require less computational resources and time. They go by various names, including time transformation, Fourier transformation, and nonlinear harmonic analysis. These methods are strictly for bladed passages. The basic idea behind them is that rather than solving the problem in time (even steady-state solutions act numerically as if they were advancing the solution through time), they convert the equations to the frequency domain, solve those new equations, and convert the results back to the time domain we humans understand.

For this to work, the analysis must "know" what fundamental frequency is of interest, so it can solve based on this frequency and its harmonics. It's up to the user to set the correct fundamental frequency, usually based on the blade passing frequency. Because they work from a selected fundamental frequency and its harmonics, these methods can miss nonharmonic effects, not picking up subsynchronous rotating stall, for example.

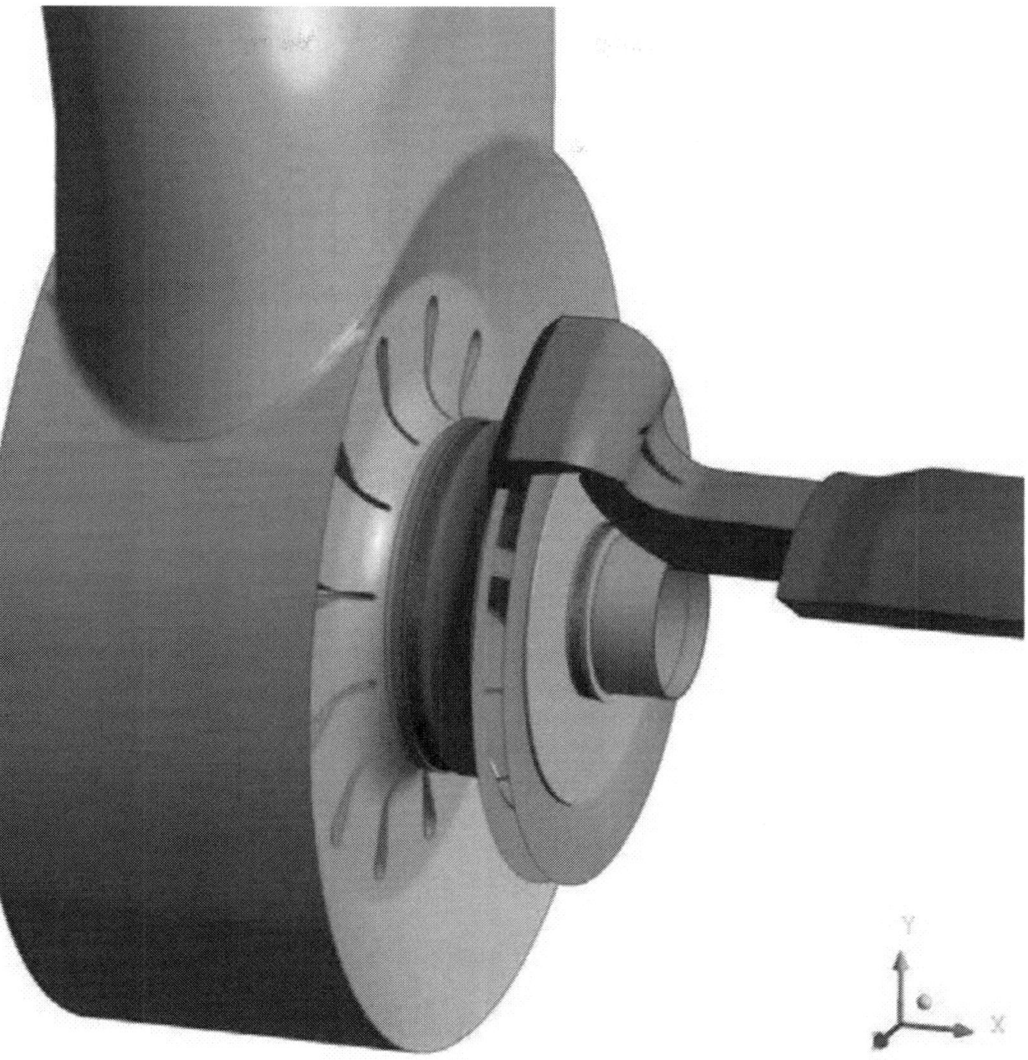

Figure 5. The model transitioned from full-circle to single-passage components after the impeller (in blue).

Pseudo-transient methods usually require only one or two blade passages for each component in a model, yet can extrapolate the flow's variation around the circle and its cyclic behavior through time.

Pseudo-transient methods can provide good answers quickly for cases where the transient effects of interest are being driven by blade interaction effects or a one-per-rev signal. Unfortunately, there's no good way to determine whether a pseudo-transient analysis is of adequate accuracy compared to a full transient except to try them both and compare results. If there are several similar cases to investigate, doing this comparison on the first case can be worthwhile to establish confidence that the pseudo-transient will be good enough for the rest of the cases.

CFD Strategy

Here are some pointers for getting the most out of CFD analyses:

- Plan on multiple analyses. If you're looking for anything more complex than a simple "no surprises" assurance that the flow field is OK, you'll probably need more than one model. Almost all cases require at least the model of interest and a baseline to compare it to (such as in Figure 6). CFD is much better at predicting differences in performance between similar models than making absolute predictions. During design work, the baseline may simply be the previous design iteration.

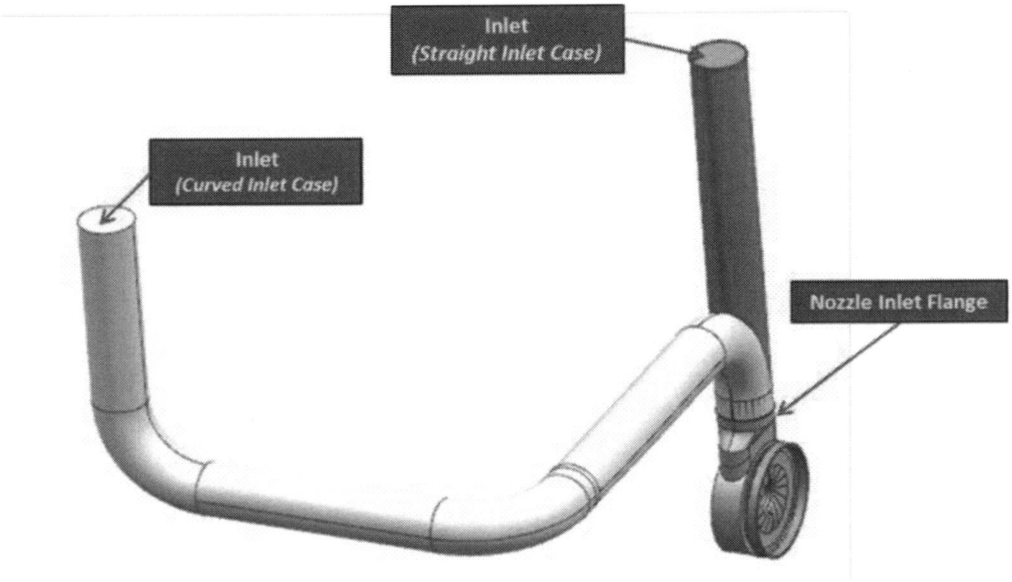

Figure 6. To find the effect of the inlet pipe bends on compressor performance, compare with a baseline case having a hypothetical straight pipe (green).

- Define the problem narrowly. Exactly what are you trying to find out? Keep the model extent as small as possible to cover just the area of interest. Leave out geometric details that are unlikely to influence the answer to the question being posed. Choose the simplest CFD approach among those discussed above that will describe the physics adequately. These choices can save considerable time and money.
- Beware of boundary conditions! Poor boundary conditions are a leading cause of poor CFD results. In the end, CFD is numerics more than physics, and it can converge nicely on a solution with boundary conditions that make no physical sense. Boundary conditions will be held no matter what, and the effects of poor boundary conditions can propagate throughout a model, as shown in Figure 7. Consider how the chosen boundary conditions could influence the flow both downstream and upstream of the constraint and whether that influence is likely to affect the answers being sought. Despite the desire to keep the model small, it may be necessary to extend the model up- or downstream to keep the imposed boundary condition far from the region of

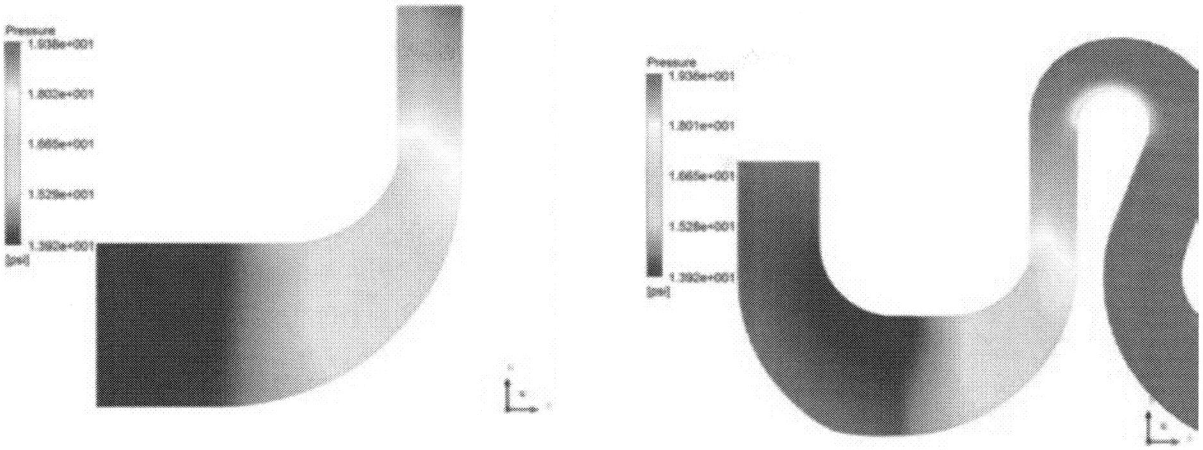

Figure 7. Adding the upstream and downstream bends changed the pressure distribution in the impeller and diffuser.

interest. Inlet turbulence, for example, is often not well known and will often be resolved to a reasonable value by the solution itself, as long as the specified inlet value is far enough upstream.

- Perform mesh sensitivity studies. Results can be affected by mesh density and also mesh topology, that is, how the blocks are structured for a structured hexahedral mesh or whether the mesh is hexahedral, tetrahedral, or of some other type. Finer mesh and higher-order numerics tend to weaken these effects. Because mesh is three-dimensional, it takes large changes in node count to affect results much: doubling density in all three coordinate directions requires increasing the node count by a factor of 8, a reasonable target for a typical mesh density check. Many CFD codes include the ability to interpolate results from one mesh onto another as a starting point for a run, which expedites mesh sensitivity studies considerably. Once experience has been gained with similar models, mesh sensitivity studies for later cases may not be necessary.
- Perform other sensitivity studies. The philosophy for mesh sensitivity studies applies to other parameters as well. Try varying any parameter that's causing doubt, such as the shape of an inlet profile, a specified turbulence value, a surface roughness value, or a time step. Turbulence values are hard to determine, and should be varied by at least a factor of two to evaluate sensitivity. Significant variations in surface roughness depend on where the roughness lies on a Moody chart, but about a factor of two is probably reasonable. A time step five to ten times smaller might bring out more flow details, especially for a transient run.
- Allocate enough time (and money) for the required analyses. A proper set of CFD analyses may require considerable time and money. If upfront planning shows that the CFD can't be completed fast enough, decide early whether to make compromises in the analyses and accept the increased risk or drop the CFD approach altogether. "Quick and dirty" may just be too "dirty" to waste time on.

If you're not the CFD analyst yourself, knowledge of these strategies can lead you to ask the analyst the right questions. It's best to discuss these up front, before spending a lot of time and money on the analyses, to make sure you'll get what you need. Answers to these questions should include explanations of why the answers were chosen.

- How will you choose the baseline case for comparison? If you feel you don't need one, why not?
- How far up- and downstream will the model extend?
- Will you run the case steady-state, pseudo-transient, or transient?
- What interfaces will you use between stationary and rotating portions of the model?
- What boundary conditions will you set? How confident are you that you really know the conditions at those locations?
- What sensitivity studies are you planning? What experience do you have to justify dropping cases?
- How long is this likely to take? Do we have enough time to get what we need, especially if transient analyses are planned? What compromises are you making for time or cost?

Expectations

With all this background, we return to the original question: what can you expect to get from CFD?

- Aerodynamic performance. CFD can predict efficiencies, loss coefficients, and similar parameters with useful accuracy. A CFD code "out of the box" will generally not predict absolute performance all that well, so its results often will not match test results. Calibrating a CFD code to test results can be a long and arduous process, undertaken by OEMs using years of experience. But the *change* in performance from one similar CFD model to another is usually predicted reasonably well.
- Mechanical performance. CFD postprocessors often include the ability to integrate pressure over specified regions, so you can find forces and torques on surfaces. Steady-state runs can give average values, and transient runs can show how these values vary. The same comments about absolute aerodynamic performance apply to mechanical performance as well.
- Frequency information. Monitoring pressures, forces, or other parameters throughout a transient run can allow you to calculate a Fast Fourier Transform to get frequency information. In some cases, simply extracting data around the circumference from a steady-state run may be sufficient.
- Details of the flow field. This may be CFD's greatest unique strength. Seeing details of the flow at every point throughout a machine can enable considerable insights for design or troubleshooting. This is also useful for evaluating test data: sampling the CFD results where the test instruments are located can show whether the values from the instruments are likely to be representative of the overall flow. Velocity vectors at various slices through a model are often a good way to get an overview of the flow field and find regions that should be explored in more depth. Zooming in on particular regions can show separations or deviations from vanes or other problems. Contour plots of total pressure can highlight regions of high loss; look for places where total

pressure falls rapidly. Many other kinds of plots can be relevant, depending on the question being investigated.

In addition, CFD can provide results for other classes of problem not discussed in this article, such as heat transfer, fluid/structure interaction, and acoustics (often in conjunction with dedicated acoustics software).

CFD will not replace a test stand and will generally not give a quick answer to anything. However, it can give useful answers to particular questions, and faster than one might expect in many cases. Plan carefully, avoid dropping into the pitfalls, and CFD analysis can be an invaluable tool for getting the real engineering done.

591159 When Disasters Happen: A Collaborative Approach Towards the Community's Recovery Process

Abstract:

The 2020 Virtual AIChE Spring Meeting & 16th GCPS Wednesday keynote "When Disasters Happen: A Collaborative Approach Towards the Community's Recovery Process" will be presented by Thomas Munoz, Emergency Manager and Homeland Security Director with Texas City. As the Emergency Manager and Homeland Security Director with Texas City, the home of 19 chemical plants, he is responsible for the safety of all its citizens along with managing the crisis communication plan with local industry and major stakeholders.

Safety and disaster planning is vital to the recovery process after a major incident. As key critical infrastructure industries, issues may arise that have not been planned for in advance, such as the recovery from an incident. Developing and maintaining partnerships with local partners in your community will support addressing unforeseen challenges. Recovery efforts include planning with key partners and understanding the phases of recovery, which also consider the mental health and well of everyone involved in the response, as well as the community impacted. In this keynote address a discussion of pre-incident planning considerations and strategies as well as lessons learned/after action improvements post-recovery will be discussed. Additionally, collaboration considerations regarding local resources to promote recovery in the event of a workplace incident will be discussed.

C0-presenter Amanda Groller, MS.Ed., LPC will discuss the mental health impacts of a disaster.

Copyright © American Institute of Chemical Engineers. All rights reserved.

590213 Making Artificial Intelligence into Augmented Intelligence: a Digitalization Approach for Ethylene Plants

Abstract:

Spanish Philosopher Jose Orega y Gasset said "Create a concept and reality leaves the room.

Creating a conceptual mantra, a monomania, makes life a lot simpler. It permits one to eliminate any interfering complexity from the intrusion of other facts or contradictory ideas or even common sense. The one idea becomes talismanic. The words in which the idea is encased immediately delivers an inner meaning, outward significance, plan of action and morality. The fact that the idea may not comport with reality is beside the point."

In today's environment as fear of falling behind and not being competitive is ever present; the concept of what digitalization can bring may be in this same category of a conceptual mantra that is so attractive that common sense is not being applied. Analyzing data for the sake of analyzing data or proving that something can be modeled does not mean it is needed. Random application of machine learning tools to Ethylene plants is occurring and the promise of digitalization is being tarnished as value is not captured.

Understanding the problems that are occurring and the financial impact of these problems are the first steps. Then determining the analytics needed to understand the problem and its potential solutions should be considered. There is tremendous potential in digitization and digitalization. The purpose should be however to be able to provide the insight and augmented intelligence needed for engineers and operators to improve operations. The goal should be to have the insight needed to take additional appropriate action.

Ethylene plants tend to have considerable room for mathematical optimization through both fundamental models and machine learning models. Ingenero applies digitalization to Ethylene plants to enable improved operations. Ingenero customizes data agglomeration, analytics, and solution visualization for augmenting the intelligence of the operators and engineers to address the specific problems encountered that limit production. Examples of appropriate application of machine learning and result visualization for increasing throughput, yield and equipment availability will be shown for both ethylene furnaces and the separation train. The value is defined upfront and then artificial intelligence is applied with the intent of augmenting intelligence for value capture.

Copyright © American Institute of Chemical Engineers. All rights reserved.

82C:

ROBOTICS AND THE ELIMINATION OF HAZOUDOUS WORK

Twain Pigott
Robotics Technology Leader
The Dow Chemical Company

Copyright retained by authors

Prepared for Presentation at the 2020 Spring National Meeting
Houston, Texas, March 30 – April 1, 2020

AIChE and EPC shall not be responsible for statements or opinions contained in papers or printed in it publications

ROBOTICS AND THE ELIMINATION OF HAZOUDOUS WORK

Twain Pigott
Robotics Technology Leader
The Dow Chemical Company

Abstract:

Mobile Robotics and UAV (Unmanned Aerial Aircraft) technology play an important role in eliminating or reducing human exposure to potentially hazardous work. Dow's Global Robotics Program has performed thousands of activities within our production facilities. The primary focus of the program is on high-risk activities such as confined space entry, elevated work and industrial cleaning; however, emerging robotic technology is also developing skilled robotics to accomplish more task and remove even more people from harm's way.

Robotics and the Elimination of Hazardous Work

The case for change:

Preventative measures for hazardous work have been firmly established in the manufacturing industry. With measures such as procedural controls, engineered design, engineered isolation and PPE, these measures have continued to improve safety and reduce worker exposure to potential risk. However, even with these controls in place, it is not always possible to fully eliminate the worker from all potential risks.

Work carried out in of confined spaces, at elevation and with high pressure blasting for industrial cleaning bring with them an increased potential for injury due to complexity. According to 2017 data collected by the U.S. Department of Labor, over 160 fatalities occurred during confined space entries and over 800 workplace fatalities were attributed to falls [1].

With the emerging UAV and robotics technologies, it is now becoming possible for tasks to be performed remotely – completely removing the worker from harm's way. In the area of maintenance and mechanical integrity inspection, there is growing momentum among end-users, mechanical integrity companies and robotic developers to expand the use of UAVs and robotics to perform tasks in these potentially hazardous environment.

Establishment of a Robotics Program:

With the continued development of the UAV and robotic technologies, we are now able to go directly after work being done in confined spaces, from temporary elevated structures and with high risk industrial cleaning activates. Many end-users in the Oil & Gas and Chemical Industry are evaluating how robotics will impact their maintenance and inspection processes, while seeking to answer how to best scale a robotics program. Industry collaborations, such as Sprint Robotics, are helping to answer these questions by bringing together end-users, inspection companies and robotic technology companies to promote the use of new tools and expand the application base by creating a strategic roadmap [2].

At Dow, a global robotics program has been established to remove workers from potentially hazardous activities utilizing UAV and robotic technology [3]. Starting in 2014, Dow deployed a UAV program to perform task to eliminate work and by 2017 Dow had a fully established robotics program with a global deployment of basic robotics kits. The success of this program has been instrumental in the elimination of over 4,000 confined space entries and over 200,000 hours of work in potentially hazardous environments. As the development and application of UAV and robotic technology continues to grow, robotics will play a critical role in Dow's target to eliminate confined space entries by 2025.

References:

1. Bureau of Labor Statistics (BLS), "National Census of Fatal Occupational Injuries in 2017," US Department of Labor (DOL), December 18, 2018, available at weblink: https://www.bls.gov/news.release/pdf/cfoi.pdf (accessed on January 6, 2019).

2. Sprint Robotics, https://www.sprintrobotics.org/

3. The Dow Chemical Company 2018 Sustainability Report, https://corporate.dow.com/en-us/science-and-sustainability.html.

589996 Gleaning Value from Digitalization in Process Manufacturing

Abstract:

The world has gone through three Industrial Revolutions over the past three centuries with the advent of steam driven machines to mass production lines and automation and is now going through a fourth with Digitalization. Digital transformation through the application of the Industry 4.0 technologies for increased global collaborative teams, augmented intelligence, and real time decisions utilizing insightful information is rapidly becoming the standard.

Digital transformation captures additional revenue and identifies opportunities using digital technologies to make higher quality information readily available. Digital transformation enables improved decision not possible with pre-digitalization paradigms. Digital transformation entails a process whereby several Digital technologies are seamlessly integrated, requiring relevant expertise and domain experience to do so. Data is the key raw material in the Digital transformation process.

There is no dearth of data available from Process manufacturing. However, direct use of data for business intelligence can be misleading. Advanced analytics and impactful presentation of data that is converted to information, is necessary to enable use of data for decision making. The Industrial Internet of Things (IIoT) is a valuable tool that can facilitate advanced analytics execution and the digital transformation process, if utilized appropriately. Digital technology software tools by themselves requires specalized engineering effort (Decision Insight Engineering), to convert it into a readily utilizable solution that could provide *Augmented Intelligence*to the manufacturer's teams to meet the digitalization objective.

The objective of any digitalization initiative is to capture revenue and value; not to implement new technology for the sake of doing so. The selection of digital technologies to implement are based on "Use Cases" whose value proposition is known up front. Defining the "Use Case" and its estimated value is the key to ensure a successful Digitalization planning process.

Such a Digitalization initiative helped a US petrochemical manufacturing major to address multiple operational challenges:

- Furnace Run-Length below Design
- High Emmissions of SOx, NOx, etc.
- Frequent Failure of Assets : Higher Equipment Downtime / Unplanned Shutdowns
- Lower Throughput

This paper will outline a roadmap for Digtal transformation in Process manufacturing, the nature and significance of the work that *Decision Insight Engineering*entails and the Digital technology tools that it helps convert to directly usable solutions, along with a case example of an application for a US ethylene manufacturer.

Copyright © American Institute of Chemical Engineers. All rights reserved.

AIChE Paper Number 124d

Reducing greenhouse gas emissions of steam cracking furnaces by oxy-fuel combustion, renewable fuels and high emissivity coatings

Stijn Vangaever
ir.
Laboratory for Chemical Technology, Ghent University, Ghent, Belgium

Yu Zhang
Dr. ir.
Laboratory for Chemical Technology, Ghent University, Ghent, Belgium

Marko R. Djokic
Dr. ir.
Laboratory for Chemical Technology, Ghent University, Ghent, Belgium

Joost Van Thielen
Mr.
CRESS B.V., Breskens, The Netherlands

John Olver
Dr.
Emisshield Inc., Blacksburg, Virginia, USA

Gilles Theis
Mr.
John Zink Hamworthy Combustion, Dudelange, Luxemburg

Kevin M. Van Geem
Prof. Dr. ir.
Laboratory for Chemical Technology, Ghent University, Ghent, Belgium

Prepared for Presentation at the 2020 Spring National Meeting
Houston, Texas (TX), March 29 – April 2, 2020

AIChE and EPC shall not be responsible for statements or opinions contained in papers or printed in its publications.

Reducing greenhouse gas emissions of steam cracking furnaces by oxy-fuel combustion, renewable fuels and high emissivity coatings

Stijn Vangaever
ir.
Laboratory for Chemical Technology, Ghent University, Ghent, Belgium

Yu Zhang
Dr. ir.
Laboratory for Chemical Technology, Ghent University, Ghent, Belgium

Marko R. Djokic
Dr. ir.
Laboratory for Chemical Technology, Ghent University, Ghent, Belgium

Joost Van Thielen
Mr.
CRESS B.V., Breskens, The Netherlands

John Olver
Dr.
Emisshield Inc., Blacksburg, Virginia, USA

Gilles Theis
Mr.
John Zink International Luxembourg, Dudelange, Luxemburg

Kevin M. Van Geem
Prof. Dr. ir.
Laboratory for Chemical Technology, Ghent University, Ghent, Belgium

Abstract: Steam cracking furnaces are responsible for the majority of the emissions of a steam cracking plant. Therefore the energy efficiency and the furnace technology will need to change drastically, preferably in a cost effective way, bearing in mind that more strict regulations will be imposed over time. Several novel technologies were tested at various technology readiness levels before implementing the most promising technologies at a commercial scale in the context of the H2020 IMPROOF project. John Zink Hamworthy Combustion (JZHC) is in the possession of a test furnace where several burners can be tested simultaneously to full scale to simulate part of an ethylene cracker furnace. Two JZHC COOLStar™ burners were tested to see the effect of biogas and oxyfuel combustion on the heat flux profile. The experimental results were validated with Reynolds Averaged Navier-Stokes Computational Fluid Dynamics simulations in ANSYS Fluent 19 using a 25 species mechanism for combustion in CO_2-rich environments. The results show that existing furnaces can easily be retrofitted to run on biogas or to implement oxy-fuel combustion. Steam cracking pilot plant experiments performed at the Laboratory for Chemical Technology show that applying an Emisshield high emissivity coating on the furnace refractory wall can lead to energy savings of approximately 5 %. These energy savings are validated using radiative heat transfer simulations in a simplified geometry. A parametric study is performed to check the effect of boundary wall emissivity on the temperatures inside a so-called tube-in-box enclosure.

Introduction: Novel technologies

Renewable fuels & oxy-fuel combustion

Biogas is the most straightforward bioenergy alternative in terms of production process and purification involved[1]. Several commercial scale biogas plants are already running successfully in different parts of the world. In these plants, biogas is produced through anaerobic fermentation of biomass, and it is mainly constituted of CH_4 (55 – 75 %) and CO_2 (25 – 45 %), with traces of other components such as H_2 (0 – 3 %), N_2 (1 – 5 %), CO (0 - 0.3 %) and H_2S (0.1 - 0.5 %)[2]. Due to the presence of carbon dioxide the caloric value of biogas is significantly lower compared to natural gas. This makes the removal or transformation of carbon dioxide, preferably combined with the purification from other contaminants, essential for increasing the calorific value of biogas in order to act as a one-to-one replacement for natural gas[3]. However, biogas upgrading and cleaning makes the process economically less feasible and puts an additional stress on the environment due to the emissions during processing. For these reasons there has been an increased interest in the direct combustion of unpretreated biogas. One of the essential questions in this work is whether the use of renewable biogas can actually be considered in steam cracking furnaces or by extension to be considered in gas-fired furnaces in general.

Oxy-fuel combustion is a promising technology where fuel is burned with pure oxygen rather than air as oxidant. Due to the absence of the nitrogen component in air, higher flame temperatures can be obtained with quasi non-existing NO_x emissions. A portion of the flue gas is often recycled to temper the flame temperature. The main advantage of the oxy-fuel combustion process is that the flue gas stream is nearly entirely composed of carbon dioxide and water instead of the large amount of nitrogen diluent present in conventional air-combustion scenarios. This means that water vapor can easily be separated from the exhaust stream via low-temperature dehydration, resulting in an near pure carbon dioxide product stream. This carbon dioxide stream can be used for carbon capture, utilization and storage. The main drawback of the oxy-fuel combustion technology is the high energy requirement related to producing high purity oxygen. For this reason, the broader perspective offered by life cycle assessment is often called upon to validate the environmental impact of oxy-fuel combustion combined with carbon capture, utilization and storage.

High emissivity coatings

Interest in high emissivity coatings originated in the aerospace industry. The association with spacecraft and NASA surrounded the research area in an aura of 'high tech' and 'mystery'[4]. Nonetheless, the underlying physical principle is straightforward. During the atmospheric re-entry of space vehicles, friction between the space vehicle and the atmospheric gases drastically heats up the surface of the spacecraft. Since the emissive properties of the radiating surface determine the amount of heat that can be dissipated back to the surroundings, increasing the emissive properties reduces the surface temperature[5]. Emisshield licensed the high emissivity coating technology from NASA in 2001 to include all applications except space and space vehicle applications. For the last decades, high emissivity coatings started to make their way into the chemical industry. High emissivity refractory and metal coatings promise to reduce downtime, minimize coking, increase productivity and to reduce the plant's overall carbon footprint by providing a uniform temperature distribution and an enhanced radiative heat transfer[6].

Methodology and procedures

Renewable fuels & oxy-fuel combustion

Test furnace

The John Zink Hamworthy Combustion test furnace, representing a section of a cracking furnace, is used to validate the feasibility of biogas and oxy-fuel combustion. The test furnace allows for full-scale tests of two floor burners simultaneously at their rated capacity with furnace temperatures up to 1623 K. The rectangular firebox has internal dimensions representative of an industrial furnace with vertical tubes acting as a heat sink. Side by side, two COOLStar™ burners are installed in the test furnace. The main benefit of a full-scale test is that emissions and flame stability with high inert content, shape and height can be studied in a detailed manner. Several openings along the furnace height allow for heat flux measurements.

Both biogas and oxy-fuel firing conditions are tested in this furnace. In both cases the fuel and oxidizer are artificially created by mixing pure CO_2, H_2O and O_2 combined with natural gas. This implies that for the biofuel tests, natural gas was gradually diluted with CO_2 to mimic biogas. For the oxy-fuel tests, the oxidant was composed of O_2 combined with CO_2 and H_2O. The ratio of CO_2 to H_2O was in correspondence with the stoichiometric combustion ratio of natural gas in air similar to operation with recycled flue gas. The O_2 content in the oxidant was initially kept at 21 vol% which is comparable to air. The main question is whether existing industrial furnaces can easily be revamped or retrofitted to run on biogas and or oxy-firing conditions.

Computational Fluid Dynamics model set-up

Three-dimensional steady-state global mass, momentum, energy and species conservation equations are solved for a compressible, reacting fluid using ANSYS Fluent 19. The RNG-based $k - \varepsilon$ two-equation model is used to model turbulence. Initially, the Westbrook and Dryer two-step combustion mechanism was used to obtain the combustion reaction rates[7]. In a later stage, the POLIMI mechanism is reduced to a 25 species mechanism with 154 reactions[8]. The abundance of CO_2 in the mixture has chemical effects resulting in a reduced mechanism that is larger than a conventional methane-in-air combustion mechanism. The Eddy Dissipation Concept is used to model the turbulence-chemistry interaction. The discrete ordinates radiation model is selected to solve the radiative transfer equation with an angular discretization of 2 by 2. A weighted-sum-of-gray gasses model accounts for the gas phase radiative properties.

High emissivity coatings

Steam cracking pilot plant

To experimentally assess the effects of applying a high emissivity coating on the energy efficiency of a gas-fired furnace, the steam cracking pilot plant at the Laboratory for Chemical Technology of Ghent University is used. A schematic representation of the steam cracking pilot plant is shown in Figure 1. For a detailed description of the unit, including all the analytical equipment, the reader is referred to previous work[9, 10]. The pilot plant is divided in different cells that can be fired independently. Twelve wall burners in each cell allow the plant operator to impose any required temperature or heat flux profile over the reactor tube length. In the first two cells 10 kg/h of propane mixed with 4 kg/h of dilution steam is preheated to the desired coil inlet temperature of 917 K. A SCOPE reactor coil, made of Centralloy® HT-E with nine fins and an internal diameter ranging from 35 to 39.1 mm, is installed in the two cracking cells[11]. At the outlet of the pilot plant a coil outlet temperature of 1111 K is maintained to guarantee the same cracking severity for each experiment, corresponding to 90 % propane conversion. At the start of every experiment, the reactor is heated with steam until the desired reactor coil temperature profile is reached. Next, the flow rate of propane and steam is set to the desired operating conditions. Cracking starts and is maintained for 6 h. Finally, the experiment is halted and the reactor is decoked. The experimental procedure, as described, is repeated two times to establish a base case with uncoated furnace walls. Next, the two cracking cells are coated with a high emissivity coating and the experiment is repeated twice. Differential pressure transmitters are installed to measure the premixed burner gas flow rate to the two cracking cells. Additionally, the total natural gas flow rate to the entire furnace is monitored. Several thermocouples are installed to measure the process gas temperature, flue gas temperature, tube metal wall temperature and refractory wall temperature, the latter both inside and outside the furnace.

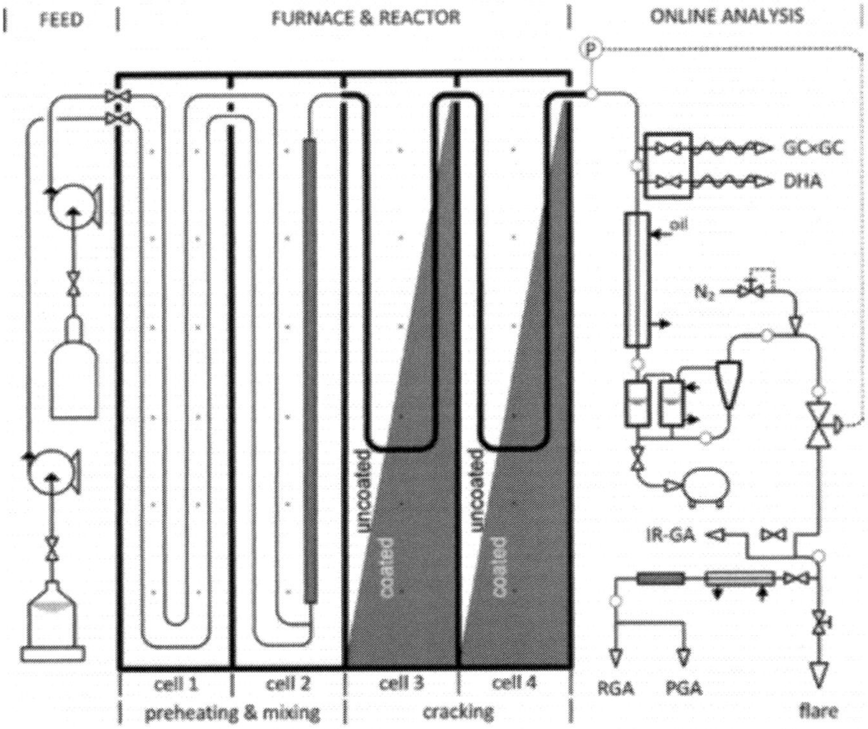

Figure 1: Schematic overview of the pilot plant setup. The pilot plant is divided into a feed section, a furnace-reactor section and an online analysis section. Cracking reactions mainly occur in cells 3 and 4. These cells are coated with a high emissivity coating. (Adapted from Van Geem et al.[10])

Radiative heat transfer tube-in-box model

In this work, a simplified two-dimensional geometry based on a horizontal cross-section of one of the pilot plant cracking cells, is studied . This so-called tube-in-box model, shown in Figure 2, consists of a central cylindrical surface, acting as a heat sink, and an enclosing adiabatic refractory surface. The enclosing adiabatic box has a length and height of 0.380 m. A circle with 0.103 m diameter, equivalent to the pilot plant heat sink area, is positioned in the center. A perfectly stirred flue gas with a uniform temperature, T_G, is located between the adiabatic enclosure and the central heat sink. The flue gas is composed of CO_2, H_2O and N_2 in the stoichiometric combustion ratio of methane in air.

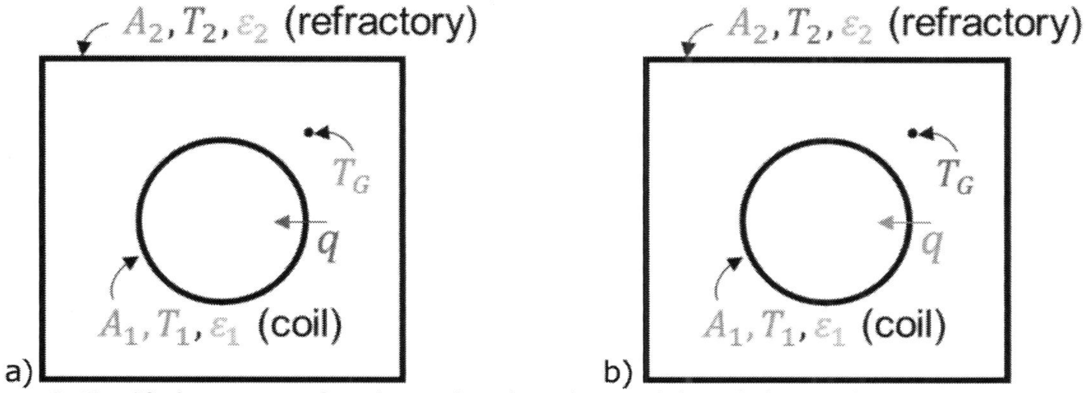

Figure 2: Simplified geometry referred to as the tube-in-box model. In a) the gas phase temperature, T_G, is assumed to be constant for different tube, ε_1, and adiabatic surface, ε_2, emissive properties. In b) the gas phase temperature, T_G, is adjusted to get the same heat flux, q.

Radiative heat transfer from the hot flue gas to the heat sink will be modeled using Hottel's zoning method[12] with varying emissive properties for both the central heat sink, ε_1, and the furnace refractory wall, ε_2.The geometry is determined by the surface areas, A_1 and A_2. The central heat sink temperature, T_1, is kept constant based on the averaged experimentally observed tube metal temperature. Initially, the gas phase temperature , T_G, is kept constant, resulting in an unknown adiabatic wall temperature, T_2, and an unknown heat flux, q, to the central heat sink, visualized in Figure 2 a). In a second simulation, shown in Figure 2 b), the gas phase temperature, T_G, is adjusted to obtain a constant heat flux, q, for varying wall emissive properties. The radiative heat transfer problem will initially be solved with a gray gas assumption. Later, this solution will be compared to a non-gray gas approach.

Results & Discussion

Renewable fuels & oxy-fuel combustion

Mechanism reduction

Conventional global combustion mechanisms such as the two-step Westbrook and Dryer mechanism[7] or the four-step Jones and Lindstedt mechanism[13] are not valid in CO_2-rich conditions[14]. The present CO_2 will compete with O_2 for atomic hydrogen leading to the formation of CO. Similarly, CO_2 might also react with the hydrocarbon radicals contributing to the formation of CO[15]. So instead of refining global combustion mechanisms to be used under oxy-fuel conditions[16], the POLIMI natural gas combustion mechanism developed by the CRECK modelling group is reduced to include the appropriate radical species[8]. The resulting mechanism consists of 25 species with 154 reactions. The mechanism has a maximum error on ignition delay time lower than 20 % in the tested range (equivalence ratio of 0.5 to 1.7 with up to 40 vol% CO_2 for biogas and under oxy-fuel conditions). Figure 3 shows a maximum error of 7 % on the laminar flame speed.

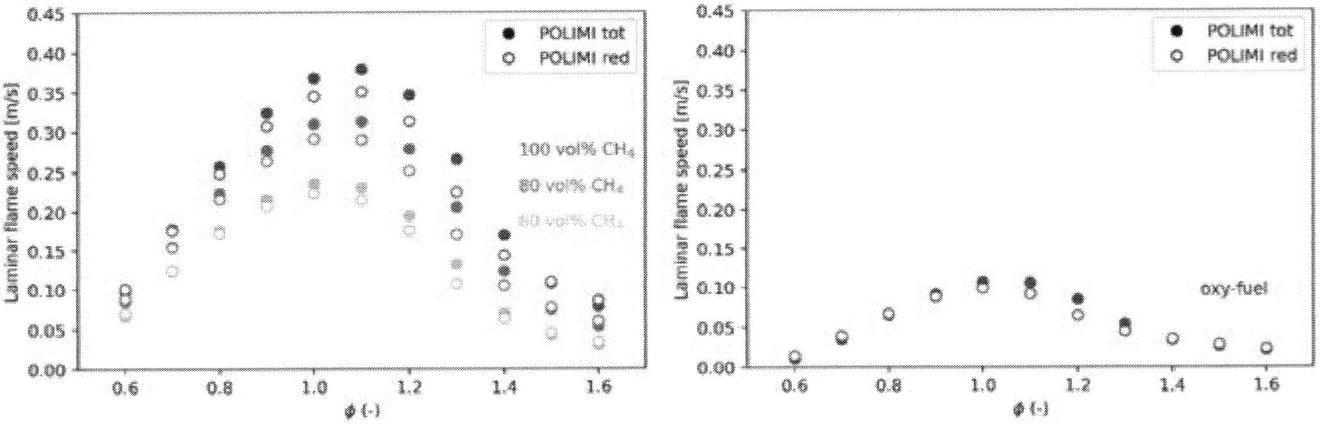

Figure 3: Laminar flame speed for the complete POLIMI mechanism versus the reduced POLIMI mechanism for biogas with varying CO_2 additions (left) or under oxy-firing conditions (right).

Flame envelop & heat flux profile

Figure 4 gives a visualization of the temperature and vertical velocity of a cross-section of the test furnace for different CO_2 dilutions. The peak flame temperature occurs at an elevation of approximately 4.5 m and increases with the addition of CO_2.

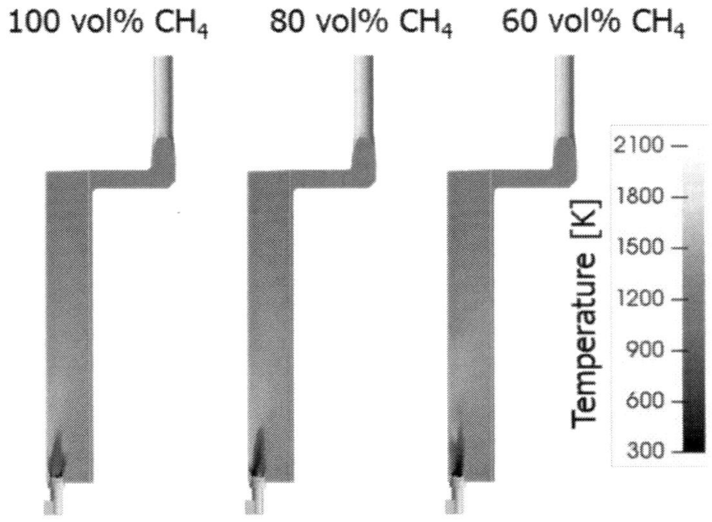

Figure 4: Temperature distribution for conventional combustion (left) and for biogas combustion with 20 vol% CO_2 (middle) and 40 vol% CO_2 (right).

The heat flux profile shows where most heat is transferred to a process coil in function of the furnace height. This visualization is important for licensors of steam cracking furnaces. In order to retrofit an existing furnace with biogas or oxy-fuel combustion, the heat flux profile should not deviate too much from the predefined shape. Experimentally, this is the case, as shown in Figure 5. The initial two-step Westbrook and Dryer combustion mechanism gives a heat flux profile that closely matches the experimental results. In a later stage, the reduced POLIMI mechanism was implemented resulting in an upward shift of the maximum heat flux. This increased flame height could eventually result in flame lift-off when the CO_2 concentration is further increased. However, Large Eddy Simulations are typically required to correctly model the flame detachment. The oxy-fuel simulations have not yet converged.

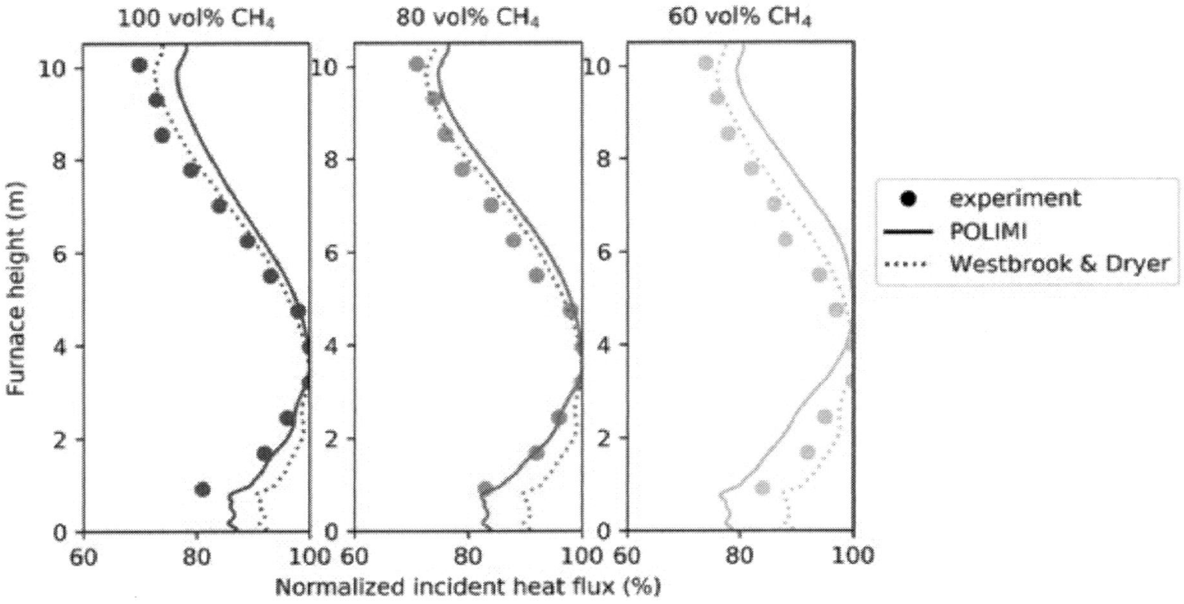

Figure 5: Normalized incident radiative heat flux to the process coils for varying CO_2 additions. The experimental measurements are compared to the numerical simulations using the simplified POLIMI mechanism and the two-step Westbrook & Dryer mechanism.

High emissivity coatings

Steam cracking pilot plant experiments

The fuel consumption of the steam cracking pilot plant before and after the high emissivity coating was applied is shown in Figure 6. For all experiments the natural gas consumption initially decreases to reach a steady-state after almost 6 h of cracking. The decrease can be attributed to the fact that the furnace needs to heat-up and needs to find an equilibrium with the surrounding environment. The fuel consumption at steady-state is gives an energy efficiency increase of 4.77 % using averaged values for the uncoated and coated runs.

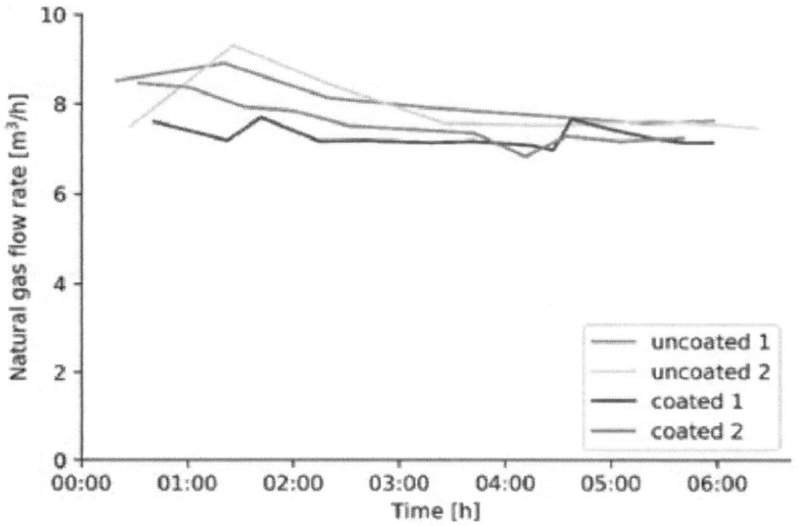

Figure 6: Natural gas flow rate to the entire pilot plant over time for the different experiments before and after applying an Emisshield high emissivity coating.

Radiative heat transfer tube-in-box model

Initially, the total surface-to-surface exchange areas with a participating gray gas and the total gas-to-surface exchange areas are derived for the tube-in-box model. The "exchange areas" approach was introduced by Hottel and Sarofim in their zonal method for radiative heat transfer[12]. The total exchange areas are functions of the geometry and radiative properties of the entire enclosure[17]. The resulting equations for a gray gas are independent of the adiabatic wall emissivity. With the gray gas approximation, all radiation coming from the gas will either be reflected or absorbed and re-emitted by the enclosing surface. Since there is no clear distinction between the 'two types of radiation', the enclosing surface can be regarded as a blackbody independent of its emissivity; redirecting all radiation towards the heat sink.

A non-gray gas model needs to be introduced to make the distinction between reflected and re-emitted radiation. An uncoated furnace, having a high probability of reflecting the incoming radiation, tends to reflect the radiation in an absorption band where it is re-absorbed by the gas. A higher emissivity will increase the probability that the incoming energy will be re-emitted in transparent window where it is more likely to heat the load.

Additionally, radiative transfer characteristics of an opaque wall can often be described with good accuracy by gray emission, absorption and reflection. The radiative properties of a molecular gas vary strongly and rapidly across the wavelength spectrum which makes that the assumption of a gray gas is never a good one[18, 19]. Hence, a non-gray gas modelling technique is certainly required to accurately capture the radiative heat transfer characteristics. In the gray-plus-clear gas model the general weighted sum of gray gasses model is reduced to the combination of one lumped absorption band where the total exchange areas derived for a gray gas are still applicable and one clear band where the gas absorbs no radiation.

Based on the pilot plant experiments, a reactor tube wall temperature of 1173 K will be used as heat sink boundary condition. For the first simulation a constant gas phase temperature of 1373 K was assumed, shown in Figure 2 a). In the next step the gas phase temperature is changed to have the same net heat flux to the inner tube, shown in Figure 2 b). The resulting net radiative heat flux for a constant gas phase temperature is shown in Figure 7 a). The gas phase temperature required to get the same heat flux is shown in Figure 7 b).The refractory emissivity was varied from 0.4 to 1.0 (y-axis) and the reactor tube emissivity was varied from 0.7 to 1.0 (x-axis) to give a complete overview of the effect of the emissive properties of the boundary surfaces on the temperature inside the geometry.

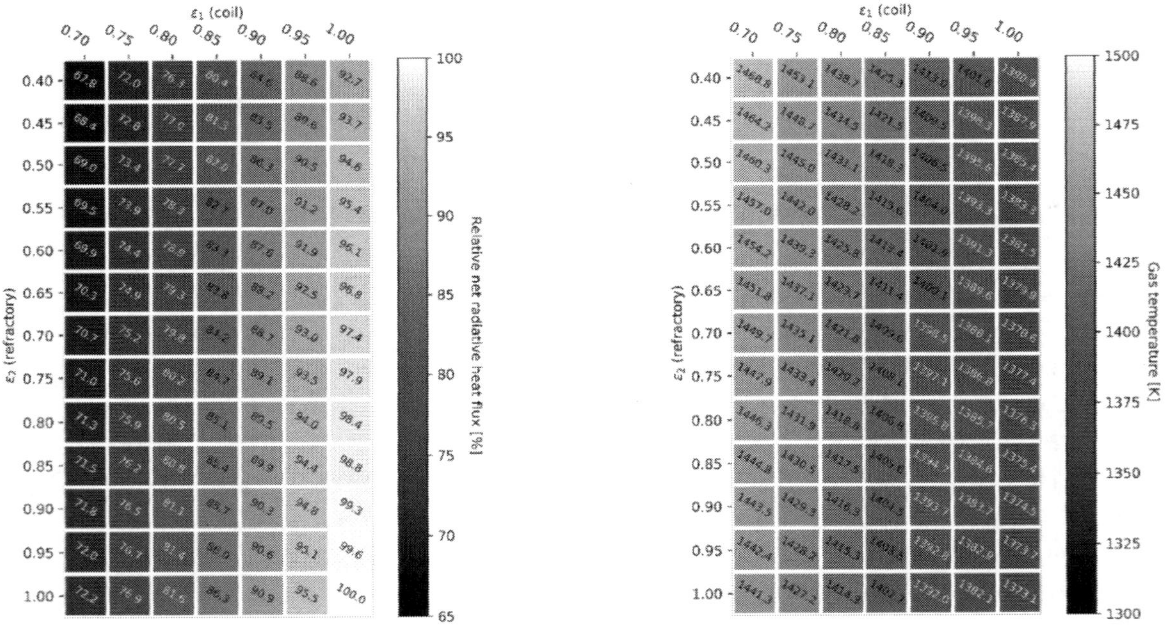

a) b)

Figure 7: Radiative heat flux towards the coil for a constant gas phase temperature a) and gas phase temperature required to get a constant net radiative heat flux b).

Conclusions

In the IMPROOF project, the effect of various novel technologies including biogas, oxy-fuel combustion and high emissivity coatings on the greenhouse gas emissions of steam cracking furnaces has been studied. Experiments performed at the John Zink test furnace shows that existing furnaces can be retrofitted to run on biogas or on oxy-fuel combustion. Numerical simulations support this claim, validating the experimentally obtained heat flux profiles. Steam cracking pilot plant experiments performed at the Laboratory for Chemical Technology show energy savings up to 5 % after an Emisshield high emissivity coating was applied. Radiative heat transfer simulations show the necessity of a non-gray gas model to accurately describe the effect of high emissivity coating in a tube-in-box enclosure. In the following step of the IMPROOF project, 3D reactor technologies and high emissivity coatings will be implemented on an industrial scale. Extensive measurements combined with a thorough life cycle analysis should be able to accurately describe the effect of different technologies on the greenhouse gas emissions of steam cracking furnaces.

Acknowledgment

This project has received funding from the European Union's Horizon 2020 research and innovation programme (H2020-SPIRE-04-2016) under grant agreement No 723706.

References

1. Gnansounou, E.; Raman, J. K., Chapter 7 - Wide Scope Environmental Assessment of Biofuels. In *Biofuels: Alternative Feedstocks and Conversion Processes for the Production of Liquid and Gaseous Biofuels (Second Edition)*, Pandey, A.; Larroche, C.; Dussap, C.-G.; Gnansounou, E.; Khanal, S. K.; Ricke, S., Eds. Academic Press: 2019; pp 163-196.
2. Karellas, S.; Boukis, I.; Kontopoulos, G., Development of an investment decision tool for biogas production from agricultural waste. *Renewable and Sustainable Energy Reviews* **2010,** 14, (4), 1273-1282.
3. Angelidaki, I.; Xie, L.; Luo, G.; Zhang, Y.; Oechsner, H.; Lemmer, A.; Munoz, R.; Kougias, P. G., Chapter 33 - Biogas Upgrading: Current and Emerging Technologies. In *Biofuels: Alternative Feedstocks and Conversion Processes for the Production of Liquid and Gaseous Biofuels (Second Edition)*, Pandey, A.; Larroche, C.; Dussap, C.-G.; Gnansounou, E.; Khanal, S. K.; Ricke, S., Eds. Academic Press: 2019; pp 817-843.
4. Stendera, J.; Bonsall, S., High Emissivity Coatings - Do They Really Work? *Refractories Worldforum* **2012,** 4, (1), 67-72.
5. He, X.; Li, Y.; Wang, L.; Sun, Y.; Zhang, S., High emissivity coatings for high temperature application: progress and prospect. *Thin Solid Films* **2009,** 517, (17), 5120-5129.
6. Emisshield website - Hydrocarbon & Chemical refractory and metal coatings. http://www.emisshield.com/industrial-applications/hydrocarbon-chemical/ (14/01/2020),
7. Westbrook, C. K.; Dryer, F. L., Simplified reaction mechanisms for the oxidation of hydrocarbon fuels in flames. *Combustion science and technology* **1981,** 27, (1-2), 31-43.

8. Stagni, A.; Frassoldati, A.; Cuoci, A.; Faravelli, T.; Ranzi, E., Skeletal mechanism reduction through species-targeted sensitivity analysis. *Combustion and Flame* **2016,** 163, 382-393.
9. Dierickx, J. L.; Plehiers, P. M.; Froment, G. F., On-line gas chromatographic analysis of hydrocarbon effluents: Calibration factors and their correlation. *Journal of Chromatography A* **1986,** 362, 155-174.
10. Van Geem, K. M.; Pyl, S. P.; Reyniers, M.-F.; Vercammen, J.; Beens, J.; Marin, G. B., On-line analysis of complex hydrocarbon mixtures using comprehensive two-dimensional gas chromatography. *Journal of Chromatography A* **2010,** 1217, (43), 6623-6633.
11. Symoens, S.; Dokic, M.; Zhang, J.; Bellos, G.; Jakobi, D.; Weigandt, J.; Klein, S.; Battin-Leclerc, F.; Heynderickx, G.; Van Thielen, J. In *IMPROOF: Integrated model guided process optimization of steam cracking furnaces*, 30th Ethylene producers' conference (30th EPC), 2018; 2018.
12. Hottel, H. C.; Sarofim, A. F., *Radiative Transfer*. McGraw-Hill: 1967.
13. Jones, W.; Lindstedt, R., Global reaction schemes for hydrocarbon combustion. *Combustion and flame* **1988,** 73, (3), 233-249.
14. Frassoldati, A.; Cuoci, A.; Faravelli, T.; Ranzi, E.; Candusso, C.; Tolazzi, D. In *Simplified kinetic schemes for oxy-fuel combustion*, 1st International conference on sustainable fossil fuels for future energy, 2009; 2009; pp 6-10.
15. Glarborg, P.; Bentzen, L. L. B., Chemical Effects of a High CO2 Concentration in Oxy-Fuel Combustion of Methane. *Energy & Fuels* **2008,** 22, (1), 291-296.
16. Andersen, J.; Rasmussen, C. L.; Giselsson, T.; Glarborg, P., Global combustion mechanisms for use in CFD modeling under oxy-fuel conditions. *Energy & Fuels* **2009,** 23, (3), 1379-1389.
17. Green, D. W.; Perry, R. H., *Perry's Chemical Engineers' Handbook, Eighth Edition*. McGraw-Hill Education: 2007.
18. Edwards, D., Molecular gas band radiation. In *Advances in heat transfer*, Elsevier: 1976; Vol. 12, pp 115-193.
19. Modest, M. F., *Radiative heat transfer*. Academic press: 2013.

Session 124: Ethylene Plant Technology Fundamentals
Paper 124f.

Separation of Olefin-Paraffin Mixtures with Optiperm™ Customized Amorphous Fluoropolymer Membranes

Dr. Hannah Murnen

Chief Technology Officer

Compact Membrane Systems

Dr. Sudip Majumdar

Membrane Expert

Compact Membrane Systems

William Charlton

VP Operations

Compact Membrane Systems

Prepared for Presentation at the 2020 Spring National Meeting

Houston, TX, April 1, 2020

AIChE and EPC shall not be responsible for statements or opinions contained in papers or printed in its publications

Separation of Olefin-Paraffin Mixtures with Optiperm™ Customized Amorphous Fluoropolymer Membranes

Hannah Murnen

Chief Technology Officer

Compact Membrane Systems

Sudip Majumdar

Membrane Expert

Compact Membrane Systems

William Charlton

VP Separations

Compact Membrane Systems

Abstract: The separation of light olefins from paraffins is one of the most energy intensive in the world. The state of the art separation method is distillation, a technology that has remained unchanged over decades. Membranes hold the promise of a more modular, energy efficient and capital efficient method for separation. However, they have historically been hampered by low performance and stability.

Compact Membrane Systems has developed a novel membrane material using a silver embedded fluoropolymer material to transport olefins across the membrane. This material has demonstrated over 2 years of longevity in laboratory operation and has shown high selectivity and permeance under realistic operating conditions. Additionally Optiperm™ membranes were trialed at the Delaware City Refining Company in over 300 days of field operations validating the robustness and capability of the technology in an operating plant environment.

During the past year, CMS has developed and installed a laboratory recycle test systems to allow for the continuous testing of spiral wound modules with areas of ~20 square feet. The recycle test unit enables CMS to test larger scale membrane modules under field operating conditions using actual process streams from end users. Furthermore the recycle rig design incorporates our innovative design for hydrating the process feed within the membrane housing, significantly reducing the complexity of the membrane systems operations.

We will share the latest learnings from our recycle rig testing as well as provide updates on the scale up of our flat sheet spiral wound membrane cartridges. We will also discuss the current state of our hollow fiber membrane configuration development and the progress on our joint project with Dow Chemical through RAPID, a manufacturing institute funded by the Department of Energy. The goal of the work with Dow is to produce a chemical grade propylene from a process feed stream with a propylene concentration of greater than 60%.

CMS will also discuss results on laboratory testing of C_4 streams. This includes mixed gas separation of n-butane and 1-butene with permeances between 200-500 GPU and selectivity greater than 20. We will share results for a number of operating conditions and will highlight several exciting C_4 separation applications including those with isomer streams. In addition, longevity results will be shared demonstrating the stability of the membrane in C_4 separations.

Background

Ethylene and propylene are major chemical industry raw materials and consume a great deal of energy in their production. The separations of these materials are some of the costliest, most energy intensive, and most technically challenging separations in the industry due to the very similar size and nature of the molecules being separated. The technology currently employed for the separation of ethylene and propylene from paraffins is distillation and is estimated to consume 250 trillion BTU/year of energy.

In addition, distillation columns to separate these molecules are very capital-intensive. A membrane based olefin-paraffin separation technology has the potential to provide substantial economic benefit to the petrochemical industry and to drastically reduce the energy required. In addition, membrane systems are scalable, allowing their implementation for smaller, stranded streams where the investment in distillation columns have not previously been installed due to limited economic value. Membrane processes utilizing facilitated transport membranes have been extensively studied for separating olefin/paraffin mixtures. While separations have been demonstrated in the laboratory, problems with membrane stability have prevented development of commercial systems.

Membrane Technology

Compact Membrane Systems (CMS) has developed Optiperm™ membrane, a customized amorphous fluoropolymer (CAF) containing silver ions that selectively transport olefin molecules from a mixture of olefin and paraffin. CMS Optiperm™ membranes are based on a silver-containing fluoropolymer material. The technology exploits a mass-transfer mechanism known as facilitated transport. Silver incorporated into the membrane polymer backbone as a binding site for the double bond in the olefins transporting them across the membrane as depicted in the cartoon below. Saturated paraffins do not have the same interaction. This difference in affinity allows for olefins to pass through the membrane at a significantly higher rate than do paraffins.

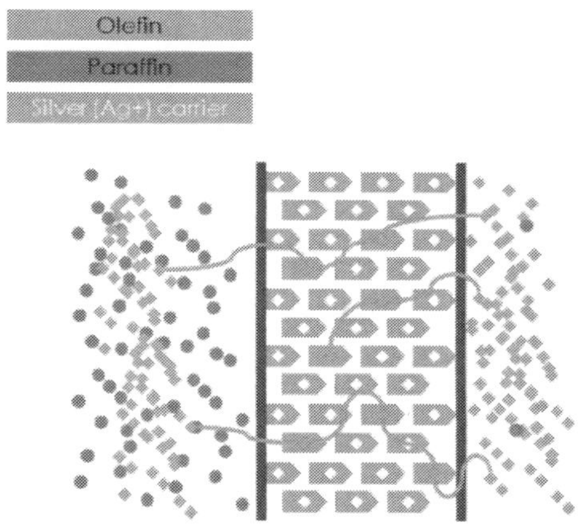

Figure 1. *Illustration of Facilitated Transport*

The two key membrane properties that characterize the effects of facilitated transport are **permeance** and **selectivity**.

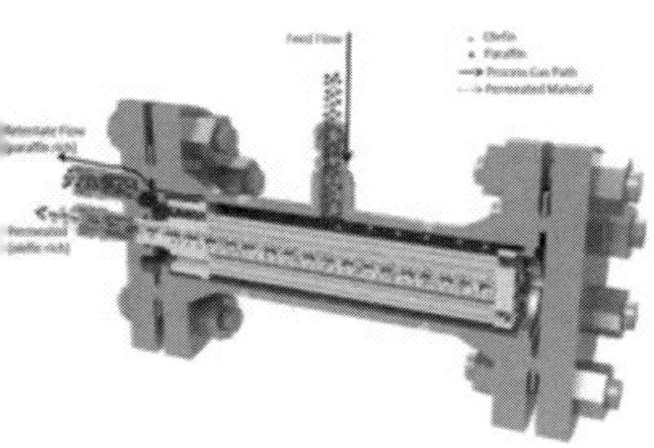

- **Permeance** is defined as the rate at which one component travels through a given surface area of membrane for a given driving force. This performance metric is typically expressed in GPU (gas permeation unit).
- The **selectivity** of one component to another is simply the ratio of the permeance of the two components. This tells you how much olefin is passing through the membrane as compared to paraffin.

Figure 2. *Illustration of Optiperm™ membrane module*

The Optiperm™ membrane has both high permeance and high selectivity, meaning high product purities can be reached with a reasonable membrane area. These characteristics yield very attractive process economics and the membrane can perform separations of olefins from paraffins over a wide range of operating temperatures, pressures, and compositions. CMS has demonstrated olefin removal at temperatures up to 70°C (158°F) and pressures up to 250 psig in the lab. For an efficient separation to occur, the olefin-paraffin mixture must be in the gaseous state. And optimal separation performance requires a highly humidified feed stream as water vapor improves the mobility of the olefin through the membrane. The membrane can remove olefin in both concentrated and dilute streams for compositions ranging 5-95% by volume of olefin. The membrane can process mixtures of C_2, C_3, and C_4 olefin and paraffin mixture.

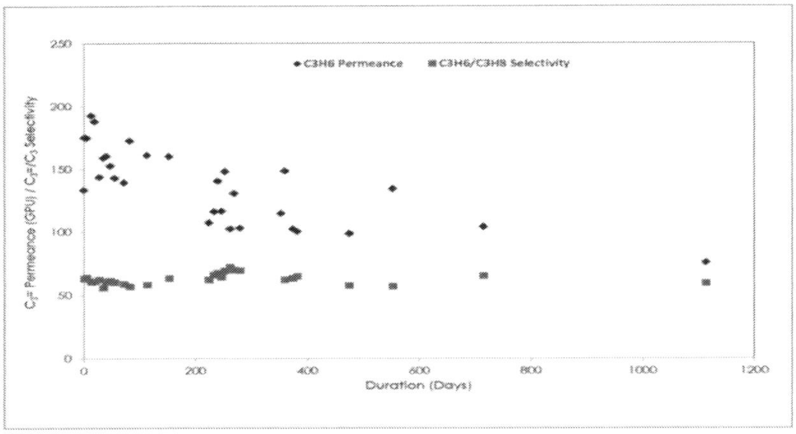

The chart displayed to the left shows a performance profile for a long term lab ageing study of an Optiperm™ membrane separating a 20% propylene /80% propane mixture at 60 psig and room temperature. As you can see, the membrane has shown stable performance over >1100 days, a vast improvement over current state of the art materials.

Figure 3. *Optiperm™ membrane long term aging study*

Pilot Units

The purpose of the CMS Pilot Units is to provide a platform for field demonstration of Compact Membrane Systems' (CMS) new Optiperm™ membrane technology for separation of olefins from paraffins in the gas state. The membranes utilize silver as a facilitating agent to improve the selective transport of olefins through the membrane relative to paraffins. These membranes provide a high olefin transport rate, which reduces the required membrane surface area, and high olefin/paraffin selectivity, which improves product split. Although various membranes for olefin/paraffin separation have been evaluated for many years, the CMS membranes are different in that they exhibit stable performance over a much longer time period than was ever achieved with earlier membranes. The improvement in membrane stability is made possible by incorporating the silver ion into backbone of the fluoropolymer which makes up the bulk of the membrane structure. As with most fluoropolymers, the fluoropolymer used in the CMS olefin/paraffin separation membrane is very resistant to chemical attack. It is this property that has been successfully utilized to create a membrane that provides stable separation performance in olefin/paraffin separations.

Prior to the refinery field trials all of the development tests done with CMS Optiperm™ membranes have been carried out in the laboratory at small scales. The DCRC Pilot Test Unit enabled testing of membranes that are an order of magnitude larger in scale with actual refinery process streams. Moreover, membrane performance could be tested at operating conditions that are difficult to achieve in the laboratory. In particular, membrane performance in the separation of propylene/propane mixtures at feed pressures above 90 psig can be readily tested.

Building upon the learnings of or membranes performance in the lab environment we completed over 300 days of operations that spanned a period of approximately 15 calendar months at the Delaware City Refining Company facility with our Optiperm™ membrane technology utilizing a paraffin-rich mixed C_3 feed stream to gauge the technical performance and economic viability of a full scale system. Prior to our field trials all of our membrane ageing and performance testing have been done in our lab facilities under controlled conditions with bottled gases and one of the things we did not know when we began the pilot trials was weather our membranes would be stable in a real process stream.

DCRC Pilot Test Findings:

DCRC Pilot tests key learnings:

- **Field membrane performance is stable**. Membrane pre-qualification and post-field performance tests confirm steady membrane performance parameters with permeance and selectivity matching the pre field testing values.
- **Membrane performance matches laboratory data** at similar conditions.
- We were able to achieve our primary goal of **producing an HD5 quality propane product**. A commercial scale system design will incorporate all of the design and control elements to continually produce on spec products including HD5 propane and propylene-rich streams.
- **Our spiral module membrane design and materials of construction performed as designed** with no mechanical failures noted in 300 days of field testing using multiple spiral wound membrane modules.

To give a little bit more detail around the first key learning, presented below in Figure 3, is field data from a membrane test conducted over a 50 day period. In this test the membrane was operating at a low stage cut with a 9% propylene feed to produce a permeate stream with a propylene concentration of 25%. Although there is some scatter in the data, the permeate stream composition is consistent over the 50 days of operation.

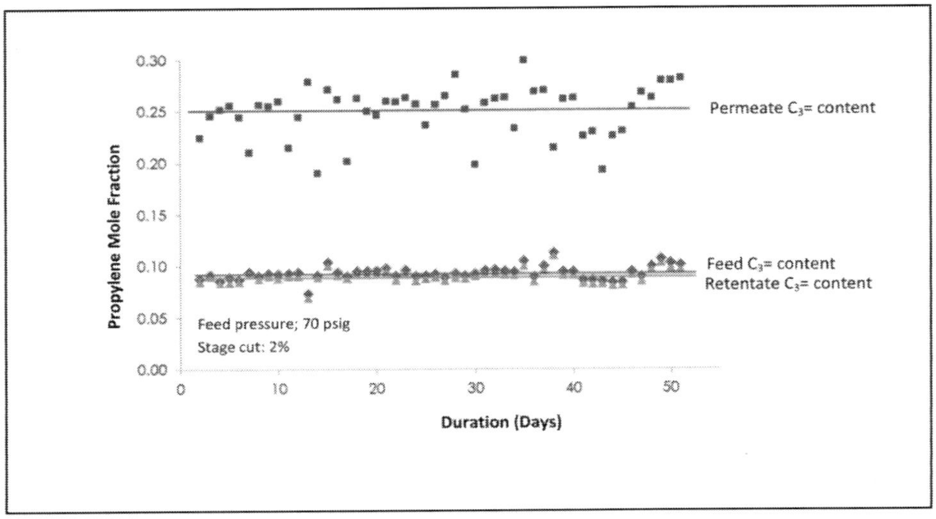

Figure 4. The results of one of our long term aging studies conducted over a 50 day period

Building upon the success of the DCRC Pilot Unit, CMS designed and built a new testing facility at our Newport, Delaware site to enable testing of larger scale spiral wound membranes at field conditions. To give a sense of scale, the testing at DCRC was done on a ~2ft^2 membrane cartridge. This test unit will allow testing of >20ft^2 cartridges. The test unit was designed as a closed loop system which allows for continual circulation, separation and recovery of a mixed olefin paraffin streams at plant operating conditions. The system will run with 24/7 autonomous operation to allow for the long term performance testing of membrane cartridges. The test system provides the research team with the capability to simultaneously test 2 large scale modules at different process conditions and has proven to be irreplaceable tool for developing and testing new membrane designs and operating concepts. Finally, the system also incorporates one our latest innovations, a novel method of humidification that ensures uniform humidification throughout the entire module which we envision to be incorporated into all pilot and commercial scale designs.

The CMS design and engineering team is currently developing three pilot scale units to be deployed in 2020 at customer sites to further enhance our understanding of the Optiperm^{TM} membranes performance in process environments.

2020 Pilot 1 - For the first project, CMS has partnered with Dow through the Department of Energy's RAPID institute to scale-up and test large scale spiral wound membrane modules at a Dow process facility. The CMS and Dow project team has designed and engineered a single stage membrane system capable of testing membrane cartridge with up to 20 ft^2 of active area. The Dow system design will also include CMS' new cutting edge hydration system that was conceptualized during the DCRC pilot trials and proven with our newly installed test systems. The pilot is scheduled to be installed in the 3rd Quarter 2020 and will be deployed in a mixed C_3 feed stream at the Dow Freeport Texas site to separate and upgrade the olefin component of the feed stream to a chemical grade propylene product.

2020 Pilot 2 - To date all of the development tests done with CMS Optiperm™ membranes have been carried out at smaller scales with membranes having less than 5 ft^2 of active surface area. These earlier separation tests were also limited to single stage membrane processes. For the second pilot trial underway in 2020, CMS is partnering with an industry major to develop a Pilot Test Unit utilizing a two stage separation process with membranes having up to 100 ft^2 of active area. Moreover, membrane performance will be tested at actual field operating conditions that are difficult to fully duplicate in the laboratory at this scale. Finally, and probably most importantly, through demonstration of the membrane technology to produce relatively high purity product streams in the field, the Pilot Test Unit will generate increased interest in the technology and promote its acceptance in streams where high purity is required. In its initial deployment, the system will be tested in the recovery of 90+% propylene and 95+% propane streams from polypropylene reactor process effluent stream containing 40% to 70% propylene. The Pilot Test Unit has been designed to enable demonstration of acceptable product split across a wide range of operating conditions (pressures, temperatures and feed compositions). The testing with the two stage pilot unit will commence in the 4th quarter of 2020.

2020 Pilot 3 - The separation and recovery of C_4 olefins is one of the most challenging separations in the petrochemicals industries as they require several integrated stepwise process to isolate the C4 isomers. These separation processes often include extractive distillation, fraction and absorption to effectively separate and concentrate the valuable components. With increasing customer inquiries around Butane separations, CMS has accelerated the study of Optiperm™ effectiveness at separating C_4 olefins from paraffins and C_4 isomers. To further investigate these separations via membrane, CMS is refurbishing and reconfiguring the pilot unit previously used at the DCRC refinery to complete field testing at a process site in Europe to demonstrate Optiperm™ capability in separating C_4 olefin paraffin streams. This will be our first test using a spiral wound module configuration outside of our lab facilities to separate these valuable olefin components of the mixed C_4 streams. Our goal is to commence field testing with C_4s early in the 3rd quarter of 2020.

Modelling of spiral wound elements

With the CMS operations team now fully focused on commercialization of our Optiperm™ membrane technology an enhanced modeling tool with predictive properties was needed to help with the design and optimization of the membrane modules in a variety of operating conditions. To accomplish this goal, CMS partnered with the University Of Minnesota Department Of Chemical Engineering[1] through the RAPID program to develop a fundamental model of spiral wound elements in olefin/paraffin applications. This work developed a mathematical model for the separation of gases with spiral-wound membrane modules. The model accounts for variation of feed and permeate-side variables in two

[1] A Mathematical Model for Gas Separations with Spiral-Wound Membranes: Validation and Design in the Presence of Uncertainty

dimensions, and is implemented with open-source software packages. The uncertainties in membrane parameters are propagated into model calculation outputs by Monte Carlo simulation. The results from the model are used to understand the transport phenomena occurring in both dimensions, and the importance of the pressure drop along the permeate channel is realized.

The charts below shows a comparison of the models predicted performance versus the actual values recorded on the CMS recycle test unit for a mixed C_3 stream with an approximate 65% propylene concentration operating at 160 psig, 40 deg F over a 6 day period. As can be seen in the graph the results achieved in this test (solid lines), are very close to that predicted by the model (dashed lines) giving good confidence that the model will enable us to predict performance at a range of operating conditions. Please note that the decrease in permeance for the final day in this test is related to a increase in feed olefin composition which results in a lower GPU due to the facilitated transport nature of this membrane.

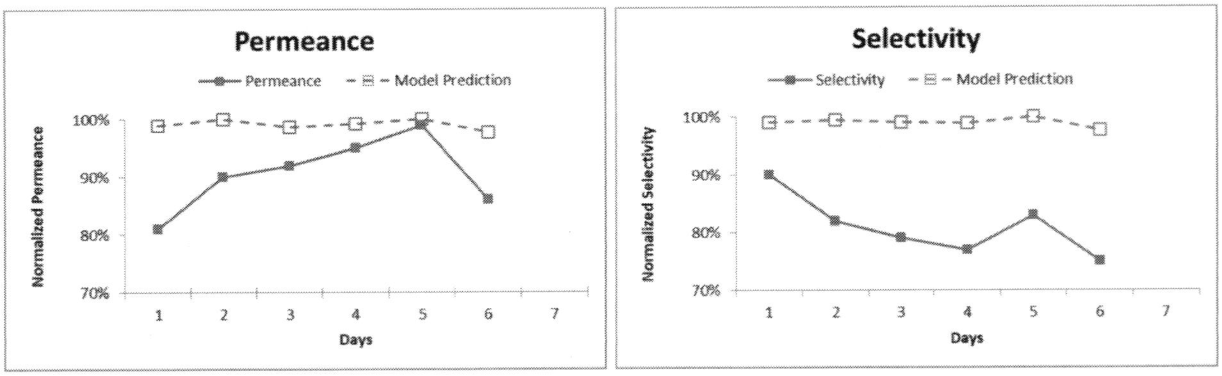

Figure 5. Comparison of the models predicted performance versus the actual values

Additional areas of development

The research team is also investigating other interesting applications of our Optiperm™ membranes both to improve process efficiencies and enable the capture and conversion of low value streams to desirable products. One example is research in the area of Enhanced Oxidative Dehydrogenation of Ethane (ODH) where the Optiperm™ membrane is partnered with an ODH reactor to produce a low cost ethylene product directly from ethane. Commercial applications of the technology may include localized small scale manufacturing of plastic components, and the direct upgrading of ethane streams to ethylene at NGL plants to name a few.

590472 Development of Novel Alloys for Use in Ethylene Cracking Furnaces

Abstract:

Cast chromia-forming austenitic stainless steels such as HP-type (25Cr-35Ni) and 34Cr-45Ni alloys are used in a wide range of industrial applications such as high temperature furnace components, radiant burner tubes, and ethylene cracking furnaces that demand high temperature microstructural stability, corrosion resistance, and creep strength. In particular, alloys used in ethylene cracking furnaces should possess good high temperature mechanical properties, coking resistance, must be fabricable using centrifugal/static casting, and must be weldable. Alloys such as cast HP-type alloys and 35Cr-45Ni achieve their environmental protection through the formation of a chromium-rich oxide scale. Recent work has shown that alloys that form a thin alumina-scale show better coking resistance in ethane steam cracking environments [1]. Although alloys that form alumina scales offer better corrosion protection and coking resistance at relevant temperatures, developing cast austenitic alloys that form a stable alumina scale and achieve creep strength comparable to existing cast chromia-forming alloys is challenging.

Cast alumina-forming austenitic stainless steels have recently been developed at ORNL for use from about 750°C to 1150°C [2]. In a recent study Muralidharan et al discussed the effect of composition on the alumina-forming capability and creep properties of cast alloys of composition Fe- 25 wt.% Ni-14 wt.% Cr-3.5 wt.% Al and with varying levels of Nb, Si, and C [3,4]. It was observed that alloys in this composition range containing about 25 wt. % Ni may be limited to operation at temperatures of about 800°C - 850°C depending on the water vapor levels in the environment due to inadequate oxidation resistance at higher temperatures. In a more recent study, Brady et al. [5] studied the ability of Fe–(25–45)Ni– (10–25)Cr–(4–5)Al–1Si–0.15Hf–0.07Y–0.01B wt% base alloys with and without Nb, Ti, and C additions to form a protective alumina scale at 1100 °C in air with 10% water vapor. It was found that Fe–35Ni–25Cr–4Al with additions of Nb, C showed promise as an alloy base for future development to achieve a cast alumina-forming austenitic alloy for use at temperatures exceeding 1100°C.

Oak Ridge National Laboratory has been working with industrial partners Duraloy Technologies, and MetalTek International and University of Wisconsin, Milwaukee to develop cast austenitic stainless steel compositions for use in ethylene cracking furnace applications. This presentation will outline the methodologies used in the development of the cast Fe-Ni-Cr-Al austenitic stainless steels, their microstructure, and highlight their high temperature oxidation resistance. In addition, the ability of these alloys to resist coking in an ethane steam cracking environment has been evaluated. This talk will outline the experimental setup, testing methodologies, and results from cyclic coking-decoking tests conducted on the developmental alloys. These results were compared to results from baseline 35Cr-45Ni alloys.

Research sponsored by ARPA-E, US Department of Energy, under Contract DE-AC05-00OR22725 with UT-Battelle, LLC.

1. E. Munoz Gandarillas, K. M. Van Geem, Marie-Francoise Reyniers, Guy B. Marin, "Influence of the Reactor Material Composition on Coke Formation during Ethane Steam Cracking," Ind. Eng. Chem. Res. 2014, 53, 6358-6371.
2. G. Muralidharan, Y. Yamamoto, and M. P. Brady, "Cast Alumina Forming Austenitic Stainless

Steels," U.S. Patent 8, 431, 072 B2, April 30, 2013.
3. G. Muralidharan, Y. Yamamoto, M. P. Brady, D. Leonard, B. A. Pint, D. Voke and R. Pankiw, (2015) "Development of Cast Alumina-forming Austenitic Stainless Steel Alloys for use in High Temperature Process Environments," NACE Paper C2015-6114, Houston, TX, presented at NACE Corrosion 2015, Dallas, TX, March 2015.
4. G. Muralidharan, Y. Yamamoto, M. P. Brady, L. R. Walker, H. M. Meyer II, and D. N. Leonard, "Development of Cast Alumina-Forming Austenitic Stainless Steels" JOM, First Online: September 6, 2016, DOI: 10.1007/s11837-016-2094-8.
5. M.P. Brady, G. Muralidharan, Y. Yamamoto, and B. Pint, "Development of 1100°C Capable Alumina-Forming Austenitic Alloys," Oxid Met, DOI 10.1007/s11085-016-9667-3.

Copyright © American Institute of Chemical Engineers. All rights reserved.

589790 Precision Procedure Planning for Executing Successful Turnarounds

Abstract:

Precision Procedure Planning for Executing Successful Turnarounds

Long term planning is becoming more recognized as key to a successful overhaul event within the petrochemical industry due to the demand for shorter outage durations driven by ever-changing market conditions. Ethylene producers need to leverage every available opportunity in performing their routine machinery maintenance and scheduled overhauls to maximize efficiency and effectiveness.

Precision Procedure Planning of these outages helps identify and mitigate safety issues, lack of detail, inefficient methods, and mechanical risks to the equipment well in advance of the execution. MCO-I understands the complexities of these major overhauls and continues to advance the detailed planning process required to make these events a success. We break the planning process into two phases, reliability and efficiency upgrades and overhaul procedures.

When developing the proper plan, eighteen months of lead time is generally needed to ensure a successful outage. The first phase of planning is to evaluate all potential reliability and efficiency upgrades. The second phase defines work scope, recognizes risks, creates detailed procedures, inventories spare parts, and identifies special tooling requirements to execute a successful outage.

Copyright © American Institute of Chemical Engineers. All rights reserved.

589739 The Use of Laser-Based Measurement Technology to Improve Critical Turbomachinery Overhauls

Abstract:

The utilization of laser-based measurement technology has rapidly spread throughout the global petrochemical industry. The evolution of both contact (laser tracker) and non-contact (laser scanning) measurement technologies has inspired a new dynamic with respect to the execution of critical turn-around activities. Rotating equipment overhauls, project planning, piping modifications, dimensional analysis and reverse engineering efforts are all poised to benefit through the integration of this technology. This paper will focus on the available technology... how it can be implemented to reduce the overhaul duration of critical equipment during major outages as well as highlight the benefits of this technology with respect to preparation and planning of future outages. The end user will provide an overview of things to consider when selecting a specialized metrology vendor as well as current / specific applications of this technology. The specialized metrology vendor will provide descriptions of both current and conceptual technologies within the field of dimensional metrology, the benefits and limitations of these technologies as well as provide prior uses / case studies of these applications.

Applications:
- Precision Alignment
- Dimensional Analysis
- Virtual Assembly
- Reverse Engineering
- Piping Modifications
- Structural Modifications
- Equipment Layout & New Installations
- Point Cloud Development for Modeling / Simulation / Virtual Reality
- Quality Assurance

Copyright © American Institute of Chemical Engineers. All rights reserved.

AIChE Paper Number: 588699

The Utilization of Laser-etched, Active Seal Face Surface Features: Maximizing Reliability in Ethylene Applications

Brian Kalfrin
Sr. Regional Engineering Manager
John Crane

Prepared for Presentation at the 2020 Spring National Meeting
Houston, Texas, August 19, 2020

AIChE and EPC shall not be responsible for statements or opinions contained in papers or printed in its publications

The Utilization of Laser-etched, Active Seal Face Surface Features: Maximizing Reliability in Ethylene Applications

Brian Kalfrin
Sr. Regional Engineering Manager
John Crane

Abstract:
The requirements for effective sealing of Ethylene in terms of the mechanical seal is a maximization of both seal face stability and interface lubrication. Mechanical seals operating in ethylene and similar light hydrocarbon services will do so with very little hydrodynamic load support due to the low viscosities in place. It is likely that the seal faces in such applications will operate in a solid to mixed friction regime. In this instance, the face materials are likely to experience higher wear rates in these regions due to increased temperature (from rubbing friction) and potential break down due to significant hydrostatic loading of the faces. The paper will highlight successful seal face component and design implementations utilized to reduce interface frictional heat and leakage, focusing on specific ethylene pump seal installations. Paper will discuss both historical laboratory testing validating performance of active face surface features along with field installations where improvements have been realized by implementing such design features. In addition, facets of how process seal design strategies coupled with an effective support system can provide improve reliability in these services will be discussed. The aspects discussed should be utilized as a template when evaluating similar problematic applications in the future.

The challenge in processing, low molecular weight hydrocarbons such as ethylene, ethane, propane, and mixtures of these fluids is their unique properties, starting with the atmospheric boiling point. As an example, ethane, ethylene, and propane are all gases at ambient temperature and pressure since their atmospheric boiling point is less than normal ambient temperature. For this reason, these fluids and others similar to them would be processed at either very high pressures or very low temperatures. For example, consider that the vapor pressure of ethylene at 0 °F is 390 PSIA, requiring that the ethylene must be pressurized greater than 390 PSIA in order to be a liquid at 0 °F. If however, the temperature is reduced to -50 F, the vapor pressure is reduced to 180 PSIA. In addition to the vapor pressure increase along with temperature, the specific gravity and viscosity would also decrease with a temperature increase. The subsequent thinning of the process fluid with this temperature increase creates specific challenges for mechanical seals in terms of maintaining adequate fluid film support between the two sliding surfaces, the seal faces.

It is likely that mechanical seal faces in such applications will operate in a solid to mixed friction regime; in these operating regions, the face materials are likely to experience higher wear rates due to increased temperature (from rubbing friction) and potential break down due to hydrostatic loading of the faces. The fundamental principle of how a mechanical seal operates and the factors that affect reliability, can be described and demonstrated by looking at the simple axial force equilibrium. It is assumed that the arithmetical sum of all axial external loads (hydraulic load from the sealed fluid, spring load and secondary seal friction) is supported by reaction forces at the sealing interface. In this simplistic approach to the problem it applies that the closing force is fully supported by the fluid film but in reality for most mechanical seals the fluid film (hydrostatic pressure and hydrodynamic effect) is not strong enough to support the full closing force and the excess is supported by asperity contact between the sealing rings. Figure 1 illustrates this relationship.

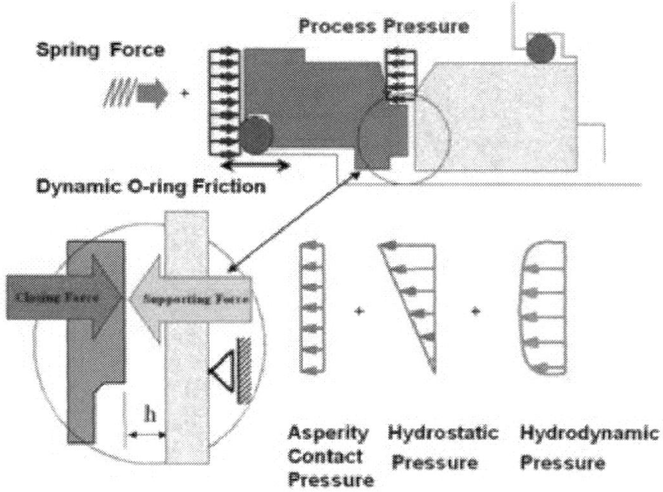

Figure 1: Seal Face Equilibrium Equation

This lubrication regime described above is called 'mixed lubrication' or if the asperity contact support is predominant 'boundary lubrication'. The sealing gap (h) itself is small, in the order of the surface asperities, and offers very high resistance of radial leakage flow.

Unfortunately, the asperity contact that minimizes the seal leakage is also the main source of frictional heat and the reason for wear. Frictional heat can cause overheating and in some cases vaporization of the fluid film, loss of seal face lubrication and severe dry running leading to premature seal failure.

Mechanical seal face contact in ethylene and similar olefin applications can produce additional challenges outside of accelerated wear of the seal faces. Asperity contact pressure and subsequent heat generated from rubbing friction in the absence of sufficient hydrodynamic load support can lead to localized interface temperatures conducive to oligomer formation. The management of heat in ethylene services, more importantly, keeping excess heat out of the sealing interface is critical to long term reliability in these instances. Once oligomer formation commences, it becomes progressively more severe and contributes to higher leakage rates from face hang-up and irregular face tracking. Figure 2 documents an example of polymerization due to localized temperatures (seal face generated heat).

Figure 2: Polymerization in Ethylene Application

An ideal mechanical seal would operate reliably, for long periods, without wear, with permanently low friction and extremely low leakage. Real mechanical seals do wear and have a limited life. They can leak appreciably, particularly when the operating conditions pressure, temperature and speed vary with time. To date, most technology solutions employed are passive methods to control the sealing gap with the emphasis on boundary lubrication or hydrostatic fluid film formation. There have also been attempts to introduce hydrodynamic seal face features on a larger scale (hydro-pads, face waviness, hydrodynamic micro-groves or thermo-hydrodynamic circulation grooves) to overcome excessive face friction and wear and with it achieving improvements in seal performance and reliability particularly for high pressure, high velocity application. However, increased fluid film thickness inevitably increases seal leakage/ emissions to unacceptably high levels. Management of seal leakage when pumping ethylene is especially critical.

When the pressure of ethylene is suddenly reduced, the temperature decreases. In the instance of leakage to atmospheric pressure, the temperature decreases to the atmospheric boiling point. For example, if ethylene at 0 °F and 500 psia leaks to the atmosphere, the temperature of the leaking gas would be -154 °F. If the leakage rate is very low, then heat transfer from the surroundings may overcome this refrigeration effect but for high leakage rates, severe icing may occur. Passive interface lubrication features used to enhance the interface lubrication by way of hydrodynamics inevitably, and sometimes unpredictably, increase the mean thickness of the sealing gap and subsequently the leakage of the seal. This increase in leakage with an increase in fluid film thickness can be substantial as the Reynolds equation suggests, in which the leakage, q is a cubic function of the face gap height, h.

$$q = -\frac{h^3}{12\mu}\frac{\partial p}{\partial r}$$

As outlined, handling of ethylene from a mechanical seal perspective is especially challenging; optimizing reliability requires effective management of both face generated heat and leakage to lower pressure regions within the assembly.

Active Seal Face Features

It becomes clear that a successful approach to increasing seal performance and reliability in challenging lubrication regimes such as ethylene applications requires full fluid interface lubrication, but also requires active leakage or fluid film control. This realization lead to the idea of creating, in combination with a lubrication enhancement mechanism, an interface pumping mechanism which returns fluid from the sealing gap to the pressurized fluid space. Figure 3 below highlights a particularly successful seal face technology feature that achieves the desired results.

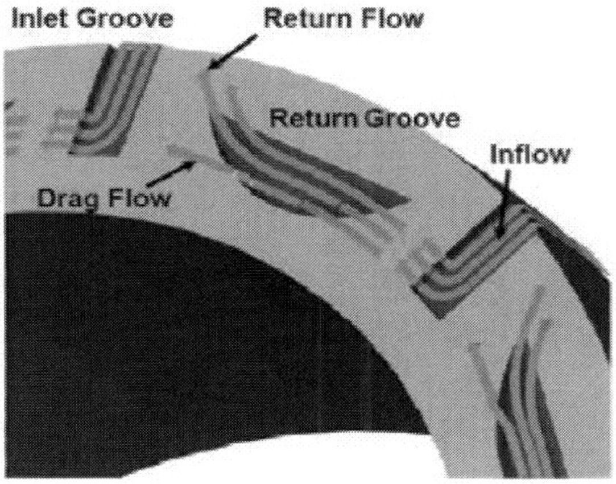

Figure 3: Active Seal Face Features

The surface features, created through advanced etching techniques, combines a pair of symmetrical precision-machined microgrooves each of which performs a different task. The square inlet-groove is connected to the pressurized fluid area and through it, the process fluid can penetrate deeply into the sealing interface. The fluid within it is dragged in tangential direction by the sliding counter face. When the fluid reaches the edge of the inlet-groove it generates a strong hydrodynamic pressure that lifts the seal faces and subsequently stabilizes the sealing gap. This hydrodynamic generated fluid film guarantees low friction and low wear performance.

As discussed, simply increasing the interface gap nets a reduced in radial leakage flow resistance of and leakage levels would usually increase substantially as the Reynolds equation suggests. The addition of semi-circular grooves, also known as 'return-grooves', control the amount of leakage passing through the sealing gap. The return grooves are located within the sealing interface and are not connected to the pressurized fluid area, therefore the groove collects the excessive fluid film and guides it along the trailing edges to the discharge end of the groove. The discharge end is very close to the high pressure edge of the sealing gap where the fluid pressure inside the groove is at a maximum and much higher than the sealed pressurized fluid. Following the path of least resistance, a majority of the fluid flows towards the higher pressure region outside the sealing gap. In addition, cavitation or vaporization of the fluid film in the expansion zones of the return-grooves block radial leakage flow from the outer seal interface diameter. Figure 4 illustrates the principle and highlights analytical models predicting pressure stability within the sealing interface.

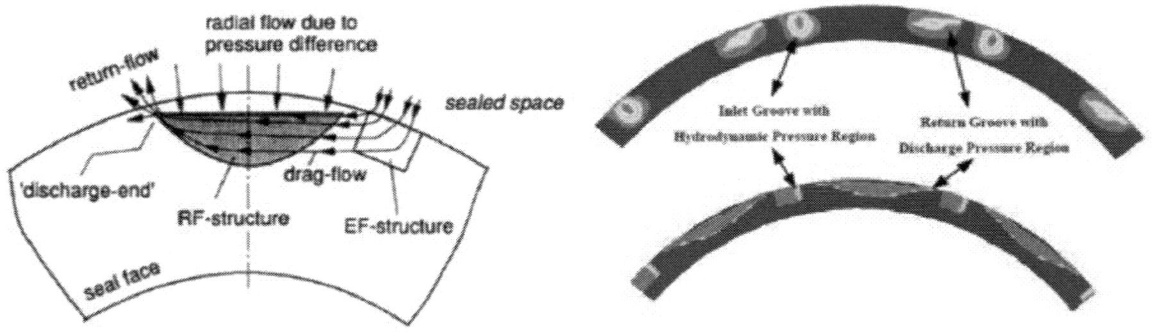

Figure 4: Principle Illustration and Analytical Results

The combination of hydrodynamic grooves and fluid film re-circulation grooves (return-grooves) gives the benefits of a seal operating with hydrodynamic lubrication and boundary lubrication, in other words, low friction and wear but together with low static and dynamic leakages. Figure 5 underlines this statement. In figure 5 the two important performance measures for a mechanical seal, leakage and face friction, are compared for a standard plain face seal, the same seal with a conventional hydrodynamic face groove (inlet groove) and a seal equipped with both entry and return-flow grooves. The bar chart on the left illustrates the comparison using good lubricating fluids and the one on the right for a more difficult duty with a flashing, poorly lubricating liquid. The performance figures are presented in relative values where 100% is always the highest recorded value from the three seal configurations.

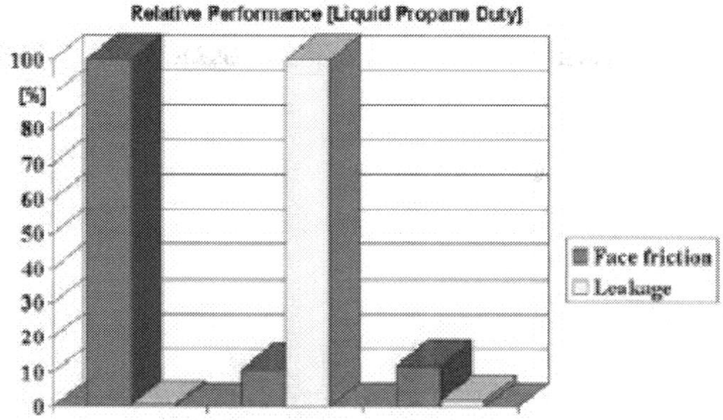

Figure 5: Comparison Results in Propane

Testing

The benefit of utilizing seal face technology such as this is best demonstrated in a comparison test of two identical seals, except that seal is equipped with entry-flow and return-flow structures on the one seal ring and the other with the conventional plain seal faces. The test duty is liquid propane at 305 psig, at a pumping temperature of 133 °F and an API flush plan 11 set at 2.6 gal / min flush flow. The mechanical seals are both 2.00" in diameter. This test duty was deliberately chosen as a more realistic example of a real-world application where temperature margins at suction pressure are minimal and the option of cooling or increasing the seal chamber pressure is limited.

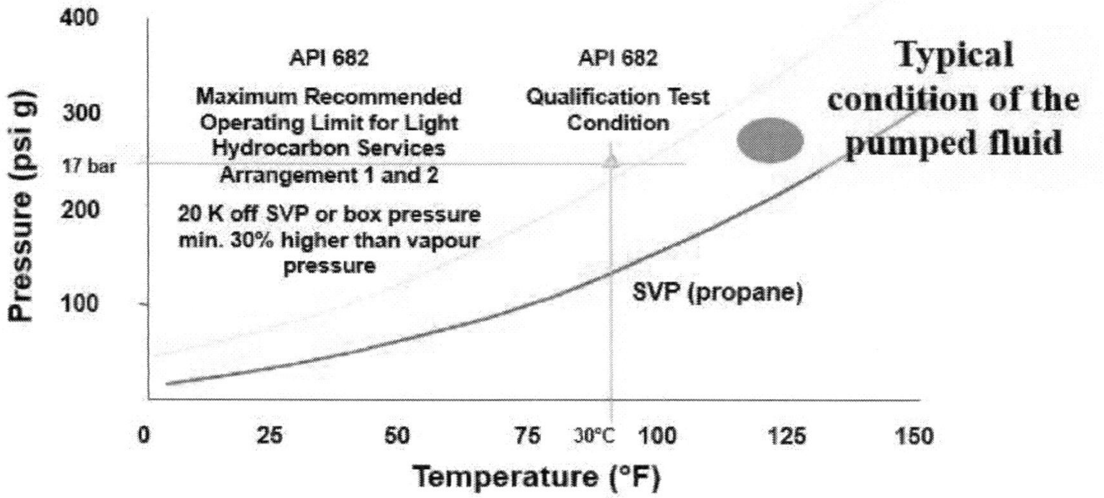

Figure 6: Graphical Representation of Real-World Application Conditions

Both mechanical seals were analyzed preliminarily through Finite Element Analysis (FEA) modeling and then subjected to testing under the noted process conditions. The conventional mechanical seal was tested initially under the specified operating parameters, key of which is the operation close to the SVP of the propane (below API 682 qualification test thresholds). After the 200 hour test, the seal faces were examined and heavy wear of the carbon face was recorded (over 0.012" in 200 hours). There was also noticeable blistering on the carbon face ID with some ID chipping. The mating ring exhibited polished wear track with moderate phonographing in the wear area. The noted face wear is indicative of marginal lubrication regimes present in the sealing interface, and suggests the majority of the face regions are operating without sufficient hydrodynamic load support. Photos of the blistering, chipping, and phonographing are shown below in Figure 7. A mechanical seal operating under these conditions would have an expected life based only on seal face wear of 3 months.

Figure 7: Plain Face Mechanical Seal – Post Test Wear

After this test, the modified seal which included the entry-flow and return-flow active face features on one seal face was tested. Post examination, dramatically different results were realized. Most noticeably, the amount of wear after the 200 hour test was measured at .0001"; more significant was the overall lack of any detectable damage to the carbon face at all. The mating ring, which included the etched face pattern, exhibited very light polishing and no visible material reduction or interface related lubrication damage. A mechanical seal operating with these design features, based on the testing, would have an anticipated life of over 274 months based solely on the amount of face wear recorded during the test. Pictures of faces post-test are included below in Figure 8.

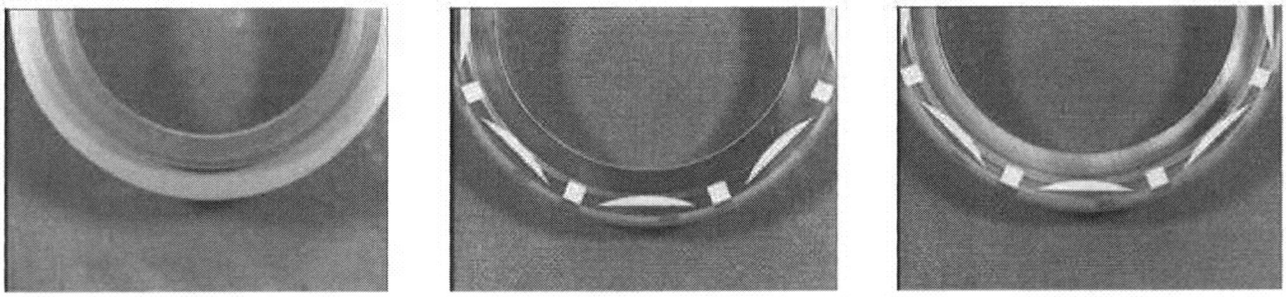

Figure 8: Etched Face Mechanical Seal – Post Test Wear

Considering the results of the design analysis and testing, potential applications for such seal face features can be considered where poor interface lubrication can lead to premature seal failure. A primary application range of focus would be light hydrocarbon services where seal chamber conditions and the fluid properties themselves can be especially volatile for the mechanical seal. A successful implementation of this technology in an ethylene service will now be discussed.

Ethylene Product Pump Case Study

A field example where active seal face features yielded significant benefits will be discussed. The application conditions concerned an ethylene product pump at a petrochemical facility on the Gulf Coast, US. The existing application conditions were as follows:

Pump:	Pacific RBIJ
Process Fluid:	Ethylene
Temperature:	-15 °F
SG:	0.431
Vapor Pressure:	280 PSIA
Suction Pressure:	293 PSIG
Discharge Pressure:	1938 PSIG
Seal Cavity Pressure:	293 PSIG
Shaft Speed:	4900 RPM

The existing mechanical seal installed in this application was a dual unpressurized cartridge seal, supported by API Piping Plan 11 / 72 / 76 / 62. The inner mechanical seal is cooled and lubricated by the process fluid, taking a bypass stream from first stage discharge pressure on the pump and injecting it into the inner seal face cavity through the seal gland connection. The outer seal in the arrangement is a non-contacting, dry running containment seal utilizing nitrogen as a purge gas to sweep inner seal vapor leakage to the plant vapor recovery system. Leakage rates and general inner seal condition are monitored through the Plan 76 monitoring panel. To provide protection against potential icing on the ID of the containment seal due to the application temperatures, a nitrogen quench is included to the cavity between the containment seal and the atmosphere, isolated by a segmented carbon bushing design. Figure 9 shows a diagram of the mechanical seal cartridge.

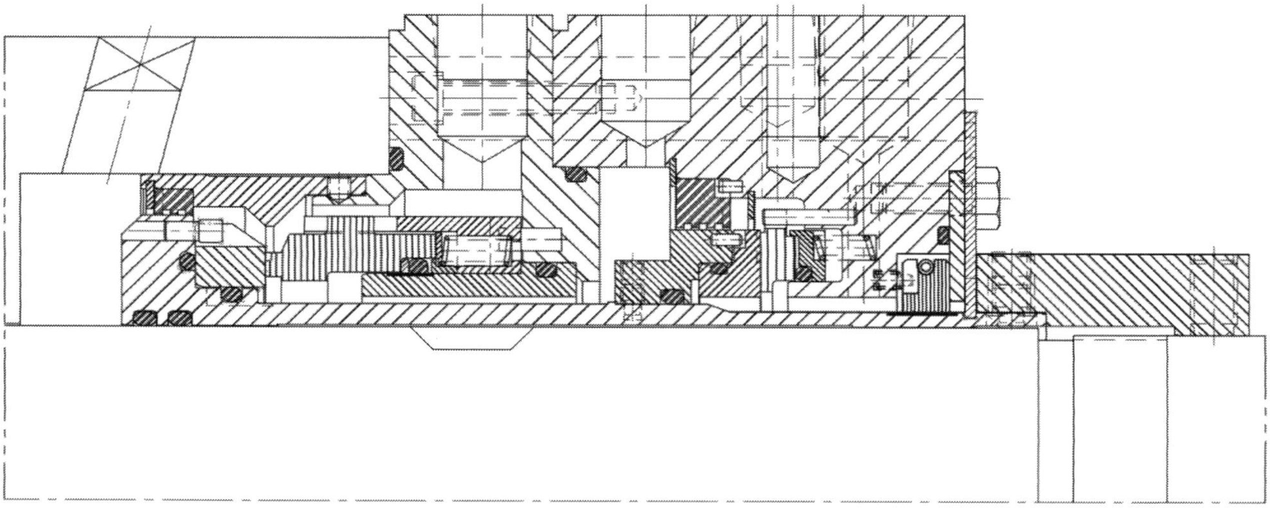

Figure 9: Original Mechanical Seal

Due to the volatility of the process, the inner seal primary ring utilized a macro-surface features incorporated to the OD of the carbon primary ring, which was the spring-loaded seal face in this particular arrangement. These recesses are sufficiently deep and allow for more fluid penetration within the sealing interface to help enhance sliding load support.

Damage to the mechanical seal faces after one repeat failure was captured in several photographs:

A & B are the photo-gallery of the as received carbon primary ring. These photos show that the carbon primary ring has severe chipping from the ID and this could be attributed to severe contact load on its seal surface. Picture C is of the Silicon Carbide mating ring. The SiC mating ring exhibits the origination of a fracture from the drive pin groove. Since, the drive pin grooves are highly stress concentrated areas the cracks originated from this location. One of the possibilities for the fracture of the SiC mating ring could be that the mating ring would have tried to spin or move from its location and hence fractured through the groove and this could be due to low lubricity condition operation which would have resulted in 'stick-slip' phenomenon. Based on these findings, there were noted concerns about the lubricity and temperature of the inner seal faces across the operating window of the pump. Elevated, localized face temperatures would not only mitigate potential fluid film support, but also potentially contribute to polymerization (oligomer) formation. In previous failures, photographic evidence supported that potentially elevated localized temperatures existed, at the inner seal faces, to support polymerization. See figure 10 below.

Figure 10: Polymerization evidence

In order to review the anticipated performance of the mechanical seal under the stated process conditions, an analysis of the design was conducted using the mechanical seal supplier's 2D finite element analysis (FEA) program. The focus of the analysis centered on the expected distortion, both from pressure and thermal effects, which the existing primary ring would be exposed to under the stated operating conditions in addition to start-up excursions, particularly a slow-roll condition of 700 RPM that had been documented. The existing primary ring design was evaluated to determine impacts on the performance in terms of overall face pressure profile, leakage, contact pressure, and temperature. The design variations considered in the analysis were the following:

1. Existing process primary ring design, startup @ 700 RPM (slow-roll condition)
2. Existing process primary ring design, startup @ 3600 RPM (turbine min-governor speed)
3. Existing process primary ring design, steady-state operation @ 4607 RPM (post min-governor)
4. Modified process primary ring design (active seal face features), startup @ 700 RPM (slow-roll condition)
5. Modified process primary ring design (active seal face features), startup @ 3600 RPM (turbine min-governor speed)
6. Modified process primary ring design (active seal face features), steady-state operation @ 4607 RPM (post min-governor)

Additional operating parameters used in the evaluation were based on field verified operating conditions in which suction pressure was recorded as 300 PSIG, discharge pressure as 2,000 PSIG, and the operating speed recorded as 4,607 RPM as noted above. Each case was modeled using two separate solvers in the FEA program, one which simulates steady-state operation and which simulates pressure profile and seal performance during initial startup with new (not worn) face profiles. Since the primary ring in all cases is carbon, there will be face contact pressure and wear in the portions of the face where the contact pressure is beyond that of the fluid film support. The utilization of the 2D FEA model is useful as it approximates the wear, distortion, and helps identify regions of the seal interface where vaporization and wear could occur. The software can be useful in an attempt to replicate seal face profiles that correlate to wear patterns and damage noted during post-failure inspection and analysis, as well as evaluating the potential influence of modifications implemented to mitigate potential detrimental effects to seal performance.

In every case, the case details assumes a constant suction pressure, but the available flush flow rate supplied to the seal under each condition is a function of available differential pressure across the pump and specifically per stage as it relates to the seal flush pressure source. Differential pressure varies with the square of a change in speed, or, differential pressure varies directly as the speed squared. In each case, the estimated available flush flow rate based on the differential pressure per stage is used in the model to determine temperature rise in the flush fluid. The associated temperature rise in the fluid around the seal components is the most commonly used criteria to determine rate of flush effectiveness in relation to generated heat. Recommended allowable temperature rise of the process fluid in the seal chamber varies based on the fluid being sealed – typical values would be 15 °F for water and low volatility hydrocarbons and 30 °F for lube oils. For volatile light hydrocarbons, the allowable temperature rise is limited to 5 °F to minimize vaporization potential around the seal faces and prevent excessive wear due to dry running. The design seal flush flow rate takes into account the temperature rise and heat generation plus a sufficient margin of safety to account for wear and also to supply sufficient velocity through the seal flush porting to move vapor bubbles that may have formed on adjacent sealing surfaces out of the chamber.

<u>Case 1 – Existing process primary ring design, startup @ 700 RPM (slow-roll condition)</u>

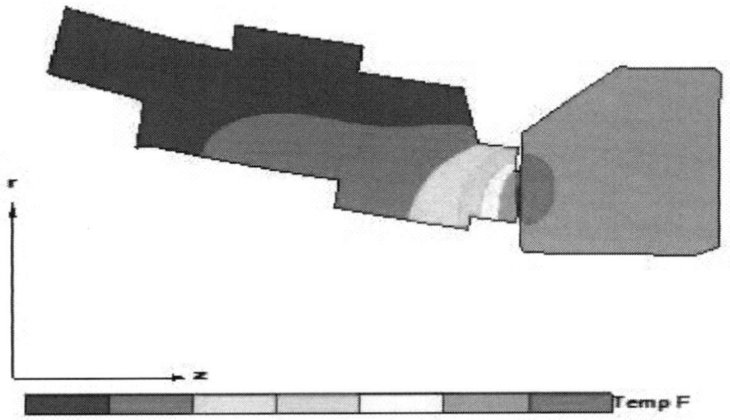

The initial profile at a startup condition of 700 RPM yields excessive face contact pressures and high average interface temperatures. Calculated flush flow rate based on available differential pressure of 3.9 GPM. Subsequent temperature rise in the sealed fluid calculated at 41.6 °F.. FEA model outputs in this conditon indicate that pressure gradient profile diverges with respect to pressure, suggesting more OD face wear.

Case 2 - Existing process primary ring design, startup @ 3600 RPM (turbine min-gov speed)

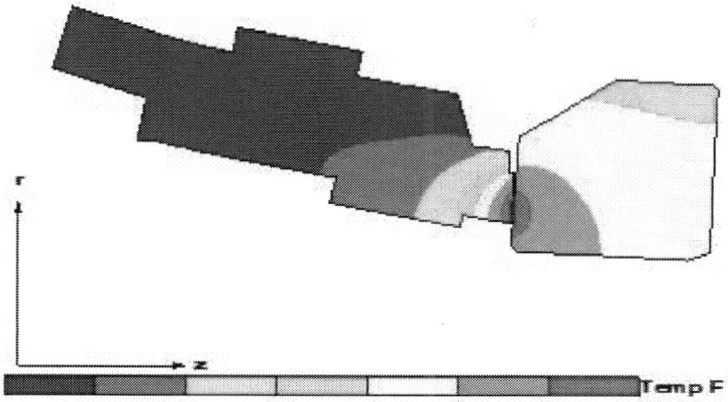

This case assumes an intial startup at 3600 RPM, which has been identified as turbine min governor speed. Contact pressure is reduced significantly when compared against the 700 RPM slow roll conditon.Calculated flush flow rate based on available differential pressure of 20.2 GPM. Subsequent temperature rise in the sealed fluid calculated at 13.6 °F. Fluid and contact pressure curves overlaid show a parabolic fluid pressure curve, which is indicative of a convergent pressure distribution in which the highest pressure drop is at the face ID, which tends to yield higher detectable leakage levels.

Case 3 - Existing process primary ring design, steady-state operation @ 4607 RPM

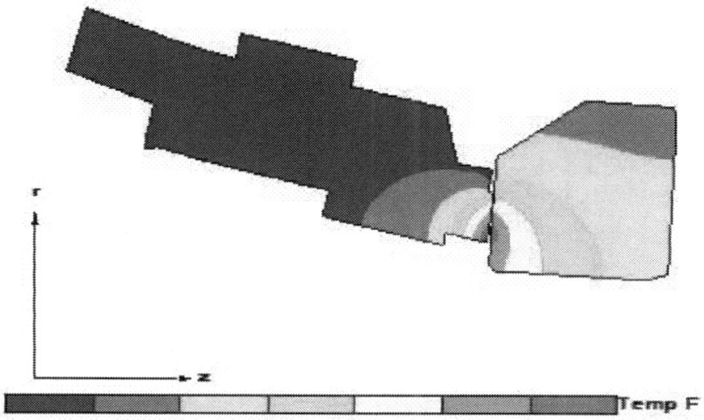

This case assumes steady-state operation at 4607 RPM post the initial condition of 3600 RPM. Further reduction in contact pressure over the previous 4607 RPM condition evaluated post slow-roll. Calculated flush flow rate based on available differential pressure of 26.0 GPM. Subsequent temperature rise in the sealed fluid calculated at 2.5 °F. Similar fluid and contact pressure curves when compared to previous cases, along with high face ID contact pressure.

Case 4 - Modified process primary ring design, startup @ 700 RPM (slow-roll condition)

Potential modifications to the seal face geometry were evaluated to determine if any improvements could be made to mitigate the effects of slow speed startup conditions. This initial case evaluates the effects of active interface features using entry-flow and return-flow grooves on the mating ring against a plain-face (non-hydropad) primary ring. The enhanced groove design utilizes a modified hydraulic balance along with the laser-etched surface structures to minimize heat, wear, and leakage. This case evaluates the same conditions as Case 1, only with a modification in face balance and inclusion of the LaserFace grooves. Calculated flush flow rate based on available differential pressure of 3.9 GPM. Subsequent temperature rise in the sealed fluid calculated at 1.3 °F. Contact pressure curve is surpressed significantly when compared the original design and calculated average contact pressure is well within materials thresholds.

Case 5 - Modified process primary ring design (active face features), startup @ 3600 RPM

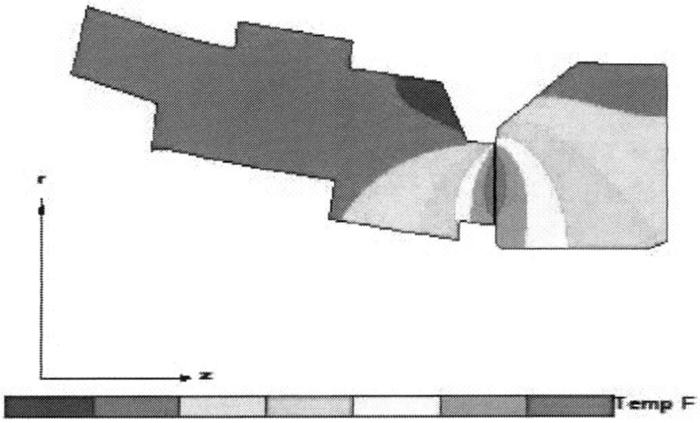

This simulation would mimic Case 2, with the modified primary ring change, at min governor condition. Calculated flush flow rate based on available differential pressure of 20.2 GPM. Subsequent temperature rise in the sealed fluid calculated at 0.4 °F. Contact pressure curve maintains a slightly linear shape, indicating pressure gradient profile is closer to parallel distribution, hence the lower contact pressure values across the interface gap and operation in a more fully hydrodynamic load support condition.

Case 6 - Modified process primary ring design, steady-state operation @ 4607 RPM

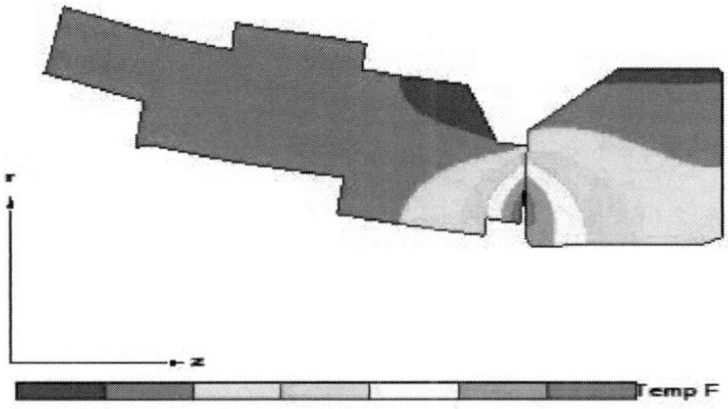

This simulation would mimics Case 3, with the modified primary ring change, post the 3600 RPM condition. Calculated flush flow rate based on available differential pressure of 26.0 GPM. Subsequent temperature rise in the sealed fluid calculated at 0.3 °F. Contact pressure curve maintains a slight linearity at the OD of the interface and there is substantial fluid film support above the contact pressure curve through the face surface. The contact pressure increase in relation to the fluid pressure curve would correlate to a contact sealing 'dam' at the very ID of the faces, minimizing leakage in this condition.

Output data comparison:

Case Number	Speed (RPM)	Flush Flow (GPM)	Avg. Face Temp (°F)	Pressure-Velocity (PSI-FPM)
1	700	3.9	41.6	186,545
2	3600	20.2	13.6	256,230
3	4607	26.0	2.5	312,461
4	700	3.9	1.3	56,110
5	3600	20.2	0.4	260,794
6	4607	26.0	0.3	246,294

Summary

After implementation of the revised mechanical seal in the application, the mean time between repair (MTBR) increase from 2 months to over 24 months after the initial replacement on the first pump. The sister pump received the identical modification to its mechanical seals. The key takeaway was that the modified mechanical seal design, implementing the active lubrication features as described was able to provide consistent and stable interface lubrication across a wide operating range. This uniformity allowed the revised mechanical seal to operate through transients without significant wear due to intermittent face contact, face generated heat, elevated leakage rates.

As outlined, the implementation of active seal face features employing deliberate entry-flow and return-flow structures in ethylene applications can yield noticeable benefits in the form of consistent interface fluid film support and stability, while at the same time managing leakage to low levels, subsequently avoiding and undesirable refrigeration effects. The minimized frictional heat generated by such designs keeps interface temperatures low, and flush fluid temperature rise differentials minimal, allowing for reliable operation in applications where very low vapor pressure margins are present. The interface features described herein should be considered as a viable option in poor performing ethylene and similar light hydrocarbon applications to manage both face generated heat and leakage. The implementation of specific face features, coupled with a reliable seal support system, has been documented to yield significant reliability improvements in these applications.

References:

1. Kalfrin, B., 2016, "Mechanical Seal and Support System Considerations for Negative Temperature Hydrocarbon Services: NGL Processing and Ethylene Production Focus," *Proceedings of the Thirty Second International Pump Users Symposium,* Turbomachinery Laboratory, Texas A&M University, College Station, Texas
2. Müller, H.K., Wallace, N.M., 1994, "Development of Low Friction, Low Leakage Mechanical Seals using Laser Technology". *Texas A&M, 11th Pump User Symposium,* USA.
3. Lebeck A.O., 1991, "Principles and Design of Mechanical Face Seals". New York, USA: John Wiley.
4. Kalfrin, B., Bagain, J., 2019, "How to Choose the Right Seal Face Technology to Improve Pump Reliability". *Plant Engineering Magazine*

Figures courtesy of John Crane

590290 Exxonmobil Baytown Olefins Plant Commissioning Safety Focus

Abstract:

ExxonMobil commissioned an ethylene production facility in Baytown in 2018. Like many ethylene plants, commissioning progressed via a series of steps including water flushing, steam blowing and air blowing. Process and personnel safety during this transient time is a major focus of the commissioning effort. Techniques and strategies for maximizing commissioning effectiveness while minimizing potential safety risks will be outlined in the presentation.

Copyright © American Institute of Chemical Engineers. All rights reserved.

590874 Improved Regeneration of Adsorbent Beds

Abstract:

Adsorbent bed treating systems are common in refineries and petrochemical manufacturing facilities and have widespread applications. They are used to dry instrument air, remove sulfur species and CO2, to meet product specifications, remove poisons from reactor feed streams, and remove water and CO2 from cryogenic fractionation trains.

Over time, changes in operating conditions may warrant a review of the treater design, operation, and regeneration procedures to improve efficiency. With changes to energy and utility costs, and improvements in desiccant technology, treaters that have been in service for a while may benefit from a fresh evaluation.

This paper discusses the operation of adsorbent beds and the typical costs associated with regenerating them. It also identifies methods for reviewing operation and regeneration techniques to reduce operating costs and emissions, using examples from an ethylene plant.

Copyright © American Institute of Chemical Engineers. All rights reserved.

PAPER#81C

USE OF DYNAMIC SIMULATION FOR SEEDING APC INPUT FOR QUICKER BENEFIT

Durgesh Soni

Manager, Process Engineering, Basell Polyolefins Ind. Pvt. Ltd, Mumbai, India

Vijay Bulusu

Process, Lead Engineer, Basell Polyolefins Ind. Pvt. Ltd, Mumbai, India

Manoj Shouche

APC/PCS Team Lead. Lyondell Chemical Company, Channelview complex, TX, USA

Prepared for Presentation at the Year 2020 AIChE Virtual Spring Meeting & 16th GCPS,

August 17 – 21, 2020

USE OF DYNAMIC SIMULATION FOR SEEDING APC INPUT FOR QUICKER BENEFIT

Durgesh Soni

Manager, Process Engineering, Basell Polyolefins Ind. Pvt. Ltd, Mumbai, India

Vijay Bulusu

Process, Lead Engineer, Basell Polyolefins Ind. Pvt. Ltd, Mumbai, India

Manoj Shouche

APC/PCS Team Lead. Lyondell Chemical Company, Channelview complex, TX, USA

Abstract

In recent years, Advanced Process Control has incorporated various new technologies and methodologies. One such advancement is Aspen DMC3 (Dynamic Matrix Control3) which has the Calibrate and Adaptive Modeling features. Calibrate mode provides Automated Step testing for which a seed model is required. One of the ways to create a seed model is through step testing Manipulated Variables (MV) in a Dynamic process simulation and collecting the data to build the model in DMC3. This approach has the advantage that no step-test is required in the real plant, thus increased savings.

LyondellBasell (LYB) used this approach in a debottleneck (DBN) Project in one of their chemical plants in the US. Existing steady state model used for DBN case, was converted in-house to Aspen Hysys dynamics and step test data was generated. The data was subsequently used prior to start up for building Advanced process control (APC) models and then implemented. This approach helped LYB save 32% cost and to gain 67% in the implementation schedule, compared to traditional step-test method.

This paper explains this methodology of step testing and model building in DMC3. Advantages of this approach over traditional plant step-test are also discussed.

Introduction:

In the current era use of software has made engineer's life easy. Now-a-days, it is possible to get very quick engineering solutions. Simulation and Control studies are widely used in practice for batch and continuous chemical process operations for the last two decades. In the last two decades there has been a remarkable increase in application of computational techniques which are readily available for engineers. Owing to these advantages' engineers can easily use more profound techniques of analysis and synthesis. Dynamics of most unit operation and process control system can be enabled effectively and are suitable for engineering purposes.

Aspen Hysys steady state and Dynamic simulators can carry out sophisticated simulations that involves multiple complex steps and scripts in a very lengthy design and analysis procedure. Dynamic simulation has the capability to define process control strategies. One can predict the behavior of the process by using fundamental engineering relationships such as, mass and energy balances, and phase and chemical equilibrium. Process simulations can run different cases likes "what if" analysis, sensitivity analysis, and to do optimization runs.

Simulations enable design of optimized plants and thus increase the profitability of existing plants by executing required number of simulation runs.

LYB performed Aspen HYSYS dynamic simulation to generate APC step test data instead of traditional plant step test. This approach is used in newly commissioned distillation columns for one of our chemical plant during DMC3 implementation. The objective of the project was to develop the step-test data which was subsequently used to generate a seed model to complete the DMC3 implementation.

Application of simulation to generate the step-test data:

Application of the Aspen suite in such exercises has unique advantage. LYB used Aspen Hysys platform, for the user to convert their existing process steady state simulation to a Dynamic simulation. Even if the existing process simulation was not available building new steady state simulation could have been a simpler task. After validation of the Dynamic simulation with plant operation and/or design data, the user can start configuration of MV's and related Control variables (CV). Multiple ways are available in Aspen Hysys to generate the step-test data e.g. use of strip chart along with event scheduler to record desired MV's and CV's trend or by using readily available utility named DMC+ 

from the palette of the Aspen Hysys.

Following sketch depict the major execution steps involved and sequence, starting from steady state simulation to DMC3 in service;

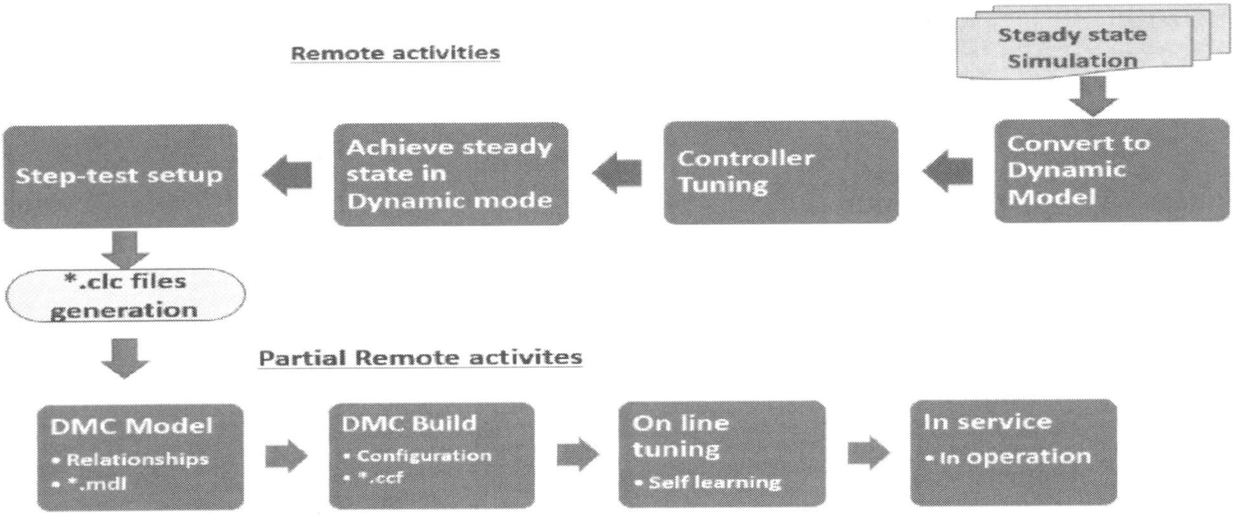

***Remote activities*:** Steady state simulation to clc file generation can be executed remotely.

Partial Remote activities: Preferably executed at process site in co-ordination with process operation and process control teams.

Abbreviations:

clc- calculation file (This contains time wise MV's & CV's value trend).

mdl- model file.

ccf- configuration file.

The developed Aspen Hysys steady state model is exported to pressure-controlled simulation in Aspen Hysys dynamics. Before exporting to Dynamics, the size and the initial specification of each equipment were defined.

LYB experienced challenges during steady-state to dynamic model conversion and subsequently in step-test, few of them are mentioned below,

- **Unit-operation complexity:**
  - Columns having multiple side draws and recycles
  - Operating under vacuum

- **Higher run time required for convergence:**
  - Due to complexity of the model, faster response time could not be achieved in dynamic simulation.
  - Speeding the Dynamic simulation run resulted in crashing of the model & bypassing few calculation steps.

- **Tuning of controllers to cover the wide range of operating conditions.**

Sketch depicting steps involved to convert steady state to dynamic simulation is given below;

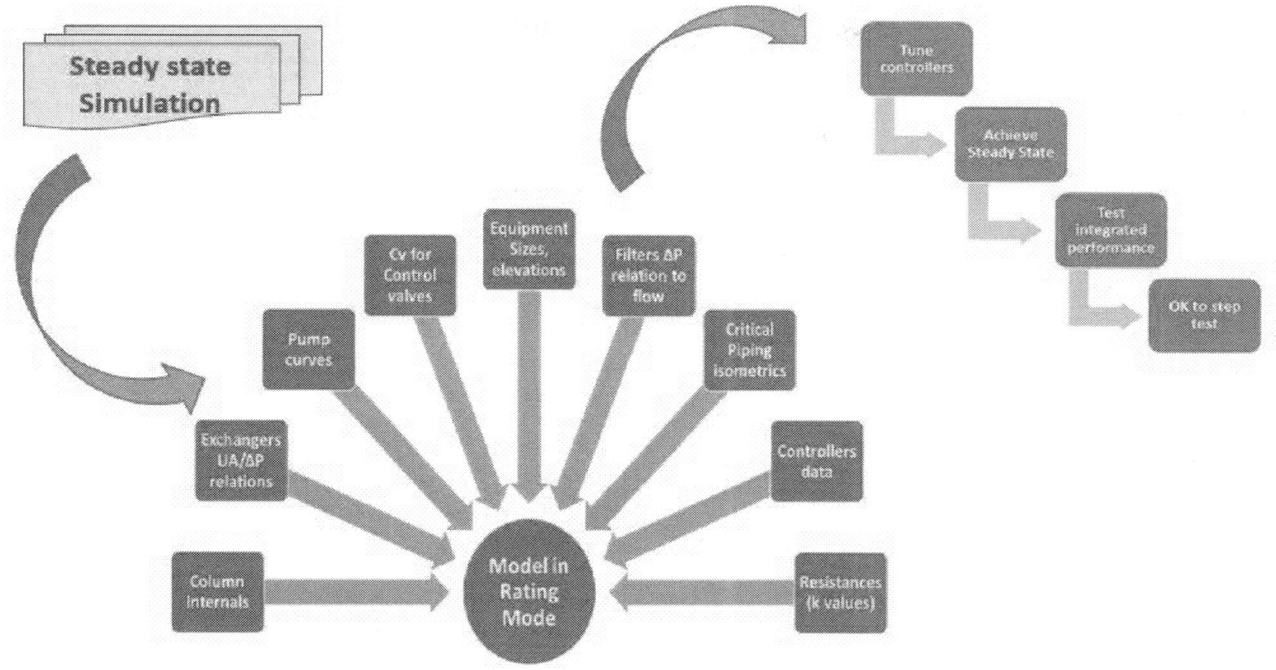

Step-test methodology:

Configuration of the MV's and CV's for step-test was executed after validation of the process dynamic simulation. Following represent steps followed for this activity

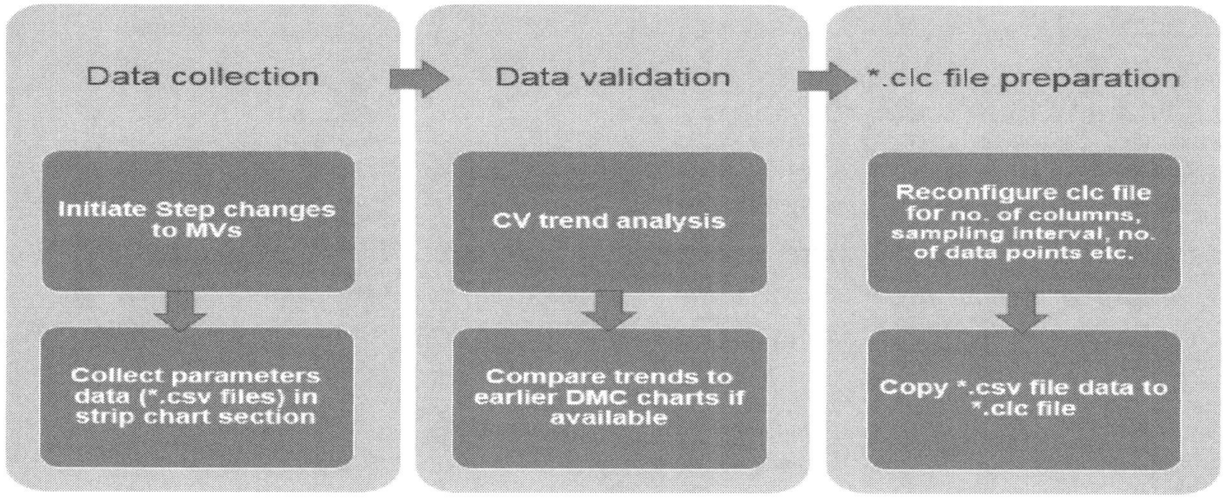

DMC3 Implementation plan:

Step-test data generated in the form of the *.csv or *.clc format is used to build the seed model.

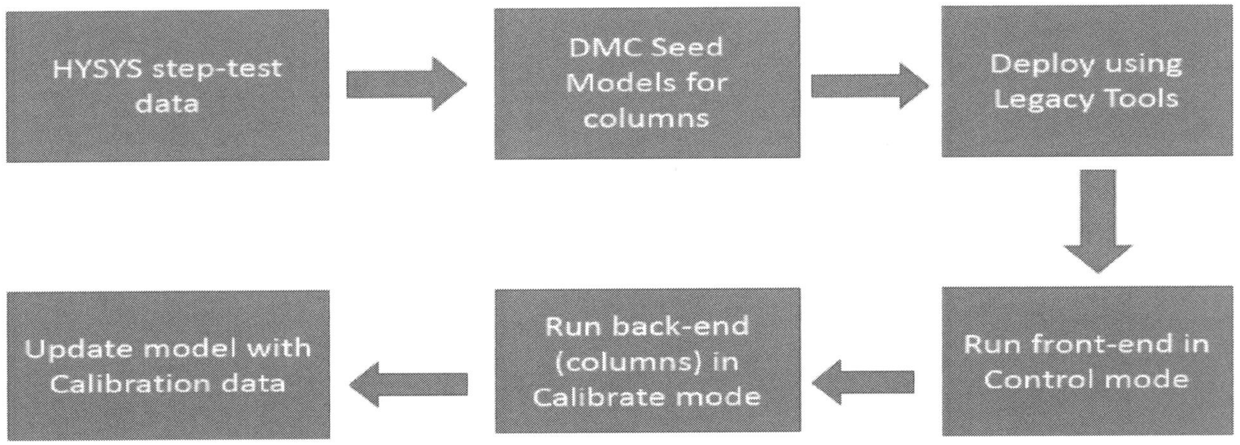

Use of first principle for process behavior:

Aspen Hysys simulation approach is based on first principles of chemical process engineering, where a mathematical model is used to simulate, a single unit operation or an entire plant. Aspen Hysys results are validated with the plant operation data in steady state mode, this way the most suitable thermodynamic behavior of the system gets chosen. Process behavior and response depends on the controller tuning used in the simulation. Hence, tuning of the process controller is the one of the critical steps.

How and when to plan these activities?

Building & solving Dynamic simulation is a process engineering specific activity where solver used in simulator to speed up the time for convergence matter to achieve steady state. Plan to develop such simulation should appropriately consider turnaround or revamp schedules. It is recommended that to plan activities such that step-test data is ready well in advance of plant commissioning. This ensures DMC3 implementation without any production loss owing to step test.

Additional benefits over the traditional way of generating the step-test data:

In traditional way, there is good amount of time required to perform the step-test which could lead to production or quality upsets. It also requires co-ordination among plant operator, process control engineer and the vendor team.

Generation of the step-test data using Aspen Hysys Dynamic simulation helped ***LYB to save 32% cost and gaining 67% in schedule*** compared to traditional method. Executing this utilizing In-house resources help company to have realistic model handy to be used in near future for any kind of "what if" analysis.

Conclusions:

Aspen Hysys dynamic simulation has been accepted across the globe among various refineries, Petrochemical & other chemical industries. The software is being widely utilized to perform various process engineering activities such as design, rating, de-bottleneck, optimization, troubleshooting etc.

Development of step-test data using Aspen Hysys and using it to build a seed model during the DMC3 implementation has proven to be a beneficial approach. This activity requires proper planning & scheduling to gain maximum benefit over the traditional approach to generate controller step-test data and use in DMC3 implementation.

The approach followed by LYB can also be applied where processes are critical/severe in nature and operation may cause delay in performing step-test in running plants.

References:

1. Aspen Hysys software
2. Aspen Support website

592114 Process and adsorbent for separating ethanol and associated oxygenates from a biofermentation system

Abstract:

Abstract for poster presentation or talk:

Abstract Title: Process and adsorbent for separating ethanol and associated oxygenates from a biofermentation system

Presenter Information: Deepak Sharma; Sr. Process Engineer, Engineering and Technology, Bayer AG

Patent number: US8658845B2

Patent Authors:

Deepak Sharma - Sr. Process Engineer, Bayer Ag, St. Louis; (Presenting Author)

Anil Oroskar – CTO, Orochem Technologies, Napervielle, IL;

David W. House – Scientist, Orochem Technologies, Napervielle, IL;

Alice M. Havill – Scientist, Lanzatech, New Zealand

The principles of process intensification if used on existing processes can lead to development and redesign of existing processes to become more sustainable. This presentation will discuss a process and an adsorbent for the separation of ethanol associated oxygenates from a dilute mixture of ethanol and associated oxygenates in water in the presence of organic compounds derived from a biofermentation process. After pretreatment, the separation is carried out in a simulated moving bed adsorption system employing a stationary phase adsorbent comprising fluorinated carbon or modified C18 silica gel selective for the adsorption of ethanol and associated oxygenates, such as 2,3-butanediol, with a mobile phase desorbent selected from the group consisting of methanol, ethanol, propanol, and methyl tertiary butyl ether. The process is useful for removing water from dilute aqueous mixtures of organic compounds comprising ethanol in dilute concentration in water and produced by fermentation, biomass extraction, biocatalytic, and enzymatic processes which are not economically recoverable by conventional distillation methods.

Keywords: Separation; Adsorption; SMB; Purification

Patent Assignee: Orochem Technologies LLC

Copyright © American Institute of Chemical Engineers. All rights reserved.

592111 Pump safety, the undiscussed topic in chemical industry leading to several fatal incidents

Abstract:

Abstract Title: Pump safety, the undiscussed topic in chemical industry leading to several fatal incidents

Presenter Information: Deepak Sharma; Sr. Process Engineer, Engineering and Technology, Bayer AG

Pumps are one of the equipment's in the process industry that require a thorough review during HAZOP studies. The constant source of energy given to the pump, if not converted to kinetic energy to flow the fluid, can lead to heating of the pumped fluid. If the fluid keeps accumulating the energy provided by the motor, in a blocked-in scenario, it can lead to very high temperatures eventually leading to loss of containment of the pump and the pump can get damaged. There have been instances in the industry where fragments of damaged pumps have flown several feet away from the point of loss of containment leading to injuries. The poster will outline a scenario where a pump is run blocked in on suction and discharge for water and a chemical which has a low thermal decomposition temperature. Arc data is used to evaluate the onset temperature of thermal decomposition. The poster will outline safeguards that can be employed to avoid catastrophic loss of containment of the pump. Outlined below are safeguards used to avoid loss of containment of a pump in case of a blocked-in scenario:

- Install PSV on suction of pump
- Employ a low amp/power interlock on the pump to turn off the pump in case of blocked in situation
- Have a high-pressure interlock on the discharge line of the pump to direct flow back to the tank in case of pump deadhead
- Install an Interlock to turn off the pump in case the feed forward valves (automatic valves) close inadvertently

Pump deadhead scenarios are not only catastrophic when in flammable service there are several salts that have low thermal decomposition temperatures. Pumps in water service left blocked in could also blow catastrophically leading to reportable incidents. I feel this is an important topic to bring to the attention of students and professionals in the chemical industry. Poster will include pictures of actual installed pumps along with required safeguards in the field along with supporting P&ID depictions.

Copyright © American Institute of Chemical Engineers. All rights reserved.

589440 New Kind of Linear Nanocomposite Applied in Improving Rheological Behavior of Crude Oil and Its Mechanism Studied through Molecular Simulation Method

Abstract:

In this work, a new kind of linear nanocomposite produced by carbon nanotube(CNT) as pour point depressant (C2-100) was developed by blending with MCNT nanoparticle and copolymer named Ethylene-vinyl acetate (EVA). Its effect of reducing yield stress and modifying the morphology of wax agglomerate were also evaluated. Besides, the synergetic effect of C2-100 nanocomposite with the magnetic field was also conducted. The mechanism of modifying rheological behavior of waxy crude oil was studied through molecular simulation method, which aims at understanding the interaction between this nanocomposite and wax molecule. The results indicated that C2-100 nanocomposite could reduce the yield stress of waxy crude oil more effectively than the copolymer (EVA) even at low concentration, as well as changing the morphology of wax crystals and agglomerate. Moreover, its effect of reducing yield stress becomes more obvious when the alternate magnetic field was applied. In addition, the results of molecular simulation also presented that the effect of improving rheological behavior of C2-100 nanocomposite is related to the adsorption behavior of wax molecules on the surface of MCNT nanoparticle, which is adsorbed with EVA molecules. In addition, the adsorption behavior of different fragments of EVA molecules can also influence the adsorption behavior of wax molecules on the surface of EVA-attached MCNT nanoparticle, which may influence the effect of C2-100 to improve the rheological behavior of waxy crude oil.

Copyright © American Institute of Chemical Engineers. All rights reserved.

589683 Solar Energy Conversion and Noise Characterization Sensing in Building Integrated Solar Photovoltaic Devices with Ventilation

Abstract:

A study on solar energy conversion and noise characterization sensing in solar photovoltaic devices with ventilation is conducted. A building integrated parallel plate solar photovoltaic (PV) device was built with a pair of PV modules, a ventilated air cavity and an insulating back panel of plywood board filled with polystyrene installed in an outdoor test-room. Photovoltaic amplification was attained with power output from a potentiometer through rotation of its circular knob. A parallel plate PV device was studied for its
electrical parameters as Resistance-Capacitance (RC) electrical analog circuit. The effect of inductive and capacitive heating losses was considered in evaluating electrical characteristics of a PV device exposed to solar radiation. The characterization of noise interference, due to power difference of two intensities in a wave on PV Device is presented. Noise characterization of a wave is obtained depending on type of wave. The difference of two power intensities is due to transmission of light, sound, heat, electricity, fluid, and fire into a parallel plate PV device. Based on analysis of a solar PV device, a definition with measurement expression and unit for an electrical noise were coined. Standard definitions of noise sources, their measurement equations, their units and their origins under limiting reference conditions were also devised. A review for sensing noise of physical agents is conducted for sensing noise of physical agents (sun, light, sound, heat, fluid, electricity and fire). With integrated environmental control, these environmental parameters are to be characterized through various environmental monitoring sensors.

Copyright © American Institute of Chemical Engineers. All rights reserved.

589945 Structural Characterization and Model Representation of Vitrinite-Rich India Coal By Using Advanced Analytical Techniques

Abstract:

The detailed investigation of the chemical structure of Karharbari coal (vitritine rich subbituminous coal from India) by utilizing advanced instrumental analytical techniques such as solid state ^{13}C nuclear magnetic resonance (^{13}C-NMR) spectroscopy, Fourier transform infrared (FT-IR) spectroscopy, X-ray diffraction (XRD), Flash-pyrolysis gas chromatography mass spectroscopy (Py-GC/MS), Laser desorption ionization time-of-flight mass spectroscopy (LD-TOF-MS), X-ray photoelectron spectroscopy (XPS), and high-resolution transmission electron microscopy (HRTEM). The Karharbari coal was an amorphous loosely organized lower rank coal with aromaticity 0.79. This was indicated by XRD results and also confirmed with ^{13}C-NMR results (aromaticity =0.84) and FTIR results (aromaticity = 0.82). The average molecular weight of the aromatic cluster is 507 amu by NMR result which is closer to the result obtained by HRTEM (MW = 530.46 amu). The presence of different heteroatoms like carbon, oxygen, nitrogen and sulphur with their functionalities, were determined by using a XPS technique. Different carbon/oxygen forms present in the Karharbari coal are (C-C) and (C-H) 68.53%, (C-O) 23.46%, (COO^-) 1.96%, (C=O) 6.05% and different nitrogen forms such as pyridine, pyrrolic, quaternary, oxidized and their distribution (mol %) are 19.36%, 45.68%, 31.25%, 3.71 respectively, and different sulphur forms were present i.e. corresponding to thiophenes, sulphones, sulphuric acid and sulphates with the molar content of 16.47%, 41.61%, 21.37%, 20.55% respectively. The construction of the individual molecules will be discussed in detail in the rest of this paper. The individual molecules were geometry-optimized and Materials Studio Amorphous Builder module was used to construct a three dimensional associated structure.

Copyright © American Institute of Chemical Engineers. All rights reserved.

589479 Life Management of High Energy Piping with Highlights from ASME B31.1

Abstract:

High energy piping (HEP) systems – or as defined by the ASME B31.1 Power Piping Code, Covered Piping Systems (CPS) – are critical assets with respect to production facilities. A failure can have catastrophic consequences from a safety perspective and in terms of equipment loss and downtime. In fact, loss of high-pressure steam due to a failure in a main run of steam supply piping could shut an entire production unit down. Because of this and the requirements of the Code, HEP maintenance and inspection programs are established to ensure that the integrity of the systems is maintained in a safe and reliable condition throughout their lifecycle. There is no one-size-fits-all for the life management of these critical assets because there is a great variety in age, design, materials, and owner objectives. Life management of these assets is beneficial in that it provides a structured, systematic approach to creating and implementing a program including evaluating the current program (whether formal or informal), what the objectives of the program are (short and long term), what equipment is in place, and its history.

A HEP program put in place for overall life management is not synonymous with an inspection program, although inspections are an important part of an overall program. Inspections should be used in conjunction with analytical tools, real time monitoring, and laboratory examinations. The overall approach is life management – not inspection management – of the equipment.

This paper describes a HEP life management approach which includes identifying locations of interest, risk-ranking those locations, implementing a plan for determining the health of the highest risk locations, and taking actions based on the system health that has been determined. This paper also presents and discusses some of the requirements of the B31.1 code, which at the highest level states that each operating company shall develop operation and maintenance procedures for piping systems deemed necessary to ensure safe facility operations based on the provisions of the code, relevant industry experience, the operating company's experience (and knowledge of its facility), and the conditions under which the piping systems are operated.

Copyright © American Institute of Chemical Engineers. All rights reserved.

590131 Optimization of C3MR LNG Liquefaction Process Using a Parallelized Simulator

Abstract:

Controlling the cost of energy-intensive processes such as C3MR (propane/mixed refrigerant) LNG liquefaction is a prime opportunity for optimization. Our work combines different C3MR models to develop an optimization that uses multiple independent variables, relaxed constraints, and flowsheet alterations to achieve an optimal operating cost. This work uses a rigorous model of the process in a process simulator rather than a reduced model of the process response. Use of a first-principles process simulator increases the accuracy of the model and therefore allows a more realistic optimization, at the cost of increased computing time. Parallelization (using multiple CPU cores) in the process simulator greatly increases the speed of the optimization calculations. CHEMCAD's optimization tools and multi-core technology make it possible to easily define the optimizations, run the optimizations efficiently, and generate global optima with a high degree of confidence.

Copyright © American Institute of Chemical Engineers. All rights reserved.

590196 Modifying Fischer-Tropsch Reaction Sequence with Addition of Zeolite to Iron-Based Activated-Carbon Supported Catalyst

Abstract:

In addition to previously presented scientific work, modified reaction sequence of Fischer-Tropsch (FT) process was studied by addition of Zeolite ZSM-5 to an iron-based activated-carbon-supported catalyst. Main reactions involved due to FT catalyst are Fischer Tropsch Synthesis (FTS) and Water Gas Shift (WGS) reactions. Main reactions due to ZSM-5 are cracking, oligomerization, isomerization, cyclization, aromatization, conversion of alcohols to hydrocarbons, and coke formation. The waxy and linear structure FT products are converted to gasoline range, lighter hydrocarbons, iso-paraffins, and aromatics using ZSM-5. The catalyst ratio of 1:2, FT catalyst: ZSM-5, is used at 300 PSIG and 300°F. Separate-bed configuration, i.e. FT and ZSM-5 are separated in a reactor, arrangement is used for the reaction sequence analysis. Alcohols with carbon-chain from 3-5 produced by FTS are converted into hydrocarbons over ZSM-5. Waxy hydrocarbons with carbon-chain greater than 23 produced by FTS are cracked and isomerized to small n-paraffins and iso-paraffins using ZSM-5. The hydrocarbon with carbon chain from 7-10 produced by ZSM-5 cracking reactions are converted to aromatic products over ZSM-5. Produced gasoline to kerosene range n-paraffins are further cracked into smaller chain (1-5) iso- and n-paraffins. Some of the iso-paraffins are oligomerized and aromatized over ZSM-5 to form more aromatics. Further, multiple aromatic rings combine to undergo cyclization and dehydrogenation reactions over ZSM-5 to form coke. Water produced as a byproduct of FTS is combined with some of the carbon monoxide, which is the WGS, to form carbon dioxide and hydrogen.

Copyright © American Institute of Chemical Engineers. All rights reserved.

590427 Lateral Retention Force Required for Sliding Saline Droplets over Silianized Silicon Surface

Abstract:

The lateral retention force required to slide a liquid drop on a surface is expected to increase with the time that the drop rests on the surface prior to sliding till it reaches a plateau. The molecular reorientation of the molecules at the periphery of the solid surface subjected to the unresolved vertical component of surface tension force causes this phenomenon. In this system, the surface tension of deionized water is increased by adding NaCl and is expected that the lateral retention force is increased proportionally with concentration. The retention force increased with increase in concentration reaching a maximum only to drop to a minimum with concentrations closer to saturation. That is, the lowest sliding force is required for both lower and higher concentrations of NaCl. This sudden decrease in observed lateral force is not a function of surface tension and the surface on which the drop slides. This phenomenon is associated with adhesion of salt to the surface and is supported by the drop sliding motion and depinning of the liquid from the solid.

Copyright © American Institute of Chemical Engineers. All rights reserved.

590433 Study of Droplet Motion across Polymeric Substrates By Centrifugal Adhesion Balance (CAB)

Abstract:

On a given surface, the contour of the drop changes accordingly based on factors of the force applied, chemical structure of the fluid, drop size, and surface topography. With increasing force that is exerted on the drop, we determine the comprehensive trend for drop motion with respect to the controlled factors mentioned above. Drop motion is categorized into four phases and the point at where the initial displacement of receding line is observed is defined as the onset of motion. In this study, the onset of motion is affected by the factors mentioned above and aids in the determination for another value correlation known as Gs, known as Interfacial Modulus. The resistance to interaction towards the droplet as observed is quantified with respect to change in surface topography. This study therefore characterized drop motion and specifically onset motion based on controlled variables (factors). Work of adhesion is also quantified across different substrates to characterize the detachment of liquid from its substrate. All these parameters such as Gs, onset of motion, contact angles were analyzed with different polymeric substrates, drop sizes, and various fluids used.

Copyright © American Institute of Chemical Engineers. All rights reserved.

590665 Population Balance Modeling and Spatially-Oriented Control for Crystal Size and Shape in Crystallization Processes

Abstract:

Crystallization is an important separation and purification process with relevant applications in the food, pharmaceutical, fine chemical and fertilizing industries. In these processes, modeling and process control for crystallization are focused on crystal size distribution (CSD) and shape of the crystals in order to make it possible to achieve the desired quality and purity of the product of interest. The control of crystal size and shape is of considerable importance for industrial crystallization processes, as crystal morphology has great influence on downstream steps and also from the point of view of product specification criteria. The consideration of crystal shape in modeling of crystallization processes is also important due to its impact on product effectiveness, such as the bioavailability and tablet stability in pharmaceuticals. Control of the crystallization process is can be facilitated with online measurements, which provide information to be used in the modeling of the process. However, if online measurements are not available, the strongly nonlinear dynamics of these processes makes difficult to establish relations between some inputs (e.g. temperature) and the final desirable characteristics of the crystal product. In this context, image analysis has been shown to be a suitable technique to monitoring the process [1].

In our present work, a population balance model-based scheme for controlling the size and shape of crystals is presented. The modeling was performed using a two-dimensional population balance (2D-PBM) in terms of the two characteristic dimensions of the crystals, describing the evolution of crystal size and shape distributions along the batch time. To identify the nucleation, growth and dissolution kinetics parameters using the 2D-PBM approach along the length and width directions, QICPIC-LIXELL's high speed dynamic image analysis equipment was used for the online measurement of the CSD, shape parameters and aspect ratio distribution of crystals during the experiments. In this way, the developed model allowed for the prediction of the crystal size and shape distributions during growth-dissolution cycles.

The developed framework was employed to obtain optimal temperature policies for controlling the crystal size and shape distributions that meet target criteria for size, morphology and productivity using the dynamic image analysis along the runs. Based on the concept of using measurements along batch crystallization processes as a trajectory in a phase space [2], a trajectory-endpoint control problem is presented with an oriented control scheme that is applied to the shape of the crystals. The scheme was used to control the batch cooling crystallization of potassium dihydrogen phosphate. The reported experimental data demonstrate the application of the control policies for temperature to produce crystals of the desired average shape and size in different batch times. Thus, using the first principle modeling coupled to the developed trajectory tracking control, the proposed framework has the potential to be applied to industrial crystallization processes for shape control and to be extended to polymorph control investigations.

References:

[1] Eisenschmidt, H.; Bajcinca, N.; Sundmacher, K. (2016) Optimal Control of Crystal Shapes in Batch Crystallization Experiments by Growth-Dissolution Cycles. *Crystal Growth & Design*, v. 16, p. 3297-2803.

[2] Griffin, D. J., Grover, M. A., Kawajiri, Y., Rousseau, R. W. (2015) Mass-count plots for crystal size control. *Chemical Engineering Science*, v. 137, p. 338-351.

Copyright © American Institute of Chemical Engineers. All rights reserved.

590450 Auditing the Process Hazard Analysis - Key Characteristics to Scrutinize in a Good-Quality HAZOP Study and Supporting Effective HAZOP Study Updates

Abstract:

Although Process Safety Management (PSM), requiring good-quality Process Hazard Analyses, was promulgated in the United States in 1992, and HAZOP Study techniques date back to the mid-1970's, the HAZOP Study technique is still not uniformly applied in a high quality fashion. The objective of this paper is to provide experience-based tips on Process Hazard Analysis auditing, and some key characteristics that the auditor can scrutinize to assure quality implementation. Analogously, a similar set of characteristics that can facilitate future HAZOP Study updates can also contribute to a quality HAZOP Study that supports the spirit of Safety Management System objectives for the identification of hazards potentially impacting plant personnel, the community, and the environment. This paper will also review tips for structuring the HAZOP Study to facilitate later updates and revalidation.

Copyright © American Institute of Chemical Engineers. All rights reserved.

590342 Development of High Performance Organic-Based Cathodes

Abstract:

For lithium-based batteries, structure/property relations in terms of cell voltage are desired for new organic-based cathode active materials. In this initial pursuit of structure/property relations, ab-initio calculations were performed on a few organic-based molecules to show the effect on the theoretical cell voltage by inclusion of nitrogen and/or oxygen. Preliminary test results are presented on a newly developed molecule and methods to potentially improve on the cell performance are given.

Copyright © American Institute of Chemical Engineers. All rights reserved.

590705 Electrochemical Kinetics of Degradation of Graphite Anodes in Lithium Ion Batteries

Abstract:

Growth of the solid electrolyte interphase (SEI) is a major driver of capacity fade in LIBs. Despite its importance, the fundamental mechanisms remain unclear, primarily because of the complicated reaction pathways involved[1–3]. SEI growth can be both electrochemical and chemical in nature[4], and thus, it is a strong function of the potential and degree of lithiation of the electrode. We model the early-stage and long-term growth of SEI by accurately capturing the potential dependence of its formation kinetics as well as long term rate limiting steps, and validating it against the world's largest open source battery cycling data, generated in-house[5]. This is done using the Multiphase Porous Electrode Theory (MPET) framework[6] on graphite (phase separating) and carbon black (non phase separating) particles.

Lithium plating is another key degradation phenomenon that has been elusive, and it becomes important while trying to fast-charge batteries, i.e., 0% - 80% state-of-charge in 30 mins. We show that lithium plating is a key function of electrode morphology, phase-separation dynamics and potential. Phase-separation in graphite is modeled in the electrode using the Cahn-Hilliard Reaction framework described by Bazant[7]. We understand the electrochemistry of the onset of lithium plating with in-situ measurements connected to real time cell potential in a phase-separating electrode for the first time[8].

Results indicate that the peak SEI-forming currents are higher for higher driving currents. Also, we find that SEI only grows during electrode lithiation, i.e. the battery only degrades while being charged. We also find that onset of lithium plating is correctly captured only when phase separation in active material is accounted for. Further, the onset of plating is delayed on electrodes with a thick SEI layer – understanding SEI/plating coupling is integral to predicting fast charging manufacturing protocols for LIBs. This work holds promise for the predictive design of procedures[9] for manufacture and formation of LIBs.

[1] Cohen, Y. S.; Cohen, Y.; Aurbach, D. Micromorphological Studies of Lithium Electrodes in Alkyl Carbonate Solutions Using in Situ Atomic Force Microscopy. J. Phys. Chem. B 2000, 104, 12282–12291. https://doi.org/10.1021/jp002526b.

[2] Horstmann, B.; Single, F.; Latz, A. Review on Multi-Scale Models of Solid-Electrolyte Interphase Formation, Current Opinion in Electrochemistry 13, 62-69 2019.

[3] Nie, M.; Abraham, D. P.; Seo, D. M.; Chen, Y.; Bose, A.; Lucht, B. L. Role of Solution Structure in Solid Electrolyte Interphase Formation on Graphite with LiPF6 in Propylene Carbonate. J. Phys. Chem. C 2013, 117 (48), 25381–25389, https://doi.org/10.1021/jp409765w.

[4] Das, S.; Attia, P. M.; Chueh, W. C.; Bazant, M. Z. Electrochemical Kinetics of SEI Growth on Carbon Black: Part II. Modeling. J. Electrochem. Soc. 2019, 166 (4), E107–E118. https://doi.org/10.1149/2.0241904jes.

[5] Attia, P. M.; Das, S.; Harris, S. J.; Bazant, M. Z.; Chueh, W. C. Electrochemical Kinetics of SEI Growth on Carbon Black: Part I. Experiments. J. Electrochem. Soc. 2019, 166 (4), E97–E106. https://doi.org/10.1149/2.0231904jes.

[6] Smith, R. B.; Bazant, M. Z. Multiphase Porous Electrode Theory, J. Electrochem. Soc. 2017, 164 (11). https://doi.org/10.1149/2.0171711jes.

[7] Bazant, M. Z., Theory of chemical kinetics and charge transfer based on nonequilibrium thermodynamics, *Accounts of Chemical Research, 46*(5), 1144–1160. https://doi.org/10.1021/ar300145c

[8] T. Gao, Y. Han, S. Das, T. Zhou, D. Fraggedakis, N. Nadkarni, C. N. Yeh, W. Chueh, J. Li, M.Z. Bazant, Interplay of lithium intercalation and plating on graphite using in-situ optical measurements, submitted.

[9] Huang, W.; Attia, P. M.; Wang, H.; Renfrew, S. E.; Jin, N.; Das, S.; Zhang, Z.; Boyle, D. T.; Li, Y.; Bazant, M. Z.; McCloskey, B. D.; Chueh, W. C. and Cui, Y.; Nano Letters 2019 19 (8), 5140-5148. DOI: 10.1021/acs.nanolett.9b01515

Copyright © American Institute of Chemical Engineers. All rights reserved.

589787 Model Development of Solid Oxide Fuel Cell Thermal Performance Using Artificial Neural Network

Abstract:

NASA's Johnson Space Center has recently begun efforts to eventually integrate air-independent Solid Oxide Fuel Cell systems, with landers that can be propelled by LOX-CH4, for long duration missions. Using landers that utilize such propellants, provides the opportunity to use such systems as a power option, especially since they will be able to process methane into a reactant through fuel reformation. To ensure fuel reformation in the systems, Steam Methane Reformation (SMR) will be employed. Various lead-up activities, such as hardware testing and computational modelling, have been initiated to assist with this developmental effort including the development of dynamic empirical models using artificial neural network. Such a modeling approach has shown to be accurate (R^2>0.9) to predict the thermal and electrical performance for SOFC systems. Hence, such models could be used for system optimization and control design.

Copyright © American Institute of Chemical Engineers. All rights reserved.

Impact of Spacer Nature on Rheological Properties of Cationic Gemini Surfactant

Abstract ID: 589650

Shams Kalam[1], Muhammad Shahzad Kamal[2], Shirish Patil[1] and S. M. Shakil Hussain[2]

(1) KFUPM, Dhahran, Saudi Arabia

(2) Center for Integrative Petroleum Research, King Fahd University of Petroleum & Minerals, Dhahran, Saudi Arabia

Surfactants play a vital role in Chemical Enhanced Oil Recovery (cEOR) techniques. Gemini surfactants are a unique style of surfactants having more water-soluble hydrophilic head and oil-soluble hydrophobic tail units. Both groups are connected by a spacer at or near to the head groups. Polymers are added into the surfactants to increase the sweep efficiency of the flooding processes. This paper presents the rheological properties of Surfactant Polymer (SP) solutions in which two recently synthesized cationic gemini surfactants with different spacer nature are used.

The rheological properties of novel SP solutions were determined using Discovery hybrid rheometer (DHR-3) from TA Instrument. The geometric configuration set for this work was concentric cylinder. All experiments were carried out at 80 °C, while the shear rate for steady shear viscosity measurements ranges from 0.001 to 1000 s^{-1}. 0.25 weight % of a polymer FO 4290 SSH was used with different concentrations of newly synthesized gemini surfactants for Rheological measurements. Two hours magnetic stirring and twenty hours retention time were provided to make uniform surfactant polymer solutions.

Both surfactants exhibited good thermal stability in seawater with no precipitation. Shear viscosity and storage modulus were measured as a function of shear rate and angular frequency respectively at different surfactant concentrations ranging from 0 to 0.5 mM, where 0 mM shows pure polymer. We discovered that counter ions also influence the rheological properties significantly. Research outcomes revealed that the novel SP solution with a benzene ring and chlorine counterion produces better performance in comparison with benzene ring and bromine counterion. Moreover, the effect enhances if we increase one more benzene ring in the spacer.

This study reports two recently synthesized SP solutions for cEOR processes in carbonate reservoirs. It presents the effect of counterion on rheological properties and improves the understanding of tuning the surfactant structure in making SP solutions to increase the flooding sweep efficiency.

590257 Real Time Energy Management Systems Implementation for Continually Optimized Performance

Abstract:

It is well known that energy management is key for the optimal operation of refining and petrochemical sites. Industrial energy systems are constituted by complex utilities networks that are interconnected and have complex interactions that in order to achieve it best performance need to comply with increasing environmental rules, push against operating constraints and adapt to altering power prices and rapidly changing equipment availability.

After the initial installation, management and continuous improvement of these systems present challenges not only from a technological but also from an organizational point of view.

The implementation of an advanced digital Energy Management System (EMS) which includes capabilities for multi-period optimized scheduling, real time optimization and monitoring, deployed either as cloud-based or on-premise solutions that could be acting on open or closed loop is known to provides large benefits.

Critical is the sustainability of the EMS in the ever-evolving energy supply and demand of refining and petrochemical sites. Commercial, asset and organizational factors all change and the EMS needs to adapt to the developing environment of each enterprise meaning the type of implementation, including appropriate management structure, culture, best practices and Key Performance Indicators (KPIs) needs to be continually improved.

The EMS implementation at the large and complex Eastman site in Longview, TX, will be presented as an example. It will provide insight on the initial successes achieved in real time energy management for both, cost optimization and energy related KPIs performance tracking. The case study evolves to discuss how technical asset improvements, once replicated in the EMS, provided alternative operating modes, large additional benefits and sustained optimal

Andrew Lammers, NN2

Eastman Longview

Pablo Montagna*, Cristian Rojas

KBC Advanced Technologies, Inc

Copyright © American Institute of Chemical Engineers. All rights reserved.

589818 Analysis of Fuel Cell System through Project Based Learning for Application of Classical Thermal Fluid Concepts in Mechanical Engineering Laboratory Course

Abstract:

A hands-on project involving analysis of air noncircular channels in Fuel Cell to measure the pressure was introduced in a thermal fluid laboratory course. Although the students studied fluid mechanics as a theoretical course, this project gave them the opportunity to build an experimental setup from scratch to help showcase their hands-on skills and to apply their understanding of relevant concepts of fluid sciences. The results of the project included a successful test of a conceptual procedure, designed to measure and calculate pressure drop and flow rates across the anode and cathode sides of a fuel cell. The students enjoyed completing and learning from the project. They also provided very useful feedback on the project by incorporating a functional experimental system for the future laboratory project. This project based learning opportunity allowed students to build their confidence in what they are learning as well as provide a glimpse of what the real world may hold.

Copyright © American Institute of Chemical Engineers. All rights reserved.

589438 Measured Versus Calculated Sour Gas Properties

Abstract:

Measured versus calculated sour gas properties
Abstract:
Recent Environmental agencies have restricted standards for sulfur emission from gas chemical processing. Sour gas can extremely impact the delivery and processing by corroding the equipment. Gas properties such as compressibility, density and viscosity are a major factor controlling prediction of future production pipelines, and production tubing and pipelines in gas industry.
Sour gas effect on all gas industry aspects usually gas properties are measured in laboratories but occasionally experimental values are not available due to complexity and time consuming hence a reliable method would be proposed.
We are assessing the accuracy of the industry standard methods in calculating sour gas properties, compressibility factor using the following correlations:
Drunchuck - Abukkasem, Hall – Yarbarough, Drunchuck- Purvis- Robinson, Londono et-al, Alanazi et-al and Azizi et-al.
Viscosity using empirical correlations such as Carr-Kobayashi Burrows, Lea-Gonzalez –Eakin and Dean- Stiel.
Density using equations of state Redlich Kwang and Soave -Redlick - Kwang.
This study shows that the best correlation for calculating compressibility factor and density of high concentration sour gas is Londono-et al, all other considered methods are not accurate. All other methods investigated include high error especially for high H_2S concentration.
However the error in estimation of Z factor for most of the considered methods has great impact on estimated sour gas density. Therefore we recommend that sour gas properties be measured experimentally at high pressure and high temperature conditions.

Copyright © American Institute of Chemical Engineers. All rights reserved.

587967 The Study of a S-CO2 Oxy-Fuel Combustion Power Plant for Low Grade Coals

Abstract:

Recently, it is very important to have reliable solutions that can maximize power generation efficiency of current power generation processes and decrease emissions of CO2 and other pollutants because there are increasing concerns about the global warming and the depletion of fossil fuels. Among the available carbon capture technologies and power generation technologies, a S-CO2 (supercritical carbon dioxide) cycle based oxy-fuel combustion process appears to be a promising mid-term solution for the production of electricity from fossil fuels. Since supercritical carbon dioxide (S-CO2) involves a rapid density change at small temperature and pressure ranges, the S-CO2 cycle can extract large amounts of energy from relatively small equipment and it is possible to minimize or simplify a power generation block and to obtain higher thermal efficiency than existing steam cycles used in power generation plants. Therefore, an indirect S-CO2 oxy-fuel combustion power plants for low grade coals was designed and analyzed by using ASPEN-PLUS. Also, the operating characteristics of the indirect S-CO2 Oxy-fuel combustion power plant were compared with those of conventional air combustion power plants.

Copyright © American Institute of Chemical Engineers. All rights reserved.

590099 Characterization and Performance Assessment of Oil Entrapment By Commercially Available Surfactants

Abstract:

Oil spills occur frequently and are a threat to the health and safety of human society and the well-being of the environment and ecosystem. Remediating oil spills is an ongoing challenge. Surfactant based dispersing agents, such as the Corexit® product line, are widely used for remediation of oil spills mainly through direct application on top of the oil slick. This work encompasses the investigation of variables that influence the emulsification process and therefore the encapsulation of oil within the surfactant micelles. Of central importance are the conditions under which oil entrapment in emulsions even occurs. A central finding is that there appears to be a threshold shear rate for emulsifying oil with surfactants, below which no emulsification occurs. Different schemes for applying shear stress in this work included simply stirring, using a high energy colloid mill, and flowing the mixture through a high shear capillary. Also, the sequence in which the oil/surfactant/water (O/S/W) components are mixed impinges strongly on whether and how much oil entrapment occurs. To try to correlate shear stress and velocity profiles throughout the sample volume with entrapment efficiency and kinetic stability of these O/S/W systems, Computational fluid dynamics (CFD) simulations were conducted. Another important factor is the ionic strength of the aqueous medium, which greatly impacts kinetics of these emulsions, but with a smaller impact on the threshold at which any emulsion is obtained at all. Surfactants tested include Corexit® EC9500A, sodium dodecyl sulfate (SDS), Triton X-100, and cetyltrimethyl ammonium bromide (CTAB). The quantification of oil entrapment was performed through either thermal gravimetric analysis (TGA) or differential scanning calorimetry (DSC).

Copyright © American Institute of Chemical Engineers. All rights reserved.

Designing a Native Model for Assessment of the Effectiveness of Advertising

Sorush Niknamian
PhD, military Medicine Department, Liberty University, USA
Email: so.niknamian@gmail.com

Abstract

Reviewing and evaluating the effectiveness of advertising can be used as a criterion to convert weaknesses into strengths and budgeting. Also, Identification of the factors influencing the effectiveness of advertising can lead to more professional design and implementation. In the banking industry, with the increasing number of banks and increased right of customers to choose, the role of advertising has become increasingly important to succeed in this field of competition; so it should be said that advertising can make the bank close to its goals when it is sufficiently effective. The current research aimed to identify and rank the effective factors on the design of a native model for evaluation of the effectiveness of advertising of the Eghtesad Novin bank in Iran. The statistical population of this research is all managers and experts of the Eghtesad Novin bank. According to the research literature, the criteria and sub-criteria affecting the effectiveness of advertising have been extracted. By using the fuzzy Delphi technique and based on the experts' opinion, the final indices were extracted. Then, using the fuzzy AHP method, the rank of criteria has been addressed. The derived criteria are demographic characteristics, individual emotions, and attitudes of the customer, apparent features of the message, message content, market characteristics, cultural, economic features, design and nature of the service, advertising methods. The results of the ranking showed that the methods of advertisement, the design, and nature of the service, the apparent characteristics of the message, the emotions and attitudes of the customer, the demographic characteristics, the content of the message, economic, market characteristics, were placed in the first to ninth ranks, respectively.

Keywords: banking services, banking services marketing, advertising, advertising effectiveness,

Introduction:

Advertising is one of the promotional mix elements of marketing mixes, which play an important role in attracting customers. Advertising is a bridge between the organization and customers and can introduce the organization and its products to customers, inform consumers about the goods and services of the organization, convince them about the benefits of the product, and ensure them to have the best choice even after the purchase when the customer is uncertain (Buck, 2016: 172).

Given the increasing competition among the brands in the marketing sector, advertising is a necessary tool to introduce the characteristics of goods and services to customers. Leong et al. (1999) pointed out the fact that consumers overcome advertising information, and many of them do not remember the advertisements that they are exposed to, and it is necessary to consider the advertising and its attractive methods to attract customers' attention. The ability to attract customers' attention is one of the main indicators of advertising effectiveness that should be taken into consideration. The attention of customers is a key component of the overall effectiveness of advertising (Pileliene et al., 2017: 2). Any professional marketing without advertising will not be effective. Despite relatively high advertising costs, advertising is mentioned as an investment rather than the cost. Since many advertisements cause that the advertising companies do not benefit their advertising because of the wrong way of advertising, the competitive companies benefit from the advertising of that company (Hartnet, 2016: 23)

Advertising should be in line with the goals and plans, as well as the strategies of the organization. Therefore, we should see to what extent the advertising has met such goals. Allocating funds without evaluating the effectiveness of past advertising is the waste of resources of the organization. By evaluating the effectiveness of advertising, changes can be made to the form, the content of messages, type of media, budget, communication channels, and even the time and conditions for conducting advertising to create more useful and effective advertising than the past (Yazgirudo-Yasteh, 2015: 359). In the current competitive world, the importance of advertising is increasing. As a result, businesses have devoted a great deal of money to customers' attraction, but it is important to see whether the company advertising has been designed and implemented in such a way to meet the goals of the company. This issue has been often neglected in our country. (Kim, 2014: 2107).

The amazing effect of advertising on the lives of individuals requires that this scientific, social and communicative branch be carefully studied and advertising is made with a scientific, artistic, specialized and expert view. Professional advertisement can be a sharp sword for sales promotion, both inside and outside the country, and otherwise, it is a waste of time, money and energy. A professional advertising lever can, with the help of new technology, penetrate humans, and cause more desire for consumption (Arribeh, 2016: 463). Today, banks have turned their attention to the widespread use of mass media advertising, given the highly competitive structure governing on them. Also, given the relationship among a wide range of businesses of organizations and households and banks through various deposit accounts, loans, and other financial services, banks can benefit from advertising in a proper way to meet their goals, namely the introduction of their services, the communication with customers, their effectiveness on them in choosing a bank to do their banking business and ultimately attracting them. Unfortunately, in most banks, despite the massive advertising and the allocation of large amounts of energy and budget, the effectiveness of such advertisements is less evaluated. Since advertising is one-way, it is necessary to receive feedback from the customers (Rabiei, 2011: 19).

Regarding marketing, the effectiveness of advertising depends on the wide range of coverage of various consumers and the cost of advertising campaigns. Economists have defined the effectiveness of advertising as a change in the profit derived from promotional activities. Therefore, the effectiveness of bank advertising is the potential quality of information reported to consumers, and in promotional campaigns, it aims at creating a bank image and informing banking products and services, as well as obtaining a specific economic result before and after the transmission of the message (Denis et al., 2012: 85).

Eghtesad Novin Bank (EN bank), which is one of the oldest banks in Iran, due to the addition of new banks, should use new tools to protect and even increase its market share for the assessment of the effectiveness of its advertising. For this purpose and given that the bank uses various advertisements, it is necessary to design a model to evaluate the effectiveness of the bank advertising; therefore, the present study seeks to answer the following questions:

- What indicators are used to assess the bank native advertising model?

- How are the indicators for bank advertising evaluation weighed and ranked?

In this research, we have tried to identify the criteria using the Fuzzy Delphi process; then we will use the Fuzzy AHP method to rank the identified criteria. In the next section of the research, the literature has been studied. Then the research design has been expressed, and in the last section the conclusion has been presented.

1-Research literature:

1-1-Marketing of banking services

The growing trend in service activities has led to the emergence of new markets and large service companies. The complexity of service industries, such as information systems, management consultancy, and logistics, has increased, as service providers have entered the market with a variety of business backgrounds and capabilities to provide a broader range of services. Facing this complexity requires knowledge and skills to understand the services. The definition of the service includes all activities that have been done (for example, writing document specifications) (Selviaridis et al., 2013: 1398).

In the industrial literature, there are various ways to define the service and product requirements. Van Weele (2010) has distinguished technical and applied definitions from each other. The first requirement expresses the exact technical specifications of the goods with the required services and inputs while it focuses on performance demands and demand-oriented performance orientations. In this regard, Axelsson and Wynstra (2002) have proposed four methods for definition of services: input definitions are based on the sources and capabilities of the service provider while the process-oriented approach focuses on how services are presented regarding activities and processes. Output-oriented definitions focus on service capabilities and performance levels, as well as the interaction between the suppliers and recipients. Ultimately, they focus on stress-oriented results in the economic value of services (Selviaridis et al., 2013: 1399).

The financial sector forms an essential part of the service sector. In this regard, banks have used marketing to improve the strategic performance of banks. In particular, since the 1980s, this trend has been confirmed and criticized. Marketing researchers and experts are interested in researching the various aspects of marketing in banking. The quality of services is essential for customers who use banking services, and the provision of accurate and high-quality services leads to the survival of banks in the competitive market. Improving the quality of services will increase the access and response to the demands of the bank's customers, which will make the bank more distinct than other competitors. Therefore, measuring the quality of services in the banking industry is essential. Providing quality services is a hard and difficult task, in this regard, the features of services in general and banking services, in particular, should be explained. Banking services require a high degree of appropriate communication between service providers and customers. If services are not provided properly, they will be at high risk.

About the five features that distinguish them from material services (intangibility, non-allocation of materials for production, diversification, non-corruption, and lack of ownership), financial services have two other specific features, that have been firstly developed by McKechnie (1997). These characteristics include trusteeship and two-way information. These special features mean that the evaluation of services is more difficult than material products, and they not only create a challenge in the development of marketing activities, but also they are the source of the problem in providing an optimal level of quality services. Banking services require a high level of communication between service providers and customers, as well as a high degree of customer participation in the process of providing services. Angurr et al. (1999) emphasized that due to the high standardization and the development of technologies in the banking sector, human factor has been considered as the most important one in making distinctions among competitors. The expertise and motivation of the employees, as well as the level of customer involvement in the provision of services, significantly affect the quality of services. Due to the wide range of banking activities, attention should be paid to cultural and social changes that affect the attitudes of customers (Durdana et al., 2015: 100).

Marketing of banking services has three components as follows: if any of them were properly executed, banking services would change the country.

1- Market study: Identification of the banking services market system, including the status of current banking services and their weaknesses, strengths, expectations, demands, ideas and opinions of customers, the conditions of environmental instruments affecting the market system, in particular government policies and governing laws, and even global changes in the banking system and successful models in this system. Indeed, market study is the foundation of the evolution of the banking system, because market study (which is the result of constant communication with the market and conducting marketing studies and research in various forms) leads to comprehensive, up-to-date, relevant and necessary information for decision-makers and bank managers (Munsung Rheea & Satish Mehrab, 2006: 508).

2. Market creation: It involves the use of a variety of simple and complex marketing techniques and competitive strategies, as well as the growth, development and market creation in the banking system; it needs to understand the internal customers (employees) and external

customers, to communicate with them and meet their needs and demands. Market creation in the competitive environment requires appropriate strategies (Moch, 2013: 2908).

3. Market protection: One of the most important tasks of marketing in the banking system of the country is to maintain market share, a stable relationship with customers, and to convert ordinary customers to loyal and supportive ones. The more intense the competition, the more valuable the loyal customers will be, as marketing research in the world has shown that the cost of finding a new customer in firms is 3 to 33 times higher than the cost of maintaining current customers regarding the type of products and services. Therefore, market protection prevents the costly advertising, and attraction of new customers and provides a kind of confidence and serenity for market survival, especially in competitive conditions (Fredriksson & Moro, 2014: 69).

1-2-Effectiveness of bank advertising

One of the indicators of success in any organization is the extent to which the organization achieves its goals. Here, the success of any organization is determined by its own goals. This approach has been called effectiveness in the management literature. The effectiveness is the degree to which the organization achieves its goals. Goals must first be defined to measure effectiveness in any organization. Then, plans are set up to fulfill these goals, and thus the plans will be implemented. In the end, the results are compared with predetermined goals (Manjegani et al., 2015: 124).

About 100 years ago, William Heketh statement, "Half of the advertisements were wasted, we just do not know which half it is" attracted everyone to the accountability of advertising. What matters to advertisers is that how much the amount of money spent on advertising has returned? Has the cost spent on making and broadcasting the advertising been effective? Has the effect of advertising been strong or weak? Knowing the answer to these questions is important to the advertiser because today companies pay much money to make and broadcast advertising. According to Jack Trout, the money we use to advertise is very much. Every second of the World War II was $ 9,000, and Vietnam cost $ 22,000 per minute. But in the American Football Final, every minute of advertising cost $ 2 million, so the advertiser has the right to know about the effectiveness of his advertising (Aksoy, Atilla, 2005, Yeni Reklamcilik, Istanbul Bilgi Universitesi, 95). It is very difficult to measure the impact of long-term advertising on many occasions. Also, academically in the world of advertising, it is impossible to create a comprehensive theory that can be used to determine the effectiveness of advertisements on people at different levels under different circumstances. The effectiveness of media is its capacity to create desired results, or special conditions in a particular position and content, that is, attaining the desired communication goals (Wery, 2012)

Evaluating the effectiveness of advertising indicates how much advertising has succeeded in achieving the goals. The evaluation of the effectiveness of advertising acts as a feedback, informs us about the weaknesses and strengths of our advertising by which we find our shortcomings in the advertising. The marketer may choose an appropriate stage regarding product life or brand position, and use it to define his advertising goals and ultimately evaluate its effectiveness (Eckler & Bolis, 2011: 2).

The effectiveness of advertising is a function of its content (message), the execution (how it is transmitted), and repetition (the number of times each consumer sees the message) (Cutler, 1997, quoted by Moorthy & Hawkins, 2005). Some of the effectiveness criteria are the call and recognition of the brand name (Park, 2006), awareness, willingness and intention to purchase (Decrop, 2007), attitude towards the ad, attitude toward the brand name and intent to purchase (Muda et al., 2014)

The control of the communication element has a direct effect on customer perception of different dimensions of the brand name, such as service quality and loyalty. It means that the elements of communication, advertising and promotion must be carefully implemented and controlled (Suvittawat, 2015). Informative advertising is one of the factors that shape the behavior of the bank customers. Studies conducted both on attracting, retaining and loyalty of customers, show the need to examine the impact of advertising on attracting and loyalty of customers (Ioannou, Boukas and Skoufari, 2014: 25).

An ideal model of world-class advertising can: 1. transfer a consistent meaning to the consumer all over the world. 2. Transfer positive global capacity for the brand. That is, the results of advertising among different peoples should produce a positive effect and transfer high-quality signals (Strebinger et al., 2018: 410).

Advertisers want to prepare consumers to like goods or services, but sometimes it harms them through unwanted and intrusive advertising. It leads to the creation of a series of negative emotions in advertising. The purpose of ads is to create positive feelings in consumers about advertised goods and services (Bell & Buchner, 2018: 1).

On the other hand, research shows that commercial advertising can be positively evaluated and considered, and this positive evaluation leads to a positive evaluation of the brand with the product. However, advertising can have negative effects that lead to the formation of negative attitudes towards that product and service. Therefore, by increasing the volume of messages, it can have positive, negative or neutral effects. Therefore, marketers should evaluate advertising messages and their behavioral outcomes. With this regard, the effectiveness of advertising, based on a hierarchical approach, should be fully investigated (Tafani et al., 2018: 125).

There are various factors which influence effectiveness of advertisement like media, time, market life stage, advertisement life stage (new or old advertisement) and also advertisement creative signs (like content and administrative elements) (Talis et.al, 2005). Also using popular cartoon personages (Ranjbarian, 2009), size (Rousat, Venous & Ebrahimi, 2004, Baltas, 2003), color (Park, 2008, Moriarty, Mitchell, Wells, 2009, Cheng & Kao, 2011), complexity or simplicity of the advertisement (Morrison & Dainoff, 1972, Di Res, 2008), using irony factor (Mukherjee, Hober, 2004), using famous artists (Rousta, Venous, Ebrahimi , 2004) and consumer features (Matrin, Phaymi, Agee, 2002, Park, 2008) are among these variables. Advertisement repetition is also a factor which helps in reminding its content to the audiences (Rousta, Venous, Ebrahimi, 2004). One of considered factors in the debate of evaluating effectiveness of advertisement is the attitude towards advertisement. Attitude towards the advertisement includes public interest or lack of interest in an advertising stimulus in encountering a special topic (Moaen & Minor, 2009). Mental imagery (B one & Elen, 1992), content transparence and emotions called through the advertisement are among variables influencing the attitude towards the advertisement. Some of these factors may lead to strong impacts while some of them may create weak impacts or don't include any impact. Awareness of the impact of each of these factors makes a considerable help in making management decisions (Talis et.al, 2005).

Fotini et.al (2009) provided a model of effective advertisements for advertising websites of big companies. The results of the study showed that measurements, activities, priorities and intentions of people have a considerable impact on evaluation and qualitative factors of relation-oriented marketing have a great impact on advertisements effectiveness indicators. The extracted criteria include:

- Advertisement policies: product information, company information, product general display, consistency with traditional advertisement

- Medium features: message features, information structure, responding time, facilitating research, beauty, interactions, communications in real-time, amusement, information control

- Relation- oriented marketing policy: communications, feedback, supporting customers

Today, organizations for competition in the business world and selling their goods and services require purposeful and efficient marketing activities. One of topics which has a very important role in the process of marketing of an organization and assign a high volume of budget to it is advertisement. Advertisement causes some changes in the knowledge, attitude and behavior level of audiences, so attention to evaluating advertisement effectiveness is a necessity for the organizations that consider advertisement as requirement of continuous life, growth and development. By assessing advertisement effectiveness and investigating consistency with the organization objectives and its expected impacts, some changes may be created in its dimensions, content and methods so that a more effective and useful advertising is performed. Regarding the research literature and internal and external background which has been stated, here a general categorization of the model proposed criteria has been provided in the following table:

Table 1- summation and extracted variables

scholar	sub standard	criterion
) Dargi & Mohammadian (2010), Ashouri (2016), Mansouri Moayed (2005), Aghapour (2016), Mirzadh & Mehrmanesh (2016), Keramforn (2011), Mac Ki- Nesbiat et.al (2011), Rosemary et.al (2012), Wang (2015), Maryion et.al(2017)	awareness of sexuality differences	individual factors
Dargi & Mohammadian (2010), Ashouri (2016), Samsam Shariat Keramforn (2011)(2007), Mirzadeh & Mehrmanesh (2016), Mac Ki- Nesbiat et.al (2011), Rosemary et.al (2012), Wang (2015), Maryion et.al(2017)	attention to age requirements	
Dargi & Mohammadian (2010), Dehdashti Shahrokh (2009)	various consumers (permanent, ordinary and case consumers)	

Khashizadeh (2010), Danaei & Dasar (2016), Pillay et.al (2012), Marion et.al (2017)	advertisement richness	
Danaei & Dasar (2016), Dargi & Mohammadian (2010), Pillay et.al (2012), Olson et.al (2012)	advertisement accuracy	
Ashouri (2016), Rabiei (2011), Taeini Harari (2014)	positive mentality of the organization	
Dargi & Mohammadian (2010), Shahmohammadi (2011), Olson et.al (2012)	honesty feeling in advertisement	
Fakhari et.al (2012)	threatening feeling	
Ashourizadeh (2016), Hamidizadeh (2013)	attitude towards brand	
madhoushi (2013), Olson et.al (2012)	being convinced	
Ahmadi (2013), Mac Ki – Nesbiat et.al (2011), Olson et.al (2012)	convicting audiences	
Ashouri (2016), Hosivar (2014), Olson et.al (2012)	mental imagery	
Siadat & Sakhi Jourashk (2014)	positive feeling	
Navaei et.al (2015), Rozmari et.al (2012)	audience being active	
Mosavi Esmaeilzadeh (2015), Olson et.al (2012)	knowledge level	
Dargi & Mohammadian (2010), Ashouri (2016), Shafiei Nikabadi & Zarei (2017)	attitude towards advertisement	
Dargi & Mohammadian (2010), Hosseinzadeh et.al (2016), Sanei et.al (2016)	familiarity with methods and forms of advertisement	message
Ashouri (2016), Olson et.al (2012), Alan (2012), Patric Ro (2014)	attention to reduction of expectation time of the user for observing advertisement	
Danaei & Dasar (2016), Forentebury & Mac Gldric (2011)	beautiful and appropriate	

	combination of colors	
Aghapuor (2016), Sanei et.al (2016)	site and location of advertisement placement	
Hassanzadeh & Danaei (2016), Sanei et.al (2016)	access to sufficient and required information	
Dargi & Mohammadian (2010), Darzban (2016)	starting the advertisement with a challenge	
Navaei et.al (2015), Salimi Fard & Pourdehghan (2016), Bargar et.al (2016)	starting the advertisement with a statement and explanation	
Dargi & Mohammadian (2010), Mirzadeh & Mehrmanesh	using helping words and terms for decision making	
Dargi & Mohammadian (2010), Daneshvar et.al (2016), Alan (2012)	using stimuli with diversions more than usual and expected extent	
Sanei et.al (2016)	using images of women and children	
Sohrabi Ranani (2016), Fortenbury & Mac Goldric (2011)	creativity in the message	
Shafei Nikabadi & Zarei (2017), Denis et.al (2012)	the text of advertisement message being a predicate sentence	

Khashizadeh (2010), Dargi & Mohammadian (2010), Sohrabi Ranani (2016), Patric Ro (2014), Abdolsalam et.al (2016)	continuation and repetition of message	
Daneshfar & Nabizadeh (2015), Fortenbury & Mac Goldric (2011)	using trustable and famous people	
Salimifard & Pourdehghan (2016)	common dialects and terms	
Daneshfar & Nabizadeh(2015), Forentnburi & Mac Goldric (2011)	loveliness of the popular person for the observer	
Daneshfar & Nabizadeh (2015), Forentnburi & Mac Goldric)(2011	familiarity of famous person with the observer	
Mansouri Moayed (2005), Aghapour (2016)	emotional music	
Khonsazadeh (2010), Hosseinzadeh et.al (2016), Aghapour (2016)	using various methods	
Siadat & Sakhi Jourashk (2014), Sohrabi Ranani (2016)	high coverage and being universal	
Navaei et.al (2016), Mosavi Esmaeilzadeh (2015), Hosseinzadeh et.al (2016), Mirzadeh & Mehrmanesh (2016), Rosemary et.al (2012), Mikit et.al (2012), Spenkar (2013)	media advertisement	
Navaei et.al (2015), Mosavi Esmaeilzadeh (2015), Hosseinzadeh et.al (2016), Ashouri (2016), Hassanzadeh & Danaei (2016), Mikit et.al (2012)	multimedia advertisement	
Navaei et.al (2015), Mirzadeh & Mehrmanesh (2016)	radio advertisement	
Navaei et.al (2015), Mosavi Esmaeilzdeh (2015), Ashouri (2016), Hassanzadeh & Danaei (2016)	cell phone advertisement	
Navaei et.al (2015), Hosseinzadeh et.al (2016), Mirzadeh & Mehrmanesh (2016)	creating TV advertisement	
Ahmadi (2013), Salimi Fard & Pourdehghan (2016), Hoang et.al (2013)	attraction of rewards	
siadat & Sakhi Jourashk (2014), Koumra et.al (2016)	being benevolent	

Darzaian (2016), Zabkouich & Louksten (2011), Abdolsalam et.al (2016)	coordination in speaking and behavior	
Ahmadi (2013), Salimi Fard & Pourdehghan(2016), Honari et.al (2015)	reality of messages	
Madhoushi (2013), Salimi Fard & Pourdehghan (2016), Houang et.al (2013)	graphic-designing	
Shahmohammadi (2011), Madhoushi (2013), Salimi Fard & Pourdehghan(2016)	psychological – geographical division	
Mirzadeh & Mehrmanesh (2016), Zabkouich & Lokesten (2011), Menroz et.al (2012), Choubhai et.al (2013), Mokan & Oltz (2014), Ayono et.al (2014)	advertisement on web	
Dehdashti Shahrokh (2009), Shahmohammadi (2011), Mirzadeh & Mehrmanesh (2016), Menroz et.al (2012), Choubhai et.al (2013), Mokan & Oltz (2014), Ayono et.al (2014), Henri et.al (2015), Yazghiro- Basteh (2015), Lin (2016)	internet & cell phone advertisement	
Shahmohammadi (2011), Sanei et.al (2016), Menroz et.al (2012), Chobai et.al (2013), Ayono et.al (2014), Henri et.al (2015)	selecting advertising media	
Shahmohammadi (2011), Sanei et.al (2016), Howang et.al (2013), Henri et.al (2015), Yazghirdo-Yasteh (2015), Martina Santa et.al (2017)	artistic aspect of advertisement	
Kashizadeh (2010), Hoshivar(2014), Pelpin et.al (2017)	gender of speaker in the advertisement	
Hosseinzadeh et.al (2016), oribeh (2016), Madra et.al (2017)	text repetition	
Shahmohammadi (2011), Hassanzadeh & Danaei (2016), Abri & Joumar (2014), Ayouno et.al (2014), Oribeh (2016), Pachman (2016), Bak (2016), Madra et.al (2017)	advertisement	
Samsam Shariat (2007), Sohrabi Ranani (2016), Daneshvar et.al (2016), Bak (2016), Kiji-Kali (2017)	factors relating to usage	environment

Fakhari et.al (2012), Mirzadeh & Mehrmanesh (2016), Howang et.al (2013), Bak (2016), kiji-Kali (2017)	factors relating to facilitation	
Dargi & Mohammadian (2010), Hamidizadeh (2013), Darzban (2016), Belman et.al (2012)	awareness of different status of the environment status	
Dargi & Mohammadian (2010), Mirzadeh & Mehrmanesh (2016)	awareness of access location of objective group to the product	
Dargi & Mohammadian (2010), Salimifard & Pourdehghan (2016)	awareness of norms	
Dargi & Mohammadian (2010), Hamidizadeh (2013), Darzban (2016), Belman et.al (2012), Koumra et.al (2016)	awareness of religions and religious values	
Dargi & Mohammadian (2010), Hassanzadeh & Danaei (2016)	awareness of method of evaluating services in relation to other services	
Hassazadeh & Danaei (2016), Radmiker (2015)	attention to security	
Samsam Shariat (2007), Mirzadeh & Mehrmanesh (2016), Belman et.al (2012)	economic beliefs	
Salimifard & Pourdehghan (2016), Pilay et.al (2012), Howang et.al (2013), Komra et.al (2016)	ethical beliefs	
Salimifard & Pourdehghan (2016)	Islamic teaching	
Samsam Shariat (2007), Daneshvar et.al (2016), Radmiker (2015)	strategic look to advertisement plan	
Dargi & Mohamadian (2010), Ansari & Rasei (2016), Abri & Joumar (2014)	market share	
Dargi & Mohamadian (2010), Darzain (2016)	obviating primary needs of audiences	

Salimi Fard & Pourdehghan (2016), Kim et.al (2012), Howang et.al (2013)	reference groups	
Sohrabi Ranin (2016), Fontini et.al (2009), Belman et.al (2012)	awareness of services	factors relating to services
Salimi Fard & Pourdehghan (2016), Abri & Joumar (2014), Radmiker (2015)	stating potential benefits of purchase ad services consumption	
Dargi & Mohamadian (2010), Hosseinzadeh et.al (2016), Fontini et.al (2009)	rate of services complexity	
Samsam Shariat (2007), Darzaian (2016), Fontini et.al (2009), Radmiker (2015)	display of expected experience	
Khashizadeh (2010), Darzain (2016)	rate of familiarity of consumer with services	
Ansari & Rasaei (2016), Belman et.al (2012)	market researches	
Hosseinzadeh et.al (2016), Mirzadeh & Mehrmanesh (2016)	record of exploiting services	
Dargi & Mohammadian (2010), Ansari & Rasaei (2016), Lousi (2012), Park (2015), Marion et.al (2017)	being unique	
Dehdashti Shahrokh (2009), Daneshvar et.al (2016), Radmiker (2015), Jankoski et.al (2016)	price, benefits	
Daneshvar et.al (2016), Fontini et.al (2009), Jankoski et.al (2016)	relation with customer	
Samsam Shariat (2007), Daneshvar et.al (201), Fontini et.al (2009), Jankoski (2016)	supporting customer	
Dehdashti Shahrokh (2009), Ansari & Rasaei (2016), Fontini et.al (2009), Radmiker (2015), Jankoski (2016)	feedback	

2- Research plan

The present study is of descriptive applied type which has been performed with survey method. The statistical population of this study includes all managers and experts of Eghtesad

Novin Bank. The statistical population of the present study contains all people who are exposed to advertisements of Eghtesad Novin Bank across the country. For selecting samples, Cochran formula is used for unlimited societies that based on this formula, the samples number were determined 400 persons for each province. The sample people were selected based on the following sampling method. At first, using stratified sampling method , the samples were randomly selected from five geographical regions of the country (north, south, east, west and center) of nine provinces (Mazandaran, Khorasan Razavi, Zanjan, Chahar Mahal Bakhtiari , Kerman, East Azarbaijan , Kermanshah, Ilam and Tehran).

3- Variables and research model

In this study, fuzzy Delphi technique has been used for identifying main indicators. For ranking the final criteria and sub standards, multi-criteria decision making models and AHP model have been used with fuzzy approach. For analysis of obtained data, Excel software and coding in the environment of this software have been used. The results obtained from data analysis have been provided in the present chapter.

3-1 Identifying criteria and sub standards of factors effective on advertisement effectiveness

At the first step, the primary indicators were identified and selected. In the first stage, by reviewing the literature and performed studies, a large number of indicators were extracted from the literature. For screening the indicators and identifying the final indicators, fuzzy Delphi approach has been used. The expert views have been collected in the field of rate of significance of main indicators. Fuzzy average and defuzzification output of rates relating to indicators are presented in table 2. The rate of defuzzification greater than 7 is acceptable and every indicator which has the score higher than 7 is confirmed (Wu& Fang, 2011).

Table 2- fuzzy average and fuzzy screening of indicators of factors effective on advertisement effectiveness

Result	Crisp	mean	U	M	L	
acceptance	7.15	(2, 7.97, 9)	9	7.97	2	Q01
acceptance	7.84	(6, 8.01, 9)	9	8.01	6	Q02

Result	Crisp	mean	U	M	L	
acceptance	7.92	(6, 8.13, 9)	9	8.13	6	Q03
acceptance	7.39	(4, 7.83, 9)	9	7.83	4	Q04
acceptance	7.70	(6, 7.8, 9)	9	7.80	6	Q05
rejection	6.68	(2, 7.28, 9)	9	7.28	2	Q06
acceptance	7.71	(6, 7.82, 9)	9	7.82	6	Q07
acceptance	7.66	(6, 7.75, 9)	9	7.75	6	Q08
acceptance	7.02	(2, 7.78, 9)	9	7.78	2	Q09
acceptance	7.27	(2, 8.15, 9)	9	8.15	2	Q10
acceptance	7.09	(2, 7.88, 9)	9	7.88	2	Q11
rejection	6.85	(2, 7.52, 9)	9	7.52	2	Q12
acceptance	7.43	(4, 7.9, 9)	9	7.90	4	Q13
rejection	6.71	(3, 7.07, 9)	9	7.07	3	Q14
acceptance	7.17	(2, 8.01, 9)	9	8.01	2	Q15
acceptance	7.13	(2, 7.95, 9)	9	7.95	2	Q16
rejection	6.85	(4, 7.02, 9)	9	7.02	4	Q17
acceptance	7.43	(4, 7.89, 9)	9	7.89	4	Q18
acceptance	7.70	(6, 7.8, 9)	9	7.80	6	Q19
rejection	6.80	(3, 7.21, 9)	9	7.21	3	Q20
acceptance	7.75	(6, 7.88, 9)	9	7.88	6	Q21
rejection	6.62	(3, 6.93, 9)	9	6.93	3	Q22
acceptance	7.61	(6, 7.66, 9)	9	7.66	6	Q23
acceptance	7.22	(2, 8.08, 9)	9	8.08	2	Q24
acceptance	7.04	(2, 7.81, 9)	9	7.81	2	Q25
acceptance	7.71	(6, 7.82, 9)	9	7.82	6	Q26
rejection	6.65	(3, 6.97, 9)	9	6.97	3	Q27
acceptance	7.98	(6, 8.22, 9)	9	8.22	6	Q28
acceptance	7.70	(6, 7.79, 9)	9	7.79	6	Q29
rejection	6.54	(3, 6.81, 9)	9	6.81	3	Q30
acceptance	7.19	(4, 7.53, 9)	9	7.53	4	Q31
acceptance	7.79	(5, 8.18, 9)	9	8.18	5	Q32

Result	Crisp	mean	U	M	L	
acceptance	7.53	(4, 8.05, 9)	9	8.05	4	Q33
acceptance	7.05	(2, 7.82, 9)	9	7.82	2	Q34
acceptance	7.43	(4, 7.89, 9)	9	7.89	4	Q35
rejection	6.92	(3, 7.38, 9)	9	7.38	3	Q36
acceptance	7.02	(2, 7.78, 9)	9	7.78	2	Q37
rejection	6.91	(3, 7.37, 9)	9	7.37	3	Q38
rejection	6.96	(2, 7.68, 9)	9	7.68	2	Q39
rejection	6.99	(2, 7.74, 9)	9	7.74	2	Q40
acceptance	7.42	(4, 7.88, 9)	9	7.88	4	Q41
acceptance	7.10	(2, 7.9, 9)	9	7.90	2	Q42
rejection	6.59	(2, 7.13, 9)	9	7.13	2	Q43
rejection	6.55	(2, 7.08, 9)	9	7.08	2	Q44
acceptance	7.77	(6, 7.9, 9)	9	7.90	6	Q45
rejection	6.91	(2, 7.62, 9)	9	7.62	2	Q46
acceptance	7.09	(2, 7.88, 9)	9	7.88	2	Q47
acceptance	7.68	(6, 7.78, 9)	9	7.78	6	Q48
acceptance	8.19	(6, 8.53, 9)	9	8.53	6	Q49
rejection	6.88	(2, 7.57, 9)	9	7.57	2	Q50
acceptance	7.75	(5, 8.12, 9)	9	8.12	5	Q51
acceptance	7.03	(2, 7.8, 9)	9	7.80	2	Q52
rejection	6.96	(4, 7.19, 9)	9	7.19	4	Q53
acceptance	7.04	(2, 7.8, 9)	9	7.80	2	Q54
rejection	6.96	(3, 7.45, 9)	9	7.45	3	Q55
acceptance	7.02	(2, 7.78, 9)	9	7.78	2	Q56
acceptance	7.92	(5, 8.38, 9)	9	8.38	5	Q57
acceptance	7.00	(2, 7.76, 9)	9	7.76	2	Q58
rejection	6.90	(1, 7.86, 9)	9	7.86	1	Q59
acceptance	7.10	(4, 7.4, 9)	9	7.40	4	Q60
acceptance	7.86	(6, 8.04, 9)	9	8.04	6	Q61
rejection	6.97	(2, 7.71, 9)	9	7.71	2	Q62

Result	Crisp	mean	U	M	L	
acceptance	7.02	(2, 7.79, 9)	9	7.79	2	Q63
acceptance	7.73	(6, 7.84, 9)	9	7.84	6	Q64
acceptance	7.01	(2, 7.76, 9)	9	7.76	2	Q65
rejection	6.65	(3, 6.98, 9)	9	6.98	3	Q66
acceptance	7.01	(2, 7.77, 9)	9	7.77	2	Q67
acceptance	7.90	(6, 8.11, 9)	9	8.11	6	Q68
acceptance	7.55	(5, 7.83, 9)	9	7.83	5	Q69
acceptance	8.02	(6, 8.28, 9)	9	8.28	6	Q70
acceptance	7.95	(6, 8.17, 9)	9	8.17	6	Q71
acceptance	7.03	(2, 7.79, 9)	9	7.79	2	Q72
acceptance	7.65	(5, 7.98, 9)	9	7.98	5	Q73
acceptance	7.00	(1, 8, 9)	9	8.00	1	Q74
acceptance	7.04	(2, 7.8, 9)	9	7.80	2	Q75
acceptance	7.03	(2, 7.8, 9)	9	7.80	2	Q76
acceptance	7.30	(4, 7.7, 9)	9	7.70	4	Q77
rejection	6.88	(2, 7.57, 9)	9	7.57	2	Q78
acceptance	7.63	(4, 8.19, 9)	9	8.19	4	Q79
acceptance	7.65	(5, 7.98, 9)	9	7.98	5	Q80
rejection	6.59	(3, 6.88, 9)	9	6.88	3	Q81
acceptance	7.09	(2, 7.88, 9)	9	7.88	2	Q82
rejection	6.83	(2, 7.5, 9)	9	7.50	2	Q83
acceptance	7.12	(2, 7.93, 9)	9	7.93	2	Q84

From the 84 indicators which were assessed, 24 indicators which have acquired the average score less than 7 are omitted. So, 60 indicators are finally confirmed (Adelazaret.al, 1995).

The remained indicators will be finally examined in the format of a 60-question questionnaire with analytical hierarchy process.

3-2 determining the priority of factors effective on effectiveness of advertisement using fuzzy AHP technique

For determining the priority of major criteria, fuzzy analytical hierarchy (FAHP) has been used. The analysis process is as follows.

1- Dual comparison of major criteria based on objective and determining the weight of major criteria

2- Dual comparison of sub standards of each criterion and determining sub standards of each cluster

3- Multiplication of the weight of sub standards in the weight of respective criterion and determining the final weight of sub standards

For dual comparison of elements, 9-degree Saaty scale has been used. 9-degree Saaty scale has been provided by Thomas Saaty , codifier of analytical hierarchy theory. Also, in this study fuzzy approach has been used for making the values quantitative. So, Saaty fuzzy range has been used.

Table 3- fuzzy range equal to 9-degree Saaty scale in AHP technique. Habibi et.al , 2014, 77

reversed fuzzy equivalent	fuzzy equivale nt	the word term of comparison status of i to j
(1,1,1)	(1, 1, 1)	Preferred Equally
$\left(\frac{1}{3},\frac{1}{2},1\right)$	(1, 2, 3)	intermediate
$\left(\frac{1}{4},\frac{1}{3},\frac{1}{2}\right)$	(2, 3, 4)	Preferred moderately
$\left(\frac{1}{5},\frac{1}{4},\frac{1}{3}\right)$	(3, 4, 5)	intermediate
$\left(\frac{1}{6},\frac{1}{5},\frac{1}{4}\right)$	(4, 5, 6)	Preferred Strongly
$\left(\frac{1}{7},\frac{1}{6},\frac{1}{5}\right)$	(5, 6, 7)	intermediate
$\left(\frac{1}{8},\frac{1}{7},\frac{1}{6}\right)$	(6, 7, 8)	very strongly Preferred
$\left(\frac{1}{9},\frac{1}{8},\frac{1}{7}\right)$	(7, 8, 9)	intermediate
$\left(\frac{1}{9},\frac{1}{9},\frac{1}{9}\right)$	(9, 9, 9)	Extremely Preferred

(source in papersL Khorshid & Zabihi, Lee my H.I; Wen-Chin Chen, Ching-Jan, Chang, 2008, 101)

The pattern of criteria and sub standards has been drawn in the figure using AHP technique. Also, the criteria and sub standards of the study have been named with numeric index so that the research flow is easily detectable.

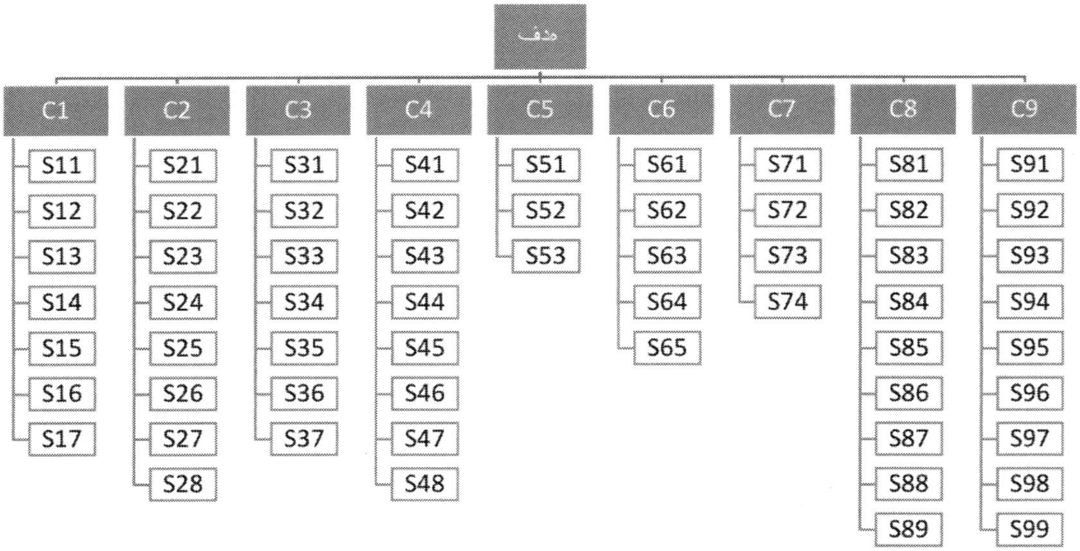

Figure 1- hierarchical model of criteria and sub standards of factor effective on advertisement effectiveness

1- Determining the priority of major criteria based on objective

In the first step, the major criteria based on the objective have been compared dually. Dual comparison is very simple and all elements of each cluster should be compared two by two. Therefore, if there is n element in a cluster, $\frac{n(n-1)}{2}$ comparisons will be performed. Since there are four criteria, so the number of performed comparisons is equal to:

$$\frac{n(n-1)}{2} = \frac{9(9-1)}{2} = 36$$

So, 36 dual comparisons have been performed from group perspective of experts. The experts view has been made quantitative using fuzzy scale. Collecting experts view has been collected with 9-degree Saaty range. So, 6 dual scales of four major four criteria have been performed based on the experts view. Using geometric average the experts views have been aggregated. Fuzzy aggregation will be calculated as follows:

Equation 1

$$F_{AGR} = (\prod(l), \prod(m), \prod(u))$$

After matrix formation of the obtained dual comparisons, the special vector has been calculated. At first, the fuzzy extension of each row is calculated. Every entry of dual comparison matrix $\widetilde{X}$

Is displayed as $\tilde{x}_{ij}$.

The fuzzy extension of every row is also displayed with symbol $\tilde{S}_i$. Therefore, the fuzzy extension of each row is calculated as follows:

Equation 2

$$\tilde{S}_i = \sum_{j=1}^{n} x_{ij}$$

Then, the fuzzy extension of every row elements will be as follows:

Fuzzy extension of row 1 (1,1,1) ⊕ (1,27,1,47,1,75) ⊕ (0.47, 0.55, 0.66) ⊕ (1.52, 1.81, 2.14) ⊕ (0.93, 1.14, 1.39) ⊕ (1.26, 1.61, 2.04) ⊕(1.25, 1.51, 1.84) ⊕(0.78, 0.97, 1.24) ⊕ (0.59, 0.75, 0.96) ⊕ (9.08, 10.8, 13.02)

Fuzzy extension of row 2 (0.57, 0.68, 0.79)⊕(1, 1, 1)⊕(0.58, 0.7, 0.89)⊕(1.84, 2.28, 2.75)⊕(0.68, 0.82, 1.02)⊕(1.42, 1.81, 2.34)⊕(1.64, 2.04, 2.53)⊕(0.71, 0.88, 1.12)⊕(0.73, fuzzy extension of row 3 (1.52, 1.82, 2.11)⊕(1.13,)0.9, 1.1)⊕ = (9.17, 11.11, 13.55 1.42, 1.72)⊕(1, 1, 1)⊕(1.89, 2.34, 2.88)⊕(0.7, 0.86, 1.09)⊕(1.45, 1.78, 2.22)⊕(1.55, 1.88, 2.29)⊕(0.57, 0.71, 0.93)⊕(0.67, 0.8, 0.96)⊕ = (10.47, 12.61, 15.2)

Fuzzy extension of row 4 (0.47, 0.55, 0.66)⊕(0.36, 0.44, 0.54)⊕(0.35, 0.43, 0.53)⊕(1, 1, 1)⊕(0.97, 1.21, 1.51)⊕(0.96, 1.33, 1.8)⊕(2.14, 2.7, 3.37)⊕(0.94, 1.21, 1.56)⊕(1.3, 1.69, 2.16)⊕ = (8.49, 10.57, 13.14)

Fuzzy extension of row 5 (0.72, 0.88, 1.07) ⊕ (0.98, 1.22, 1.48)⊕(0.92, 1.17, 1.43)⊕(0.66, 0.82, 1.03)⊕(1, 1, 1)⊕(0.67, 0.83, 1.04)⊕(0.23, 0.29, 0.36)⊕(0.28, 0.32, 0.39)⊕(0.3, 0.38, 0.48)⊕ = (5.76, 6.91, 8.29)

Fuzzy extension of row 6 (0.49, 0.62, 0.79)⊕(0.43, 0.55, 0.7)⊕(0.45, 0.56, 0.69)⊕(0.55, 0.75, 1.04)⊕(0.96, 1.2, 1.48)⊕(1, 1, 1)⊕(0.34, 0.41, 0.51)⊕(0.27, 0.34, 0.44)⊕(0.24, 0.31, 0.41)⊕ = (4.74, 5.75, 7.08)

Fuzzy extension of row 7 (0.54, 0.66, 0.8)⊕(0.4, 0.49, 0.61)⊕(0.44, 0.53, 0.64)⊕(0.3, 0.37, 0.47)⊕(2.81, 3.5, 4.3)⊕(1.95, 2.43, 2.96)⊕(1, 1, 1)⊕(0.34, 0.42, 0.54)⊕(0.29, 0.36, 0.47)⊕ = (8.06, 9.76, 11.78)

Fuzzy extension of row 8 (0.8, 1.03, 1.28)⊕(0.89, 1.14, 1.41)⊕(1.08, 1.41, 1.76)⊕(0.64, 0.82, 1.06)⊕(2.55, 3.1, 3.63)⊕(2.25, 2.97, 3.72)⊕(1.86, 2.39, 2.94)⊕(1, 1, 1)⊕(0.58, 0.69, 0.83)⊕ = (11.66, 14.56, 17.64)

Fuzzy extension of row 9 (1.04, 1.33, 1.69)⊕(0.91, 1.11, 1.37)⊕(1.04, 1.25, 1.5)⊕(0.46, 0.59, 0.77)⊕(2.08, 2.63, 3.3)⊕(2.42, 3.23, 4.1)⊕(2.15, 2.74, 3.43)⊕(1.2, 1.46, 1.72)⊕(1, 1,)1)⊕ = (12.29, 15.35, 18.88

Then, the fuzzy sum of total priorities column elements is calculated

Equation 3

$$\sum \tilde{S}_i = \sum_{i=1}^{n} \sum_{j=1}^{n} x_{ij}$$

The sum of elements of the priorities column of major criteria will be as follows:

$$\sum_{i=1}^{4} \sum_{j=1}^{4} M_g^j = (79.72, 97.41, 118.59)$$

For normalization of each criterion priorities, that criterion total rates sum should be divided into sum of all priorities (column elements), since the rates are fuzzy, so fuzzy sum of every row is multiplied in summation reverse. The summation reverse should be calculated.

Equation 4

$$if\ \tilde{F} = (l, m, u)\ then\ \tilde{F}^{-1} = \left(\frac{1}{u}, \frac{1}{m}, \frac{1}{l}\right)$$

Then, based on the equation 4, we will have

$$\left(\sum_{i=1}^{n} \sum_{j=1}^{n} M_g^j\right)^{-1} = (0.01, 0.01, 0.01)$$

$$P_k = \sum_{i=1}^{n} M * \left(\sum_{i=1}^{n} \sum_{j=1}^{n} M_g^j\right)^{-1}$$

So, the results obtained from normalization of obtained rates will be as follows:

$\widetilde{W}_{C1} =$ (0.08, 0.11, 0.16)

$\widetilde{W}_{C2} =$ (0.08, 0.11, 0.17)

$\widetilde{W}_{C3} =$ (0.09, 0.13, 0.19)

$\widetilde{W}_{C4} =$ (0.07, 0.11, 0.16)

$\widetilde{W}_{C5} =$ (0.05, 0.07, 0.1)

$\widetilde{W}_{C6} =$ (0.04, 0.06, 0.09)

$\widetilde{W}_{C7} =$ (0.07, 0.1, 0.15)

$\widetilde{W}_{C8} =$ (0.1, 0.15, 0.22)

$\widetilde{W}_{C9} =$ (0.1, 0.16, 0.24)

Each of the obtained rates is fuzzy and normalized weight relating to major criteria. There are various methods for defuzzifcation of the obtained rates. In this study, Bojadziev proposed method has been used for defuzzification.

Table 4- defuzzification and computation of final and normal weight of criteria of factors effective on advertisement effectiveness

Normal	Deffuzy	X3max	X2max	X1max	Crisp
0/111	0/117	0/113	0/115	0/117	C1
0/114	0/120	0/115	0/118	0/120	C2
0/103	0/137	0/133	0/135	0/137	C3
0/108	0/113	0/112	0/113	0/113	C4
0/070	0/073	0/072	0/073	0/073	C5
0/060	0/063	0/062	0/063	0/063	C6
0.101	0/107	0/103	0/105	0/107	C7
0.149	0/157	0/153	0/155	0/157	C8
0.158	0/167	0/163	0/165	0/167	C9

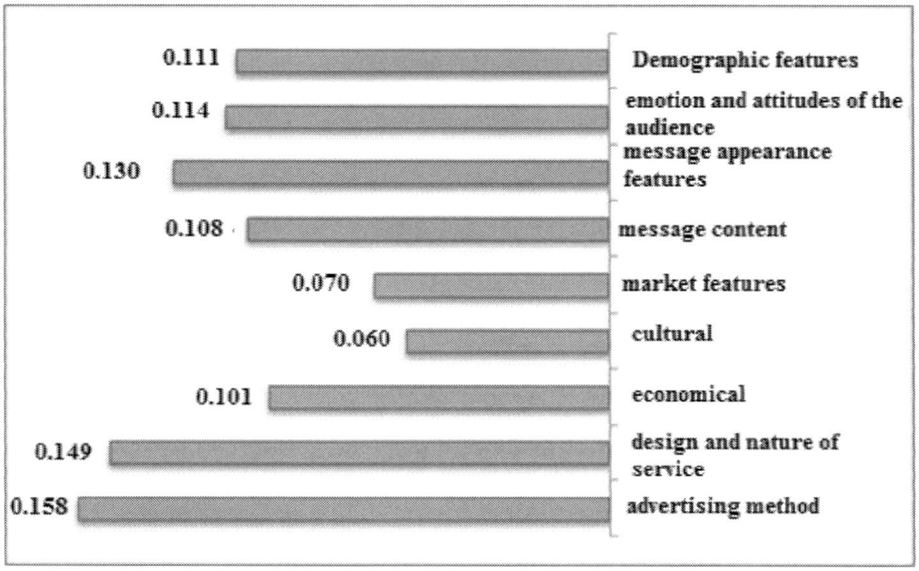

Figure 2- graphical display of significance of criteria of factors effective on advertisement effectiveness

Based on the obtained special vector:

- Demographic features with weight 0.111 have been located in rank five

- Individual feelings and attitude of the addressee with weight 0.114 has been located in rank four.

- The message surface features with weight 0.13 has been located in rank three.

- The message content with weight 0.108 has been located in rank six

- The market feature with weight 0.07 has been located in rank eight.

- Cultural with weight 0.06 has been located in rank nine.

- Economic with weight 0.10 has been located in rank seven.

- Plan and nature of services with weight 0.149 has been located in rank second.

- Advertisement methods with weight 0.158 has been located in rank first.

The inconsistency rate of performed comparisons has been obtained 0.030 which is smaller than 0.1 and so the performed comparisons is sufficed.

2- Comparison and determining priority of sub standards

At first, the experts' views have been collected in the dual comparison of sub standards. Then, the performed calculations for fuzzification of the experts views average have been provided for determining the priority of sub standards.

Priority of sub standards of demographic features

The audience job with weight 0.151 has been located in the third rank.

The audience income with weight 0.093 has been located in the seventh rank.

The audience religion with weight 0.16 has been located in the second rank.

Social class with weight 0.141 has been located in the fifth rank.

Awareness of gender differences with weight 0.173 has been located in the first rank.

The individual knowledge level with weight 0.135 has been located in the sixth rank.

Priority of sub standards of the audience individual feeling and attitude

Positive mentality with weight 0.124 has been located in the fifth rank.

Feeling of honesty in advertisements with weight 0.089 has been located in the eighth rank.

Threatening feeling with weight 0.14 has been located in the third rank.

Mental image with weight 0.102 has been located in the seventh rank.

Attitude towards brand with weight 0.121 has been located in the sixth rank.

Convincing audiences with weight 0.149 has been located in the first rank.

Attitude towards advertisement with weight 0.143 has been located in the second rank.

The individual perception of accuracy of advertisement with weight 0.133 has been located in the fourth rank.

Priority of sub standards of the message surface features

Then, based on the results of performed dual comparisons:

Beautiful and appropriate combination of colors with weight 0.143 has been located in the fourth rank.

Attention to reduction of the user waiting time with weight 0.159 has been located in the second rank.

Graphics- design with weight 0.131 has been located in the sixth rank.

Creativity in the message with weight 0.137 has been located in the fifth rank.

Emotional music with weight 0.099 has been located in the seventh rank.

Text repetition with weight 0.153 has been located in the third rank.

Access to required and sufficient information with weight 0.178 has been located in the first rank.

Determining the priority of sub standards of the message content

Benevolent with weight 0.093 has been located in the eighth rank.

Coordination in speech and behavior with weight 0.125 has been located in the fourth rank.

Reality of the messages with weight 0.119 has been located in the sixth rank.

Using women and children images with weight 0.131 has been located in the third rank.

Gender of speaker in the advertisement with weight 0.106 has been located in the seventh rank.

Using trustful and famous people with weight 0.16 has been located in the first rank.

Loveliness of famous people for the observer with weight 0.122 has been located in the fifth rank.

Familiarity of the famous person with the observer with weight 0.144 has been located in the second rank.

Priority of sub standards of the market features

Then, based on the results of the performed dual comparisons:

Factors relating to its application in the market with weight 0.326 have been located in the second rank.

Awareness of different aspects of status and environment with weight 0.283 has been located in the third rank.

Awareness of evaluation method of service receiver with weight 0.391 has been located in the first rank.

Priority of cultural sub standards

Then, based on the results of performed dual comparisons:

Awareness of norms with weight 0.199 has been located in the third rank.

Awareness of religion and religious values with weight 0.238 has been located in the first rank.

Ethical beliefs with weight 0.18 have been located in the fourth rank.

Obviating primary needs of the audiences with weight 0.164 has been located in the fifth rank.

Reference groups with weight 0.219 have been located in the second rank.

Determining the priority of economic sub standards

Then, based on the results of the performed dual comparison:

Economic beliefs with weight 0.261 have been located in the second rank.

Strategic look at an advertising program with weight 0.215 has been located in the fourth rank.

Market share with weight 0.274 has been located in the first rank.

Attention to economic security with weight 0.251 has been located in the third rank.

Priority of the plan and services nature sub standard

Awareness of services with weight 0.104 has been located in seventh rank.

Expressing potential benefits of purchase and services consumption with weight 0.117 has been located in the third rank.

Services complexity rate with weight 0.089 has been located in the ninth rank.

Uniqueness with weight 0.108 has been located in the sixth rank.

Rate of familiarity of the services consumer with weight 0.12 has been located in the second rank.

Price and allowances with weight 0.114 has been located in the fourth rank.

Relation with the customer with weight 0.111 has been located in the fifth rank.

Displaying the expected experience with weight 0.098 has been located in the eighth rank.

Supporting customers with weight 0.139 has been located in the first rank.

Priority of sub standards of the advertising methods

Beginning the advertisement with a challenge with weight 0.079 has been located in the ninth rank.

Beginning an advertisement with a statement and explanation with weight 0.107 has been located in the fifth rank.

Using stimuli with diversions more than usual and expected with weight 0.097 has been located in the seventh rank.

The advertisement message text being predicating with weight 0.126 has been located in the third rank.

Using diverse methods with weight 0.088 has been located in the eighth rank.

Media advertisement with weight 0.113 has been located in the fourth rank.

Dimensions of TV advertisement with weight 0.151 have been located in the first rank.

Psychological, geographical division with weight 0.107 has been located in the fifth rank.

Advertising media selection with weight 0.132 has been located in the second rank.

Determining the final priority of major criteria with FAHP technique

For determining final priority of major criteria using FAHP technique, the weights relating to major criteria (W1) and weight of indicators based on each criteria (W2) are needed. The results of comparison of the research sub criteria relating to them forms matrix W2. For determining the final priority of indicators with AHP technique, it is sufficient that the weight of indicators based on every criterion (W2) is multiplied in the major criteria (W1). Any of these matrices has been calculated in the previous steps. Based on the obtained output, it is obvious that the market share item, one of economic factors with weight 0. 277 is the most important indicator in the existing indicators. Item of awareness of evaluation method of service receiver in relation to other services with weight of 0.0273 has the second priority. Item of economic believes, one of economic factors, has the third priority.

So, the research model is as follows:

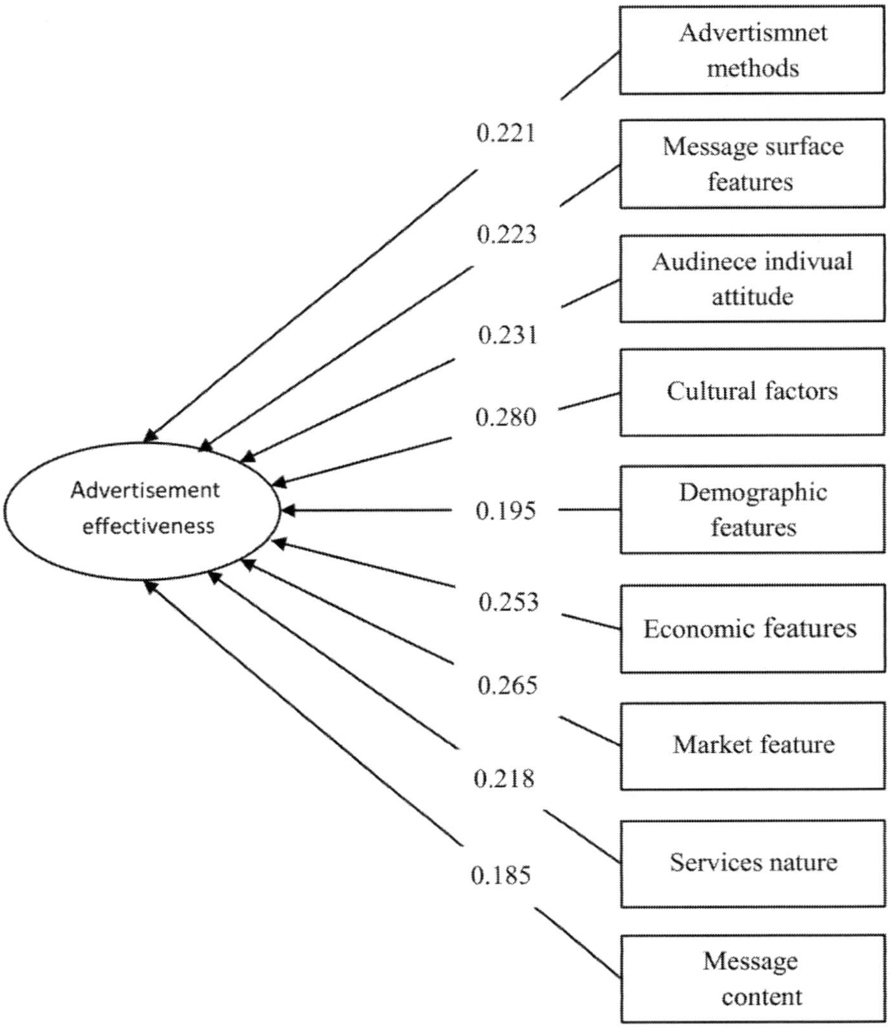

4- Conclusion:

In this study, identification and ranking of factors effective on designing the native model of advertisement in Eghtesad Novin bank has been addressed. In this first step, the primary indicators were identified and selected. In the first stage, by reviewing the literature and performed studies, a great number of indicators were extracted from the literature. For screening the indicators and identifying final indicators, fuzzy Delphi approach has been used. The experts' views have been collected in the field of significant rate of major indicators. From 84 indicators which were assessed, 24 indicators which have acquired score less than 7 are omitted. So, 60 indicators were finally confirmed. Then, regarding the performed calculations, the final weight of any of the model indicators have been calculated with fuzzy AHP. Based on the obtained output, it is obvious that the market share of economic factors with weight of 0.027 is the most important indicator in the existing indicators. The item of awareness of evaluating services in relation to other services is one of market features factors with weight 0.0273 which has second priority. The item of economic beliefs with weight 0.0264 is in the third priority.

Today, considering extension of competition in the international level and formation of world competition, all countries of the world may be successful in this arena to coordinate them with this trend and create more value for their customers by services and products they provide. In the today modern world, almost all people are exposed to advertisements and other communicational activities. Effectiveness of advertisement is one of important issues in the advertisement domain. In this regard, Mansouri Moyed (2005) showed that attention attraction, commercial advertisements of Maskan Bank are effective but in respect of other variables, that is, creating interest, stimulation of tendency and measurement to purchase, commercial advertisements of Maskan Bank are not effective. Samsam Shariat (2007) showed that TV channels and cartoon animations , attractive package, colors relating to the product and innovation, noticing safety feeling and associated symbols, informing of quality, background and price, dialects and common phrases, diversity in the message content, in TV, broadcasting from 18-23 and before beginning programs (in media) have been effective factors in effectiveness of commercial effectiveness from people viewpoint. Mosavi Esmaeilzadeh (2015) showed that oral advertisements (in 4 dimensions) and after that TV advertisement (just in 2 dimensions) are effective. Also, radio, internet and billboard advertisement are not effective. And also Friedman test was used for ranking advertisement tools. Salimi Fard & Pourdehghan (2016) showed that among factors effective on attitude to cell phone advertisement, components of general attitude towards advertisement, knowledge level and value have respectively the greatest significance and components of time of presentation and reference groups with the least significance in effecting the individuals attitude towards cell phone advertisement. Ansari & Rasaei (2016) showed that the creativity of advertisement message , market research, competition, market share, uniqueness and relation with customer are effective on the effectiveness of brand advertisement. Louis (2012) showed that advertising in the urban public spaces will have a great impact in the consumers' mentality. Since some advertisements are transitory and are often displayed in a short period of time and just in several spaces, but advertisement in public spaces exist everywhere and is similar to advertising through print and media, but since they have lower cost for production and distribution, have lower cost in this respect. Zabckowich & Loukston (2011) showed that in necktie advertisements use more of sensory requests and information signs. Findings showed that in western and European countries, people establish more contact due to globalization increase. They declared that content of printed advertisements, includes visual and text advertisements which are effective in establishing relation with many audiences. Also, ethnic identity as an important cultural variable increases attention to consumer purchase behavior. Besides that, they found out that the image of people, models, ceremonies and managers constitute an important part of images and background image plays an important role in effectiveness of visual factors in advertisement. Mac Ki- Nesbiat et.al (2011) showed that younger adults remind emotional messages specially negative ones better than others and older adults more prefer positive and logical messages to negative and emotional ones, but attitude towards advertisement among younger adults is not different. Also, they found that older adults have the most positive attitude towards advertisements.

Perzouti (2015) showed that advertising models are different in various ages and each age is consistent with a special model and using various and consistent models with people conditions may make the advertisement more effective. Abri & Joumar (2014) showed that advertising messages are effective on the consumer behavior and the advertiser may use various media for communicating consumer and media schedule is effective on the advertisement success. Attention to environmental factors where the customer lives are important factors in the advertisement success. Madra et.al (2017) showed that the advertisement text, text repetition impact the advertisement effectiveness. Maryion et.al (2017) showed that simultaneous use of several media like TV and internet may increase effectiveness of the advertisement and considering two variables of gender and multimedia create a better individual attitude towards the message effectiveness.

In ranking the identified factors, advertisement methods have assigned the first rank among factors. It is suggested that in designing advertisement and selecting advertising instruments, the dimensions of TV advertising, selecting median and predicative message text should be noticed. Using advertising instrument like internet advertisement, adverting on public vehicles like buses and metro due to great number of people who use them can increase stability of the advertisements in the customers mind.

In codifying and designing bank advertisements, the design and services nature should be noticed. It is proposed that supporting customer, familiarity of consumer with services are considered and during the advertisement process, it is required that the potential benefits of purchase and services consumption are expressed.

In designing the advertisement and content of bank advertising message, it is required to use famous and trustful people for introducing the services. Also, children as future customers and audiences should be noticed since they are future potentials. Total advertisements provided by the bank through various media should move in a unit line so that the audience can establish a logical relation between the observed advertisements in various media that is total advertising efforts lead to synergy of advertisement effectiveness. This issue won't be acquired unless through a media planning and efficient advertisement.

In advertisements, the message surface features should be noticed. It is required that in advertisement, access to required and enough information for the customers, attention to reduction of expectation time of user for observing the advertisement and suitable repetition of the text are considered. Providing large information in an advertisement which has a short time and providing information that require special mathematical analysis should be avoided.

References:

Hosseinabadi, Maryam, Mahmoud Nematian, Mohammad Besharati, investigating the relation between mixed elements of marketing and customers satisfaction with Tehran

Sarmayeh bank branches, management and entrepreneurship studies, fall 2016, 2nd period, No. 3 (page 162-169)

Dargi Parviz, 2008, subjects and topics of marketing management with Iran market attitude, Tehran, Rasa publication

Sarmad, Zohreh, Bazargan Abbas, Hejazi, Elaheh, 2014, research methods in behavioral sciences, Tehran, Agah. 27-13

Salimi Fard, Khodakaram, Pourdehghan, Adel, 2016, identifying and ranking factors effective on attitude towards cell phone advertisement using AHP, Brand management, period 3, no. 6, summer 2016, page 133-160

Seyed Akbar Manjgani, Vahideh Tabasi Lotfabadi, 2015, factors effective on effectiveness of advertisement on financial stoop of legal clients of Mashhad City with AHP approach, finance journal, fall 2015, No. 75 (page 121-146)

Sanayei, Ali, Majid Mohammad Shafiei, Maedeh Amini Valashani, 2016, effectiveness of bank advertisement in customer attraction, commercial management searches, spring and summer 2016, No. 15 (page 185-210)

Mahmoudian, Mahmoud, Rabiei, Ali, Jamili, Bita. 2011, evaluation of effectiveness of Parsian bank advertisement and identifying the most important factors in increasing its effectiveness in Tehran city, modern marketing studies, summer 2011, No. 2 (page 17-40).

Manjagani, Seyed Akbar, Tabasi Lotfabadi, Vahideh, 2015, factors effective on advertisement effectiveness on financial stoop of legal clinets of Mashhad city with AHP approach, finance journal, fall 2015, No. 75, 1-121

Venous, Davar, Safaeian, Mitra, (2005), marketing bank services in Iran and bank services features, Tehran, Negah Danesh, 4th edition

Angulo, N. (2016). 10 digital marketing trends to watch in 2016 and beyond. Retrieved from: http://www.marketingdive.com/news/10-digital-marketing-trends-to-watch-in-2016-./and-beyond/411457

Bell Raoul, Buchner, Axel.(2018). Positive Effects of Disruptive Advertising on Consumer Preferences, Journal of Interactive Marketing 41 (2018) 1–13

Corvi. Elisabetta.(2010).The effectiveness of advertising: a literature review, 10th Global Conference on Business & Economics, ISBN : 978-0-9830452-1-2

Cramphorn Michael F.,(2011), Forum Gender effects in advertising, International Journal of .Market Research Society Vol. 53 Issue 2

Cronin, J. J. & Fox , G. L. (2010). "The implications of third party customer complaining for .advertising efforts". Journal of advertising. 39(2): 21-23

Decrop, A. (2007). The influence of message format on the effectiveness of print advertisements for tourism destinations. International Journal of Advertising, 26(4), 505–.525

Denis .A. Shakhov and ,Anna A. Panasenko.(2012).Evaluating Effectiveness of Bank Advertising in the Internet: Theory and Practice, World Applied Sciences Journal 18 (Special Issue of Economics): 83-90, 2012 ISSN 1818-4952

Denis 1,2 A. Shakhov and 3Anna A. Panasenko.(2012).Evaluating Effectiveness of Bank Advertising in the Internet: Theory and Practice, World Applied Sciences Journal 18 (Special Issue of Economics): 83-90, 2012

Durdana Ozretic-Dosen Ines Zizak , (2015),"Measuring the quality of banking services targeting student population", EuroMed Journal of Business, Vol. 10 Iss 1 pp. 98 - 117

Eckler, P., and Bolls, P. (2011). Spreading the Virus: Emotional Tone of Viral Advertising and Its Effect on Forwarding Intention and Attitudes. Journal of Interactive Advertising, .11(2): 1-11

Eric W.T. Ngai Ka-leung Karen Moon S.S. Lam Eric S. K. Chin Spencer S.C. Tao , (2015),"Social media models, technologies, and applications", Industrial Management & Data Systems, Vol. 115 Iss 5 pp. 769 – 802

Fotini Patsioura Maro Vlachopoulou Vicky Manthou, (2009),"A new advertising effectiveness model for corporate advertising web sites", Benchmarking: An International Journal, Vol. 16 Iss 3 pp. 372 - 386

Frederic Crane, , G. (2010). «Marketing FO Entrepreneurs: ConceptandApplication for new .Venture. New York: SAGE Pablication, Inc

Frolova. Svetlana.(2014).THE ROLE OF ADVERTISING IN PROMOTING A PRODUCT, Thesis CENTRIA UNIVERSITY OF APPLIED SCIENCES ,Degree Programme in Industrial Management May 2014, p"1-61

Henry Boateng Abednego Feehi Okoe , (2015),"Consumers' attitude towards social media advertising and their behavioural response", Journal of Research in Interactive Marketing, Vol. 9 Iss 4 pp. 299 -312

Henry Boateng Abednego Feehi Okoe , (2015),"Consumers' attitude towards social media advertising and their behavioural response", Journal of Research in Interactive Marketing, Vol. 9 Iss 4 pp. 299 -312

Huang. Jinsong, Song Sub, , Liuning Zhouc, , , Xi Liu .(2013).Attitude Toward the Viral Ad: Expanding Traditional Advertising Models to Interactive Advertising, Journal of Interactive Marketing, Volume 27, Issue 1, February 2013, Pages 36–46

Idris, I., M.S.A. Yajid and A. Khatibi, 2009. Personal and Psychological Factors-Does it .Impact the Choice of Advertising Medium? Journal of Social Sciences, 5(2): 104-111

Ioannou Myria, Nikolaos Boukas, Evi Skoufari .(2014).Examining the role of advertising on the behaviour of co-operative bank consumers, Journal of Co-operative Organization and Management, Volume 2, Issue 1, June 2014, Pages 24-33

Jankowski. Jarosław, Przemysław Kazienko, Jarosław Wątróbski, Anna Lewandowska, Magdalena Ziolo.(2016).Fuzzy multi-objective modeling of effectiveness and user experience in online advertising, Expert Systems with Applications, Volume 65, 15 December 2016, Pages 315-331

Kim, C., Park, S., Kwon, K., & Chang, W. (2012). An empirical test to measure the effectiveness of online advertising in online marketplaces using a hierarchical Bayes model. .Expert Systems with Applications (39), 117–128

Kim, ji-Hern., Hyun, yong j., (2010). " A model to investigate the influence of marketing – mix efforts and corporate image on brand equity in the IT software sector " . industrial . Marketing Management. No 40,2011, pp 424-438, Emerald Group Publishing Limited

Kim, S., & Jun, J. (2016). The impact of event advertising on attitudes and visit intentions. .Journal of Hospitality and Tourism Management, 29, 1-8

Kirby, J. and Marsden .(2006). Connected marketing : the viral, buzz and word of mouth .revolution, Oxford : Butterworth-Heinemann, 2006

Madera. Quetzali, Oscar Castillo ∗, Mario García-Valdez, Alejandra Mancilla..(2017). A method based on Interactive Evolutionary Computation and fuzzy logic for increasing the effectiveness of advertising campaigns, Information Sciences 414 (2017) 175–186

Marion Garaus and Udo Wagner , Anna-Maria B¨ack.(2017). The Effect of Media Multitasking on Advertising Message Effectiveness, Psychology & Marketing, Vol. 34(2):)138–156 (February 2017

Martín-Santana J.D., E. Reinares-Lara , P. Reinares-Lara .(2017).Influence of radio spokesperson gender and vocal pitch on advertising effectiveness: The role of listener gender, Spanish Journal of Marketing – ESIC, Volume 21, Issue 1, February 2017, Pages 63-71

McKay-Nesbitt, J., V. Manchanda, R., C. Smith, M., & A. Huhmann, B. (2011). Effects of age, need for cognition, and affective intensity on advertising effectiveness. Journal of Business Research (64), 12–17

Moch, Nils (2013). Competition in fragmented markets: New evidence from the German banking industry in the light of the subprime crisis. Journal of Banking &; Finance, Volume 37, Issue 8, Pp 2908–2919

Moorthy, S. and Hawkins, S. (2005). Advertising repetition and quality perception. Journal of .Business Research, 58, 354– 360

Mucan. Burcu, Özeltürkay. Eda Yaşa.(2014).Social Media Creates Competitive Advantages: How Turkish Banks Use This Power? A Content Analysis of Turkish Banks through Their Webpages, Procedia - Social and Behavioral Sciences, Volume 148, 25 August 2014, Pages 137-145

Muda, M.; Musa, R.; Niana-Mohamed, R. and Borhan, H. (2014). Celebrity Entrepreneur .Endorsement and Advertising Effectiveness. Social and Behavioral Sciences, 130, 11-20

Oladele. Kehinde Joseph, Ogunnaike, Akinbode Mosunmola, Aka Deborah .(2016). Effective Advertising: Tool For Achieving Client-Customer Relationships, Researchjournali's Journal of Media Studies Vol. 2 | No. 1 February | 2016

Olsen, G., Pracejus, J. W., & O'Guinn, T. (2012). Print advertising: White space. Journal of .Business Research (65), 855–860

Park, M. (2006). The compensatory effects of pictorial and verbal information for haptic information on consumer responses in non-store shopping environments. Unpublished Ph.D, .Ohio, Ohio state University

Park, T. (2008). Effective design of advertisement in handheld devices (Doctoral dissertation .)PURDUE UNIVERSITY

Park, T. (2008). Effectiveness Design of Advertisement in Handheld Devices. Dissertation for .doctoral degree, Purdue University, West Lafyette India

Pergelova, A., D. Prior and J. Rialp, 2010 Assessing Advertising Efficiency Does the Internet Play a Role? Journal of Advertising, 39(3): 39-54

Pilelienė Lina Viktorija Grigaliūnaitė , (2017)," The effect of female celebrity spokesperson in FMCG advertising: neuromarketing approach ", Journal of Consumer Marketing, Vol. 34 - .Iss 3 pp

Rosemary J. Averya, Matthew D. Eisenbergb, Kosali I. Simon .(2012).The impact of direct-to-consumer television and magazine advertising on antidepressant use, Journal of Health Economics, Volume 31, Issue 5, September 2012, Pages 705–718

Selviaridis. Kostas, Martin Spring b, Luis Araujo b .(2013). Provider involvement in business service definition: A typology, Industrial Marketing Management 42 (2013) 1398–1410

Strebinger .Andreas, Xiaoling Guo, Ferdinand Klauser, Peter Grant-Hayd .(2018).Is Multi-Ethnic Advertising a globally viable strategy for a Western luxury car brand? A mixed-method cross-cultural study, Journal of Business Research 82 (2018) 409–416

Suvittawat, A., (2015). "Commercial Bank Service Quality: Chonburi Province, Thailand". .International Journal of Management Sciences and Business Research. 4(2): 2226-8235

Tafani. Eric, Elyette Rouxa, Rainer Greifeneder .(2018).In the mood for action: When negative program-induced mood improves the behavioral effectiveness of TV commercials, Journal of Business Research 84 (2018) 125–140

Te'eni-Harari, T. (2014). "Clarifying the relationship between involvement variables and .advertising effectiveness among young people". Journal of Consum Policy. 37: 183–203

Thomas, S. and Kohli, S.C. (2011). Can Brand Image Move Upwards after Sideways? A .Strategic Approach to Brand Placements. Business Horizons, 54 (1): 41-49

Yahyaoui . Hamdi, Own .Hala S..(2018). Unsupervised clustering of service performance behaviors, Information Sciences 422 (2018) 558–571

Yuksel, A., Yuksel, F. & Bilim, Y. (2010) "Destination Attachment: Effects on Customer Satisfaction and Cognitive, Affective and Cognitive Loyalty", Journal Of Tourism Management, Vol.31, pp: 274-284.

STUDY OF NATURAL ANTIOXIDANT INHIBITORS FOR CORROSION PROTECTION OF MILD STEEL IN ACIDIC MEDIUM

Ms. Ibtihal Hassan AL-Bahrani, National University of Science and Technology, Sultanate of Oman.

ABSTRACT

Corrosion is a natural phenomena and it is a very big loss for structure as well as business; so corrosion minimization is one of the biggest areas of research. In this project; corrosion inhibition efficiency of berries juice extract was tested in mild steel sample against the sulfuric acid medium which was prepared from four types of berries; blackberries, blueberries, cranberries, and raspberries.

The experiment was designed by using Design Expert V 11 where immersed time, acid and berries juice extract concentrations were set as factors and a limited range was selected. The corrosion rate was set as a response. 20 runs were under taken and the experiments were done in the lab with maximum care.

The corrosion rate of 1020.374 mpy minimized to 597.659 mpy in presence of 5ml of extract and the efficiency of berries juice extract reached to 41.427%. The RSM gave 232.022 mpy corrosion rate as a minimum value and the optimum equation given by RSM was used for Genetic Algorithm analysis which gave 188.866 mpy corrosion rate at the optimizing conditions.

The surface and chemical characterization of the mild steel sample in presence and absence of berries juice extract were analyzed along with the relations between temperature, immersed time, acid and berries juice extract concentrations with response. The mixing tank was designed, calculations and drawing were presented.

KEYWORDS

Corrosion, Mild steel, Berries, Inhibitor, Weight loss, Sulfuric acid, RSM, SEM.

INTRODUCTION

Mild steel is used in most of our daily life applications as well as in industrial applications. In other hands, mild steel has low corrosion resistance which corrodes easily once it reacts with the surrounding environment; corrosion is a phenomenon where the mild steel reacts with the deteriorates and surrounding environment. Corrosion of mild steel is a continuous and constant problem where it severe financial and life loses as it leads to loss in the product, efficiency of equipment and failures in some cases (Roberge,2012).

The corrosion phenomenon controlled by adding a small amount of component to the corrosive environment which is a corrosion inhibitor. There are organic and inorganic inhibitors. The inorganic inhibitors are used in limit amount as they are toxic such as nitrites, molybate, nitrate, silicate, chromate, and phosphate (Cafferty,2010). Now a day, the corrosion engineers looking for cheap, renewable, friendly, readily available and ecologically acceptable inhibitors which are organic inhibitors which comes from the plants such as antioxidant components consisting of tannins, saponins, flavonoids, alkaloids and amino acids. Organic inhibitors consist of some heteroatoms such as oxygen, nitrogen, multi-bound, and sulphur atoms with the conjugated system which allows the absorption in the metal surface. Those atoms attract the absorption of some components on the metal surface and that done by the interaction of metals and by the electron of the conjugated system or by forming a bond with the free electron of heteroatom which forms a layer in the metal surface, as a result, it decreases the corrosion rate (Roberge,2012).

Berries are small pulpy fruits which have up to 9.2 mmol of antioxidant per 100g of berries. Berries can be used as a corrosion inhibitor as they are rich in antioxidant. The studies show that the berries juice extract of blueberries, cranberries, raspberries, and blackberries rich with antioxidant where the antioxidant capacity of them reach to 10 mg/ 100 g of blueberries (Michalska&Lysiak,2015), 58 mg/100g of cranberries (Blumberg,J.,et al,2013), 26.2mg/ 100 g of raspberries (Bobinaite,R.,Viskelis&Venskutonis,2015), and 21 mg/100 g of blackberries (Kaume,L.,et al,2012).

OBJECTIVES

To prepare anticorrosive extract (berries), to study corrosion in sample in presences and absences of the inhibitors, to determine the optimum conditions to inhibit corrosion and to design mixing tank.

METHADOLOGY, DESIGN AND EXPERIMENTATION

PREPARATION OF MILD STEEL SAMPLES

The mild steel sheets were cut into 2.5*3 cm^2 by using steel cutter and cleaned by surface cleaning machine to remove any corroded products in the metal surface then 4mm hole done by using drilling machine. Bunches for all metal samples (1-20) were made to differentiate between them. The drilled samples were washed by distilled water then rinsed with benzene and acetone respectively and dried with a hot air dryer. They were weighted using sensitive balance and kept in the desiccator until the time of usage.

PREPARATION OF CORROSIVE MEDIUM

The corrosive medium of different concentrations of H_2SO_4 prepared by using 36.8 N of the sulfuric acid.

PREPARATION OF BERRIES JUICE EXTRACT

137g, 209g, 282g, and 200g of raspberries, cranberries, blueberries, and blackberries respectively have been used to produce 309.5 ml of berries juice extract. The different types of berries washed by cold water, crashed in the electrical crasher and compressed mechanically by hand then filtered by cloth filter. The filtrate filtered again by using filter paper. The berries juice extract kept in the freezer at 0°C till the time of usage.

The efficiency of berries juice extract has been tested by running three experiments. The cleaned and weighted mild steel samples immersed in 0ml(blank), 2ml and 5ml of berries juice extract respectively with 50 ml of 1M H_2SO_4. After 2 hours; the samples have been taken out and cleaned by distilled water, benzene and acetone then dried by air dryer and weighted again. The corrosion rate and inhibitor concentration have been calculated by using equations (1) and (2) respectively.

$$Cr(mpy) = 22300\times\frac{\Delta W}{\rho\times A\times t} \quad \textbf{equation(1)}$$

$$\mathrm{IE}(\%) = \frac{\text{initalcorrosionrate-finalcorrosionrate}}{\text{initalcorrosionrate}}\times 100 \quad \textbf{equation(2)}$$

DESIGN THE EXPERIMENT CONDITIONS

The experiments conditions have been designed by using Design Expert software-version 11 and following Response Surface Module(RSM); the range of the parameters have been given as inputs to the software which are inhibitor concentration (5-15 ml), immersed time (180-240 min) and acid concentration (1-2 M) and the response set as corrosion rate in mpy. Table(1) shows the experiment conditions given by the software.

Table(1): Designed runs given by RSM.

Run.No	Factor(A): inhibitor, ml	Factor(B): Time, min	Factor(C): Acid concentration, M
1	10	180	1
2	15	240	2
3	5	120	1
4	5	240	1
5	10	180	1.5
6	15	180	2
7	10	120	1.5
8	5	120	1.5
9	10	180	2
10	10	180	1.5
11	10	180	1.5
12	15	240	1
13	15	120	1
14	5	240	2
15	10	180	1.5
16	15	180	1.5
17	5	120	2
18	10	180	1.5
19	10	180	1.5
20	10	240	1.5

APPLYING THE DESIGNED RUNS IN THE LABORATORY

The weighted mild steel sample has been hanged by using a plastic thread in a 50 ml of sulfuric acid in presence of berries juice extract of known concentration and contact time as per the software; the batch has been kept in the shaker. After the contact time the mild steel sample cleaned by using 120 grade emery paper to remove any corroded products, washed by distilled water, rinsed by benzene and acetone respectively and dried by hot air dryer, weighted by using sensitive balance. The corrosion rate in mpy was calculated by using equation(1) where the weight losses results have been used.

STUDYING THE IMPACT OF TEMPERATURE IN CORROSION RATE

To study the impact of temperature in the corrosion rate; mild steel samples weighted and hanged for 2 hours in 50 ml of 1M sulfuric acid with 10ml of berries extract at different temperatures(30, 40, 50 and 60 °C) as shown in Table(2); each batch has been kept in the water bath shaker with heater to provide perfect contact between the solution and mild steel sample and to maintain the required temperature. Corrosion rate has been calculated by equation(1).

Table(2): Experiment conditions of temperature into corrosion rate.

SI.NO	Temperature (°C)	Inhibitor concentration (ml)	Acid concentration (M)	immersed time, (hours)
1	30	10	1	2
2	40	10	1	2
3	50	10	1	2
4	60	10	1	2

SURFACE AND CHEMICAL CHARACTERIZATION ANALYSIS

The mild steel surface of blank, absence and presence of berries juice extract samples have been studied by using Scanning Electron Microscopy (SEM) and Energy Dispersive Spectroscopy (EDS).

OPTIMIZATION STUDY BY GENETIC ALGORITHM

Matlab has been used for optimization where the genetic algorithm method has been followed. The actual equation form Design expert(RSM) has been used where "Optimtool" followed. The same limit of the conditions from RSM has been selected. The same previous procedure has been followed in the lab with maximum care by using those optimum conditions which given by the software, 1.001M of H_2SO_4 with 15 ml of berries juice extract immersed for 237.208 minutes.

DESIGN OF MIXING TANK

Design of mixing tank is a part of the project where the industrial scale mixing tank has been designed, the tank will be formed from high resistance material such as S.S to be able to resist the corrosive medium of sulfuric acid which will be mixed with berries juice extract. It is a container which used to mix various material (heterogeneous materials) together to form homogenous materials even the different viscosity materials can be mixed easily which provides continuous process, good control, easy cleaning, and simplest construction. It can be made from S.S material which provides a smooth surface and it can be easily cleaned (Petropedia,2018).

Parameters to be considered to design mixing tank are length, torque, weight, stiffness and hydraulic loading.(Dynamixianc,2015). The important factors in design mixing tank are tank type, tank volume, specific gravity, process and environment.(Ibryamoval,Y.et al,2009)

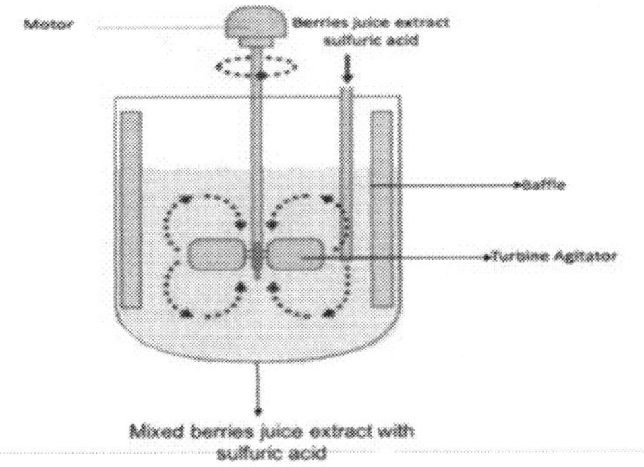

Figure(1): Mixing Tank(*Mixing Tank,*2018)

From the graph in Figure(2) Turbine agitator has been selected as it is suitable with the viscosity of the fluid and the tank volume. (Sinnot&Towler,2009)

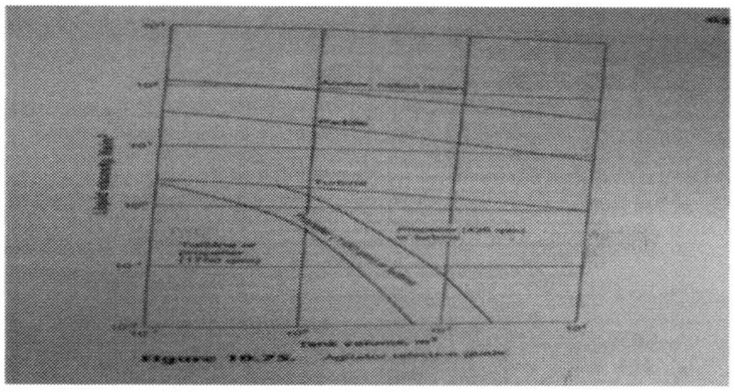

Figure(2): Graph of liquid viscosity Vs tank volume. (Sinnot&Towler,2009)

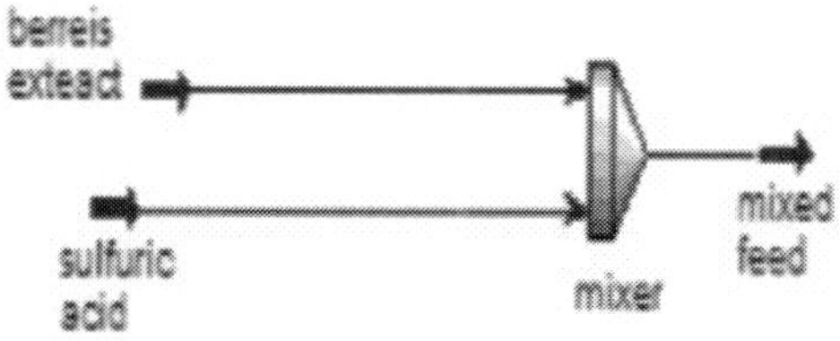

Figure(3): Result of Mixing Tank by Aspen Hysys.

Table(3): Calculated parameters for designing mixing tank

Tank diameter	m	3.5
Tank Height	m	1.81
Across sectional area	m^2	9.62
Volume of the tank	m^3	17.5
Impeller diameter	m	1.16
Length of paddle	m	0.29
Width of paddle	m	0.23
power	watt	204085
Reynolds Number	-	136486.6391

RESULTS AND ANALYSIS

EFFECT OF BERRIES JUICE EXTRACT CONCENTRATION INTO CORROSION RATE AND INHIBITOR EFFICIENCY

As shown in Table(4) the corrosion rate in the absence of berries juice extract inhibitor reach to the maximum (1020.374mpy); the corrosion rate has been minimized to 898.117mpy and 597.659mpy in presence of 2 and 5 ml of berries juice extract inhibitor respectively which means that berries juice extract inhibits the corrosion as the concentration of the berries juice extract increases. As well as the inhibitor efficiency reaches to 41.427% which obtained at 5ml of berries juice extract inhibitor in 1M of sulfuric acid immersed for 2 hours. The adsorption of the organic compounds such as ascorbic acid that present in the berries juice extract is blocking the active sites on the mild steel samples as it is forming a protective film on the metal surface as what kumar.A,2012 mention in his studies so the corrosion rate keeps decreasing as the amount of the berries juice extract is increasing. (Kumar.A,2012) Those compounds have heteroatoms such as oxygen which is a center of adsorption. In addition; It forms a coordinate bond as it has the ability to transfer the electrons to the vacant d-orbital of mild steel from the heteroatoms. (Ostovari, L. et al,2009)

Table(4): Experimental results of testing the efficiency of berries juice extract.

SI.NO	Inhibitor concentration (ml)	Corrosion rate (mpy)	Inhibitor efficiency (%)
1	0	1020.374	-
2	2	898.117	11.976
3	5	597.659	41.427

EFFECT OF IMMERSED TIME INTO CORROSION RATE AND INHIBITOR EFFICIENCY

The maximum corrosion rate reaches to 765.005mpy which obtained at 5ml of berries juice extract in the presence of 50 ml of 2 M of H_2SO_4 at 2 hours. Since the minimum corrosion rate of 232.022mpy obtained in presence of 5ml of berries juice extract with 50 ml of 1 M of H_2SO_4 immersed for 4 hours. The effect of immersed time into corrosion rate can be observed at low acid concentration and in the presence of berries juice extract, the corrosion rate is minimized at high immersed time. The decrease in the corrosion rate with the increase in immersed time is related to the adsorption of the corrosion product ($FeSO_4$) as shown in equation(3); According to Xianghong,L. et al ; the corrosion product of $FeSO_4$ becomes thicker with the immersed time, so it form a protective film (barrier) in the surface of mild steel (Xianghong, L. et al,2009).

$$Fe_{(s)}+H_2SO_{4(aq)} \rightarrow FeSO_{4(aq)}+H_{2(g)}$$, **equation(3)**

EFFECT OF ACID CONCENTRATION INTO CORROSION RATE

The effect of acid concentration into corrosion rate has been studied by using 1, 1.5 and 2 M of H_2SO_4 at different inhibitor concentrations and immersed time where the corrosion rate increase with the acid concentration at low inhibitor concentration and immersed time which reach to 765.005mpy at 2M H_2SO_4, 5ml of berries juice extract which immersed for 120 minutes as a maximum corrosion rate as shown in Table (3). In presence of the corrosive medium of H_2SO_4 with less amount of berries juice extract; the mild steel surface was exposed to H_2SO_4 medium where the corrosion reaction started as shown in equation(3) and $FeSO_4$ was formed with H_2 bubbles.

EFFECT OF TEMPERATURE INTO CORROSION RATE

From Table(5) can be noted that the rate of corrosion increases with temperature where it reaches to maximum 2407.994mpy at maximum temperature of 60°C. According to Ambrish, S. et al, at high temperature the berries juice extract molecules were partially absorbed in the mild steel surface so it was exposed to the corrosive medium of H_2SO_4, as a result, increasing the corrosion rate (Ambrish,S. et al,2012).

Table(5): Results of studying the effect of temperature into corrosion rate

SI.NO	Temperature,° C	Corrosion rate, mpy
1	30	701.507
2	40	1666.404
3	50	2008.861
4	60	2407.994

ANALYSIS BY RSM

Once the experimental results and corrosion rate calculated the RSM analysis has been done where the corrosion rate has been given to the software as a response. The response must be modified where the results fit with linear. The fit summary shows that the model F-value and P-value of 12.66 and 0.0001 respectively imply that the model is significant where A, B, C, and B^2 are the significant terms in this model only. 1.48 is a lack of fit P-value which indicates that the fit is not significant as there are pure errors.

0.5562 of the predicted R^2 is a reasonable agreement with 0.7105 of the adjusted R^2. The signal to noise ratio of 13.7252 of Adeq precision that the ratio grater than 4 is desirable. (Mycourses,2016)

The process variables are studied using ANOVA. The estimate coefficients in term of coded factors are A-Inhibitor, B-time, C-Acid and B^2. It is creating an accompanying condition for the sample with 1.0000 correlation coefficient. The coded and actual equations given by software are shown in equation(4) and (5) respectively.

- Corrosion rate(mpy)= 370.52-40.52(A)-126.97(B)+ 80.85(C) +66.58$(B)^2$ **equation(4)**
- Corrosion rate(mpy)= 1189.15487-8.10479(Inhibitor) -8.77421(Time) +161.69302 (Acid Conc)+0.018495 $(\text{Time})^2$ **equation(5)**

The various relations can be studied from one factor graph; it can be noted that the corrosion rate increases with acid concentration and decreases with the increase in the amount of berries juice extract and immersed time as can be seen from the plots in Figure(3):Plot pf corrsoion rate Vs inhibitor concntration, Figure(4): plot of corrosion rate Vs acid concentration and Figure(5):Plot of corrosion rate Vs immersed time respectively.

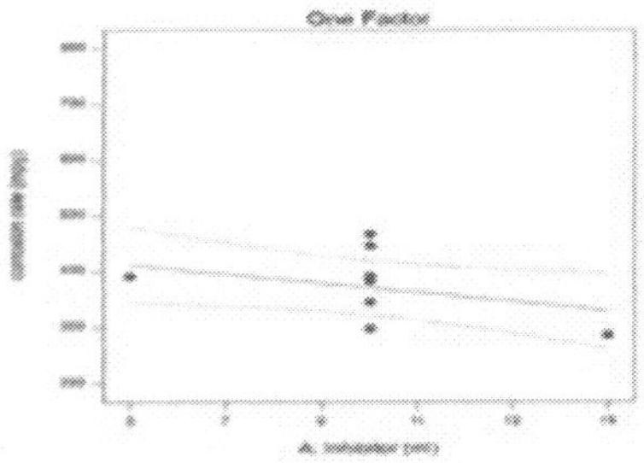

Figure(3)

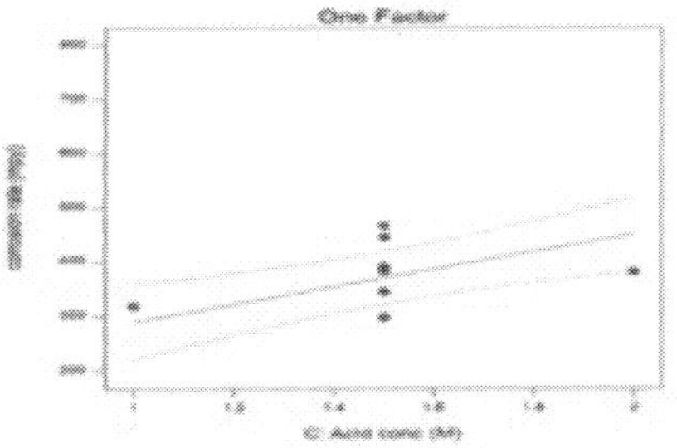

Figure(4)

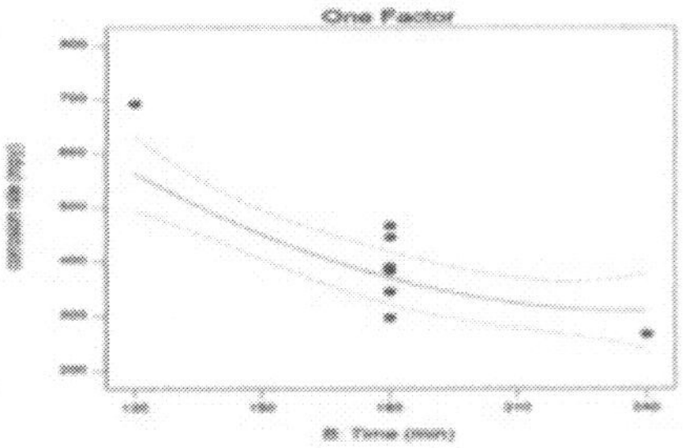

Figure(5)

The effects of two factors into the corrosion rate can be studied from the 3D plots as shown in Figure(6): 3D plot of corrosion rate Vs inhibitor concntration and acid concntration and Figure(7): 3D plot of corrosion rate Vs inhbitor concntration and immersed time.

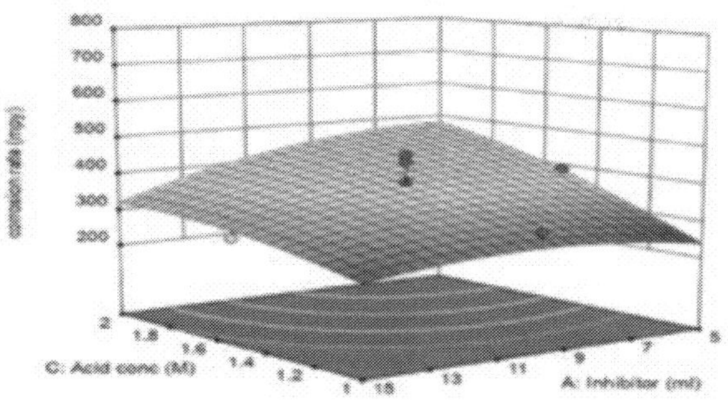

Figure(6)

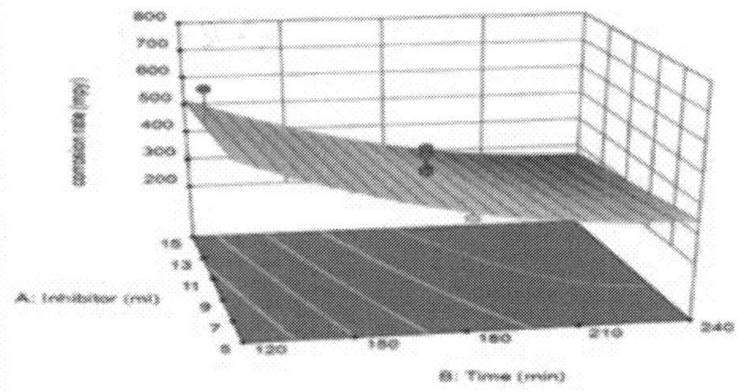

Figure(7)

The optimum conditions given by the RSM of 1M of H_2SO_4 with 5ml of berries juice extract inhibitor runs for 240 minutes gives 213.402mpy as a minimum corrosion rate but the experimental results at the same conditions gives 232.022mpy which means that they are too close as the suggested solution considered as an excellent optimum conditions.

SEM ANALYSIS

The SEM has been done for the blank mild steel sample for facilitating comparison as shown in Figure(8) which shows smooth mild steel surface compare to others. Rough surface in absence of berries juice extract is shown in Figure(9); where corrosion starting with the formation of H_2 bubbles as per the equation(3) which means that the surface of mild steel is exposed to the H_2SO_4 medium. In presence of 15 ml of berries juice extract, the surface of mild steel is protected by the adsorption of the organic compounds that are present in the berries juice extract where fewer cracks and pits can be noted and smooth surface is shown in Figure (10), which means that the mild steel surface is protected. (Raja, P. et al, 2008)

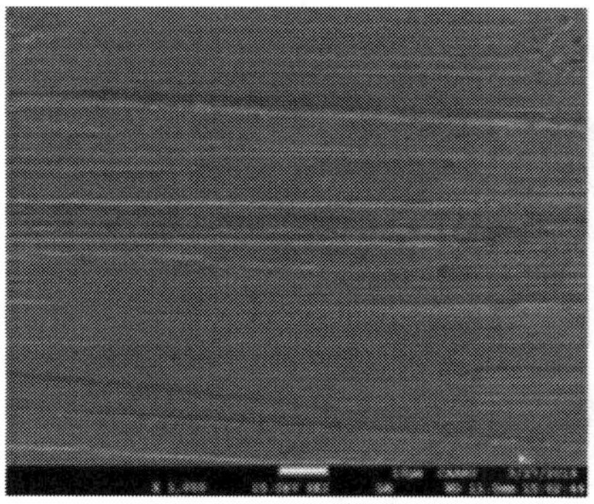

Figure(8)

Figure(9)

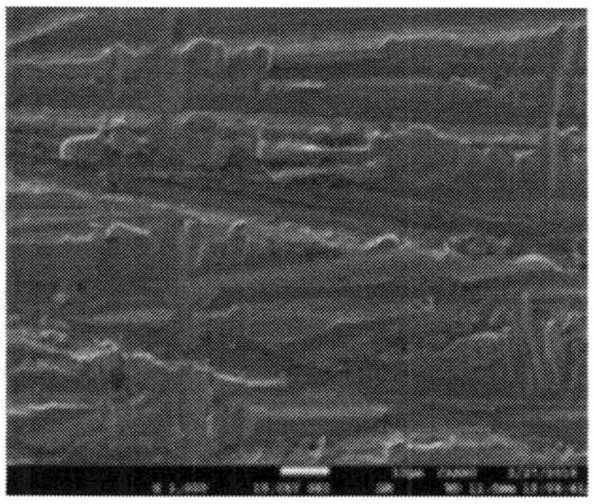

Figure(10)

EDS ANALYSIS

EDS has been used to study the chemical characterization or elemental analysis; Figure (11) shows the blank mild steel sample before being immersed in 1M H_2SO_4 in absence of berries juice extract where high Fe peak is shown, which indicates the presence of high amount of iron which reaches to 88 Wt% in the mild steel surface. Figure (12) shows the mild steel surface that has been immersed for 2 hours in 50 ml of 1M H_2SO_4 in absence of berries juice extract; the mild steel surface contains low amount of Fe (87.3 Wt%) compared to the blank mild steel sample. In Figure (13) the mild steel sample immersed for 2 hours in 50 ml of 1M H_2SO_4 in the presence of 5ml of berries juice extract, it shows high spark of Fe where the mild steel surface contains lower amount of Fe (84.8 Wt%) compared to Figure(11) and (12) which indicate lower iron dissolution rate in 1M H_2SO_4 medium containing berries juice extract (Alaneme, K. et al,2016). A high amount of oxygen and carbon are observed in presence of berries juice extract compared to the absence of berries juice extract. Si is shown in the absence and presence of berries juice extract which means that's it comes out to the surface during the reaction with sulfuric acid.

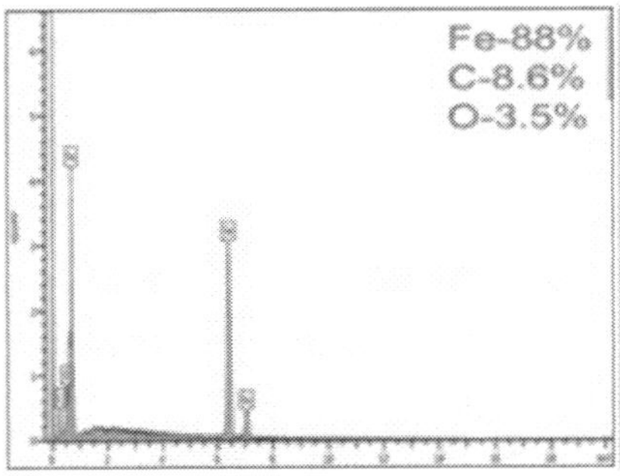

Figure(11)

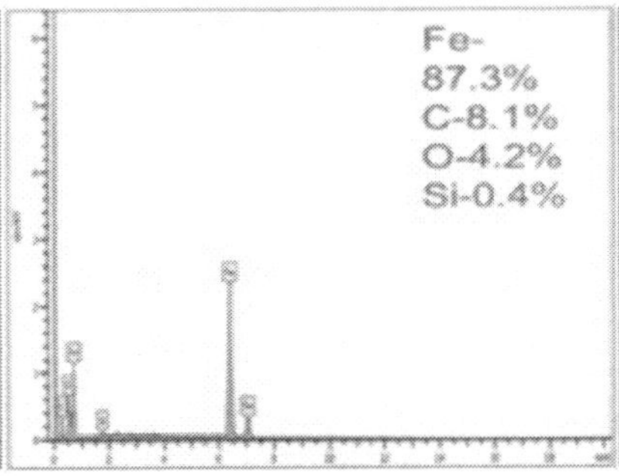

Figure(12)

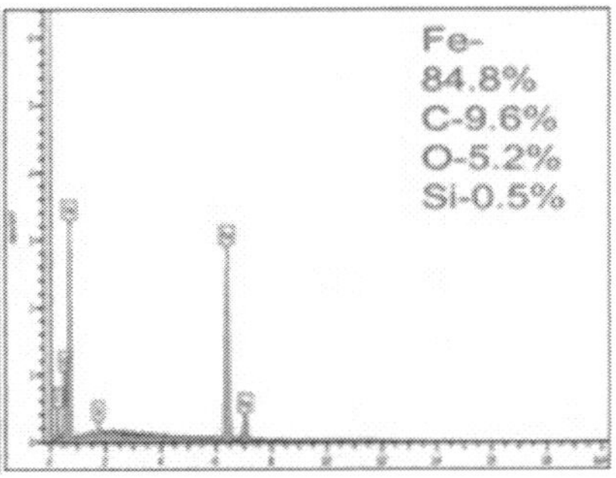

Figure(13)

GENETIC ALGORITHM OPTIMIZATION

The experimental result gives 288.367mpy as a corrosion rate. The software gives the minimum corrosion rate value of 188.866mpy from 50 runs as shown in Figure(14). A small difference can be noted as the genetic algorithm is not the exact fit with linear equations so the experimental and software results are too close to each other.

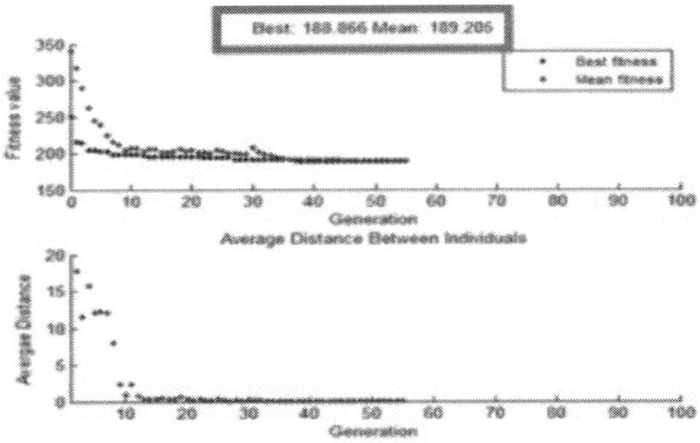

Figure(14): Genetic Algorithm optimization.

CONCLUSION

In this project, all the objectives have been achieved by designing and experimenting.

Firstly; the anticorrosive extract (berries) has been prepared, the prepared extract which contains 4 types of berries namely blueberries, blackberries, cranberries, and raspberries inhibit the mild steel corrosion in sulfuric acid medium with 41.427% efficiency in presence of 5 ml of berries extract.

Secondly, the mild steel corrosion in the presence and absence of berries juice extract were studied where the corrosion rate reaches to 1020.374mpy in absence of berries juice. By using 5 ml of berries juice extract the corrosion rate minimized to 232.022mpy.

Thirdly, the optimum conditions to inhibit the mild steel corrosion were studied by Design Expert V11. At the optimum conditions of 1M of H_2SO_4. 5ml of berries juice extract, 240 minutes and at room temperature the corrosion rate reaches to 232.022mpy as a minimum value.

The actual equation obtained from Design Expert V11 was later optimized using Genetic Algorithm. Genetic Algorithm shows that at 1.001M of H_2SO_4, 15 ml of berries juice extract and 237.208 minutes at room temperature, the corrosion rate is minimized to 188.866mpy.

Finally, mixing tank has been designed with 1.18m height and 2.5m diameter and a turbine agitator has been selected.

ACKNOWLEDGEMENT

I would like to thank god for giving me patience and power to work in my most important part of my life, my father, mother, my brother and my friends for their help and support. I would also like to express my sincere gratitude to Dr. Priy Brat Dwivedi for his support and help in developing my project idea, Mr. M.J.Varghese for his help in RSM analysis, Mr. Hatem Yazidi and Dr. Moammar Ali AlTubi for providing me the technical knowledge in Matlab optimization, Mr.Mohammed Salim Adeeb in providing me the technical help in the laboratory, Ganesh Nagraj Patil, Dr.Lakkim Setty

Rao, and Dr.Santosh Madhav Walke for their help in the calculations of design mixing tank. Last but not the least, I would like to thank National University of Science and Technology-College of Engineering for giving me this opportunity to complete the project in their laboratories and supporting me by providing me with the required chemicals.

REFERENCES

1.Alaneme,K. et al .,2016.Corrosion Inhibitory Properties of Elephant Grass(Pennisetum Purpureum) Extract Effect on Mild Steel Corrosion in 1M HCl Solution :*Alexandria Engineering Journal*.[e-journal]. 55 (8). p. 1069-1076. Available from: https://www.sciencedirect.com/science/article/pii/S1110016816300242.[Accessed: 8th May 2019].

2.Ambrish,S. et al .,2012. The Effect of Environmentally Bening Fruit Extract of Shahjan(Moringa Olifera) on the Corrosion of Mild Steel in Hydrochloric Acid Solution:*Journal of Chemical Engineering Communications*.[e-journal]. 199 (6). p. 63-77. Available from:https://www.tandfonline.com/doi/abs/10.1080/00986445.2011.570390.[Accessed: 6th May 2019].

3.Blumberg,J.,et al.,2013.Carnberries and Their Bioactive Constituents in Human Health:*Journal of Advances in Nutrition*.[e-journal]. 12 (7). p. 618-632. Available from: https://academic.oup.com/advances/article/4/6/618/4595566.[Accessed: 1th May 2019].

4.Bobinaite,R.,Viskelis,P.&Venskutonis,P.,2015.Chemical Composition of Raspberry Cultivars:*International Journal of Molecular Sciences*.[e-journal]. 16 (8). p. 18642-18663. Available from: https://www.researchgate.net/publication/306158828_Chemical_Composition_of_Raspberry_Rubus_spp_Cultivars.[Accessed: 1th May 2019].

5.Cafferty,E.,2010.*Introduction to Corrosion Science*.United States:Springer.
Roberge,P.,2012.*Handbook of Corrosion Engineering*.2nd edition.United States:McGraw-Hill company.
6.Dynamixianc,2015.*Optimal Mixer Design* .[Online]. Available from: https://www.dynamixinc.com/optimal-mixer-design. [Accessed: 14th December 2018]

7.Ibryamoval,Y. et al .,2009.Design of Mixers for Chemical and Biochemical Industry: *Trakia Journal of Sciences*.[e-journal]. 8 (6). p. 6-11. Available from: http://tru.uni-sz.bg/tsj/Vol8.Suppl.3.2010/Y.Ibryamova.pdf.[Accessed: 13th December 2018].

8.Kaume, L. et al.,2012. The Blackberry Fruit A Review on Its Composition and Chemistry, Metabolism and Bioavailability and Health Benefits:*Journal of Agricultural and Food Chemistry*.[e-journal]. 37 (7). p. 5716-5727. Available from: https://www.ncbi.nlm.nih.gov/pubmed/22082199.[Accessed: 6th May 2019].

9.Kumar.A,2012. Inhibition of Mild Steel Corrosion in HCl Solution Using Pipali(Piper Longum)Fruit Extract:*Arabian Journal for Science and Engineering*.[e-journal]. 38 (9). p. 85-97. Available from: https://link.springer.com/article/10.1007/s13369-012-0409-9.[Accessed: 9th May 2019].

10. Michalska,A.& Lysiak.G .,2015.Bioactive Compounds of Bluberries:*International Journal of Molecular Sciences*.[e-journal]. 16 (8). p. 18642-18663. Available from: https://www.ncbi.nlm.nih.gov/pmc/articles/PMC4581264/.[Accessed: 1th May 2019].

11.*Mixing Tank*,2018.[Online]. Available from: https://www.conceptdraw.com/examples/agitator-tank-symbol.[Accessed: 19th November 2018].

12. Mycourses,2016. *Response Surface Modelling-RSM*.[Online]. Available from:https://mycourses.aalto.fi/pluginfile.php/364340/mod_resource/content/1/BOS_Lecture3_2016.[Accessed: 4th May 2019].

13. Ostovari, L. et al.,2009.Corrosion Inhibition of Mild Steel in 1M HCl Soluition by henna extract:*Journal of Corrosion Science*.[e-journal]. 32 (9). p. 1935-1949. Available from: https://www.sciencedirect.com/science/article/pii/S0010938X09002157. [Accessed: 6th May 2019].

14. Petropedia,2018. *Mixing Tank*.[Online]. Available from: https://www.petropedia.com/definition/7742/mixing-tank.[Accessed: 11th December 2018].

15.Raja,P. et al.,2008.Inhibitive Effect of Black Pepper Extract on the Sulphuric Acid Corrosion of Mild Steel:*Journal of Materials Letters*.[e-journal]. 62 (9). p. 1977-2979. Available from: https://www.sciencedirect.com/science/article/abs/pii/S0167577X08001316.[Accessed: 8th May 2019].

16. Sinnot,R.&Towler,G.,2009.*Chemical Engineering Design*.5th edition.India:Elsevier.

17. Xianghong,L. et al.,2009.Inhibition Effect of 6-benzylaminopurine on the Corrosion of Cold Rolled Ssteel in H_2SO_4 Soluition:*Journal of Corrosion Science*.[e-journal]. 51 (9). p. 620-634. Available from: https://www.sciencedirect.com/science/article/pii/S0010938X09000080. [Accessed: 6th May 2019].

590434 Verification of Firefighters' Heuristics and Predictive Model Development through Big Data Analysis

Abstract:

The heuristics accumulated in the field activities of firefighters are reviewed through big data analysis of fire occurrences and emergency responses in Gyeonggi-do Province, Korea and expanded as predictive models to be utilized for proper fire prevention activities according to time, day, and target through quantitative modeling. Empirical rules with high sympathy are collected through direct interviews with firefighters. Among these, three types of high-priority rules are compared and analyzed for big data such as fire occurrence and damage history in Gyeonggi-do in 2018. Big data comparison analyses are conducted, including the number of fires and damages that occurred in Gyeonggi-do in 2018. Furthermore, fire occurrence patterns by region, day of the week, time of day, and building type are derived. In addition to using empirical rules that have been validated through research, relatively inexperienced firefighters also can make better decisions by relying on refined quantitative predictive models replacing empirical rules, including local government and time-based factors that reflect big fire occurrence data.

Copyright © American Institute of Chemical Engineers. All rights reserved.

CONVERTING SHELTER-IN-PLACE LOCATIONS TO ZERO VULNERABILITY SAFE HAVENS

J. A. Rau
MineARC Systems America, LLC
4850 W. Ledbetter Drive
Dallas, Texas 75236
USA

A. Young
MineARC Systems, Pty Ltd
274 Welshpool Rd
Welshpool, WA 6106
Australia

M. McDermott
High Performance Building Solutions
7415 Folger Dr.
Charlotte, NC 28226
USA

CONVERTING SHELTER-IN-PLACE LOCATIONS TO ZERO VULNERABILITY SAFE HAVENS

ABSTRACT

While it is not uncommon to find shelter-in-place or safe haven locations designated across chemical facilities, what varies is the level of protection offered at these sites. Functionality can range from merely a muster point (shelter-in-place) to a positively pressurized room (safe haven). Either way, when determining the room's acceptable performance criteria, the industry generally applies ERPG-3[1] or a vulnerability factor. An acceptable vulnerability factor is often deemed the probability of a fatality being less than one percent.

In 2017, MineARC Systems and High Performance Building Solutions (HPBS) began working with a leading energy company to convert five shelter-in-place spaces to safe havens, which included additional life support systems that would ensure zero percent vulnerability. The objective of this paper is to examine the process to harden the space against contaminated air intrusion and what critical life support systems are essential for converting a shelter-in-place location to a safe haven.

INTRODUCTION

Emergency shelter-in-place (SIP) forms an integral part of any plant's broader emergency response plan. Explosions, fires, and the release of smoke or other toxic gases are some of the types of incidents that can occur despite high levels of planning and safety precautions[2].

Numerous government agencies and other organizations recommend shelter-in-place as a measure to reduce harm in the event of a chemical release. Any building may be used as a temporary measure to reduce health risk from exposure to the toxic materials by simply closing windows and doors and turning off ventilation fans and air conditioning/heating systems.

Three factors govern the effectiveness of a SIP location: 1) the tightness of the building or leakage rate, 2) the concentration of the toxic gas, and 3) the release duration. The tightness of the building is primarily based on building construction; however, temperature differential and wind speed also play a factor. Identification of the maximum toxic concentration occurs during a Facility Siting Study (FSS). FSSs are a hazard analysis that defines Maximum Credible Event (MCE) scenarios using a consequence-based approach to produce a quantitative risk assessment. Determining potential consequences is a necessary step in the process of developing a comprehensive safety plan. The FSS determines the expected release duration based on stored inventories of toxic materials. The higher the toxic concentration and longer the release duration, the less effective sheltering in place becomes.

While occupants may feel safe sheltering-in-place from the external toxic hazard, typical buildings have high leakage rates. A typical wood framed constructed building can have 3-5 air changes per hour (ACH50[3]), rendering it ineffective against most chemical releases. Additionally, sealing to an acceptable airtightness can be time consuming and expensive and even with professional air sealing it is very difficult to air seal existing structures to an acceptable tightness.

A case study for the conversion of five shelter-in-place spaces to zero percent vulnerability safe havens is presented. There are considerations for effectively sealing existing infrastructure and ensuring positive pressure to ensure contaminants do not infiltrate safe haven locations along with the critical element of life support. The critical elements of life support that are of primary importance to human survival in a safe haven are:

- Positive pressure
- Breathable air supply
- Supplemental O_2
- Removal of accumulated CO_2

[1] Emergency Response Planning Guidelines (ERPGs) are values developed by the AIHA Guideline Foundation's Emergency Response Planning Committee. ERPG-3 is the maximum airborne concentration below which it is believed nearly all individuals could be exposed for up to 1 hour without experiencing or developing life-threatening health effects.

2 The CSB is an independent federal agency charged with investigating industrial chemical accidents. Over the last 10 years they have completed investigations into 15 separate chemical release incidents.

[3] ACH50 is the number of times the air volume in a building changes per hour at 50 Pascal of pressure. Commercial air leakage rates are commonly reported at 75 Pascals, but this leakage study was conducted at 50 Pascals based on the targets included in the FSS report.

- Cooling and dehumidifying

CASE STUDY

In 2017 MineARC Systems and its project partner HPBS began working alongside a major energy company to convert five shelter-in-place locations into zero percent vulnerability safe havens.

The process started with the energy company completing its FSS, which included plume modeling to define the chemical release concentrations at each building location along with ideal shelter-in-place locations[4] and internal minimum positive pressure[5]. This information was used by MineARC to calculate indoor concentrations and occupant vulnerability to highlight that shelter-in-place alone would result in the vulnerability of greater than zero percent.

From there, HPBS conducted air infiltration testing following ASTM E779-10 to determine an air leakage baseline (see Table 6) for rooms that were designated to be sealed (pre-weatherization). A site assessment was then completed to review each location and determine its suitability for sealing and conversion to a safe haven. Firstly, the measured baseline infiltration rates were compared to the FEMA 453 estimates based on construction type[6] (see Figure 2). All surfaces of the room were examined for ease of conventional air sealing and mechanical equipment sealing. Rooms that included false floors, large amounts of electrical conduit, cabinetry, computer servers or windows etc. were deemed difficult to seal (i.e. control rooms). Secondly, the building or rooms open floor space was checked to ensure it could accommodate the specified number of people in an emergency. FEMA 453 section 1.6 recommends 20ft^2 per person for durations of less than 24 hours. With the FSS recommending 4 hours, it was deemed acceptable to have an unrestricted floor space of 6ft^2 per person.

Manual weatherization (air sealing) of the rooms included installing a cleanroom suspended ceiling system with gaskets and tiles that met the air permeance rate for an air barrier material, cleanroom airtight light fixtures with gaskets, sealing around electrical and plumbing, air sealing all visible air pathways to make sure the interior gypsum board was functioning as the air barrier, installing low-leakage HVAC dampers, installing an emergency shut off for the mechanical system and adding door weather-stripping and sweeps. Post-weatherization air leakage tests were conducted to determine the air infiltration rate of the SIP spaces after completion of the manual air sealing, cleanroom suspended ceiling system and HVAC work. The weatherization air sealing methods improved the air barrier integrity for all the proposed SIP spaces by between 57% and 65% (refer to Table 6).

Aerosol-based envelope sealing was used to reduce leakage rates even further. AeroBarrier® is a building envelope sealing technology that simultaneously measures and seals buildings. AeroBarrier® works by forcing an atomized sealant into orifices under pressure gradually sealing leaks. Rooms were sealed during night shift with personnel able to re-access the rooms by morning. The AeroBarrier® air sealing methods improved the air leakage for all the SIP spaces between 7% and 88% over the post manual weatherization leakage rates (refer to Table 7). AeroBarrier® is not effective for sealing large holes so manual air sealing is required for certain locations prior to using AeroBarrier®.

MineARC engineers once again calculated infiltration concentration rates along with the vulnerability rate. As the probability of a fatality was still higher than zero, MineARC then specified its AirBANK Pressure System and necessary carbon dioxide and oxygen control equipment.

SAFE HAVEN LIFE SUPPORT EQUIPMENT

Primary Life Supporting Atmosphere

A safe haven needs to provide a reliable breathable atmosphere independent of external toxic gas concentrations. Human breathing consumes oxygen (O_2) and expels carbon dioxide (CO_2), carbon monoxide (CO), water vapor, and heat. At complete rest, an average person exhales air at 17% O_2 and 3.2% CO_2. This breathing rate results in a 'resting' O_2 uptake of 0.3L/min of O_2 and a resting CO_2 discharge of 0.24L/min (Brake, Fellows, & Bates, 1999). Taking into consideration the above values, a person in a closed system such as a shelter-in-place succumbs to CO_2 poisoning well before O_2

4 Shelter-in-place location is beyond the scope of this paper, but should be determined based on the chemical hazard, facility layout, and location of personnel.

5 Positive pressure requirements are based on average wind speed at the facility. As wind speed increases so does the SIP leakage rate.

6 The FEMA 453 estimates are intended to estimate filter sizing when ASTM E779 air leakage testing is not conducted.

asphyxiation. Venter et al. (1998) found an average oxygen consumption for 12 entrapped individuals of 0.44 liters per minute, at STP, over 24 hours (including sleeping). A figure of 0.5 liters per minute over a non-sleeping entrapment period of 12 hours is a prudent, conservative design value. This corresponds to a metabolic rate of about 160W or $80W/m^2$ for a typical person with a body skin area of about $2m^2$, which is equivalent to a breathing rate of about 12.5 liters per minute of fresh air.

Consequently, life support systems are needed within a safe haven that can ensure positive pressure keeps contaminants outside, along with replacing the O_2 consumed, and removing exhaled CO_2, moisture, and heat to maintain the atmosphere within acceptable threshold limits.

Positive Pressure

Positive pressure is integral to prevent the ingress of airborne contaminants that may compromise the breathable atmosphere for occupants. Establishing a Class 1 Safe Room (FEMA 453) requires the sufficient release of filtered air from an external source, to form a positively pressurized zone.

Buildings are typically constructed to provide natural or forced ventilation to allow permanent occupancy and potentially combined with large exposed windows and surfaces which can present challenges for ensuring protection during a chemical release. When determining adequate positive pressure within a space, several structural factors need to be considered, including:

- Shelter volume
- HVAC (Heating, Ventilating, and Air Conditioning)
- Door and window tightness
- Existing opening such as vents and elevator shafts
- Indoor and outdoor temperature
- Wind speed and building orientation
- The external screening of the structure

Building design and multiple ingress points suggest the conversion of a smaller interior room within a building can provide a higher level of protection, than the whole building, as it is less subject to wind and buoyancy forces which induce the infiltration of external hazards.

Breathable air from compressed air cylinders venting directly into the SIP forms a positive internal pressure. Determining the amount of air required to reach the necessary level of positive pressure involves measuring the areas internal volume and the relationship between the internal and external atmospheric conditions.

After reaching required ACH (air changes per hour) or tightness, a positive-pressure management system must be sized for the expected shelter duration.

Breathable Air Supply

If the shelter-in-place is airtight, a supply of breathable air is necessary to support occupants over the required shelter duration. Unventilated SIPs that are tightly sealed cannot be occupied for long periods without the risk of high carbon dioxide levels. A secondary source of breathable air must be circulated within the shelter to meet occupancy over the required amount of time.

Clean, breathable air is supplied to a chemical, biological and radiological (CBR) safe room in a few ways: by trapping air inside the room and minimizing the air exchange, passing contaminated air through a filter (not practical for high concentration chemical releases or for chemicals that cannot be filtered) to purify it as it is supplied to the room, and the supplemental supply of oxygen through high pressure compressed air cylinders.

The air supply must contain enough oxygen in its mix, and not contain contaminants like carbon monoxide, carbon dioxide, water, or any other toxins that poison safe haven occupants during entrapment.

Table 1 lists international standards for flowrates for breathable compressed air for self-contained breathing apparatus (SCBA), supplied air respirators (SAR) and refuge chambers.

Table 1 – Standards for acceptable air supply requirements

Air supply requirements	Relevant Standard
Open-circuit positive-pressure SCBA, including all components and accessories except the air cylinder (shell) must resist the permeation and penetration of HD and GB chemical agents when tested for a period of 6 hours on an upper-torso manikin connected to a breathing machine operating at an airflow rate of 40 liters per minute (lpm), 36 respirations per minute, 1.1 liters tidal volume.	NIOSH 42 CFR Part 84 Respiratory Protective Devices
The flow rate should be at least 85L/min (3cfm) per person to remove expired CO2 and CO but at least 255L/min (9cfm) to provide adequate cooling and maintain comfortable conditions. The air inflow is assumed to be at a temperature of 26.5°C and 65% relative humidity	ITA Report No. 14 – Guideline for the provision of refuge chambers in tunnels under construction, May 2014
For manually operated blower type air-hose respirators, the blower shall be capable of delivering air to the space enclosed by the facepiece at a flow rate of at least 85 L/min.	AS/NZS 17:16:2003 Respiratory Protective Devices

The NIOSH standard of 40L/min is for SCBA apparatus (i.e. mouthpiece or face mask) where dilution of CO_2 is not a concern. Conversely, the ITA and Australian standards specify 85L/min and are concerned with both oxygen consumption and carbon dioxide dilution. It is important to note that three times the amount of air is required to maintain comfortable conditions inside a refuge chamber, when compared to simply breathing (Health & Safety Executive UK, 2006).

Unless the positive pressure system is supplying breathable air at a rate that maintains oxygen and compressed gas levels within acceptable ranges (dependent on safe haven occupancy and duration), a CO_2 scrubber and supplementary oxygen supply are required.

Supplemental O_2

O_2 is an essential requirement for cellular activity, and therefore, the absence of O_2 for relatively short periods results in death. Humans within an enclosed environment consume O_2 and if not replaced, would result in an irrespirable atmosphere. An O_2 concentration of less than 11% poses a high risk of death.

Retaining the oxygen levels between acceptable limits is critical. Various international standards outline acceptable O_2 limits for a breathable environment; examples are shown in Table 2.

Table 2 - Standards for allowable O_2 levels

Minimum O_2 Content (% by volume)	Maximum O_2 Content (% by volume)	Relevant Standard
18.5%	23.0%	Mines Safety and Health Administration 30 CFR Parts 7 and 75 Refuge Alternatives for Underground Coal Mines; Final Rule
19.5%	23.5%	Occupational Safety and Health Administration 29 CFR 1910.146 Permit-Required Confined Spaces for General Industry
19.5%	23.5%	Australian Standard AS 2865–2009 Safe working in a confined space

OSHA and Australian Standards for working in a confined space specify an acceptable lower limit of 19.5%, which is consistent with many other standards around the world. The slightly lower MSHA limit of 18.5% is based on a person at rest, not working, therefore, not likely to have trouble in breathing; conditions that are likely to be present in a refuge chamber.

O_2 itself is not flammable but supports and accelerates combustion. OSHA defines O_2 enriched mixtures or atmospheres as those containing more than 23.5% O_2 by volume (OSHA, 2013). In O_2-enriched atmospheres, the reactivity of O_2 significantly increases the risk of ignition and fire.

Within a safe haven, pure O_2 must be utilized in conjunction with a CO_2 removal system.

Removal of Accumulated CO_2

Protection within a safe haven is substantially increased by filtering CO_2 from the air. CO_2 is being produced continuously by the cells of our body and is a byproduct of respiration. The inability to expel this gas or an environment with high concentrations of CO_2 leads to CO_2 poisoning. Breathing air with more than 3% CO_2 can induce dizziness, headaches, and vision disturbance. Prolonged exposure at greater than 6% could result in unconsciousness and death. Listed in Table 3 are examples of international standards of satisfactory CO_2 removal requirements for a breathable environment.

Table 3 – Acceptable CO_2 removal requirements

CO_2 Removal Requirements	**Relevant Standard**
CO_2 removal components shall remove CO_2 at a rate of 1.08 ft^3 (30.5 L)/h/person.	Mines Safety and Health Administration 30 CFR Parts 7 and 75 Refuge Alternatives for Underground Coal Mines; Final Rule
OSHA has established a Permissible Exposure Limit (PEL) for CO2 of 5,000 parts per million (ppm) (0.5% CO2 in air) averaged over an 8-hour workday (time-weighted average or TWA.)	Occupational Safety and Health Administration
5.2.8 Atmospheric monitoring. Atmospheric monitoring There should be the capability to monitor temperature continuously, and for carbon monoxide, carbon dioxide, and oxygen in the chamber during use in both externally supported or standalone modes. The concentration of carbon dioxide should not exceed 10000 ppm.	ITA Report No. 14 – Guideline for the provision of refuge chambers in tunnels under construction, May 2014

The CO_2 that is exhaled by the occupants within the safe haven must be removed by, constant flushing by an external breathable air source (e.g. compressed air cylinders used for positive pressure) or by the regeneration of the internal atmosphere with a CO_2 absorbent (scrubber).

Different regulations around the world, state varied human production rates between 30–50L per hour per person. When considering an acceptable production rate; it must be noted that the rate of breathing is primarily triggered by the CO_2 content of inspired air with higher CO_2 levels triggering faster respiration rates (Brake, Fellows, & Bates, 1999) (i.e. if the maximum allowable CO_2 concentration inside the safe haven is 1%, a lower production rate is acceptable compared to a 3% CO_2 limit).

If CO_2 removal mechanisms are inadequate, the occupants can develop CO_2 toxicity by re-breathing their own exhaled CO_2.

Cooling and Dehumidifying

Thermal tolerability with a safe haven should be addressed during the design phase. Assumptions must be made in preparation for safe haven design in regard to the number of personnel, activity level, available hydration and heat production. The minimum airflow to maintain a comfortable environment is set by taking these assumptions into account as well as the input temperature and humidity of supplied air.

The healthy human body maintains its internal temperature around 37°C. An essential requirement for continued normal body function is that the deep body temperature is maintained within the narrow 1°C limit (Epstein, Moran, 2006). A change of body temperature exceeding 1°C occurs only during illness or when environmental conditions surpass the body's ability to cope with extreme temperatures.

When a person is subject to extreme heat and humidity over some time, the person may experience a very rapid progression of symptoms or exhibit a few of the less significant symptoms before falling victim to more severe forms of heat-related illness.

The acceptable cooling requirements for underground refuge chambers are defined in various international standards, examples of which are listed in the below table.

Table 4 – Standards for acceptable cooling requirements

Cooling Requirements	Relevant Standard
Simulated emergencies, in which a full complement of people has occupied a refuge chamber for a significant period, indicate that humidity and temperature can increase very rapidly to potentially heatstroke-inducing levels. Refrigerated air conditioning is strongly recommended for both externally supported and stand-alone refuge chambers to counter this potentially dangerous problem. Inevitably, this places a heavy demand on the stand-alone power supply, but there are systems available that can cope, and at an acceptable cost.	Department of Mines, Industry Regulations and Safety, 2005 Refuge Chambers in Underground metalliferous mines – guideline: Safety and Health Division, Department of Industry and Resources Western Australia, 32 pp
Apparent temperature in the fully occupied refuge chamber must not exceed 95 °F.	Mines Safety and Health Administration 30 CFR Parts 7 and 75 Refuge Alternatives for Underground Coal Mines; Final Rule
It should be possible to maintain an internal temperature not exceeding 30 °C wet bulb globe temperature for at least 24 hours	ITA Report No. 14 – Guideline for the provision of refuge chambers in tunnels under construction, May 2014

MSHA regulations specify a maximum internal apparent temperature[7] of 95°F (35°C). NIOSH specifies an equivalent maximum wet bulb temperature of 28.9°C (NIOSH, 2007).

The temperature maintained inside the safe haven is determined by the metabolic heat liberated by the occupants, dissipated to the external environment by the heat transfer characteristics of the room or building. The external temperature primarily influences the heat transfer while the humidity is the combination of moisture produced by the occupants (via exhaled air and sweat) and water liberated by CO_2 scrubbing chemicals.

It is important to remember that in most cases, before sheltering-in-place, the air conditioning system will need to be shut down (unless it is a split system). For short durations, artificial cooling is not a critical requirement unless external temperatures are extraordinarily high, and/or the safe haven's volume is small compared to its occupancy.

INFILTRATION TESTING

Background

High Performance Building Solutions Inc. (HPBS) was contracted to perform an air infiltration study for spaces identified as candidates to be used as shelter-in-place (SIP) in the event of an accidental chemical release. This SIP strategy uses the existing structure and its indoor atmosphere to separate individuals from a hazardous outdoor atmosphere temporarily.

A series of pre-weatherization air leakage tests were conducted to determine the baseline air infiltration rate of potential SIP spaces. Manual air sealing, a cleanroom suspended ceiling system and low-leakage HVAC dampers were installed to reduce the overall air infiltration of the SIP spaces and to close off air leakage pathways from non-protected spaces in the building through the HVAC system. Post-weatherization air leakage tests were conducted to determine the air infiltration rate of the SIP space after the completion of weatherization and HVAC work. The SIP spaces were additionally air sealed using the proprietary AeroBarrier® system to further reduce infiltration rates.

Scope

Air sealing work was completed, and low-leakage HVAC dampers were installed to reduce the overall air infiltration of the SIP spaces and to isolate the HVAC from non-protected spaces in the building. Post-weatherization air leakage tests were conducted to determine the air infiltration rate of the SIP space after the completion of air sealing, suspended ceiling and HVAC work. AeroBarrier® air sealing was conducted on SIP spaces as a pilot project on September 18 through September 21, 2017. Post AeroBarrier® air leakage tests were conducted on September 21 and September 22, 2017, to determine the air infiltration rate of the SIP space after the completion of the AeroBarrier® sealing.

7 Apparent temperature is the temperature equivalent perceived by humans, caused by the combined effects of air temperature and relative humidity.

Following ASTM E 779-10, pressurization air leakage tests were conducted on the following spaces recommended in the Facility Siting Study (FSS).

Table 5: SIP specifications

SIP Space	Occupancy	Floor Area (Square Feet)	Volume (Cubic Feet)
SIP Space 1	40	990	10,481
SIP Space 2	6	955	8,210
SIP Space 3	5	391	3,917
SIP Space 4	10	1,697	16,968
SIP Space 5	40	293	2,637

Air Leakage Test Results

Positive pressure air leakage test results are reported in four different ways:

CFM4 is the quantitative volume of air passing through an enclosure under a pressure difference of 4 Pascals (equivalent to a 5mph wind). This is a measurable physical quantity and the value recommended in the FSS to eliminate toxic gas ingress into the SIP spaces.

CFM25 is the quantitative volume of air passing through an enclosure under a pressure difference of 25 Pascals (0.1-in w.g.). FEMA 453 requires Class 1 Safe Havens be pressurized to 25 Pascals. This is a measurable physical property.

CFM50 is the quantitative volume of air passing through an enclosure under a pressure difference of 50 Pascals (0.2-in w.g.).

ACH50 is the quantitative volume of air passing through an enclosure under a pressure difference of 50 Pascals (0.2-in w.g.) normalized by the enclosure volume. ACH is the abbreviation for air changes per hour. It is the number of times the air volume in a building changes per hour at 50 Pascals of pressure.

Table 6 below shows the CFM50 and ACH50 for each of the SIP spaces pre-weatherization and post-weatherization along with the percentage reduction in leakage. SIP Space 1 and SIP Space 5 did not have pre-weatherization results as initially the whole building was tested. Smaller spaces were selected as the whole building was too large to feasibly convert to a SIP space (re-testing of the new SIP spaces was not completed). The control room, SIP space 4, was abandoned in favor of a smaller adjacent room due to high leakage post manual weatherization and the computers that could not be shut down, which prevented the use of AeroBarrier® in the space.

Table 6: Post-weatherization leakage results

Safe Haven Space	CFM50		ACH50		% Reduction
	Pre-Weatherization	Post-Weatherization	Pre-Weatherization	Post-Weatherization	
SIP Space 1	N/A	1,871	N/A	10.7	N/A
SIP Space 2	3,623	1,560	26	11.4	57%
SIP Space 3	385	135	6	2.1	65%
SIP Space 4	N/A	N/A	N/A	N/A	N/A
SIP Space 5	N/A	655	N/A	19.4	N/A

The AeroBarrier® air sealing was prescribed a target. According to the AeroBarrier® simultaneous air leakage calculations, all the spaces met and surpassed this airtightness target; AeroBarrier® sealing was halted at this point. Table 7 shows the CFM50 and ACH50 for each of the SIP spaces as tested by HPBS post AeroBarrier® along with the percentage reduction. The HPBS results were utilized to calculate the amount of positive pressure required to convert the rooms to positive pressure safe havens.

Table 7: Post AeroBarrier® leakage results

Safe Haven Space	CFM50		ACH50		% Reduction
	Post-Weatherization	Post AeroBarrier®	Post-Weatherization	Post AeroBarrier®	
SIP Space 1	1,871	455	10.7	2.6	76%
SIP Space 2	1,560	182	11.4	1.3	88%
SIP Space 3	135	125	2.1	1.9	7%
SIP Space 4	N/A	173	N/A	3.3	N/A
SIP Space 5	655	138	19.4	3.1	79%

The AeroBarrier® air sealing methods improved the air barrier integrity for all the proposed SIP spaces between 7% and 88% over the post manual weatherization leakage rates. Air leakage was still observed behind many electrical penetrations including HVAC thermostats, and fire and smoke alarms. These were covered with plastic to protect them from the AeroBarrier® Sealant but had hidden holes behind them.

Figure 1 displays the post AeroBarrier® leakage rate in comparison to the FEMA 453 estimates for air leakage rates based on construction types. FEMA 453 estimates are reported in square feet of floor area at 25 Pascals and do not take into consideration the ceiling height and volume of the space.

Figure 1: Post AeroBarrier® results compared to FEMA 453 estimates

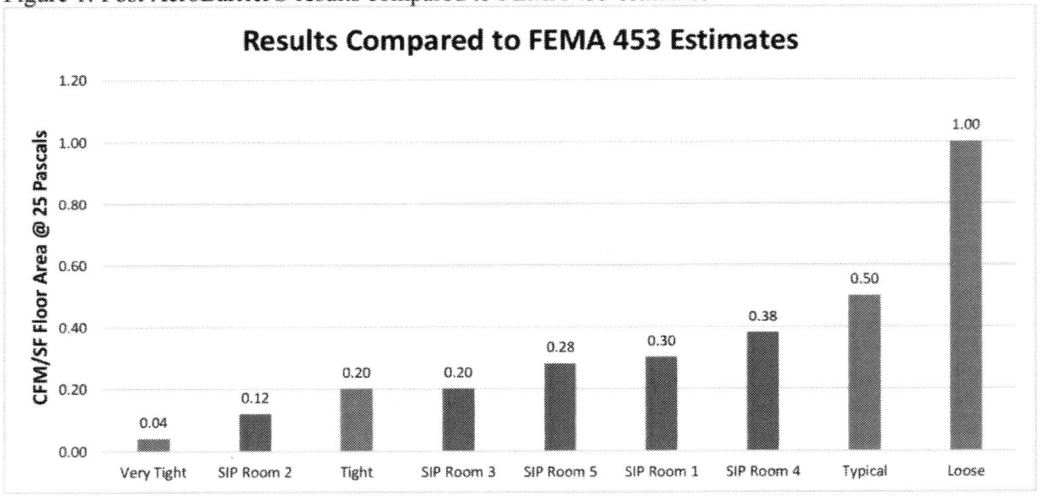

Figure 2 is an estimated leakage rate based on construction type to estimate the filtration and positive pressure system size under the FEMA 453 standard. All the SIP locations were between tight and typical except for SIP Room 2, which was a small prefabricated metal constructed portable building.

Figure 2: Estimated leakage rate per square foot of floor space to achieve 0.1 iwg (25Pa) overpressure

Construction Type	cfm per square foot of floor area
Very tight: 26-inch thick concrete walls and roof with no windows	0.04
Tight: 12-inch thick concrete or block walls and roof with tight windows and multiple, sealed penetrations	0.20
Typical: 12-inch thick concrete or block walls with gypsum wall board ceilings or composition roof and multiple, sealed penetrations	0.50
Loose: Wood-frame construction without special sealing measures	1.00

Table 8 shows the combined AeroBarrier®, and manual air sealing methods improved the airtightness for all the proposed SIP spaces between 68% and 95% over the original leakage rates.

Table 8: Air infiltration improvement for SIP spaces

Safe Haven Space	ACH50			% Reduction
	Pre-Weatherization	Post-Weatherization	Post AeroBarrier®	
SIP Space 1	N/A	10.7	2.6	76%
SIP Space 2	26.5	11.4	1.3	95%
SIP Space 3	5.9	2.1	1.9	68%
SIP Space 4	N/A	N/A	3.3	N/A
SIP Space 5	N/A	19.4	3.1	84%

Figure 3 compares the ACH50 pre-weatherization, post-weatherization and AeroBarrier® air infiltration results.

Figure 3: Pre-weatherization, post-weatherization and AeroBarrier® comparison

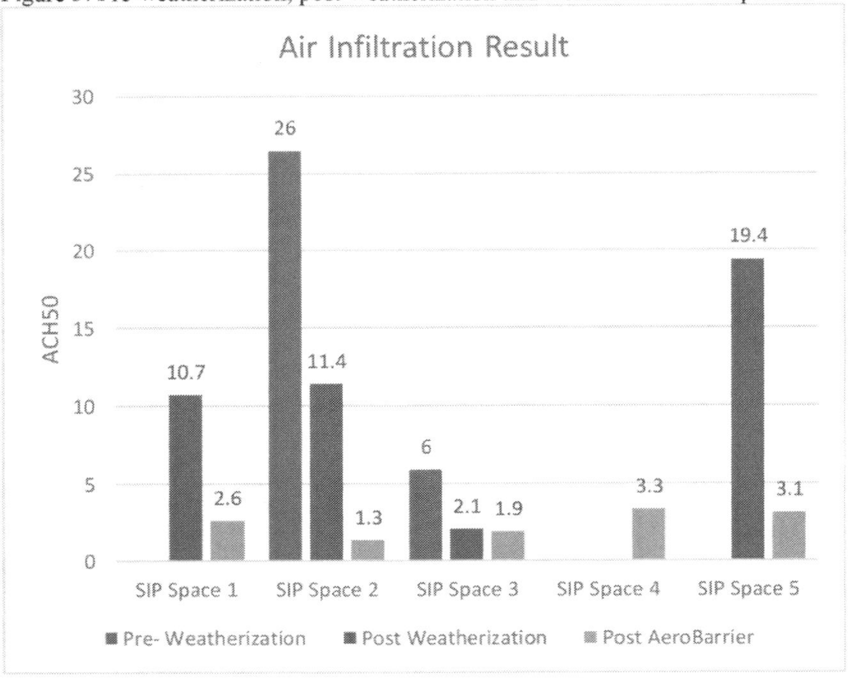

Life Support Systems

MineARC re-calculated the vulnerability for each SIP space based on the post AeroBarrier® leakage rates. All SIP spaces still had a vulnerability of 100% and therefore required the installation of independent life support systems to be deemed zero percent vulnerability safe havens.

Figure 4 shows the toxic gas concentration, duration of the release, and vulnerability for each SIP space. As the largest MCE was a full tank rupture the toxic concentrations were extremely high, but the duration was short at one minute.

Figure 4: SIP space vulnerability post AeroBarrier®

Safe Haven Location	TOXIC GAS				
	Worst Case Threat Scenario	Toxic External [ppm]	Duration of Reslease [min]	ACM	Vulnerability
SIP Space 1	Tank Rupture	585,000	1	0.032	100%
SIP Space 2	Tank Rupture	652,000	1	0.042	100%
SIP Space 3	Tank Rupture	660,000	1	0.007	100%
SIP Space 4	Tank Rupture	412,000	1	0.042	100%
SIP Space 5	Tank Rupture	600,000	1	0.028	100%

The leakage rate at 4 Pascals of pressure difference is slightly more than would occur under a 5 mile per hour wind used in the dispersion model in the FSS report based on climatic data for the facility (to maintain positive pressure and neutralize air infiltration). Post-weatherization infiltration rates in cubic feet per minute at 4 Pascals of pressure for each of the SIP spaces is shown in Table 9.

Table 9: Post AeroBarrier® air leakage results at CFM4

Safe Haven Space	Volume (Cubic Feet)	CFM4
SIP Space 1	10,481	99
SIP Space 2	8,210	32
SIP Space 3	3,917	24
SIP Space 4	3,110	45
SIP Space 5	2,637	21

The air leakage test results provide a mathematical formula that can be used to calculate the amount of breathable compressed air cylinders (437ft^3) that must be added to each candidate SIP space to maintain a 4 Pascal positive pressure to overcome a 5mph design wind coincident with a toxic release.

MineARC Systems was contracted to install life support systems to convert the SIP spaces to positive pressure safe havens. The MineARC AirBANK Positive Pressure System provides rapid pressurization, which is activated and maintained using the AirBANK Control via a simple human-machine interface (HMI) touch screen. MineARC's integrated Aura-FX Gas Monitor ensures breathable air automatically remains within acceptable limits. Additionally, the installation of supplementary oxygen and carbon dioxide scrubbing systems in safe havens, are as needed.

Figure 5 shows the number of 437 cubic feet cylinders required for each safe haven location. The minimum breathable air per person is 2.5cfm (70L/min) per person based on MineARC's experience and testing within the refuge chamber industry.

Figure 5: Quantity of cylinders required for 4Pa positive pressure

	SAFE HAVEN SPECIFICATIONS		POSITIVE PRESSURE (@ 4Pa)						
Safe Haven Location	Occupancy	Duration	CFM	CFM per Floor Space	ACH	No. Cylinders	No. Cylinder Racks	Breathable Air per Person [CFM]	Breathable Air per Person [LPM]
Safe Haven Space 1	40	4	99	0.10	0.6	55	5	2.5	70.8
Safe Haven Space 2	6	4	32	0.03	0.2	18	2	5.4	152.9
Safe Haven Space 3	5	4	24	0.06	0.4	14	2	4.8	135.9
Safe Haven Space 4	10	4	45	0.14	0.9	25	3	4.5	127.4
Safe Haven Space 5	40	4	21	0.07	0.5	12	2	0.6	17.0

The maximum CO_2 concentration is at 1% and O_2 between 19.5% and 23.5% based on OSHA limits (refer to Table 2). Due to the volume and number of people in Safe Haven Space 5, it required supplemental oxygen and carbon dioxide scrubbing to maintain the internal gas concentrations within specified limits. Three sodium chlorate oxygen candles and two AirGEN CO_2 scrubbers were included in the design.

Figure 6: Internal gas concentrations in safe havens

	SAFE HAVEN SPECIFICATIONS		POSITIVE PRESSURE (@ 4Pa)			02 & CO2 CONCENTRATIONS (@ 4Pa)			
Safe Haven Location	Occupancy	Duration	CFM	CFM per Floor Space	ACH	Indoor O_2 %	No. O_2 Candles	Indoor CO_2 %	No. AirGEN Scrubbers
Safe Haven Space 1	40	4	99	0.10	0.6	20.45	0	0.50	0
Safe Haven Space 2	6	4	32	0.03	0.2	20.84	0	0.17	0
Safe Haven Space 3	5	4	24	0.06	0.4	20.77	0	0.23	0
Safe Haven Space 4	10	4	45	0.14	0.9	20.66	0	0.31	0
Safe Haven Space 5	40	4	21	0.07	0.5	18.67	3	2.04	2

In 2018 MineARC completed installation and training on the five safe haven spaces. Each safe haven was tested during the commissioning process to ensure it could maintain acceptable positive pressure.

CONCLUSIONS

The conversion of shelter-in-place (SIP) locations to zero percent vulnerability safe havens is a four-step process consisting of the following:

1) Identifying the correct SIP location based on ease of sealing and small volume based on the required occupancy.
2) Manual air sealing of SIP spaces.
3) AeroBarrier® sealing of SIP spaces.
4) Installation of life support equipment.

Identifying SIP spaces that can seal from the outset is crucial to a successful result. In the case study, three of the five original locations were abandoned in favor of rooms in adjacent locations due to high leakage rates or reducing the size to make the project more cost effective. Control rooms pose the most difficulty in trying to seal due to the false floor and a myriad of cable tray and pass-throughs installed for the computer systems. The smaller the room and the more robust the construction (i.e. prefabricated metal or concrete buildings), the better it seals.

Manual air sealing needs to be overseen by someone with familiarity in building tightness. While most of the tasks are general construction, standard contractors are not trained to work at the level of precision needed to ensure proper sealing. Even a tiny orifice or ceiling tile not sealed correctly can ruin a leakage test result. HPBS and MineARC Systems have developed a system for ensuring airtightness in construction methodology and by avoiding un-sealable spaces or spaces that are not cost effective to seal.

For future construction projects, incorporation of SIP spaces should occur during the design phase. Air barrier commissioning is standard in new construction and should include the following tasks at a minimum: design review of the plans with specifications for the air barrier and mechanical system - specifically as these relate to spaces designated for use as safe havens; quality assurance inspections during construction to ensure a continuous air barrier is in place; and final air leakage testing of the continuous air barrier to ensure the leakage rate of the safe haven spaces are within acceptable limits.

Complementary life support equipment is essential to sustain life within an airtight environment. For personnel to remain within the sealed safe haven, the environment must be habitable. Air is to be free of toxins such as carbon dioxide and acceptable levels of oxygen maintained. Positive pressure systems reduce the ingress of toxins and provide some or all of the required source of breathable air. Additional equipment such as compressed air, oxygen cylinders and air scrubbers are used to meet the shortfall. Furthermore, internal conditions must consider the human condition, supporting workable body temperature and humidity.

Reducing the risk of harm to personnel is a priority for emergency response planning. While the likelihood of an MCE occurring may be low, the risk to on-site personnel in the event of an incident is high. Converting shelter-in-place (SIP) locations to zero vulnerability safe havens improves emergency response.

REFERENCES

ASTM E779-10, *Standard Test Method for Determining Air Leakage Rate by Fan Pressurization*, ASTM International, West Conshohocken, PA, 2019, www.astm.org

Brake, Fellow, & Bates (1999). *Criteria for the design of emergency refuge stations for an underground metal mine.* Aus. IMM Proceedings, No.2, 1999, pp 1–7.

Department of Mines and Petroleum, 2013, *Refuge chambers in underground mines — guideline: Resources Safety*, Department of Mines and Petroleum, Western Australia, 41 pp.

Epstein, Y., & Moran D. (2006). *Thermal Comfort and the Heat Stress Indices,* Industrial Health, 44, 388–398.

FEMA 453. (2006). *Design Guidance for Shelters and Safe Rooms.* Federal Emergency Management Agency

Health & Safety Executive UK (2008). *Guidance and Information on the role and design of safe havens in arrangements for escape from mines.* Retrieved from http://www.hse.gov.uk/pubns/mines08.pdf

ITA Working Group 5 (2016). *Provisions for refuge chambers in tunnel projects under construction.* ITA Report N°014 - V2 / March 2018

Mines Safety and Health Administration (2008). *30 CFR Parts 7 and 75 Refuge Alternatives for Underground Coal Mines; Final Rule.*

National Institute for Occupational Safety and Health (2007). *Research Report on Refuge Alternatives for Underground Coal Mines.* Retrieved from http://www.cdc.gov/niosh/docket/archive/pdfs/NIOSH-125/125 ResearchReportonRefugeAlternatives.pdf

Occupational Safety & Health Administration (2013). *Permit-required confined spaces.* Retrieved from http://www.osha.gov/pls/oshaweb/owadisp.show_document?p_table=STANDARDS&p_id=9797

Suming Du. (2014). *On the effectiveness of shelter-in-place as a measure to reduce harm from atmospheric releases. Journal of Emergency Management, Vol. 12, No, 3, May/June 2014.*

Venter. J. (1998). *Portable Refuge Chambers: Air or Tomb in Underground Escape Strategies.* School for Mechanical and Materials Engineering, Potchefstoom University, Republic of South Africa

Worksafe Australia National Standard (2005). *Australian Standard AS: NZS 1716:2003 Respiratory protective devices.*

Yantosik. G. (2006). *Shelter-in-Place Protective Action Guide.* The Chemical Stockpile Emergency Preparedness Program (CSEPP). U.S. Department of Energy

589008 How the Field of Human Factors Helps Solve the Riddle of Serious Injuries and Fatalities

Abstract:

Process and occupational safety are often treated as distinct endeavors, with the former tending to focus on facility system function and the latter on individual worker wellbeing. Though other differences certainly exist between the two, both endeavors are generally purposed to prevent incidents that can cost lives, health, time, and money, in addition to employee morale and public trust. The overarching thesis of this paper is that the field of Human Factors is uniquely positioned to effect positive change in both of these major safety categories.

The backdrop for our argument is the recent finding that while injury and illness incidence rates published by the Bureau of Labor statistics have decreased in recent years, the occurrences of serious injuries and fatalities have declined at a significantly slower rate. We discuss possible explanations for this finding but note that regardless of the preferred explanation, the trend should motivate scientists, engineers, and other safety professionals to revisit long-held accident causation theories holding that near misses – these are incidents that could have resulted in serious injuries and fatalities but did not – are precursors of future serious incidents. While this idea still has merit, the just-mentioned statistics warrant reconsideration of the relations between serious and less-serious incidents to explain why the rates of the two are not decreasing in lockstep. We suggest these data indicate additional (human) factors are involved.

To address this riddle and to support our general thesis, we (a) review core tenets of process safety, occupational safety, and the field of Human Factors, (b) point out that the human is always an element common to both process and occupational safety contexts, and (c) discuss how preventing or mitigating human error can benefit both. We offer specific examples from both safety contexts to show how Human Factors perspectives and techniques can help. Examples touch on standards, policies, and other administrative controls (e.g., procedures), human-machine interfaces of both workplace hardware (e.g., valves) and software (e.g., SCADA systems), environments (e.g., physical, regulatory, and cultural), and communications between personnel (e.g., elements of team resource management). We conclude that fundamental Human Factors considerations can help prevent and mitigate human performance gaps and therefore increase the rate of reduction in serious injuries and fatalities in both process and occupational safety.

Copyright © American Institute of Chemical Engineers. All rights reserved.

589233 Initial Conditions Used in Atmospheric Dispersion Models:Issues with Commonly Used Assumptions for Gas Jets and Plumes

Abstract:

When simulating jets and plumes, all atmospheric dispersion models require initial conditions that appropriately parameterize the behavior of a gas release to the atmosphere. Often, process conditions are such that loss of primary containment of a gas will result in choked flow conditions. A great deal of experimental and theoretical consideration has been given to the flow from stagnation conditions to choked flow conditions, but less experimental work has centered on describing the flow through the depressurization process to atmospheric pressure where atmospheric dispersion models would be applicable. Traditional modeling of the depressurization process has taken the simplified approach of assuming a (depressurized) "top hat" profile (laterally constant) with application of the principles of conservation of mass, momentum, and energy. Such assumptions are known to result in predicted depressurized velocities that exceed the (sonic) velocity at the point of loss of containment, and because of the energy balance, the associated temperatures are significantly less than temperatures at the choked conditions. Physically, these effects could possibly self-cancel because as jet velocity slows, the decrease in kinetic energy of the jet will increase the jet temperature. However, no atmospheric dispersion model used for routine hazard assessment purposes accounts for kinetic energy in the energy balance. The traditional modeling approaches of the depressurization process are reconsidered based on analysis of available data, and an alternative approach to modeling the depressurization process is proposed. Example cases showing the important differences between the two approaches are discussed.

Copyright © American Institute of Chemical Engineers. All rights reserved.

Fault Tree Uncertainty Analysis

Raymond "Randy" Freeman
S&PP Consulting
12303 Lake Shore Ridge
Houston, TX 77041
Cell: 713 408 0357
Rafree@yahoo.com

For Presentation at the 2020 Spring National AIChE Meeting
Houston, Texas

March 30, 2020

ABSTRACT

Fault tree analysis (FTA) is a widely used methodology in the process industries. FTA is used for the development of failure mechanisms, computation of failure frequencies and the determination of the probability of failure on demand of safety systems. Much of the data used in a FTA study are uncertain. For example, the failure rate of a pump is often not known with great precision. Likewise the failure rates of instrumentation are often known only within some defined limits. The common practice, used by analysts in the quantification of a fault tree, is to use the most likely or best guess as to the needed failure rate data. The use of best guess values as data inputs to the quantification of a fault tree creates uncertainty in the computed results.

This paper presents a general methodology for the determination of the impact of uncertainty on the results of a fault tree study. The general methodology is based on the mathematics of propagation of error and variance contribution analysis. For a fault tree that contains 10 event minimal cut sets, the methodology will compute the variance in the probability (or frequency) of the top event with a maximum error of 5%. For 5 event minimal cut sets the error in variance of the top event probability (or frequency) will be no more than 2%. An example is presented to illustrate the application of the fault tree uncertainty analysis methodology to a real world problem.

Keywords

Quantitative Risk Assessment, Fault Tree Analysis, Uncertainty Analysis

1. Background

Fault tree analysis is a widely used method for representation of failure mechanisms in chemical plants. The methodology is described in the CCPS CPQRA guideline book [Ref. 1] and is extensively discussed in Chapter 9 of Lees [Ref. 2]. Lacking in these discussions is a systematic method to quantify the uncertainty of the results of the fault tree calculations.

Fault tree analysis has also been used in the aircraft and aerospace industries, nuclear power plant review and is recognized by the OSHA Process Safety Management (PSM) standard [Ref. 3] as a method for process hazards analysis.

2. Review of Error Propagation and Variance Contribution Analysis (VCA) Methodology

The mean and variance of a function of random variables can be approximated using the method described by Haugen [Ref. 4] and applied by Freeman [Ref. 5, 6, 7]. Define an arbitrary function of a set of random variables, x_i, as:

Let

$$Y = F(x_i) \qquad \text{(Eq 1)}$$

The mean of Y can be estimated using the following approximation:

$$E(Y) = F[\, E(x_i)] \qquad \text{(Eq 2)}$$

Where:

$E(Y)$ = expected value of random variable Y = mean of Y

$E(x_i)$ = expected value of random variable x_i = mean of x_i

The variance of Y can likewise be estimated as:

$$V(Y) = \sum_{i=1}^{n} \left[\frac{\partial Y}{\partial x_i} \right]^2 V(x_i) \qquad \text{(Eq 3)}$$

Where:

$V(Y)$ = variance of random variable Y as defined above in Equation 1

$V(x_i)$ = variance of random variable x_i as defined above in Equation 1

Note that the variance is simply the square of the standard deviation. Using the variance will simplify the mathematics that is described below. The contribution of each independent random variable to the overall variance in the function is:

$$V(Y \text{ from } x_i) = \left[\frac{\partial Y}{\partial x_i}\right]^2 V(x_i) \qquad \text{(Eq 4)}$$

The relative contribution of each term to the overall variance V(Y) is a measure of the importance in the uncertainty in the particular random variable, x_i. In effect, this is a sensitivity analysis combined with a uncertainty evaluation. The variance contribution combines the sensitivity in the answer to changes in the uncertain random variable, x_i, with a measure of the uncertainty in the random variable, x_i. The overall variance in Y is found by summing the sensitivity weighted variances from each random variable.

Numerical Estimate of the Sensitivity

Let us return to the fundamental definition of the derivative.

As before, set y = F(xi)

$$\frac{\partial y}{\partial x} = \lim_{\Delta xi \to 0}\left[\frac{F(xi+\Delta xi) - F(xi)}{\Delta xi}\right] \qquad \text{(Eq 5)}$$

We are interested in a numerical estimate of the derivative. Therefore, we will make the following approximation.

$$\frac{\partial y}{\partial x} \sim \left[\frac{F(xi+\Delta xi) - F(xi)}{\Delta xi}\right] \qquad \text{(Eq 6)}$$

Linear Functions

Let us examine the approximation for a simple linear function.

Let Y = AX + B (Eq 7)

What is the sensitivity of Y to a change in variable X? By simple calculus:

$$\frac{\partial y}{\partial x} = A \quad \text{(Eq 8)}$$

We can also find the sensitivity using the numerical approximation as:

$$\frac{\partial y}{\partial x} \sim \left[\frac{F(xi+\Delta xi) - F(xi)}{\Delta xi}\right] \quad \text{(Eq 9)}$$

$$\frac{\partial y}{\partial x} \sim \left[\frac{(A(xi+\Delta xi)+B)-(A(xi)+B)}{\Delta xi}\right] \quad \text{(Eq 10)}$$

$$\frac{\partial y}{\partial x} \sim \left[\frac{(A(xi+\Delta xi))-(A(xi))}{\Delta xi}\right] \quad \text{(Eq 11)}$$

$$\frac{\partial y}{\partial x} \sim \left[\frac{((A\,xi+A\,\Delta xi))-(A(xi))}{\Delta xi}\right] \quad \text{(Eq 12)}$$

$$\frac{\partial y}{\partial x} \sim \left[\frac{A\,\Delta xi}{\Delta xi}\right] \quad \text{(Eq 13)}$$

$$\frac{\partial y}{\partial x} = A \quad \text{(Eq 14)}$$

Note that for a simple linear equation the numerical result for the sensitivity is exactly the same as the analytical expression. There is no restriction on the size of the perturbation (Δxi) used in the calculations. Previously, Freeman [Ref 7] has suggested using a 10% perturbation in xi to do the sensitivity calculations. However any size of perturbation will work and generate the correct sensitivity.

This is a general result. As long as the function can be expressed as a linear function of a number of variables, xi, the sensitivity can be directly calculated by a simple perturbation of the function. The resulting sensitivity will be exactly the same as would be obtained using analytical methods.

Non-Linear Expressions

In the quantification of a fault tree, the top event frequency is computed using failure rates of the component devices and time that measures the interval between system validations. In a minimal cut set of two or more basic events, the time variable may be the same. For example the test interval for a valve may be the same for a pressure transmitter. A minimal cut set involving both the valve and the pressure transmitter would create a 2rd order term in the test interval. If the test interval is uncertain, how do we compute the sensitivity of the top event frequency with respect to the test interval? We will now explore how to deal with this type of problem when completing the uncertainty analysis.

As before, let

$$y = F(xi) = Axi^n + B \quad \text{(Eq 15)}$$

What is the sensitivity of Y to a change in variable X? By simple calculus:

$$\frac{\partial y}{\partial x} = nAxi^{n-1} \quad \text{(Eq 16)}$$

Let us evaluate the numerical approximation for this case. If n=2, the function of interest is:

$y = F(xi) = Axi^2 + B$ (Eq 17)

From basic calculus, the exact analytical expression for the sensitivity is:

$$\frac{\partial y}{\partial x} = 2\ A(xi) \quad \text{(Eq 18)}$$

If xi = 1

$$\frac{\partial y}{\partial x} = 2\ A \quad \text{(Eq 19)}$$

Again the approximation of the derivative is:

$$\frac{\partial y}{\partial x} \sim \left[\frac{(A(xi+\Delta xi)^2+B)-(Axi+B)}{\Delta xi}\right] \quad \text{(Eq 20)}$$

$$\frac{\partial y}{\partial x} \sim \left[\frac{F(xi+\Delta xi)-F(xi)}{\Delta xi}\right] \quad \text{(Eq 21)}$$

Inserting the equation for F(xi) we obtain:

$$\frac{\partial y}{\partial x} \sim \left[\frac{(A(xi+\Delta xi)^2+B)-(Axi+B)}{\Delta xi}\right] \quad \text{(Eq 22)}$$

$$\frac{\partial y}{\partial x} \sim \left[\frac{(A\ (xi^2+2\Delta xi+\Delta xi^2)\ +B)-(Axi+B)}{\Delta xi}\right] \quad \text{(Eq 23)}$$

$$\frac{\partial y}{\partial x} \sim \left[\frac{A\ (xi^2+2\Delta xi+\Delta xi^2)\ -Axi}{\Delta xi}\right] \quad \text{(Eq 24)}$$

$$\frac{\partial y}{\partial x} \sim A\left[\frac{(xi^2+2\Delta xi+\Delta xi^2)\ -xi}{\Delta xi}\right] \quad \text{(Eq 25)}$$

Setting xi =1, and using a 10% perturbation as $\Delta xi = 0.1\ xi = 0.1$

$$\frac{\partial y}{\partial x} \sim A\left[\frac{(1^2+(2)(0.1)1+0.1^2\ )\ -1}{0.1\ (1)}\right] \quad \text{(Eq 26)}$$

$$\frac{\partial y}{\partial x} \sim A\left[\frac{(2)(0.1)+0.01}{0.1}\right] \quad \text{(Eq 27)}$$

$$\frac{\partial y}{\partial x} \sim A\left[\frac{0.21}{0.1}\right] \quad \text{(Eq 28)}$$

$$\frac{\partial y}{\partial x} = 2.1\ A \quad \text{(Eq 29)}$$

Note that the numerical estimate (Eq 29) is approximately 5% greater than the analytical value (Eq 19). As shown in Table 1, for higher order equations the error will increase as the size of the perturbation becomes more important. Table 2 shows the impact of reducing the perturbation size to 1%. For 10th order variables (say a minimal cut set with 10 basic events) the error is 5%. For almost all engineering evaluations, a maximum error of 5% should be adequate.

For cases where the test frequency, repair time, or mission time appears in a minimal cut set 3 or more times, I suggest that the following perturbation in time be used to compute the sensitivity of the top event:

$$\Delta time = 0.01\ time \qquad \text{(Eq 30)}$$

The sensitivity of the top event frequency to all uncertain variables can be computed using a perturbation of 1%.

$$\Delta xi = 0.01\ xi \qquad \text{(Eq 31)}$$

3. Fault Tree Analysis

Fault trees are often used to analyze potential failures of an engineered system. A fault tree is a failure logic diagram that shows the relationship of device or system failures leading to an accident. Fault tree analysis is covered in detail in the CCPS CPQRA book [Ref. 1]. A fault tree is constructed by starting with a top event or condition and asking the question:

What conditions lead to this outcome?

Each condition in the next layer down is then analyzed by asking the same question. What conditions lead to this outcome? The logic of how these intermediate conditions (termed intermediate events) are related is expressed using AND and OR logic gates. An AND gate output is true if and only if (iff) all of the inputs to the AND gate are all true simultaneously. An OR gate output is true iff one of the inputs to the OR gate is true.

Once the fault tree logic structure is created, numerical data can be used to calculate the frequency or probability of the top event of concern occurring. For fault trees where a basic event is found in multiple branches of the tree, such as power failure, the logic structure of the tree must be simplified to remove the impact of the repeated events on the top event of concern. This simplification is done using Boolean Algebra. Each resulting grouping of basic events can cause the top event to occur. These groupings of basic events are called minimal cut sets. As a simple example consider the Boolean Algebra representation of a fault tree :

$$T = BE11 + BE1 \bullet BE2 + BE1 \bullet BE3 + BE2 \bullet BE9 + BE3 \bullet BE7 \qquad \text{(Eq 32)}$$

Where:

T = Fault Tree Top Event

BE11 = Basic Event 11

BE1 = Basic Event 1

BE2 = Basic Event 2

BE3 = Basic Event 3

BE7 = Basic Event 7

BE9 = Basic Event 9

+ = Boolean Algebra Addition Symbol

• = Boolean Algebra Multiplication Symbol

For this example the top event T, the minimal cut sets are:

BE11

BE1 BE2

BE1 BE3

BE2 BE9

BE3 BE7

For example the existence of basic events BE1 and BE3 guarantees that the top event T will occur. Publically available standard computer software such as SAPHIRE [Ref. 8] may be used to find the minimal cut sets of a large fault tree.

The representation of the top event in equation 32 can now be used to compute the probability or frequency of occurrence of the top event. If the devices describe by the basic events are non-repairable and are tested after a period of time TI, the probability of top event, T, may be calculated as:

$$\text{Prob (T)} = \lambda_{11}\,\text{TI}/2 + (\lambda_1\,\text{TI}/2)(\lambda_2\,\text{TI}/2) + (\lambda_1\,\text{TI}/2)(\lambda_3\,\text{TI}/2) + (\lambda_2\,\text{TI}/2)(\lambda_9\,\text{TI}/2) + (\lambda_3\,\text{TI}/2)(\lambda_7\,\text{TI}/2) \quad \text{(Eq 33)}$$

Where:

λ_{11} = failure rate of device 11 in basic event BE11, failures per unit time

λ_1 = failure rate of device 1 in basic event BE1, failures per unit time

λ_2 = failure rate of device 2 in basic event BE5, failures per unit time

λ_3 = failure rate of device 3 in basic event BE3, failures per unit time

λ_7 = failure rate of device 7 in basic event BE7, failures per unit time

λ_9 = failure rate of device 9 in basic event BE9, failures per unit time

TI = test interval, time

Equation 33 assumes that the device failure rates are small and the product λ TI < 0.1. The other major assumption is that the values of all of the failure rates and test intervals are known. In reality the failure rate data are not known exactly with only a range or probability distribution representing the state of knowledge.

4. Example Problem

The following example is taken from the ISA technical report on fault tree analysis [Ref. 9]. This example should not be considered as a recommendation by either ISA or AIChE. The example is presented to allow for the demonstration of the methods that can be used in a fault tree study to quantify the uncertainty in the resulting calculated failure frequency or failure probability.

Consider the interlock on the intermediate storage tank T101 shown on Piping and Instrument Diagram in Figure 1. The interlock block diagram is shown in Figure 2. The interlock is intended to prevent an abnormal condition in the tank which could lead to a uncontrolled release of the tank contents. The process hazards analysis team has recommended that the interlock be designed to safety integrity level 2 (SIL 2) with a target probability of failure on demand (PFD) of no greater than 1E-2 or a risk reduction factor (RRF) of 100. Does the proposed interlock shown on Figure 1 meet the SIL 2 target? What is the uncertainty in the predicted PFD of the interlock?

A fault tree for the failure of this interlock is shown in Figure 3. A minimal cut set analysis has been completed and the resulting minimal cut sets are presented in Table 3. We can now write the Boolean Algebra equation that represents this fault tree as:

$$T = PE + TS1 \bullet TS2 + LS1 \bullet LS2 + FT1 \bullet FT2 + FT2 \bullet FT3 + FT1 \bullet FT3 + BV1 \bullet BV2 + BV1 \bullet SOL1 + BV2 \bullet SOL2 + SOL1 \bullet SOL2 + PT1 \bullet PT2 \qquad \text{(Eq 34)}$$

Assuming that the interlock is non-repairable until tested at time TI and using lamda-time (λT) approximation of the failure rate, the Boolean representation of the fault tree may now be converted to a failure probability model as:

$$\begin{aligned} Prob(T) = {} & Prob(PE) + (\lambda_{TS1}\, TI/2)(\lambda_{TS2}\, TI/2) + (\lambda_{LS1}\, TI/2)(\lambda_{LS2}\, TI/2) + (\lambda_{FT1}\, TI/2)(\lambda_{FT2}\, TI/2) + \\ & (\lambda_{FT2}\, TI/2)(\lambda_{FT3} TI/2) + (\lambda_{FT1}\, TI/2)(\lambda_{FT3}\, TI/2) + (\lambda_{BV1}\, TI/2)(\lambda_{BV2}\, TI/2) + \\ & (\lambda_{BV1}\, TI/2)(\lambda_{SOL1} TI/2) + (\lambda_{BV2}\, TI/2)(\lambda_{SOL2}\, TI/2) + \\ & (\lambda_{SOL1}\, TI/2)(\lambda_{SOL2}\, TI/2) + (\lambda_{PT1}\, TI/2)(\lambda_{PT2}\, TI/2) \qquad \text{(Eq 35)} \end{aligned}$$

Where:

Prob(PE) = failure probability of electronic logic solver, assumed to be constant at 5E-3.

λ_{TS1} = failure rate of temperature switch 1, failures per unit time

λ_{TS2} = failure rate of temperature switch 2, failures per unit time

λ_{LS1} = failure rate of level switch 1, failures per unit time

λ_{LS2} = failure rate of level switch 2, failures per unit time

λ_{FT1} = failure rate of flow transmitter 1, failures per unit time

λ_{FT2} = failure rate of flow transmitter 2, failures per unit time

λ_{FT3} = failure rate of flow transmitter 3, failures per unit time

λ_{BV1} = failure rate of block valve 1, failures per unit time

λ_{BV2} = failure rate of block valve 2, failures per unit time

λ_{SOL1} = failure rate of solenoid valve 1, failures per unit time

λ_{SOL2} = failure rate of solenoid valve 2, failures per unit time

λ_{PT1} = failure rate of pressure transmitter 1, failures per unit time

λ_{PT2} = failure rate of pressure transmitter 2, failures per unit time

TI = test interval, time

The electronic logic solver PE is assumed to have a constant failure probability of 5E-3.

Failure rate data for the other devices are present in Table 4. The mode (most likely value) of the failure rate is set equal to the point value taken from the ISA example [Ref. 9]. The triangular probability distribution (Appendix A) was assumed to represent the failure rate data for the equipment items. The upper and lower limits are based on the failure rate data of Appendix 4 of Smith [Ref. 10].

Using the device failure rates represented by the Table 4 column labeled as the mode and the minimal cut sets presented in Table 3, the ISA technical report [Ref.9, page 35] computes an interlock PFD as 7.5E-3 (RRF of 133). This would satisfy the requirement for a SIL 2 interlock {PFD of 1E-2 or a RRF of 100}. Now what is the uncertainty in the interlock PFD?

The first thing to remember is that the PFD should be calculated based on the mean value of the failure rate lambda (λ), not the mode of the failure rate. For a data set, the mean is the best single point representation of the data set. Table 5 presents the PFD of each minimal cut set using the mean of the device failure rate lambda (λ). The total of the PFD becomes 1.47E-2 (RRF of 68). This would not satisfy a SIL 2 interlock requirement.

Now we will compute the variance of the predicted probability of top event, T. To compute the variance of the probability of the top event, T, we will need to compute the sensitivity of the top event to each of the variables.

Sensitivity Using Numerical Perturbation Calculation

We will use the ISA interlock example previously analyzed by Freeman (Ref. 7). Using public data sources Freeman found that the probability of failure on demand (PFD) was 1.49E-2. To use the perturbation method for the evaluation of the sensitivities requires evaluating the impact of small changes in the uncertain input parameters. To enable these calculations to be completed numbers using 8 significant digits were used. The calculations were completed using Microsoft Excel which uses double-precision floating-point arithmetic compliant with IEEE 754 specification. Excel nominally carries 15 significant figures in calculations. We will complete the calculations with Excel displaying 5 significant digits and round the result at the end of the calculations to 4 significant digits. We will illustrate the calculations by looking at the Flow Transmitter, FT1.

a. First we recalculate the PFD of the example and find it to be 1.4594E-02. This is the base number that we will use in the perturbation calculations.
b. We now make a small change in mean failure rate of FT1. The mean failure rate is 8E-6 failures/hour. A 1% perturbation is used or an increase of 0.08E-6 failures/hour.

c. The perturbed value of the failure rate of FT1 becomes (8 + 0.08) * 1E-6 failures/hour or 8.08E-6 failures/hour.
d. The perturbed value of the failure rate of FT1 is now used to re-calculate the PFD of the interlock. The resulting re-calculation finds that the PFD is 1.4618E-02
e. The change in the PFD from the base number due to the perturbation of the failure rate of FT1 is now computed as:

   delta PFD = 1.4618E-02 - 1.4594E-02 = 2.4556E-05

f. The sensitivity of the PFD to a change in the failure rate of FT1 is now computed by dividing the change in the PFD by the change FT1 failure rate as:

   sensitivity = (2.4556E-05)/(0.08E-6 failures/hour) = 307.0 hours
   Result is rounded to 4 significant digits.

We now note that in the paper by Freeman (Ref. 7) that equation 32 presents the sensitivity for FT1 as:

$$\frac{\partial Prob(T)}{\partial \lambda FT1} = 307.0 \text{ hr} \qquad \text{(Eq 32 of Ref 7)}$$

The answers are exactly the same. The sensitivity calculations for all of the devices are presented in Table 6. Note that in every case the sensitivity found using the analytical expression and the numerical estimate are the same. This is a general result which will be true for all basic event equipment failure rates found in a fault tree minimal cut set.

Once found, the sensitivities can now be used to compute the variance contribution of each particular device to the overall variance of the top event PFD in the same manner as Freeman previously presented.

The sensitivities of the top event probability are summarized in Table 6. We may now compute the variance contribution of each uncertain variable using Equations 1 – 4 in the methodology section of this paper. The relative variance contribution is expressed as a percent of the total variance of the top event probability. The top event variance is found to be 8.24E-6 (failures per hour)^2 or equivalently the top event standard deviation is 2.78E-3 failures per hour.

In this example the variance (measure of uncertainty) is dominated by the flow transmitters. The flow transmitters account for almost 2/3 of the uncertainty in the top event probability. Failures of the flow transmitters appear in 3 of the minimal cut sets. Since the uncertainty in the transmitter failure rates is large, the flow transmitters should contribute a significant amount to the variance of the fault tree top event.

We can use the variance in the top event to define the likelihood of achieving SIL 2 performance of the interlock. The calculations are done using the normal probability distribution. The normal distribution is tabulated as the standard normal distribution using a normalization factor Z.

The standard normal factor, Z, [Ref. 11] is defined as:

$$Z = \left[\frac{xi - E(x)}{\sigma}\right] \qquad \text{(Eq 36)}$$

Where:

σ = standard deviation. Note that the variance of a random variable is the square of the standard deviation of the random variable.

E(x) = Expected value of the random variable x_i

For the example, the expected value of the probability of the top event of concern was calculated as

E[Prob(T)] =1.47E-2 (Eq 37)

From Table 4, the standard deviation in the probability of the top event is

σ_T = 2.78E-3 (Eq 38)

The target PFD for a SIL 2 interlock must not be worse than 1E-2 (RRF of 100).

We compute the normal distribution Z factor at the SIL 2 target value as

Z = (0.01 – 0.0147)/2.78E-3 (Eq 39)

Z = -1.69065 (Eq 40)

Tabulations of standard normal distribution are presented in most statistics books such as Meyer [Ref. 11]. For this value of Z, the corresponding probability that the interlock will achieve SIL 2 performance is 4.55 percent. We can also compute the risk reduction that the proposed interlock design will achieve.

What is the interlock probability of failure on demand (PFD) that we can be 95% certain that the interlock will provide? At the 95% level, the corresponding Z factor is 1.65 [Ref 11]. We now compute the corresponding PFD of the interlock that corresponds to this Z factor.

$$Z = \left[\frac{x95\% - E[\text{Prob}(T)]}{\sigma}\right]$$ (Eq 41)

In this case:

$x_{95\%}$ = the PFD that we can be confident that the interlock will achieve

σ_T =2.78E-3 (Eq 42)

E[Prob(T)] =1.47E-2 (Eq 43)

Z= 1.65 (Eq 44)

Rearranging equation 40 to find $x_{95\%}$we obtain:

$x_{95\%}$ = E[Prob(T)] + Z σ_T (Eq 45)

$x_{95\%}$ = 1.47E-2 + (1.65) (2.78E-3) = 0.0193 (Eq 46)

or a 95% certain risk reduction factor of

RRF = 1/PFD = 1/0.0193 = 52 (Eq 47)

By the same logic there is 5% chance that the risk reduction is a PFD of 0.01 or an RRF of 99. In summary, this interlock is likely to perform as a mid-range SIL 1 interlock, not the SIL 2 interlock desired by process hazards analysis team.

5. Methodology for Fault Tree Uncertainty Analysis

We can now generalize the methods used in the above example to a proposed method for the evaluation of the uncertainty in a fault tree. The following step-by-step procedure is modified from that that previously published by Freeman (Ref 7) to indicate the use of numerical methods to compute the sensitivity. Step 8 shown in *italics* indicates the modification..

1. Create the fault tree using standard methods outline in the CPQRA book [Ref 1].
2. Determine minimal cut sets by hand or using standard computer software such as SAPHIRE [Ref 12].
3. Define the needed failure rate data for each basic event.
4. Define those basic events that are considered to be uncertain.
5. For the basic events that are uncertain, define the probability distribution and associated parameters needed to numerically define the probability distribution. Appendix B of this paper presents the needed description of 4 commonly used probability distributions. In many cases the triangular distribution will be selected. The minimum, maximum and mode (most likely) parameters will be needed for the triangular distribution.
6. Compute the mean of each of the uncertain variable
7. Compute the probability or frequency of the fault tree top event of interest using the mean value for each uncertain variable.
8. *Compute the sensitivity of the top event probability or frequency using the numerical methods outlined in this paper.*
9. Compute the variance of each of the uncertain variables
10. Compute the variance contribution of for each of the uncertain variables to the top event probability or frequency using equation 6.
11. Compute the total variance of the top event of interest probability or frequency by summing all of the contributions determined in Step 10.
12. Compute the variance contribution percent of each uncertain variable by dividing the variable contribution (Step 10) by the total variance of the top event of interest (Step 11).
13. Define the level of risk the project is willing to take. What chance will the project management accept for potential failure of the interlock to achieve the desired risk reduction? In the above example, I have used a 5% risk of failure or a 95% certainty that the interlock will achieve a risk reduction factor of 52. In the example, there is a very low probability that the interlock will achieve a SIL 2 PFD of 0.01 or an RRF of 100.

Alternately you can report the 90% range for the top event of interest. For the example the 90% range starts at the 5% RRF of 99 and there is a 95% certainty that that an RRF of at least 52 will be achieved. The expected risk reduction for this interlock is an RRF of 68.

6. Conclusions

A general procedure has been presented to quantify the uncertainty in calculations of failure frequencies using the fault tree methodology. The uncertainty analysis procedure is based on the application of propagation of error and variance contribution analysis techniques to the minimal cut sets created during a fault tree study. The procedure is simple and can be incorporated into standard fault tree analysis programs such as SAPHIRE [Ref 8]. This procedure is based on perturbation calculation of the frequency of the top event from changes in the uncertain variables. Incorporation of a numerical perturbation method into standard fault tree software such as SAPHIRE would be easy to implement and would provide the means for evaluation of the uncertainty in the quantified fault tree results.

For the example interlock presented in this paper, the uncertainty in the device failure rates degrades the predicted interlock SIL from meeting SIL 2 requirements to becoming a mid range SIL 1 interlock. This is a general result. Uncertainty in design parameters will reduce the likelihood that the equipment will achieve desired results. The design engineer has two choices:

- Accept the conclusion that the interlock will not perform as SIL 2
- Re-design the interlock using more reliable components to achieve the SIL 2 target

The variance contribution analysis shown in Table 7 indicates that the major contributor to the uncertainty in the interlock performance is the Flow Transmitters. Flow transmitters certified to SIL-2 or SIL-3 performance with lower failure rates could improve the calculated PFD of the interlock.

7. References

1. Center for Chemical Process Safety , *Guidelines for Chemical Process Quantitative Risk Analysis – 2rd edition*, American Institute of Chemical Engineers, New York, 2000.
2. Frank P Lees, Loss Prevention in the Process Industries – 2rd Edition, Butterworth, ISBN 0-7506-1547-8, Oxford, UK, 1996
3. US Department of Labor, Occupational Safety and Health Administration, CFR 1910.119 - Process Safety Management of Highly Hazardous Chemicals, 1910.119(e)(2)(vi).
4. Edward B. Haugen, *Probabilistic Approaches to Design*, John Wiley, New York, 1968
5. Raymond. Freeman and Angela Summers, "Evaluation of Uncertainty in Safety Integrity Level Calculations," *Process Safety Progress,* Published online in Wiley Online Library (wileyonlinelibrary.com) DOI 10.1002/prs. 11805, 2016
6. Raymond. Freeman, "General Method for Uncertainty Evaluation of Safety Integrity Level Calculations – Part 2 Analytical Methods," *Process Safety Process*, Published online in Wiley Online Library (wileyonlinelibrary.com) DOI 10.1002/prs 111915, October 2017
7. Raymond. Freeman ,"Error Propagation and Uncertainty Analysis: Application to Fault Tree Analysis", Process Safety Progress, Published Wiley Online Library (wileyonlinelibrary.com/journal/prs) DOI: https://doi.org/10.1002/prs.12080, June 29, 2019

8. US Nuclear Regulatory Commission, "Systems Analysis Programs for Hands-on Integrated Reliability Evaluations (SAPHIRE) – Version 8," NUREG/CR-7039, Vol 3, June 2011.
9. ISA, "Safety Instrumented Functions (SIF) – Safety Integrity Level (SIL) Evaluation Techniques Part 3: Determining the SIL of a SIF via Fault Tree Analysis, ISA Technical Report ISA-TR-84.00.02-2002 – Part 3, Approved 17 June 2002.
10. David J. Smith, *Reliability Maintainability and Risk – 8th Edition*, ISBN 978-0-08-096902-2, Elsevier, London, 2011
11. Paul L. Meyer, Introductory Probability and Statistical Applications, John Wiley, New York, 1972

Table 1. Comparison of Analytical Sensitivity with Numerical Approximation with Perturbation of 10% in xi for the equation: $y = F(xi) = Axi^n + B$

Order of Equation n	Equation for Analytical Sensitivity $\frac{\partial y}{\partial x} = nAxi^{n-1}$	Sensitivity with xi = 1 and A=1	Numerical Sensitivity with $\Delta xi = 0.1\ xi$	Error %
1	$\frac{\partial y}{\partial x} = 1A(xi)^0$	1	1	0
2	$\frac{\partial y}{\partial x} = 2A(xi)^1$	2	2.1	5
3	$\frac{\partial y}{\partial x} = 3A(xi)^2$	3	3.31	-10%
4	$\frac{\partial y}{\partial x} = 4A(xi)^3$	4	4.64	-16%
5	$\frac{\partial y}{\partial x} = 5A(xi)^4$	5	6.11	-22%
6	$\frac{\partial y}{\partial x} = 6A(xi)^5$	6	7.72	-29%
7	$\frac{\partial y}{\partial x} = 7A(xi)^6$	7	9.49	-36%
8	$\frac{\partial y}{\partial x} = 8A(xi)^7$	8	11.44	-43%
9	$\frac{\partial y}{\partial x} = 9A(xi)^8$	9	13.58	-51%
10	$\frac{\partial y}{\partial x} = 10A(xi)^9$	10	15.94	-59%

Error = (Analytical- Numerical)/Analytical

Table 2. Comparison of Analytical Sensitivity with Numerical Approximation with Perturbation of 1% in xi for the equation: $y = F(xi) = Axi^n + B$

Order of Equation n	Equation for Analytical Sensitivity $\frac{\partial y}{\partial x} = nAxi^{n-1}$	Sensitivity with xi = 1 and A=1	Numerical Sensitivity with $\Delta xi = 0.1\ xi$	Error %
1	$\frac{\partial y}{\partial x} = 1A(xi)^0$	1	1	0
2	$\frac{\partial y}{\partial x} = 2A(xi)^1$	2	2.01	-1%
3	$\frac{\partial y}{\partial x} = 3A(xi)^2$	3	3.03	-1%
4	$\frac{\partial y}{\partial x} = 4A(xi)^3$	4	4.06	-2%
5	$\frac{\partial y}{\partial x} = 5A(xi)^4$	5	5.10	-2%
6	$\frac{\partial y}{\partial x} = 6A(xi)^5$	6	6.15	-3%
7	$\frac{\partial y}{\partial x} = 7A(xi)^6$	7	7.21	-3%
8	$\frac{\partial y}{\partial x} = 8A(xi)^7$	8	8.29	-4%
9	$\frac{\partial y}{\partial x} = 9A(xi)^8$	9	9.37	-4%
10	$\frac{\partial y}{\partial x} = 10A(xi)^9$	10	10.46	-5%

Error = (Analytical- Numerical)/Analytical

Table 3. Basic Events in Minimal Cut Sets For Example

Cut Set No	BE 1	BE 2
1	PE	-
2	TS1	TS2
3	LS1	LS2
4*	FT1	FT2
5*	FT2	FT3
6*	FT1	FT3
7*	BV1	BV2
8*	BV1	SOL1
9*	BV2	SOL2
10*	SOL1	SOL2
11	PT1	PT2

* Minimal cut Set with same basic event as another cut set

Table 4. Lamda (λ) Failure Rate Data With Uncertainty Limits

Device	Minimum Failures per 1E6 Hours	Mode Failures per 1E6 Hours	Maximum Failures per 1E6 Hours	Mean Failures per 1E6 Hours	Lamda Variance (Failures per 1E6 Hours)^2
Flow Transmitters FT1, FT2 and FT3	1.0	2.9	20.0	8.0	18.3
Pressure Transmitters PT1 and PT2	1.0	2.3	20.0	7.8	18.8
Temperature Switch TS1 and TS2	3.0	7.6	20.0	10.2	12.9
Level Switch LS1 and LS2	2.0	4.6	20.0	8.9	15.8
Block Valves BV1 and BV2	0.2	2.3	10.0	4.2	4.4
Solenoid Valves SOL1 and SOL2	1.0	2.3	8.0	3.8	2.3

Table 5. Probability of Minimal Cut Sets Based on Mean Failure Rates

Cut Set No	BE 1	BE 2	PFD based on mean λ
1	PE	-	5.00E-03
2	TS1	TS2	2.00E-03
3	LS1	LS2	1.52E-03
4	FT1	FT2	1.23E-03
5	FT2	FT3	1.23E-03
6	FT1	FT3	1.23E-03
7	BV1	BV2	3.38E-04
8	BV1	SOL1	3.38E-04
9	BV2	SOL2	3.38E-04
10	SOL1	SOL2	2.77E-04
11	PT1	PT2	1.17E-03

Table 6. Sensitivity of Example System PFD to Basic Event Failure Rates Based on a 1% Perturbation

Basic Event Label	Device Type	Mean Failures per 1E6 Hours	Failure Perturbation of 1% of Mean Failures per 1E6 Hours	Perturbed Failure Rate Failures per 1E6 Hours	Total System PFD Using Perturbed Failure Rate of Basic Event	Change in Total System PFD	Sensitivity of PFD to Change in Failure Rate Hours
FT1	Flow Transmitter	8	0.08	8.08	1.461848E-02	2.455603E-05	307.0
FT2	Flow Transmitter	8	0.08	8.08	1.461848E-02	2.455603E-05	307.0
FT3	Flow Transmitter	8	0.08	8.08	1.461848E-02	2.455603E-05	307.0
PT1	Pressure Transmitter	7.8	0.078	7.878	1.460560E-02	1.167179E-05	149.6
PT2	Pressure Transmitter	7.8	0.078	7.878	1.460560E-02	1.167179E-05	149.6
TS1	Temperature Switch	10.2	0.102	10.302	1.461389E-02	1.995945E-05	195.7
TS2	Temperature Switch	10.2	0.102	10.302	1.461389E-02	1.995945E-05	195.7
LS1	Level Switch	8.9	0.089	8.989	1.460912E-02	1.519596E-05	170.7
LS2	Level Switch	8.9	0.089	8.989	1.460912E-02	1.519596E-05	170.7
BV1	Block Valves	4.2	0.042	4.242	1.460037E-02	6.445958E-06	153.5
BV2	Block Valves	4.2	0.042	4.242	1.460037E-02	6.445958E-06	153.5
SOL1	Solenoid Valves	3.8	0.038	3.838	1.459976E-02	5.832058E-06	153.5
SOL2	Solenoid Valves	3.8	0.038	3.838	1.459976E-02	5.832058E-06	153.5

Table 7. Variance of Top Event Probability Due to Uncertain Device Failure Rates

Device No	Device Type	Label	Sensitivity of Top Event Probability to Device Failure Rate $Hour^{-1}$	Variance of Device Failure Rate (Failures per 1E6 $Hour)^2$	Variance Contribution to Top Event Variance	Contribution to Top Event Variance %
1	Logic Solver	PE	-	0	0.00E+00	0%
2	Temperature Switch	TS1	195.7	12.9	4.94E-07	6%
3	Temperature Switch	TS2	195.7	12.9	4.94E-07	6%
4	Level Switch	LS1	170.7	15.8	4.60E-07	6%
5	Level Switch	LS2	170.7	15.8	4.60E-07	6%
6	Flow Transmitter	FT1	307	18.3	1.72E-06	21%
7	Flow Transmitter	FT2	307	18.3	1.72E-06	21%
8	Flow Transmitter	FT3	307	18.3	1.72E-06	21%
9	Block Valve	BV1	153.5	4.4	1.04E-07	1%
10	Block Valve	BV2	153.5	4.4	1.04E-07	1%
11	Pressure Transmitter	PT1	149.6	18.8	4.21E-07	5%
12	Pressure Transmitter	PT2	149.6	18.8	4.21E-07	5%
13	Solenoid valve	SOL1	153.5	2.3	5.42E-08	1%
14	Solenoid valve	SOL2	153.5	2.3	5.42E-08	1%

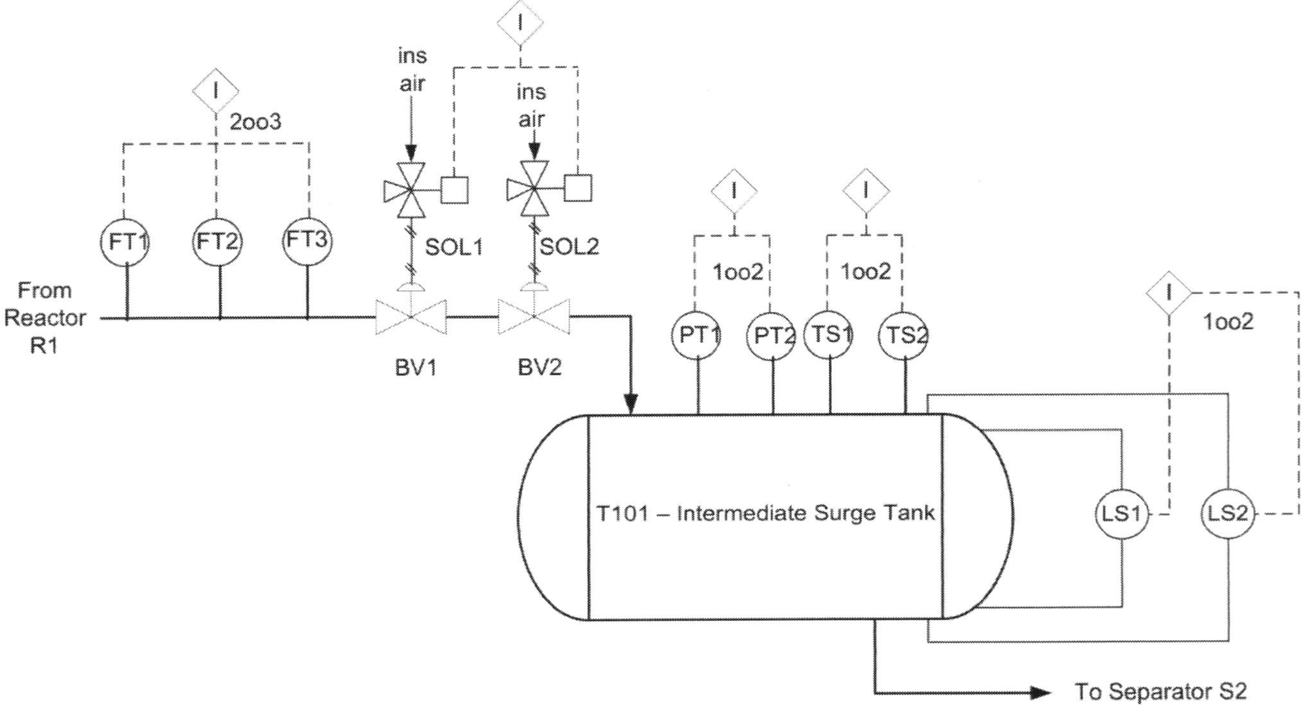

Figure 1. P&ID Diagram for Example

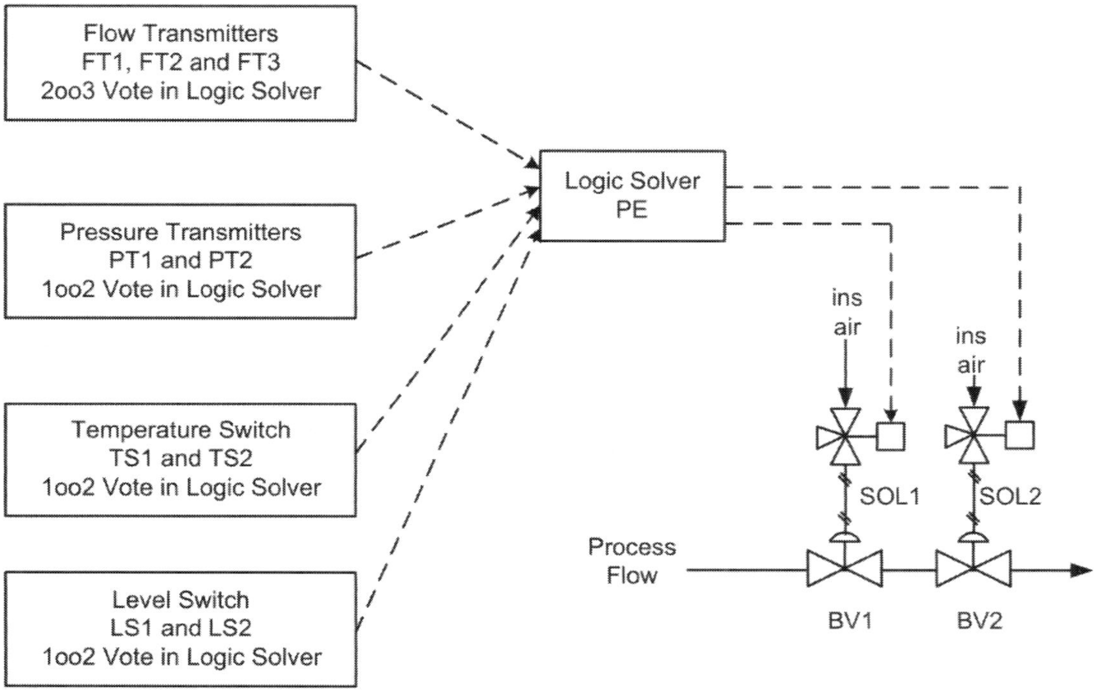

Figure 2. Interlock Block Diagram

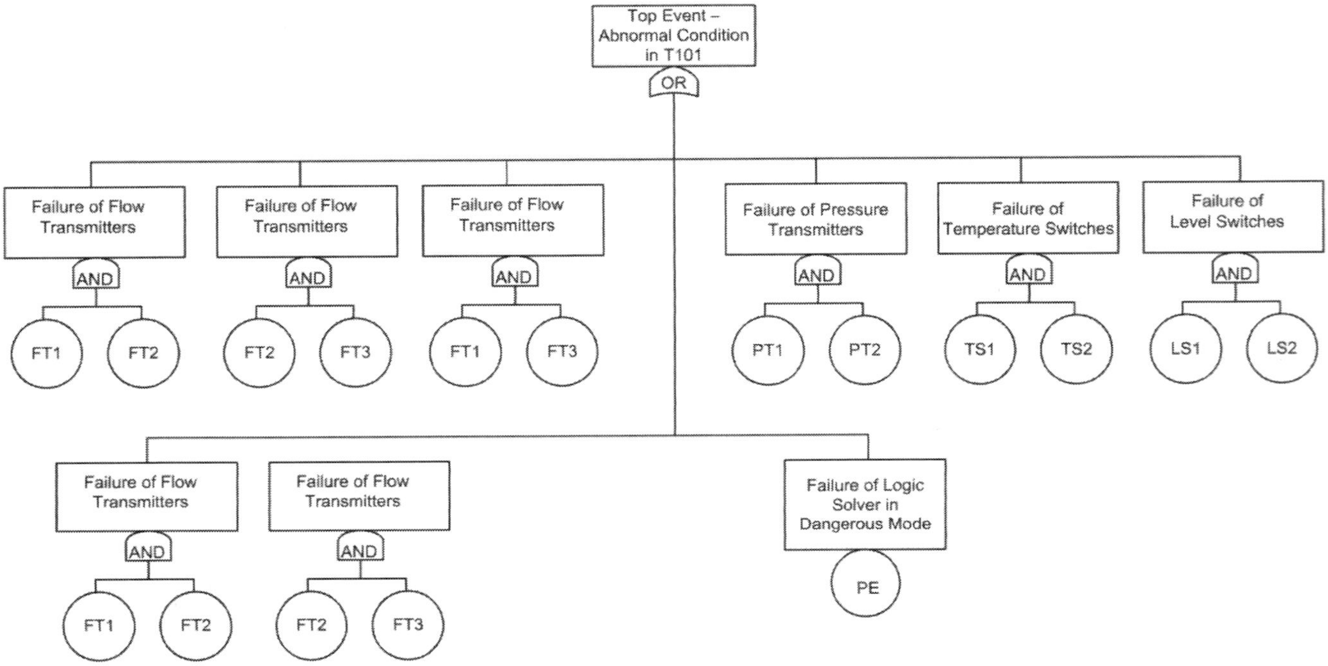

Figure 3. Fault Tree for Example Interlock

Appendix A – Commonly Encounter Probability Distributions

There are four probability distributions used to represent the failure rate of equipment items. These are:

1. Uniform Distribution
2. Triangular Distribution
3. Normal Distribution
4. Log-Normal Distribution

This appendix reviews the properties of these four probability distributions.

1. Uniform Distribution

The uniform distribution [Ref. A-1, pg 74; A-2, pg 687-688] is used when a variable is only known within a defined range. Figure A-1 presents a plot of the probability density function (pdf) for the uniform distribution. The distribution is defined using two parameters, A and B.

Where:

A = minimum value of the random variable

B = maximum value of the random variable

The mean or expected value of a random variable, x, that follows the uniform distribution is:

Mean = Expected Value – $E(x) = (A+B)/2$ (Eq A-1)

The variance of a random variable, x, that follows the uniform distribution is:

Variance = $V(x) = (B-A)^2/12$ (Eq A-2)

The standard deviation of a random variable, x, that follows the uniform distribution is:

Standard Deviation = $(\text{Variance})^{1/2} = (B-A)/(12)^{1/2}$ (Eq A-3)

2. Triangular Distribution

The triangular distribution [Ref. A-2, pg 686-687] is used when a variable is known to lie within a defined range and a "best guess" or estimate can be made as to the most likely value to the variable. Figure A-2 presents a plot of the probability density function (pdf) for the triangular distribution. The distribution is defined using three parameters, A. B and C.

Where:

A = minimum value of the random variable

B = maximum value of the random variable

C = most likely (mode) of the random variable

The mean or expected value of a random variable, x, that follows the triangular distribution is:

Mean = Expected Value – $E(x) = (A+B+C)/3$ (Eq A-4)

The variance of a random variable, x, that follows the triangular distribution is:

Variance = $V(x) = [A^2 + B^2 + C^2 - AB - AC - BC]/18$ (Eq A-5)

The standard deviation of a random variable, x, that follows the triangular distribution is:

Standard Deviation = $(\text{Variance})^{1/2} = \{[A^2 + B^2 + C^2 - AB - AC - BC]/18\}^{1/2}$ (Eq A-6)

3. Normal distribution

The normal distribution [Ref A-2, pg 665-666] is used when a variable when data analysis finds the normal distribution is the best model to describe the spread in the measured variable. Figure A-3 presents a plot of the probability density function (pdf) for the normal distribution. The distribution is defined using two parameters, μ and σ.

Where:

μ = mean of the measured data for variable

σ = standard deviation of the measured data for variable

The expected value of a random variable, x, that follows the normal distribution is:

Expected value = mean = $E(x) = \mu$ (Eq A-7)

The variance of a random variable, x, that follows the normal distribution is:

Variance = square of standard deviation = $V(x) = \sigma^2$ (Eq A-8)

The standard deviation of a random variable, x, that follows the normal distribution is:

Standard deviation = σ (Eq A-9)

4. Lognormal Distribution

The lognormal distribution [Ref. A- 2, pg 658-659] is used when a variable when data analysis finds the lognormal distribution is the best model to describe the spread in the measured variable. Figure A-4 presents a plot of the probability density function (pdf) for the lognormal distribution. Many variables in nature are lognormally distributed (for example the height of people). The log normal distribution is used when the variable of interest has a known physical lower limit of zero. A variable, x, is lognormally distributed if ln(x) is normally distributed. The lognormal distribution is defined using two parameters, μ_y and σ_y.

Let the original data be defined by the variable x with mean and standard deviation as:

μ_x = mean of the measured data for variable x

σ_x = standard deviation of the measured data for variable x

Define a new variable y as:

$y = \text{Ln}(x)$ (Eq A-10)

Then variable x is said to be lognormally distributed if y is normally distributed. We can now write the mean and standard deviation of y (μ_y and σ_y.) [Ref. A-3, A-4 and A-5] as

$$\mu_y = \mathrm{Ln}\left[\mu_x^2 / ((\sigma_x^2 + \mu_x^2)^{1/2})\right] \quad \text{(Eq A-11)}$$

$$\sigma_y = \left[\mathrm{Ln}(\sigma_x^2 / \mu_x^2 + 1)\right]^{1/2} \quad \text{(Eq A-12)}$$

Where:

Ln is the natural logarithm (base e) of the argument

Appendix A References:

A-1. P. L. Meyer, *Introductory Probability and Statistical Applications*, 2nd Edition, Addison-Wesley, Reading Mass, Library of Congress Catalog no. 75-104971, 1972, pp. 74

A-2. D. Vose, *Risk Analysis–A Quantitative Guide*, 3rd edition, Wiley, Chichester, West Sussex, England, ISBN 978-0-470-51284-5, 2008, pp. 686-687

A-3. "Lognormal mean and variance - MATLAB lognstat". www.mathworks.com. Retrieved 27 December 2018

A-4. M. Mood, F. A. Graybill, and D. C. Boes. *Introduction to the Theory of Statistics.* 3rd ed., New York: McGraw-Hill, ISBN 0-07-042864-6, 1974. pp. 117.

A-5. E.L. Crow and K. Shimizu, Lognormal Distributions – Theory and Applications, Marcel Dekker, New York, ISBN 0-8247-7803-0, 1988, pp. 9

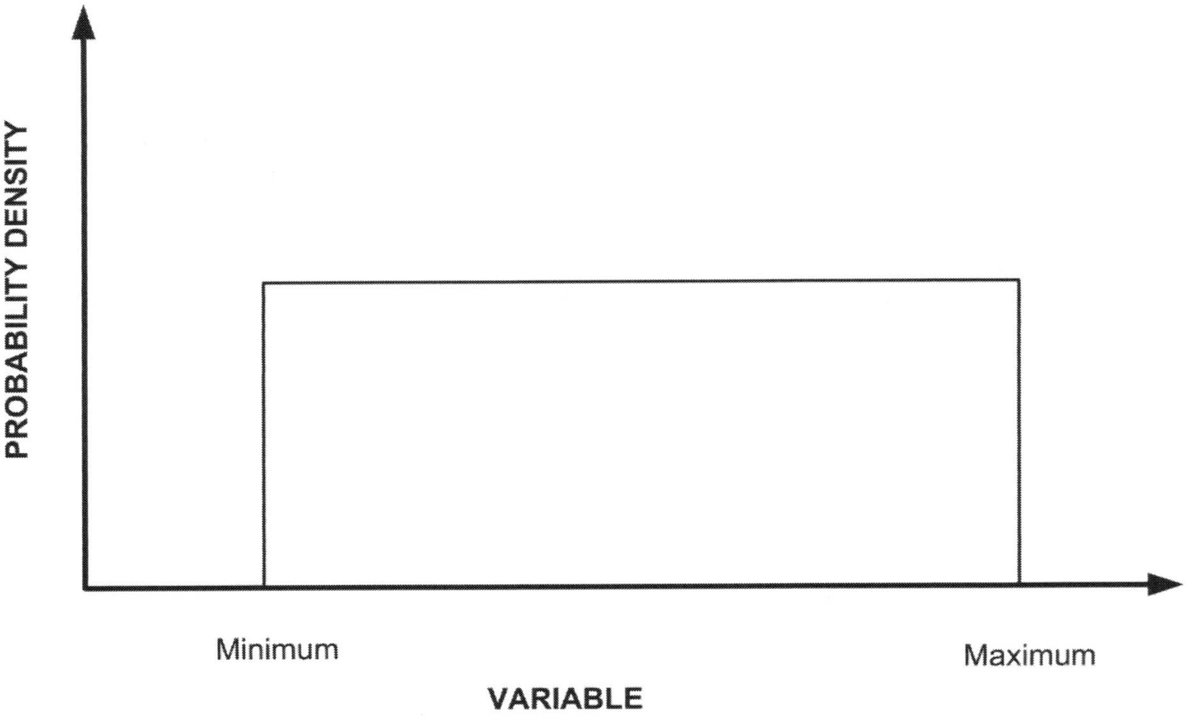

Figure A-1. Uniform Probability Distribution

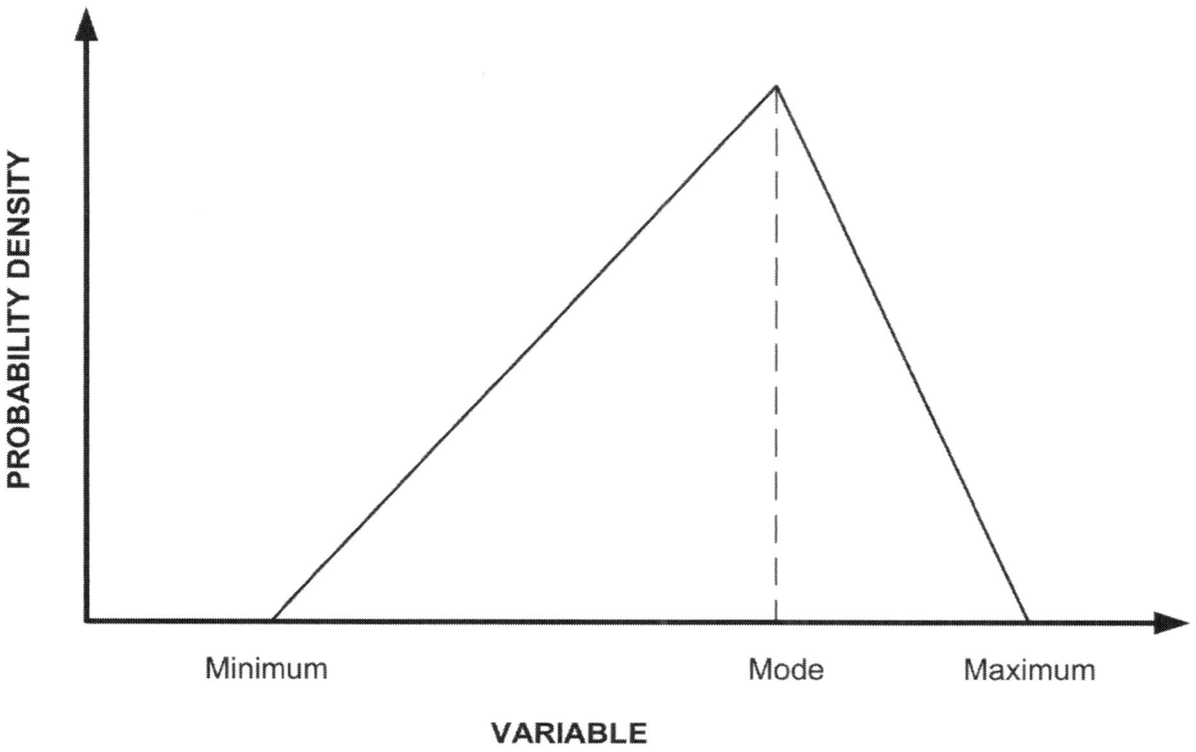

Figure A-2. Triangular Probability Distribution

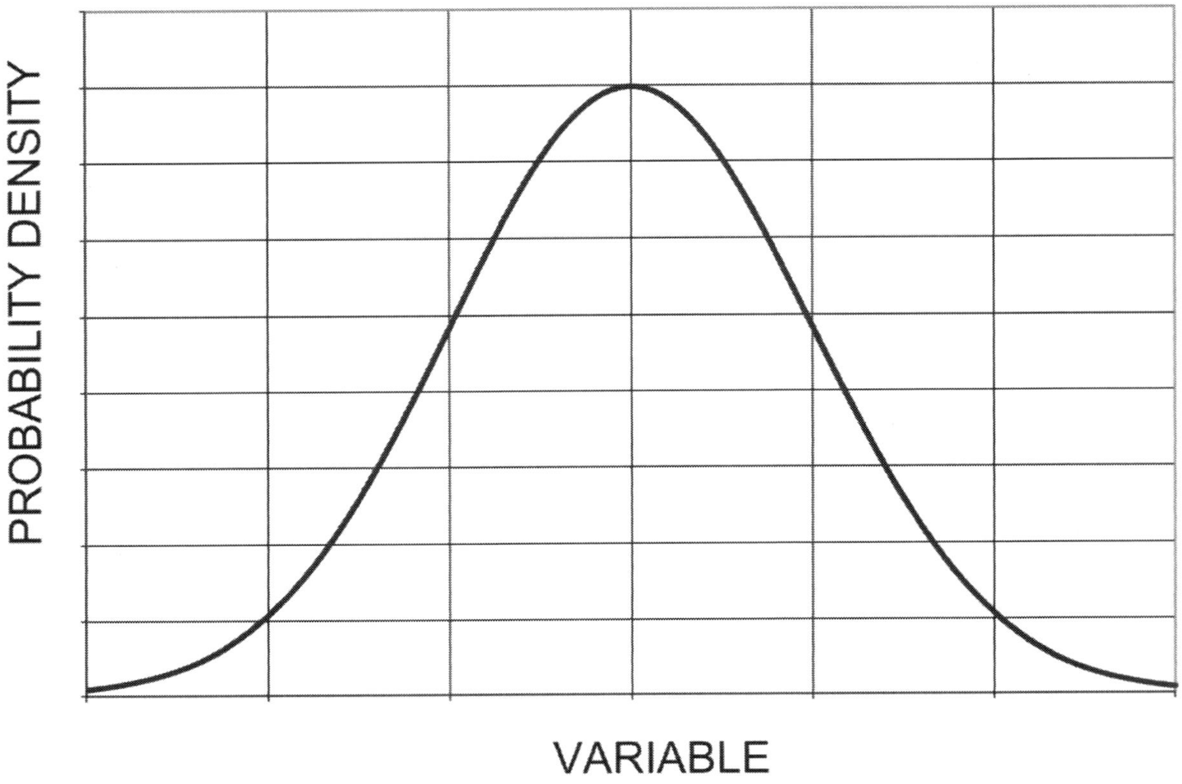

Figure A-3. Normal Probability Distribution

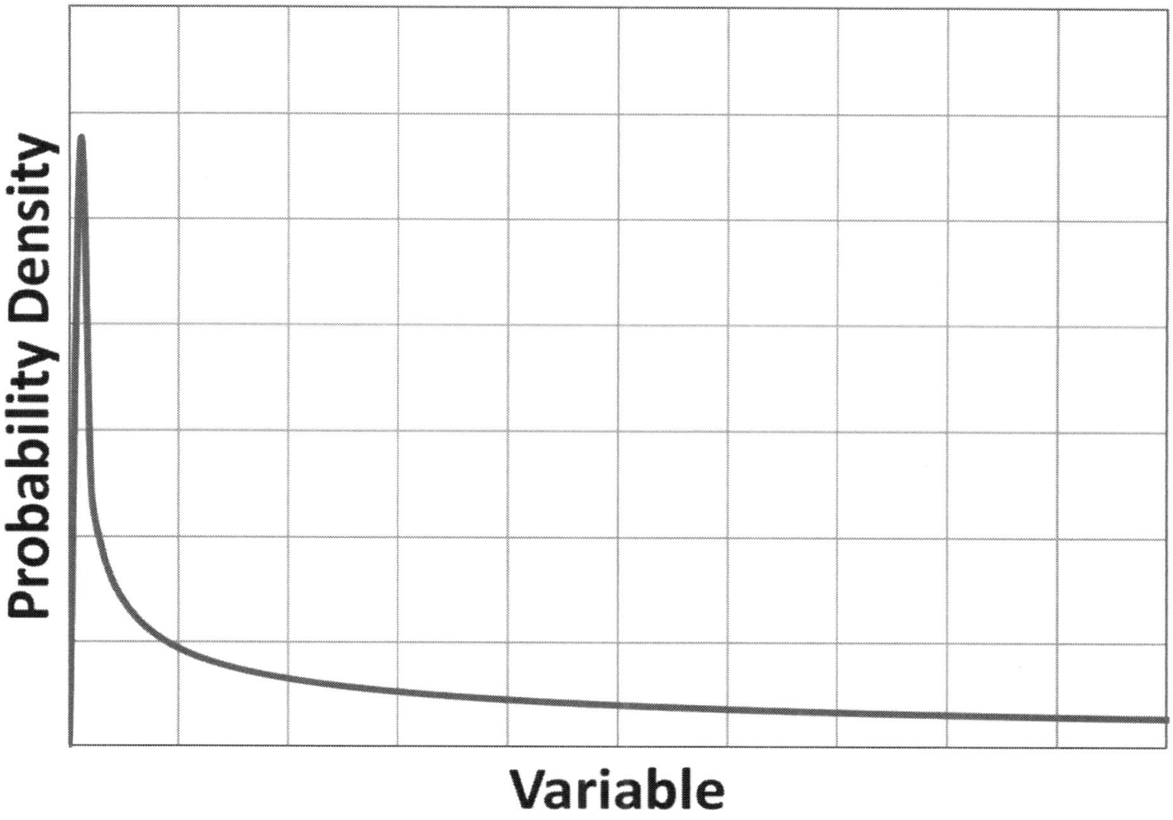

Figure A-4. Lognormal Probability Distribution

590424 Making the Most of Your PHA

Abstract:

A Process Hazard Analysis (PHA) is required to address OSHA Process Safety Management (PSM) and EPA Risk Management Program (RMP) regulations. Understanding the potential process hazard is widely recognized as a foundational element of PSM and RMP, and in addition to simply meeting regulatory requirements, can be a critical mechanism for flushing out potential vulnerabilities early in the design process for capital projects. Although an important activity, PHA is also recognized as a sizeable investment in time and energy for an Owner/Operator. So why not make the most of it? Other important activities at plants such as Mechanical Integrity, Alarm Rationalization, and Management of Change are able to stay focused and possibly be streamlined (i.e., reduced cost) based on the ability to use the results of a quality PHA having the necessary information and usability of documentation. So why not structure the PHA and documentation to be able to capitalize on potentially significant cost-savings and possible program integration?
The thrust of this paper and conference presentation will be to identify tips for conducting and documenting the PHA to maximize its ability to be used for supporting other important programs such as MOC, MI, and Alarm Rationalization. The objective will be to allow PHA Coordinators to be able to readily apply these tips to bring about improvements and cost-savings on these interrelated programs.

Copyright © American Institute of Chemical Engineers. All rights reserved.

589814 Theoretical Analysis of the Viability of Isopropanol-Cyclohexane Solvent Systems for Biomatter Lipid Extraction

Abstract:

A theoretical alternative solvent combination for the extraction of lipids from biomatter is the Water, Isopropanol, and Cyclohexane instead of the more common combination used in the Bligh and Dyer Method (Water, Methanol, and Chloroform). While the Bligh and Dyer Method is considered effective, toxicological concerns have been raised over its use of both Chloroform and Methanol. The alternative method utilizes solvents that do not pose as high as a toxicological concern while theoretically maintaining a similar level of practical viability. The alternative method's extraction efficacy was judged against the performance of the Bligh and Dyer method using the simulation software Aspen Plus V10®. The equilibrium data generated from Aspen Plus V10® was then graphed into tertiary diagrams for data analysis using the graphing software SigmaPlot.

Copyright © American Institute of Chemical Engineers. All rights reserved.

592166 Techno-Economic Analysis of the Conversion of CH4 and CO2 Intoacetic Acid Using Microwave Plasma

Abstract:

Methane and carbon dioxide together represent 90% of greenhouse gases [1], and one way of diminishing their emissions is through the development of pathways that convert them into added value chemicals. Acetic acid is one of such chemicals, and efforts at the laboratory scale have been carried out to obtain alternative ways of producing it. However, there are very few studies that evaluate the possibility of implementing these processes at a larger scale. In this paper, techno-economic and sustainability analyses were used to assess the feasibility of upscaling two pathways to produce acetic acid: (1) by direct conversion of methane and carbon dioxide and (2) by indirect conversion, using syngas (obtained from dry methane reforming) as an intermediate. For the direct method a microwave non-catalytic plasma reactor was considered as the main driver in the process simulation [2]. For the indirect method, a zeolite catalytical single step reaction unit was considered [3]. APEN Plus was used to simulate the reaction and purification systems of both methods, and alongside literature values and plant cost estimation techniques, the capital and operational expenses were calculated. Finally, the feasibility of both projects is assessed on carbon sequestration mass balances and economic metrics. The main sensitivity variables that need to be improved in order to obtain a minimum ROI threshold value were also evaluated.

Key words: carbon management, process design, microwave plasma reactor

[1] EPA. *Overview of Greenhouse Gases*. United States Environmental Protection Agency. 2017.

[2] Wang, L., Yi, Y., Wu, C., Guo, H., Tu, X. *One-Step Reforming of CO2 and CH4 into High-Value Liquid Chemicals and Fuels at Room Temperature by Plasma-Driven Catalysis*. Angewandte Chemie. 2017.

[3] Lopez, L., Velasco, J., Montes, V., Marinas, A., Cabrera, S., Boutonnet, M., Jaras, S. *Synthesis of Ethanol from Syngas over Rh/MCM-41 Catalyst: Effect of Water on Product Selectivity*. Catalysts. 2015.

Copyright © American Institute of Chemical Engineers. All rights reserved.

589668 Model Predictive Control Design for the Allam Power Plant Integrated with an Air Separation Unit

Abstract:

Since electricity and heat generation account for around 25% of this greenhouse gas emission, high efficiency power generation technology such as the Allam cycle which emits essentially no CO_2 and NOx is much needed. Being an oxy-fuel power cycle, the Allam power plant is inherently integrated with an air separation plant (ASU). In this study Aspen Plus Dynamics was used to model the integrated Allam power plant/Air Separation Unit (ASU) with a high degree of heat and work integration based on the steady state work of Net Power. In our previous work, regulatory and advanced PID controls were implemented and tested for various set point tracking and disturbance rejection scenarios. However, PID controllers have limitations for various reasons. Consequently, a model predictive control (MPC) was designed in this study to optimize and improve the performance of the integrated unit and to test its performance in terms of dynamic behavior and robustness to model uncertainties.

Copyright © American Institute of Chemical Engineers. All rights reserved.

589672 Explore the Function of Ion-Exchange Membrane (IEM) in Membrane Capacitive Deionization (mCDI) Cell, an Alternative Technique in Brackish Water Desalination Field

Abstract:

In the arid west, the freshwater supply of many communities is limited, leading to increased interest in tapping brackish water resources. Although reverse osmosis is the most common technology to upgrade saline waters, there is also interest in developing and improving alternative technologies. Here we focus on Capacitive Deionization (CDI), which has attracted broad attention as an energy efficient desalination technology, especially for ion extraction from low salinity brackish water. The objective of this work is to explore the function of ion-exchange membrane (IEM) in enhancing the performance of CDI cell. In this study, a fully coupled two-dimensional model describing ion transport and adsorption behavior in Membrane Capacitive Deionization (mCDI) cell was developed and solved numerically, taking advantage of Darcy's Law, Nernst-Planck equation, and Donnan equilibrium considering hydraulic dispersion in channel, non-ideal membrane, and modification of ion diffusivity in porous media. The model behavior was supported by the experimentally observed variations in transient effluent concentration under different operating conditions. The results showed that ion flux was significantly improved because of the IEM, leading to a faster desalination rate. The equilibrium adsorption amount is increased due to the permselectivity of the IEM, resulting in extra ion accumulation in electrode's macropores during desalination process. The cut-off time (optimal time to switch polarity to generate highest energy efficiency) of mCDI was shorter with a higher salt removal percentage compared to CDI under the same operating conditions. A sensitivity study of the thickness and the fixed charge density (starting from 500mM) of the IEM was done separately, and neither of them were found to be the dominant influencing factors under a low concentration range of brackish water (below 100mM) in this study.

Copyright © American Institute of Chemical Engineers. All rights reserved.

590103 An Integrated Process for the Production of Propylene Glycol from Shale Gas

Abstract:

Propylene glycol (PG) is produced on a large scale primarily for the production of environmentally safest and least toxic industrial and domestic products such as polyester resins, thermoset plastics, polyurethanes, food additives and antifreeze. In this research, a novel integrated plant is developed for the production of PG from shale gas. This integrated approach has the benefit of safer operating conditions since the intermediate propylene oxide, which is explosive, does not need to be stored and transported. Furthermore, there are potential economic benefits from integration. The overall plant is simulated in the Aspenplus environment, and a variety of process conditions are tested at steady-state to optimize the production of PG. Heat-integration tools are utilized for energy-saving and capital cost reduction opportunities. A comparative economic assessment based on existing plant information indicates that the use of process integration techniques has the potential to reduce costs significantly.

Keywords: Process Integration, Heat Integration, Propylene Glycol

Copyright © American Institute of Chemical Engineers. All rights reserved.

592456 Monitoring Methods for Fault Detection and Diagnosis: Algorithms and Applications

Abstract:

Many industrial processes collect an abundance of data from different sensors. These sensors often measure a wide variety of physical properties in order to ensure that these parameters are adhering to expected values. This is essential to ensure plant and operator safety, increase economic benefits, and maintain product quality.

A number of algorithms have been developed in order to improve existing fault detection and diagnosis performance. These algorithms integrate a number of different data-driven driven tools and methods. Multiscale wavelet-based representation of data can be used in order to handle data that is autocorrelated, non-Gaussian, and noisy. Hypothesis testing methods such as the Generalized Likelihood Ratio (GLR) technique can be used in order to provide the best possible detection for a fixed false alarm rate. Moreover, certain model-based methods have been developed in order to monitor process drifts and degradations in the process model, even when a process is operating under control.

This work will discuss the different tools utilized in development of the algorithms, their features, and practical applications, including the benchmark Tennessee Eastman Process (TEP).

Copyright © American Institute of Chemical Engineers. All rights reserved.

589902 Selective Separation of Industrial Relevant Metal Ions from High Tds Water Using Wafer-Enhanced Electrodeionization (WE-EDI)

Abstract:

There is a need for an efficient and economical process of ion-selective separation. In hydraulic fracturing, for instance, several million gallons of water are used in a site, most of which are disposed of by deep well injection taking the water from the ecosystem and causing earthquakes. However, due to the high amount of ions in wells at fracking sites, pressure driven processes for water recovery are impossible to use. This project will demonstrate an emerging environmentally friendly Wafer Enhanced Electrodeionization (WE-EDI) technology which attempts to eliminate the shortcomings that come with normal EDI and separates effluent streams into concentrated ions and clean water. In WE-EDI, ion-exchange resins are replaced by wafers that consist of cation and anion exchange beads and a polymer binding agent. In this study, selective separation of these ion exchange wafers and the effect of electroosmotic drag on the membrane structure have been performed for various anions and cations (Na^+, K^+, Ca^{2+}, Cl^-) and improved selectivity and higher water recoveries have been observed while significant modeling has taken place to predict the selective separation as a strong function of the equilibrium constants between the ions and the beads. This project is supported by the Membrane Science and Technology (MAST) Center and is conducted with the collaboration and guidance of industry mentors.

Copyright © American Institute of Chemical Engineers. All rights reserved.

593303 Influence of Particles on Amine Losses during CO_2 Capture: A Process Simulation Coupled Aerosol Dynamics Model _

Abstract:

Influence of Particles on Amine Losses during CO_2 Capture: A Process Simulation Coupled Aerosol Dynamics Model

David Dhanraj and Pratim Biswas

Aerosol and Air Quality Research Laboratory

Department of Energy, Environmental and Chemical Engineering

Washington University in St. Louis

St. Louis, MO 63130, USA

Efforts to reduce the concentration of atmospheric CO_2 have been initiated due to escalating concerns of global warming and climate change. Amine-based post-combustion CO_2 capture offers an economically viable retrofit for power-plants in combating these emissions. However, solvent losses have been a major challenge to commercializing this technology. Recent experiments[1] show that vapor-based and aerosol-based solvent losses are the key mechanisms of amine solvent loss during CO_2 capture. While vapor-based amine losses can be abated by employing techniques like water-washing, there has been little work to curtail aerosol-based solvent losses. This is primarily because the underlying mechanisms governing these losses are still not well understood. Although there are a few numerical models proposed for this system, there are no models that consider aerosol phenomena such as coagulation or particle loss. Moreover, these models have not been fully validated through experiments.

Furthermore, there are a few techniques that are being developed to tackle solvent losses. It is known that reducing particle number concentration in the flue gas will reduce amine emissions. Conventional Electrostatic Precipitators (ESP) are however inefficient in capturing particles in the size range 0.1 – 1.0 μm. Amongst different approaches demonstrated, introducing a pre-charging stage and a photo-ionization stage has shown potential in increasing the charging efficiency.

Therefore, the broad objectives of this work are primarily focused on two facets of the amine loss problem. The first is the development of a model that integrates aerosol phenomena with a process engineering model that accounts for the reactive mass transfer process in a packed-bed absorber. The second is the design, fabrication and testing of a bench-scale Photoionization Enhanced Two-staged Electrostatic Precipitator (PI-ESP) at Abbott Power Plant. In essence, with the PI-ESP aimed at reducing the particulate concentration in the flue gas, and the model relating the influence of particulate concentration in the flue gas to amine emissions, this work aims to illuminate the underlying mechanisms governing amine emissions, and demonstrate a solution to tackle the problem.

References

1. International Journal of Greenhouse Gas Control, vol. 85, pp. 86-99, 2019.
2. International Journal of Greenhouse Gas Control, vol. 19, pp. 138-144, 2013.

Copyright © American Institute of Chemical Engineers. All rights reserved.

AIChE Paper Number: 81D

TDLAS ANALYSIS OF AMMONIA (NH_3) IN HIGH-PURITY ETHYLENE

Greg Lankford
Marketing
SpectraSensors, Inc.

W. Gary Engelhart
Marketing Consultant
SpectraSensors, Inc.

Prepared for Presentation at the 20202 Spring National Meeting
Houston, TX, March 29 – April 2, 2020

AIChE and the EPC shall not be responsible for statements or opinions contained in papers or printed in its publications.

TDLAS ANALYSIS OF AMMONIA (NH_3) IN HIGH-PURITY ETHYLENE

Gary Lankford
Marketing
SpectraSensors, Inc.
W. Gary Engelhart
Marketing Consultant
SpectraSensors, Inc.

Abstract:

Ethylene plants utilize molecular sieves and adsorbents to remove polar contaminants (H_2O and NH_3) from ethylene to achieve polymer-grade specifications. Polymer-grade ethylene has stringent specifications for NH_3 content, which is measured in ethylene production plants and at custody transfer points. Pipeline specifications for high purity ethylene set the maximum NH_3 concentration at < 0.25ppm. Because catalysts used in polyethylene production processes are highly sensitive to NH_3 and other contaminants that poison and reduce catalyst activity, out-of-spec ethylene may be rejected, or sent to flare.

The subject of this paper is the use of Tunable diode laser absorption spectroscopy (TDLAS) analyzers to monitor sub-ppm levels of ammonia (NH_3) in high-purity ethylene gas streams. The high sensitivity and exceptionally fast response time of TDLAS analyzers to changes in NH_3 concentration is a key factor in the adoption of the technology for monitoring ethylene purity at pipeline custody transfer points and in feed streams to polyethylene polymer reactors.

Data demonstrating detection sensitivity at sup-ppm concentrations of NH_3, long-term measurement stability and repeatability, and results obtained from field installations will be presented.

Introduction

Catalysts used in polyethylene polymerization processes, and gas-phase processes in particular, are highly sensitive to the presence of contaminants in ethylene gas introduced to reaction vessels. Accordingly, ethylene plants must purify their ethylene product using molecular sieves and adsorbents to supply polymer-grade ethylene gas to polyethylene plants.

Metallocene catalysts are utilized in gas-phase polyethylene processes because of their high activity. These catalysts are capable of supporting one ethylene monomer insertion every millionth of a second. One gram of metallocene catalyst can produce 1,000 Kg of polymer before becoming inactive.

The basic structure of a metallocene catalyst is a transition metal atom sandwiched between two aromatic rings as depicted in Figure 1 (1). This structure was discovered by X-ray crystallography and spectroscopy in 1952. Ernst Fischer and Geoffrey Wilkinson were awarded the Nobel Prize in 1973 for the discovery of metallocene structures.

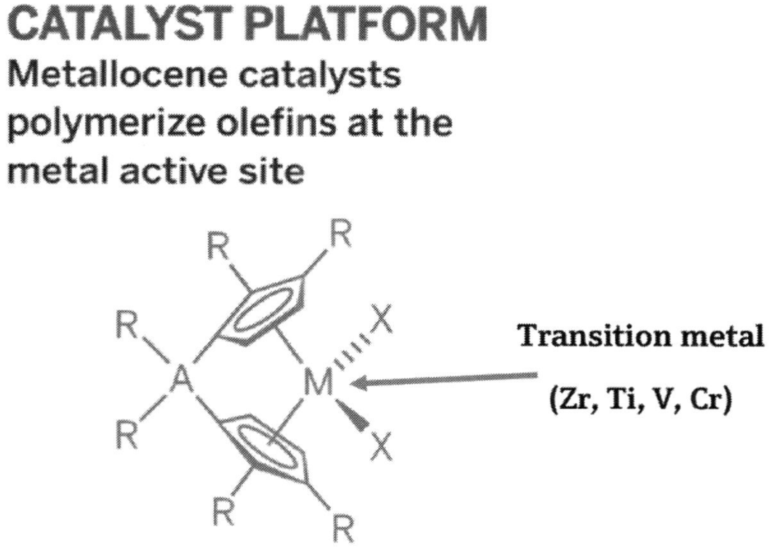

Figure 1. Basic Chemical Structure of a Metallocene Catalyst

The sensitivity of metallocene polymerization catalysts to deactivation by contaminants is specifically cited in patents for the application. U.S. Patent 3,709,853 (2), states; "However when the bis-(cyclopentadienyl) chromium II compound is contacted with an inorganic oxide support having a surface area in such a manner that it becomes adsorbed on the inorganic oxide, there is formed an unusually effective catalyst, in the absence of moisture and of air, for the polymerization of ethylene over a wide range of reaction conditions.

In some polymerization processes, ethylene, the nitrogen purge gas used to eliminate oxygen, hydrogen, and co-monomer gases undergo further purification to optimize catalyst performance and avoid premature deactivation as depicted in Figure 2. Water and ammonia, (with molecular sizes of 2.6 and 2.89 angstroms respectively), can be removed from ethylene using 3Å molecular sieves (3).

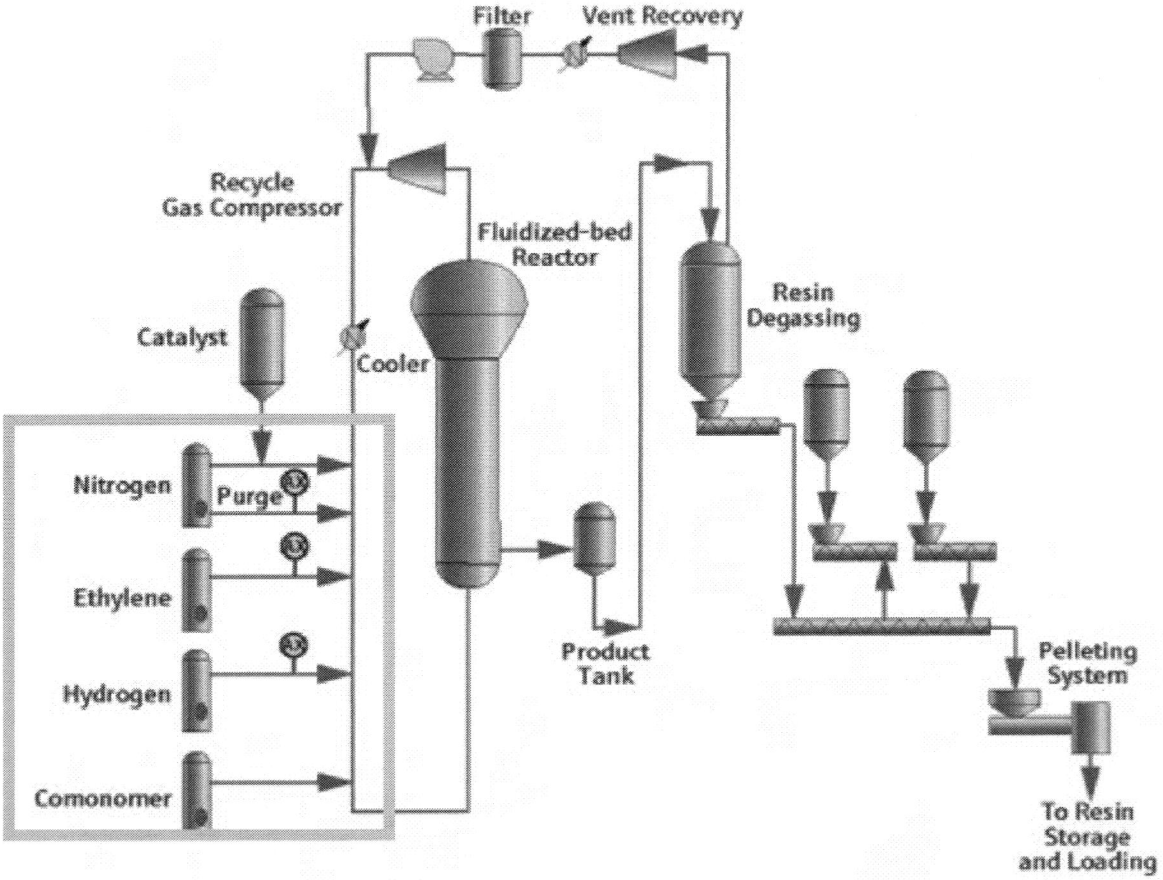

Figure 2. Purification of Ethylene, Nitrogen Purge Gas, and Hydrogen Prior to Introduction to a Fluidized-bed Polyethylene Reactor

On-Line Monitoring of NH_3 in Ethylene

The concentration of ammonia in ethylene is monitored at several points along the ethylene value chain as depicted in Figure 3. Pipeline tariff agreements include specifications defining maximum contaminant levels for water and ammonia. Likewise, polymer and petrochemical plants receiving the ethylene will have specifications on maximum contaminant levels.

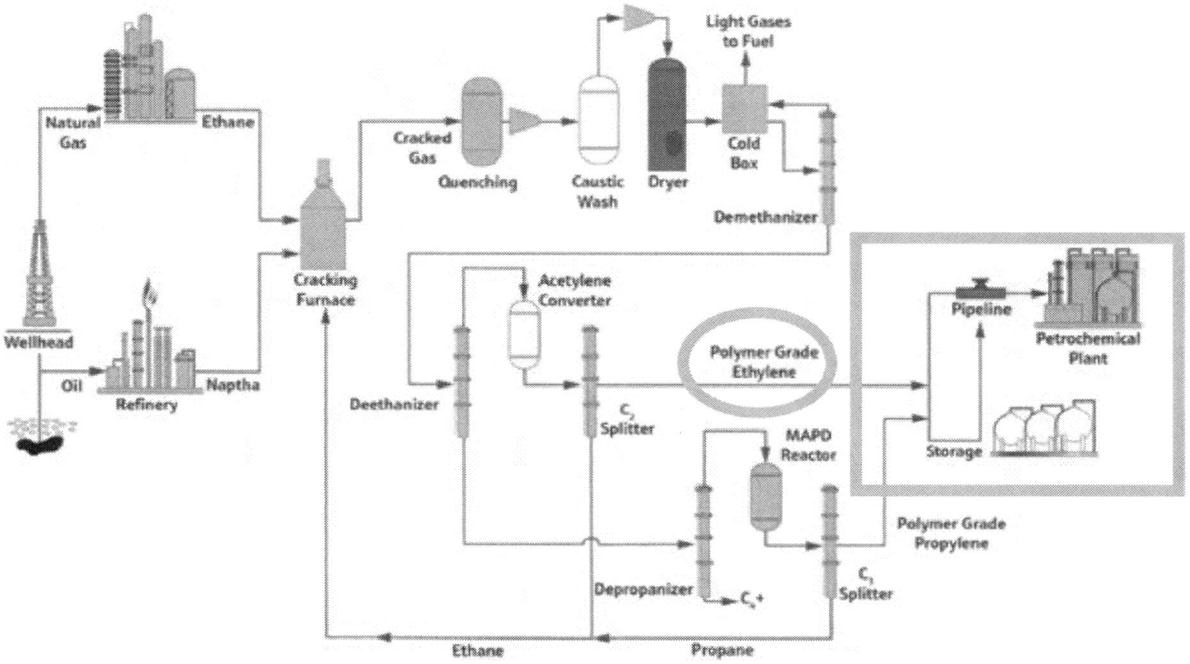

Figure 3. Simplified Diagram of Unit Operations in the Ethylene Production Process Highlighting Locations Where Ammonia Levels are Monitored

Technologies for On-line Monitoring of NH_3 in Ethylene

Several technologies have been employed for monitoring ammonia in ethylene and propylene, including; Ion Mobility Spectrometry (IMS), Dry Colorimetry, and Gas Chromatography before Tunable Diode Laser Absorption Spectroscopy analyzers became available.

Analyzers for on-line monitoring of ammonia in ethylene require certain performance characteristics to provide reliable data. An analyzer must be capable of performing accurate and repeatable measurements of ammonia at sub-ppm levels. Because the concentration of ammonia may be at or near zero for extended periods of time, it is important that the analyzer is capable of performing automated validation checks to ensure the analyzer is responding properly during these intervals. Unattended operation with little to no analyzer maintenance and support is important for analyzers intended for continuous on-line monitoring ethylene pipelines.

Ion Mobility Spectrometers are capable of performing sub-ppm measurements of ammonia in ethylene with good accuracy. However, many IMS analyzers deployed in the past for this measurement application required use of an isotopic ionization source, such as Tritium or ^{63}Ni. A site permit, with regular safety checks may be required for analyzer installations in some locations. Some external means must also be devised and installed to conduct validation checks of IMS analyzers or GC systems using an IMS detector. Manufacturers of IMS analyzers have moved away from supplying and servicing analyzers for process applications, preferring to focus their attention on airport security, and chemical weapons detection applications (4).

Dry Colorimetry Tape Analyzers are also capable of performing sub-ppm measurements of ammonia in ethylene. These analyzers use a chemically treated tape as a dry reaction substrate. The tape is mechanically fed to expose it to the process gas. When the tape is exposed to gas containing the target analyte, the tape will change color in proportion to the concentration of the analyte. A photo-optical system reads the intensity of the stain and compares it to a set of pre-programmed stain test results to provide the analyte concentration. Dry Colorimetry tape analyzers use a proprietary sensing tape, that must be continually replaced. The mechanical tape feed system must be also be regularly checked and maintained. While these analyzers are capable of low-level measurements, they are generally not equipped and able to perform automated validation checks. For pipeline custody transfer situations, where continuous 24-7 uninterrupted operation is required, dry colorimetry tape analyzers may not meet a user's analyzer availability (uptime) requirements.

TDLAS Analyzers for On-line Monitoring of NH_3 in Ethylene

TDLAS analyzers possess design and performance characteristics that make them especially well-suited for monitoring trace-level (low ppm to sub-ppm) concentrations of NH_3 in hydrocarbon gas streams.

The high-resolution laser used in TDLAS NH_3 analyzers emits a wavelength of near-infrared (NIR) light that is selective and specific for NH_3 molecules in a gas stream facilitating concentration measurements based upon molar absorptivity.

The solid-state laser and detector components in a TDLAS analyzer are isolated by a window from the process gas flowing continuously through the measurement cell as depicted in Figure 4. This design allows measurements to be performed with absolutely no contact between the process gas (and entrained contaminants) and these critical analyzer components.

The non-contact laser spectroscopy principle of TDLAS analyzers avoids measurement biases and errors and provides better analyzer availability.

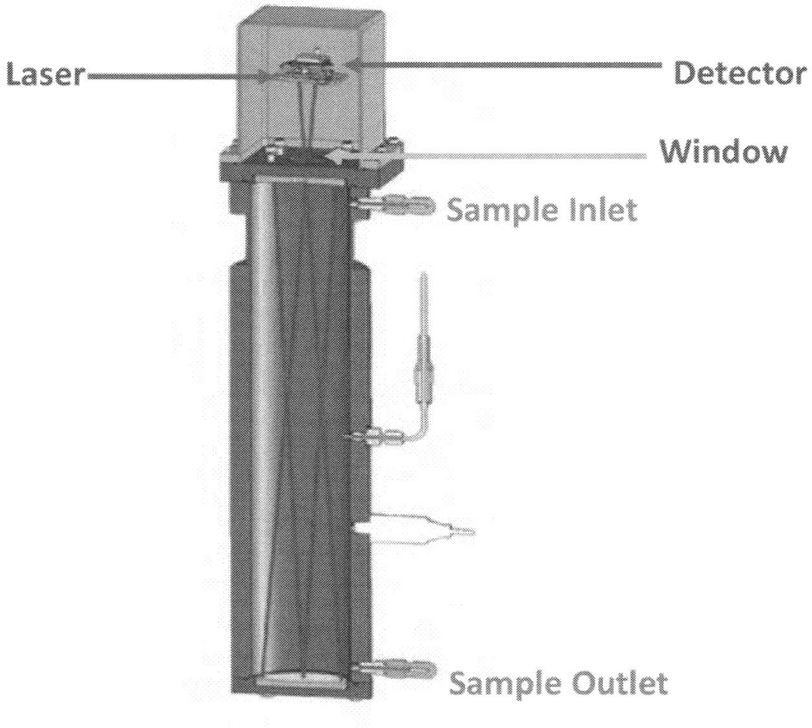

Figure 4. TDLAS Analyzer Flow Cell for Trace NH_3 Measurements Showing Positions of Laser, Detector, and Window

Molecular sieve purification is a highly efficient means of removing NH_3 from ethylene, and propylene. Under normal operating conditions gas exiting a molecular sieve vessel contains no detectable NH_3, or only trace-level (sub-ppm) concentrations of NH_3.

Measuring trace levels of NH_3 in pure ethylene or propylene requires high sensitivity detection because the background gas absorbs some of the NIR energy at or near the wavelength used to measure NH_3.

To achieve optimum sensitivity SpectraSensors developed and patented a spectral subtraction technique (differential spectroscopy) that enables trace-level measurements of NH_3

(and H_2O or H_2S) to be made when a process gas stream contains very low levels of an analyte and background interferences (5,6,7,8).

The differential spectroscopy technique is depicted in Figure 5. In operation a TDLAS analyzer performs a sequence of steps to obtain a "zero" spectrum and "process" spectrum that are used to calculate analyte concentration by spectral subtraction. The zero spectrum is obtained by passing the process gas through a high-efficiency scrubber which selectively removes the trace NH_3, without altering the process gas composition and background absorbance. The analyzer records the resulting NH_3-free spectrum of the process gas and automatically switches the sample gas flow to bypass the scrubber and collect the process gas spectrum with NH_3. Subtraction of the recorded zero spectrum from the process spectrum generates a differential spectrum of the trace analyte which is free of background interferences. The concentration of NH_3 is calculated from the differential spectrum.

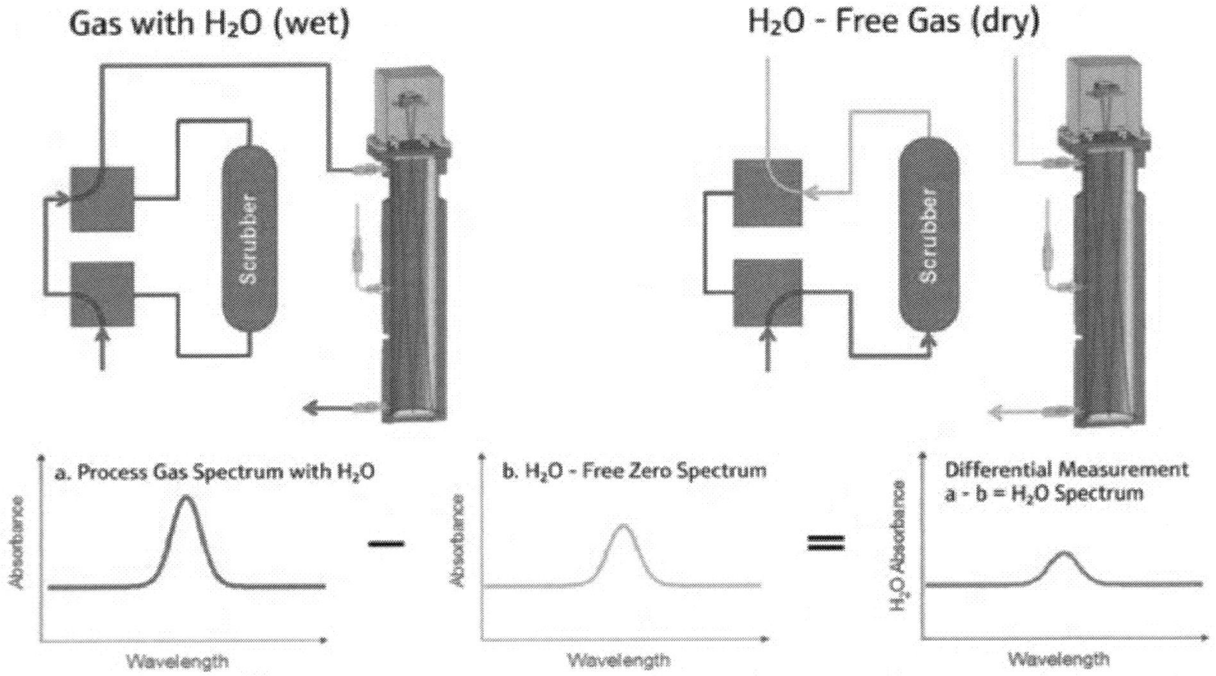

Figure 5. Depiction of Differential Spectroscopy Using Spectral Subtraction

Differential spectroscopy is used to perform trace-level measurements of NH_3 in ethylene gas with a calibration range of 0 – 10 ppm. Figure 6A shows the dry and wet spectra of NH_3 in ethylene gas, and Figure 6B shows the differential spectrum derived by spectral subtraction.

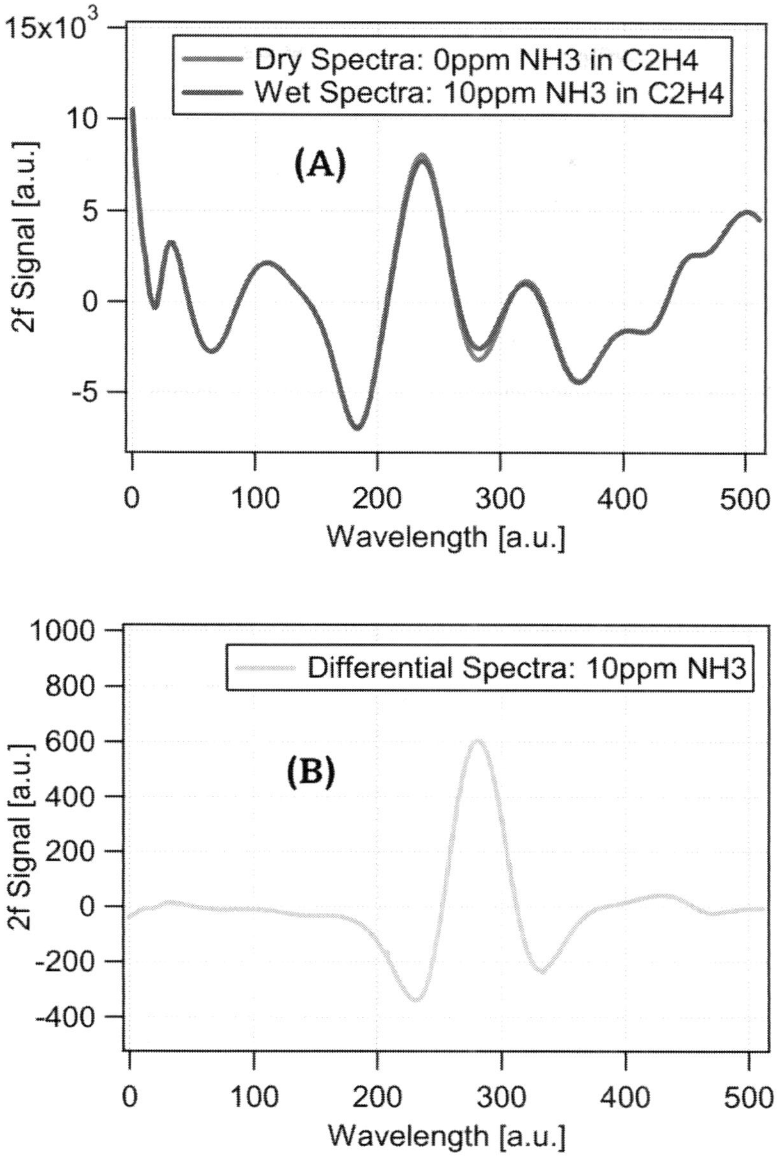

Figure 6. Wet and Dry Spectrum (A) and Differential Spectrum (B) of NH_3 in Ethylene

Factory Calibration, Performance Testing, and Field Validation

Every TDLAS trace NH_3 analyzer is factory tested and calibrated using a test mixture blended to simulate the ethylene gas stream composition furnished by the end user. The dilution ratio of the standard and mixing ratio of the background stream gases are controlled by digital mass flow controllers with NIST certifications.

The solid-state laser and detector components used in TDLAS analyzers are intrinsically stable, so no field calibration is required over the lifetime of the analyzer. Users perform periodic validation checks to verify the analyzer is operating properly within its factory-certified calibration range and ensure measurements are accurate for process control purposes.

Validation of analyzers for trace-level measurements of NH_3 in high-purity ethylene is particularly important because NH_3 will not be present for extended periods of time under normal operating conditions. The analyzer will read at or near zero ppm.

Cylinders of NIST-traceable certified reference gases are not commercially available for trace-level (low to sub-ppm) concentrations of NH_3 because they would not be stable. To address this situation TDLAS analyzers for trace-level NH_3 measurements are equipped with a permeation tube system to perform automated validation checks at user-designated time intervals. Because reliable NH_3 measurements in pipeline systems and polyethylene polymer plants are critical to uninterrupted operation, TDLAS analyzer validation checks are commonly performed daily in these systems.

The results of a long term (9-hour) factory repeatability test of a TDLAS analyzer conducted with 150 ppb in ethylene are shown in Figure 7. The 2σ repeatability observed during this test period was ± 14 ppb, well within the analyzer specifications of ± 40 ppb represented by the green shading on the graph.

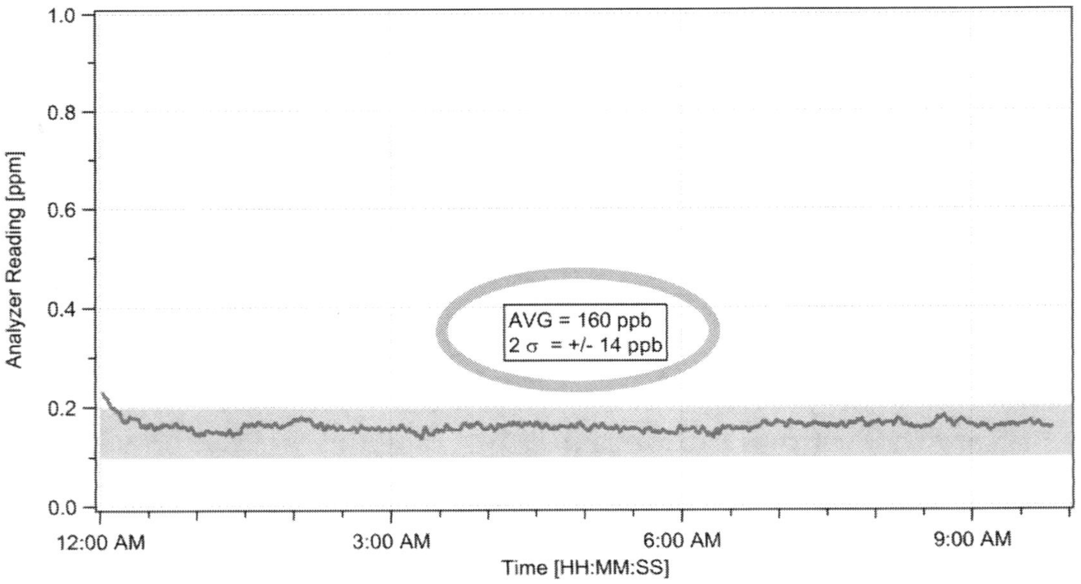

Figure 7. 9-Hour Repeatability Test of TDLAS Analyzer at 150 ppb NH_3 in Ethylene (2σ Repeatability is ± 14 ppb Well within Analyzer Specification of ± 40 ppb Represented by Green Shading)

Figure 8 shows the results of a concentration step change test from 5 ppm down to 0 ppm NH_3 in ethylene, demonstrating the exceptionally fast response of the analyzer to changes in NH_3 concentration.

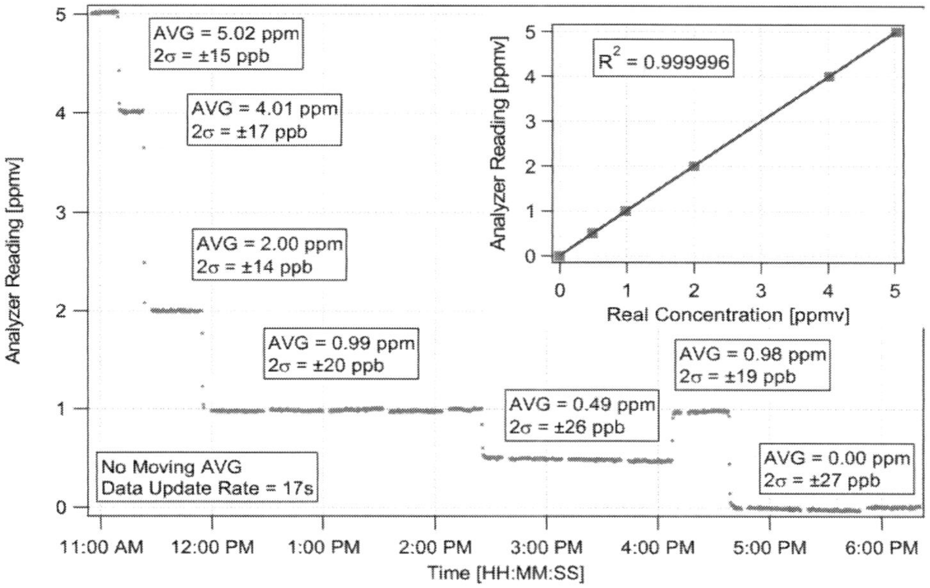

Figure 8. Concentration Step Change Test 5 – 0 ppm NH_3 in Ethylene

TDLAS Analyzer Field Results

Figure 9 shows two TDLAS analyzers installed on a rack for monitoring NH_3 and H_2O in an ethylene pipeline. The NH_3 analyzer has a measurement range of 0 – 5 ppm_v, for monitoring pipeline gas with a normal concentration of < 150 ppb_v. The H_2O analyzer has a measurement range of 0 – 10 ppm_v, for pipeline gas with a normal concentration of < 0.2 ppm_v.

Figure 9. Two TDLAS Analyzers for Monitoring NH_3 and H_2O in an Ethylene Pipeline

The exceptionally fast response time of TDLAS analyzers to changes in NH_3 concentration allow a single analyzer to be used to monitor multiple ethylene pipeline gas streams. Figure 10 shows a TDLAS analyzer for monitoring NH_3 in ethylene (cabinet on right) with stream switching (cabinet on left) for 3 pipeline gas streams.

Figure 10. TDLAS Analyzer for Monitoring NH_3 in Ethylene (Cabinet on Right) with Stream Switching (Cabinet on Left) for 3 Pipeline Streams. Gold Arrow Shows Tubing from Stream Switching Module Entering the Analyzer Housing

Figure 11 shows field data from a TDLAS analyzer monitoring NH_3 in ethylene over an 8-hour time period demonstrating the LDL (lowest detection limit) and repeatability of the analyzer are < 20 ppb.

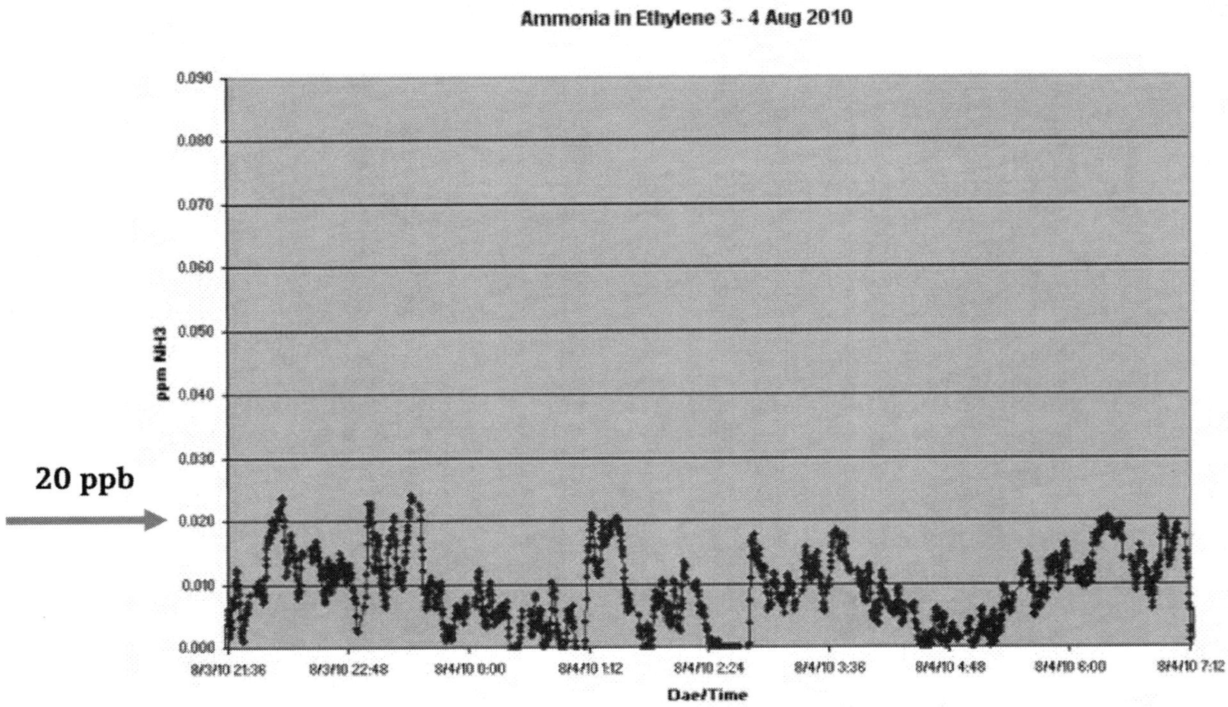

Figure 11. Field Data from 8-Hour Test Period Demonstrating Analyzer LDL (Lowest Detection Limit) and Repeatability are < 20ppb

Figure 12 shows the response of a TDLAS analyzer monitoring NH_3 in ethylene over a 7-day period. The average concentration of NH_3 is < 140 ppb during this time period. The spikes on the data readout are the automated analyzer validation checks performed with the on-board permeation tube.

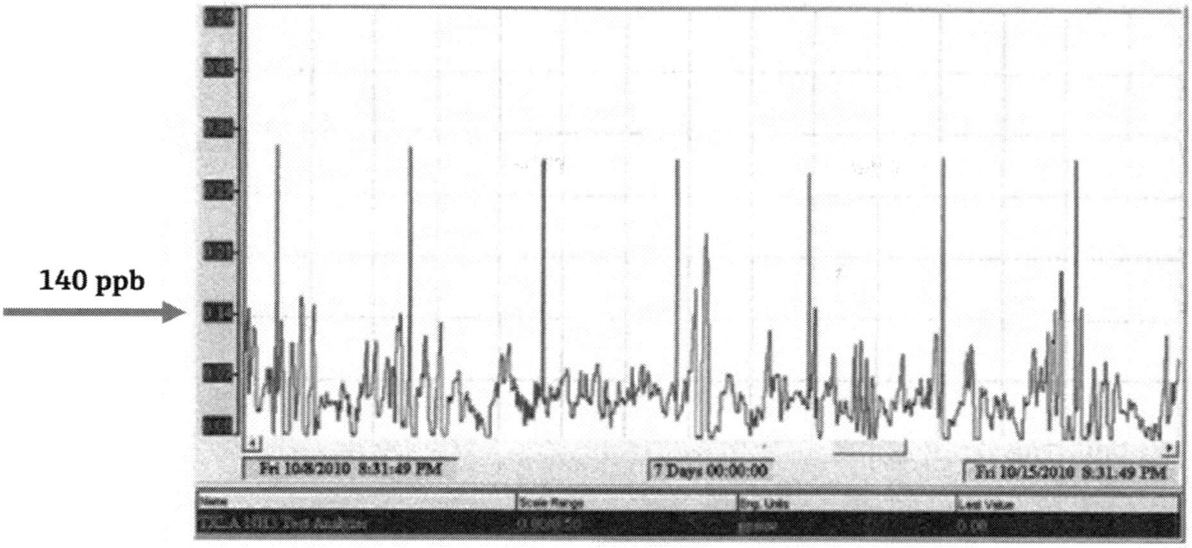

Figure 12. Response of a TDLAS Analyzer Monitoring NH_3 in Ethylene Over a 7-Day Period. The Average NH_3 Concentration is < 140 ppb. Spikes are Automated Analyzer Validation Checks Performed with the On-board Permeation Tube

Conclusions

On-line monitoring of ammonia (NH_3) in high-purity ethylene is of critical importance in pipeline systems and in the gas introduced into reactors of gas phase polyethylene processes. TDLAS analyzers possess design and performance characteristics that are well-suited for monitoring trace-levels of NH_3 in polymer-grade ethylene.

Differential spectroscopy enables detection and quantitation of sub-ppm concentrations of NH_3 in high purity ethylene gas streams. The fast response of TDLAS analyzers to NH_3 concentration changes helps avoid contamination of ethylene pipeline gas and protects expensive, high-activity polyethylene polymerization catalysts from poisoning and premature inactivation.

The on-board permeation tube system enables users to program automated analyzer validation checks to ensure a TDLAS analyzer is operating and responding properly during the extended time periods when little or no NH_3 is present in the ethylene gas stream.

The non-contact, laser-based measurement principle of TDLAS analyzers does not require use of a radioisotope ionization source, or proprietary tape colorimetry tape. The lower OPEX, maintenance and service requirements, and higher analyzer availability of TDLAS analyzers has led to increasing use of the technology along the ethylene production and supply chain.

References:

1. American Chemical Society, www.ACS.org
2. G.L. Karapinka, "Polymerization of Ethylene Using Supported Bis-(Cyclopentadienyl) Chromium II Catalysts", US Patent 3,709,853, Jan. 9, 1973
3. J. Gorawara, "Use of Molecular Sieves and Adsorbents for Contaminant Removal in Ethylene and Propylene Production", AIChE 2009.
4. J. Puton, and J. Namiesnik, "Ion Mobility Spectrometry: Current Status and Application for Chemical Warfare Agent Detection", Trends in Analytical Chemistry, 10.1016, 2016.
5. Zhou, X., Liu, X., Feitisch, A., and Sanger, G.M., "Reactive Gas Detection in Complex Backgrounds", US Patent 7,704,301 B2, Apr. 27, 2010.
6. Zhou, X., Liu, X., Feitisch, A., and Sanger, G.M., "Reactive Gas Detection in Complex Backgrounds", US Patent 7,819,946 B2, Oct. 26, 2008.
7. Zhou, X., Liu, X., Feitisch, A., and Sanger, G.M., "Reactive Gas Detection in Complex Backgrounds", US Patent 8,500,849 B2, Aug. 6, 2013.
8. Liu, X., Ji, W., and Feitisch, A., "Development of H_2S, H_2O, NH_3, and C_2H_2 TDL Analyzer for Petrochemical Applications in Optically Interfering Hydrocarbon Streams", ISA-AD, Apr. 2010

589894 Olefin Plant Wastewater Treatment Issues - Unique Challenges and Their Solutions

Abstract:

The ethylene/propylene production industry has been profoundly changed due to the frac gas production. A large number of world class gas steam crackers are being designed or under construction in the Gulf Coast Region and beyond. Although the wastewater production of these units is not as complex compared with integrated refinery/naphtha cracker, it has unique characteristics which will demand careful evaluations. These issues are:

1. Spent caustic treatment
2. Quench water pretreatment
3. Large amount of cooling tower blowdown handling
4. Biological wastewater treatment due to inorganic nutrient deficiency
5. Others

The presentation will consist of:

1. A general discussion of the ethane cracking process and wastewater generation;
2. Characteristics of effluent streams generated from gas cracking operations;
3. Challenges handling, treatment and management of gas cracking wastewater;
4. Wastewater treatment processes – biological treatment operational problems encountered in ethane crackers.
5. Case histories on the challenges and their mitigation at three crackers globally will be presented.
6. The remediation steps to control these biological processes deficiencies.

Copyright © American Institute of Chemical Engineers. All rights reserved.

81F

STUDY ON 1,3-BUTADIENE PRODUCTION ISSUES ARISING WITH THE INCREASING TREND OF LIGHTER FEEDSTOCK

Namit Tripathi
PhD Student
Lamar University

Srinivas Palanki
Professor
Lamar University

Qiang Xu
Professor
Lamar University

Prepared for Presentation at the 2020 Spring National Meeting
Houston, TX, March 31, 2020

AIChE and EPC shall not be responsible for statements or opinions contained in papers or printed in its publications

STUDY ON 1,3-BUTADIENE PRODUCTION ISSUES ARISING WITH THE INCREASING TREND OF LIGHTER FEEDSTOCK

1, 3-butadiene is used in the production of various materials and chemicals for applications in a wide range of industries. The majority of 1,3- butadiene is produced from steam cracking processes. However, the sustained production of 1,3-butadiene is facing big challenges as the cracking feedstock around the globe is trending towards lighter feedstocks, which not only causes a decrease in the butadiene supply, but also forces various butadiene facilities to run at high turndown capacities. In this paper, two relevant issues will be addressed. First, the optimal operating strategy has been studied for an existing 1, 3-butadiene process operated at a high turndown ratio. Both steady-state and dynamic process models have been developed for the virtual examination. The simulation results show that the optimal solvent to feed ratio is in the range of 6-7 when the plant is running at normal throughput; however, its necessary change the solvent to feed ratio in the range 10-11 when the plant is operating at its turndown capacity. Dynamic simulations also indicate that the effect of fluctuations in feed flowrate on product purity can be minimized via the use of ratio controller to change the solvent flowrate and a composition controller to change the side-draw flowrate. Next, potential bio-feedstock and technologies to make up the shortage of current 1,3-butadiene supplies have been studied. Three different routes that utilize biomasses to produce 1,3-butadiene and associated co-products ethylene and propylene from lignin are explored, where relevant process simulation models have been developed. The study shows that all three technological routes are all feasible and might be considered by the olefin industry. It also suggests that the comprehensive utilization of all olefin products (ethylene, propylene, and 1,3-butadiene) from biomasses based on process system integration might be a promising and profitable solution to deal with 1,3-butadiene shortfall issues in current olefin plants.

Keywords: Dynamic simulation, 1,3-Butadiene, Lignin, Extractive Distillation.

1. Introduction

1,3-Butadiene is an important feedstock in the production of rubbers and plastics, such as styrene butadiene rubber, polybutadiene rubber, and styrene butadiene latex.[1] Almost all 1,3- butadiene is produced based on steam cracking processes in olefin plants. Heavier cracker feedstock produces more butadiene than the lighter ones. As the cracker feedstock around the globe is trending towards lighter feedstock (mainly ethane) due to the shale gas booming, the sustained production of 1,3-butadiene in olefin plants is facing a grand challenge. In particular, most of current butadiene production facilities have to run at high turndown capacities. Projected by the current market trends in North America and Europe, 1,3-butadiene production would continue to decrease for the near future[2], which motivates our concerns on 1,3-butadiene plant operating issues under high turndown capacities. Thus, two relevant research aspects are addressed in this paper based on our research findings: (i) the optimal operating strategy has been studied for existing 1, 3-butadiene process operated at a high turndown ratio. In this aspect, both steady-state and dynamic models have been developed with Aspen Plus V10[3] to study how operation at turndown ratios of feed rates affects the production of 1,3-butadiene using N-methyl-2-pyrrolidone (NMP) as a solvent in an olefin plant. (ii) Potential bio-feedstock and technologies to make up the shortage of current 1,3-butadiene supplies have also been studied. In this aspect, three different routes that utilize biomasses to produce 1,3-butadiene and associated co-products ethylene and propylene from lignin are explored, where relevant process simulation models have been developed. The study shows that all three technological routes are all feasible and might be considered by the olefin industry. It also suggests that the comprehensive utilization of all olefin products (ethylene, propylene, and 1,3-butadiene) from biomasses based on process system integration might be a promising and profitable solution to deal with 1,3-butadiene shortfall issues in current olefin plants.

2. Process Description

2.1. 1,3-butadiene extraction process description

Figure 1 shows the process flow diagram of the extraction process for separating 1,3-butadiene from C_4 mixtures. The process consists of three main sections, which are extractive distillation section, ordinary distillation section and degassing section. The liquid C_4 feed first goes to the feed surge drum to minimize fluctuations in the flow to the process. The C_4 mixtures feed then goes into the main washer, which is first extractive distillation column. The required flow rate of C_4- mixture feed entering the main washer is measured in the feed line connecting the surge drum with main washer and adjusted with the help of a flow controller. Vaporized C_4-mixture enters the main washer from the bottom along with the process gas from the top of the rectifier and is washed in a counter-current operation with NMP, which enters the main washer from the top[4]. 1,3-Butadiene is almost completely absorbed in this tower. The overhead product from the main washer mainly contains propane, propene, propadiene, butanes and butanes that are present in the C_4-feed stock to the plant.

The solvent (NMP) with approximately 8.3wt% of water, enters the main washer from the top. The flow of the solvent is also controlled by a flow controller. The solvent withdrawn from the bottom of the main washer is pumped to the top of the rectifier. The level controller on the first extractive distillation column (main washer) controls the flow of the solvent. The bottom product from the first extractive distillation column, containing 1,3-butadiene, vinyl acetylene, ethyl acetylene and some butenes dissolved in the solvent, enters the rectifier from the top. In the upper part of the rectifier, the less soluble and more volatile butenes are stripped from the solvent stream through a counter-current vapor stream that contains more soluble butadiene, which is rising from the bottom of the rectifier. The gaseous mixture of 1,3-butadiene and butenes leaves from the top of the rectifier and returns to the main washer entering from the bottom. The mass fraction of 1,3-butadiene in the vapor rises to its maximum in between the upper section and lower section of the rectifier. From this location in the rectifier, 1,3-butadiene along with small amount of acetylene compounds, 1,2-butadiene, propyne and heavy C_4 and C_5 compounds are withdrawn as a side stream and are fed into the second extractive distillation column (after washer). The bottom product of the rectifier, containing C_4 acetylenes, and C_5 hydrocarbons dissolved in NMP, is preheated and sent to the degassing unit for stripping. The hot-stripped solvent from the bottom of the degasser passes through a train of heat exchangers and is recycled to both of the extractive distillation columns.

The rest of the butadiene along with C_4-acetylenes is withdrawn as a gaseous side-stream from the rectifier and fed into the bottom of the after-washer. In this column, vinylacetylene and ethylacetylene (1-butyne) are separated from 1,3-butadiene. Since C_4 - acetylenes are more soluble in NMP and heavier than 1,3-butadiene, they are removed counter-currently with fresh solvent that enters from the top of the column. The C_4 - acetylenes absorbed in the solvent are removed from the bottom of the second distillation column and fed to the rectifier and water which is present in the overhead vapor of this tower is separated from the bootleg of an accumulator.

The bottom product from the rectifier is heated up again and flashed into the degassing tower. The hydrocarbons dissolved in this solvent from the rectifier bottom have to be separated completely before the solvent is recycled to the extractive distillation columns. The stripped butadiene is recycled to the rectifier which gets withdrawn through the side-stream for further processing in the simple distillation section.

Propyne (methyl acetylene) is withdrawn with traces of 1,3-butadiene as an overhead product in propyne column through simple distillation. In the 1,3-butadiene distillation column, a mixture containing 1,2-butadiene and heavy C_4-hydrocarbons is separated as the bottom product. The pure 1,3-butadiene product is withdrawn from the

1,3-butadiene distillation column as an overhead product[4]. As can be seen from Figure 1, the liquid phase side-stream containing 1,3- butadiene and solvent is withdrawn from middle section of the rectifier and fed into the after washer (second extractive distillation). The withdrawal of the side stream from the rectifier provides lower energy costs than does the processes with other solvents[5].

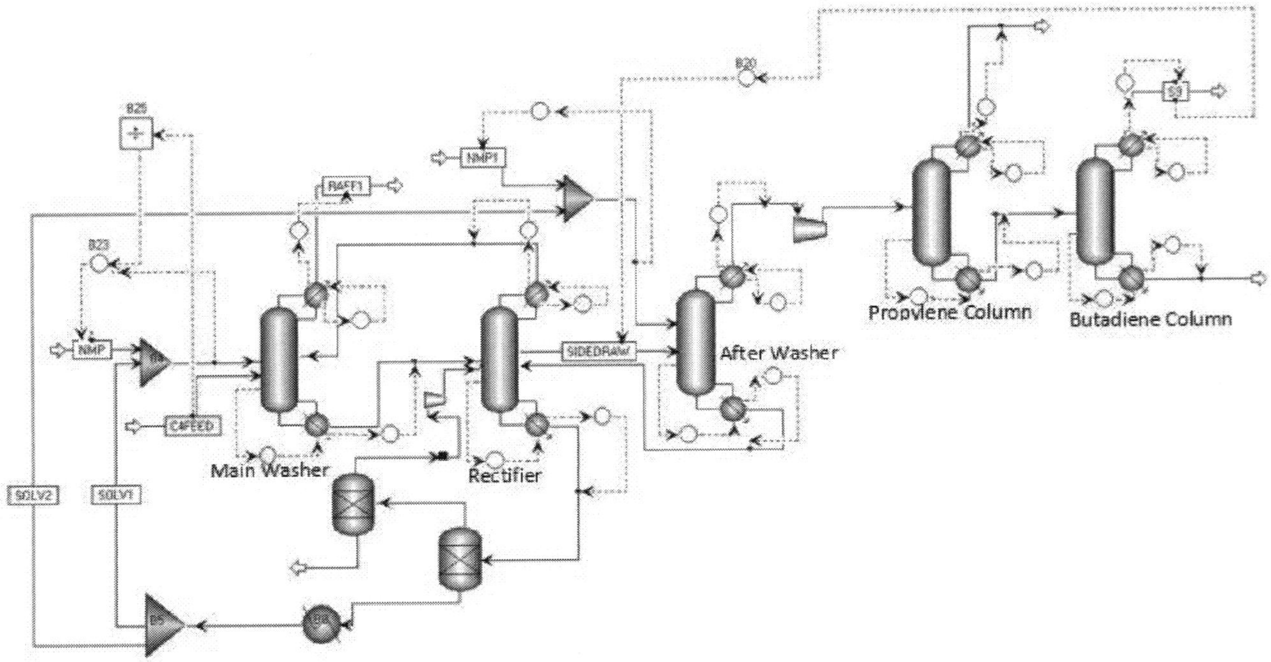

Figure 1. Process flow diagram of 1,3-butadiene extraction process.[25]

2.2. Process Description of proposed lignin-based processes

Production of syngas from biomass has been investigated in the literature using various types of gasifiers[6-9]. Syngas plays an important role as an intermediate in the production of Fischer–Tropsch liquids and its production from biomass exhibits a promising prospective22. Several biomass to syngas demonstration projects have been developed recently, such as the Hynol project in the United States, the BioMeet and Bio-Fuels projects in Sweden and the BGMSS project in Japan[10-12]. In this paper, lignin is first converted to syngas in a high-temperature, low-pressure gasifier. Then, three alternative routes are considered to convert the syn-gas to 1, 3-butadiene and associated co-products. Each of these routes involves multiple steps (reaction and separation) and the process conditions for each step are taken from the literature as described in the succeeding subsections. The first route converts lignin to 1, 3-butadiene and associated co-products via syn-gas, dimethyl ether (DME) and light oils (LO). The second route converts lignin to 1, 3-butadiene and associated co-products via syn-gas, methanol (MeOH) and light oils (LO). The third route converts lignin to 1, 3-butadiene and associated co-products via syn-gas and light oils (LO). In particular, we have focused on the use of the commercially available and most commonly used SAPO-34 catalyst for the conversion to 1,3 butadiene, ethylene and propylene. A brief description of each process is given below.

2.2.1 Gasification

In the present study, syngas with high CO and H_2 content is obtained via a gasification step. The gasification unit is based on the Thermoselect process for producing syngas from waste biomass[9]. A mixture of lignin and water is utilized as the feedstock for the reactor operating at 1500 K temperature and 0.1 MPa pressure. The following reaction occurs26:

$$Lignin + 8.7H_2O \rightarrow 9CO + 14.5H_2 \quad (1)$$

The syngas produced in the gasification unit is cooled and CO_2 in the cleaned syngas is separated in an absorption tower. This gasification step is common to all three routes considered in this research.

2.2.2 Route 1

Figure 2 shows a process flow diagram of Route 1. The Dimethyl ether (DME) synthesis unit is based on the process described by Holdings[13]. The following reactions occur:

$$CO + 2H_2 \rightarrow CH_3OH \ (methanol\ synthesis) \quad (2)$$

$$CO + H_2O \rightarrow CO_2 + H_2 \ (water - gas\ shift) \quad (3)$$

$$2CH_3OH \rightarrow CH_3OCH_3 + H_2O \ (methanol\ dehydration) \quad (4)$$

The synthesis of DME from syngas is favored at higher pressures. The syngas stream is pressurized to 3.3 MPa and cooled to 553.15 K and then supplied to a reactor[14]. The CO conversion[15] is 36%. The product mixture leaving the reactor is cooled 223.15 K, and then separated into gas and liquid phases. The liquid mixture is supplied to the DME-to-LOs reactor, which is based on a report by Park et al.[16] The following reaction occurs at 553.15 K:

$$CH_3OH + CH_3OCH_3 \rightarrow CH_3CH_2OCH_3 + H_2O \quad (5)$$

$$CH_3CH_2OCH_3 \rightarrow CH_3OH + C_2H_4 \quad (6)$$

$$C_2H_4 + CH_3OH \rightarrow C_3H_6 + H_2O \quad (7)$$

$$C_3H_6 + CH_3OH \rightarrow C_4H_8 + H_2O \quad (8)$$

The product leaving the reactor is pressurized to 2.1 MPa pressure and cooled to 223.15 K temperature and then supplied to the first distillation tower. It is necessary to utilize cryogenic distillation conditions because of the low boiling points of propylene and butenes. The distillate, which is predominantly propylene, is collected as the top product in the tower. The bottoms product of the first distillation consists mainly of butenes and water and is fed to the second distillation tower. The bottoms product of this column is mainly water. The distillate, which is predominantly butenes, is collected as the top product in the second distillation tower and is supplied to the Iso/Dehyd unit, which is used for the production of 1,3-butadiene.

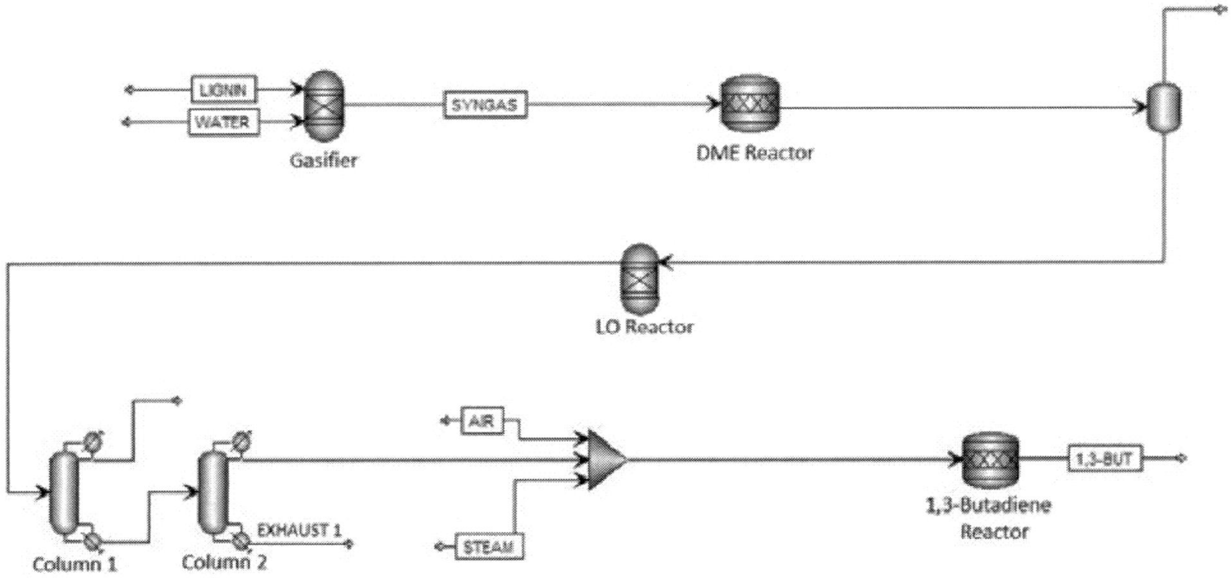

Figure 2. Process flow diagram for Route 1.[26]

2.2.3 Route 2

Figure 3 shows a process flow diagram of Route 2. In the H2/CO/CO2 adjustment unit, the syngas is mixed with H2 to achieve a H2/CO ratio 5:1 before it is supplied to the reactor[15]. The following reaction occurs:

$$CO + CO_2 + 5H_2 \rightarrow 2CH_3OH + H_2O \quad (9)$$

The MeOH synthesis unit is designed based on the operation conditions for a commercial plant[17]. The syngas is pressurized to 5.1 MPa and cooled to 573.15 K before being supplied to the reactor. The product mixture leaving the reactor is cooled to 273.15 K and the excess hydrogen is separated as gas phase. The liquid phase, which consists mainly of methanol, is supplied to the MeOH-to-LOs unit. The following reactions occur in the MeOH-to-LOs reactor:

$$2CH_3OH \rightarrow C_2H_4 + 2H_2O \quad 10)$$

$$3CH_3OH \rightarrow C_3H_6 + 3H_2O \quad (11)$$

$$4CH_3OH \rightarrow C_4H_8 + 4H_2O \quad (12)$$

The MeOH-to-LOs unit is designed based on a report by Kumita et al.[18]. The feed mixture is supplied to the reactor at 0.1 MPa pressure and a temperature of 273.15 K.

The product stream is pressurized to 2.1 MPa pressure and cooled to 223.15 K temperature, and then supplied to the first distillation tower. The distillate, containing predominantly propylene, is collected as a top product in the tower. The bottoms product of the first distillation consists mainly of butenes and water and is fed to the second distillation tower. The bottoms product of this column is mainly water. The distillate, which is predominantly butenes, is collected as the top product in the second distillation tower and is supplied to the Iso/Dehyd unit, which is used for the production of 1,3-butadiene. The details of this unit are the same as that used for Route 1.

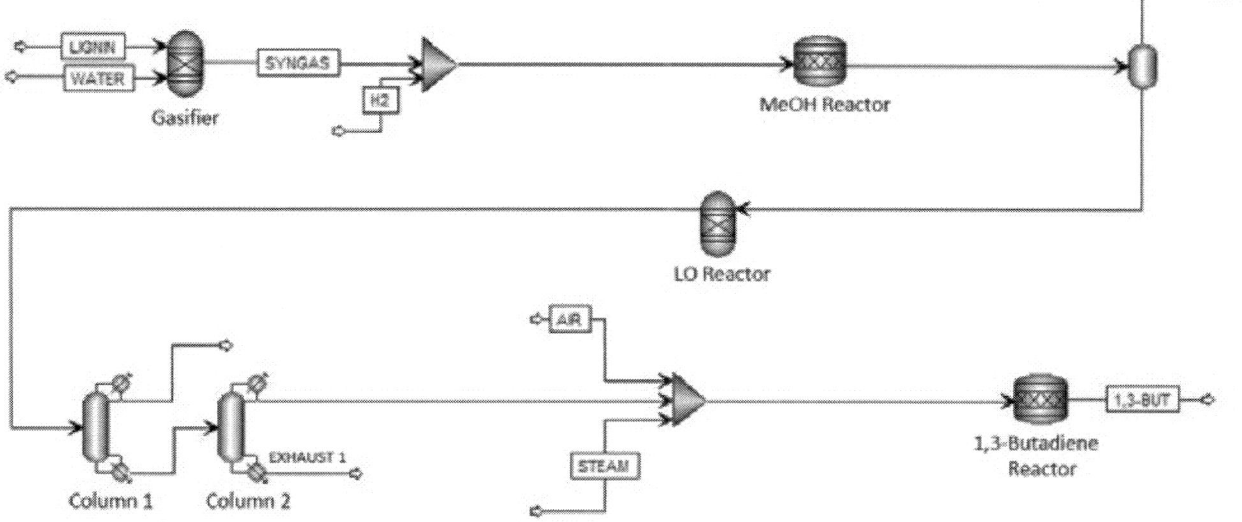

Figure 3. Process flow diagram for Route 2.[26]

2.3.4 Route 3

Figure 4 shows a process flow diagram of Route 3. The unit for the direct synthesis of LOs from syngas is designed based on a report by Schulte et al.[19]. The syngas is pressurized to 2.2 MPa and cooled to 613.15 K and then supplied to the reactor. The following reaction occurs:

$$11CO + 20H_2 \rightarrow CH_4 + C_2H_4 + C_3H_6 + C_4H_8 + CO_2 + 9\,H_2O \quad (13)$$

The product leaving the reactor is cooled to 223.15 K temperature and a pressure of 2.1 MPa. The liquid phase, which consists of propylene, butenes and water, is sent to the first distillation tower and the gas phase, which contains primarily ethylene is collected for further processing of ethylene. The distillate, containing predominantly propylene, is collected as a top product in the tower. The bottoms stream, which consists of butenes and water is fed to the second distillation tower. Water is removed as the bottom product. The distillate, which is predominantly butenes, is collected as the top product in the second distillation tower and is supplied to the Iso/Dehyd unit, which is used for the production of 1,3-butadiene. The details of this unit are the same as that used for Route 1 and Route 2 and are described below.

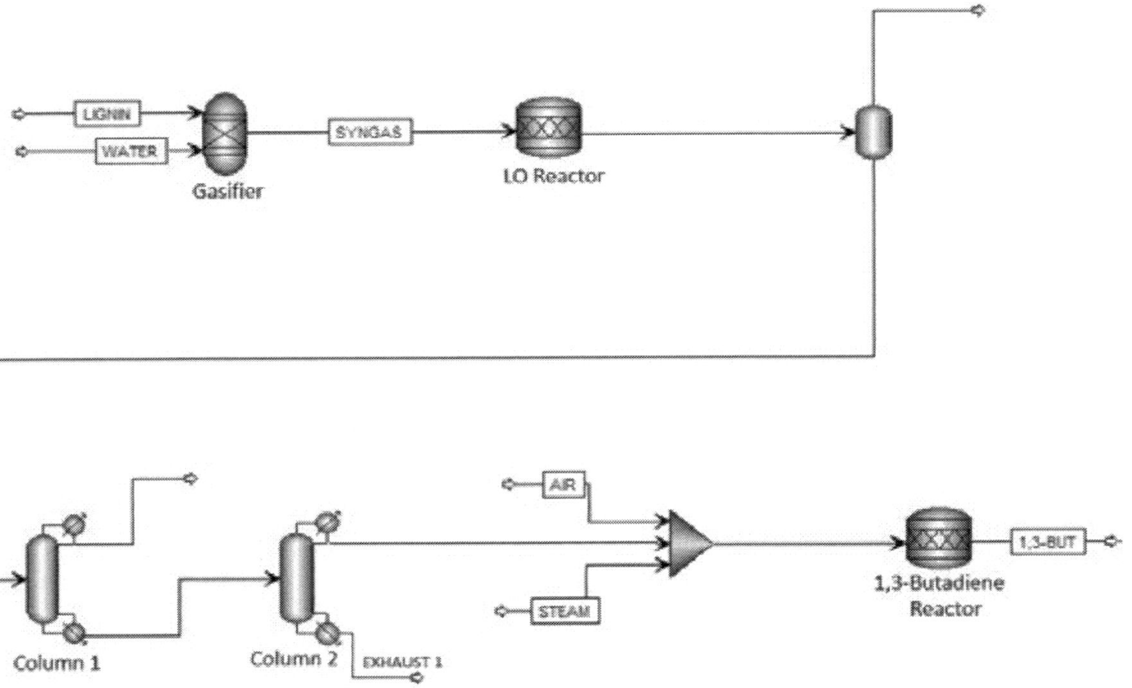

Figure 4. Process flow diagram for Route 3.[26]

Note that for all three routes, the Iso/Dehyd unit is designed based on a Mitsubishi patent[20]. This process is used to supply 1,3-butadiene commercially from n-butene obtained through the steam cracking of naphtha. The hydrocarbon mixture containing n-C4H8 is mixed with air, and steam, and then heated to the desired temperature and supplied to the reactor. The composition of the air/steam/hydrocarbon mixture used as a feed gas is 13.1:11.2:3.6[15]. The product mixture leaving the reactor is supplied to the distillation towers, which in turn provides 1,3-butadiene as the final product.

3. Process Simulation

3.1. Steady-state Simulation

A steady state model is developed with Aspen Plus V10[3] based on the plant PFD and operating conditions acquired from the literature[21,4,5]. Radfrac is used to model the azeotropic distillation columns. The Redlich-Kwong (RK) equation of state is employed for predicting vapor fugacity coefficient, while NRTL model of state is used for liquid phase in extractive distillation process. For the simple distillation section, Peng-Robinson equation of state model is used. Binary parameters for NRTL model are obtained from the literature and estimated using the UNIFAC method[22]. Thermodynamic properties of lignin are not directly available in the standard ASPEN PLUS properties database. A methodology to compute the properties of lignin in ASPEN PLUS was recently developed and verified experimentally by the National Renewable Energy Laboratory35 and this methodology is used in this research. The composition of the feed and solvent is taken from existing data[5]. Typically, the 1,3-butadiene content in C_4 mixtures received from naphtha crackers is between 40-55wt%. The solvent used for extraction contains 8.3 wt% of water and 91.7 wt% of n-methylpyrrolidone (NMP). Water is mixed with NMP because NMP-water mixtures have higher selectivity for unsaturated hydrocarbons than the pure NMP solvent[21]. Equipment configurations and operating parameters for each column have been decided based on the information available in the literature[21,4]. The composition of lignin and feed is taken from existing data[14,23]. Equipment configurations and operating parameters for each equipment are based on the information available in the literature[15].

3.2. Dynamic Simulation

The steady state model simulation file for 1,3-butadiene extractive distillation process is exported to the Aspen Dynamics and dynamic model is developed with the help of Aspen Plus V10[3]. Additional data on column diameter, tray spacing and weir height for the six columns needed for running the dynamic simulation is available in literature[25]. Conventional PID controllers are used for the ratio controller (B23) to maintain a constant solvent to feed ratio and the composition controller (B20). The pressure in all the columns is controlled by the flowrate of the vapor leaving from the top of the column. The liquid level in the reflux drum is controlled by reflux flowrate. The pressure in the reflux drum is controlled by manipulating the vapor flowrate. The reboiler level of the column is controlled by bottom product flowrate. The reboiler temperature is controlled by manipulating the reboiler heat duty. Controllers are tuned by trial-and-error by running several dynamic simulations and the detailed controller parameters is available in literature[25].

4. Results and Discussion

4.1 1,3 butadiene extractive distillation process

Table 1 shows the comparison between the simulated results and the plant data of the after washer (second extractive distillation column) overhead product, which further goes to the distillation section, where the final butadiene product is produced, and it is observed that the simulated stream composition is in good agreement with published literature[5]. Table 2 presents the comparison of both of simulation result and plant data composition of the final 1,3-butadiene product from the overhead of butadiene column. The results obtained from the simulation are in good agreement with the plant data provided in the literature[5].

After verifying the fidelity of the steady-state simulation results with published literature, dynamic simulations are conducted to study the effect of changing the solvent rate at normal and turn-down capacities. Figures 5 and 6 show the effect of excess solvent flow in the main washer to the purity of the final 1,3-butadiene product when the plant is running at normal throughput. Since the selectivity of the n-methylpyrrolidone (NMP) for 1,3-butadiene is high as compared for butanes and butenes, an excess of NMP solvent flow forces 1,3-butadiene to come out from the bottom of the main washer as the bottom product. Selectivity of NMP is also high enough to separate 1,2-butadiene from 1,3-butadiene. A range of solvent to feed ratio of 5-20 has been provided in the literature[21] but based on the dynamic simulation results from Figure 5, it is clear that the product purity of over 99% is achieved in a relatively narrow range of 6-7 for solvent to feed ratio. However, when this solvent to feed ratio in the range 6-7 is implemented on a plant operating at turndown capacity, there is a significant reduction in final product quality from 99.2% to 96.49% as shown in Figure 6. Dynamic simulation results indicate that the optimal solvent to feed ratio for the plant operating at turndown conditions is 10.0-11.0.

Table 1. Comparison between simulated results and plant data for after washer overhead product [25]

Output Section	Component	Simulated Results (wt%)	Plant Data[24] (wt%)
Overhead	1,3-Butadiene	99.04	99.08
Overhead	Ethylacetylene	0.00	0.00
Overhead	Vinylacetylene	0.00	0.00
Overhead	1,2-Butadiene, Others	0.96	0.92

Table 2. Comparison between simulated results and plant data for 1,3-butadiene column overhead product [25]

Output Section	Component	Simulated Results (wt%)	Plant Data[24] (wt%)
Overhead	1,3-Butadiene	99.21	99.70
Overhead	Isobutene	0.0215	0.045
Overhead	Trans-2-butene	-	0.21
Overhead	Cis-2-butene	-	0.20
Overhead	Others	0.7685	-

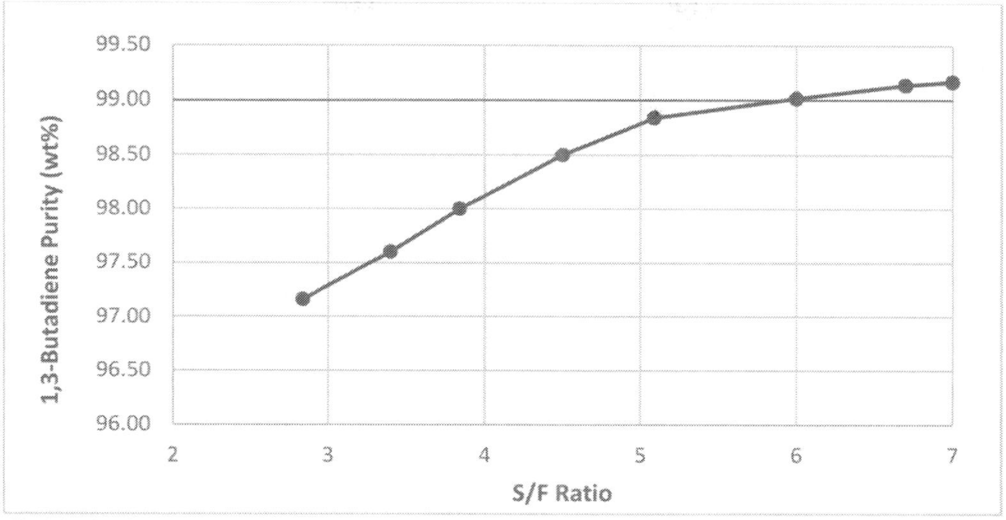

Figure 5. Variation in 1,3-butadiene product purity by changing solvent to feed (S/F) ratio in main washer at normal throughput.[25]

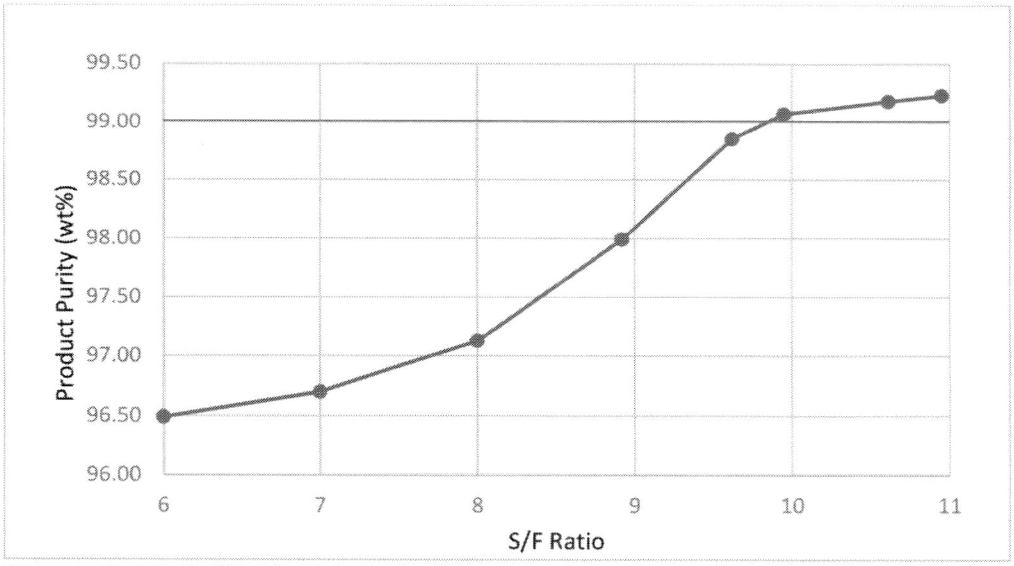

Figure 6. Variation in 1,3-butadiene product purity by changing solvent to feed (S/F) ratio in main washer at turndown capacity.[25]

The effect of fluctuations in the flow of the side draw stream is also studied. The side draw stream, which mainly contains 1,3-butadiene with small amounts of 1,2-butadiene, n-butane, butene, isobutene and propyne is stripped from the butanes and butenes in second distillation column. A study of this offset condition was simulated both at normal throughput and turndown capacity operating conditions in the dynamic simulation. It is observed from Figure 7 that a side draw of at least 40,500 kg/hr is required to maintain the desired product purity under normal operations. At turndown capacities, dynamic simulation results shown in Figure 8 indicate that the desired product purity of 99% is maintained for side draw flowrates between 39,000 kg/hr and 43,000 kg/hr. In particular, the final product purity changes from 99.23% to 99.09% when the side draw flowrate is increased from 39,000 kg/hr to 43,000 kg/hr because the entrainment of butanes and butenes in the final 1,3-butadiene product stream increases as shown in Figure 9.

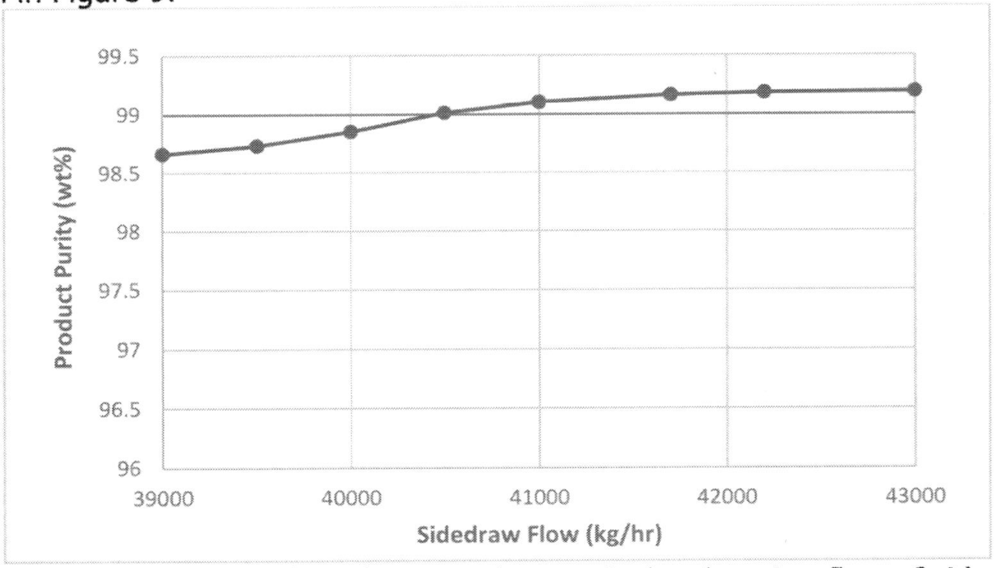

Figure 7. Variation in 1,3-butadiene product purity by changing flow of side draw stream at normal throughput.[25]

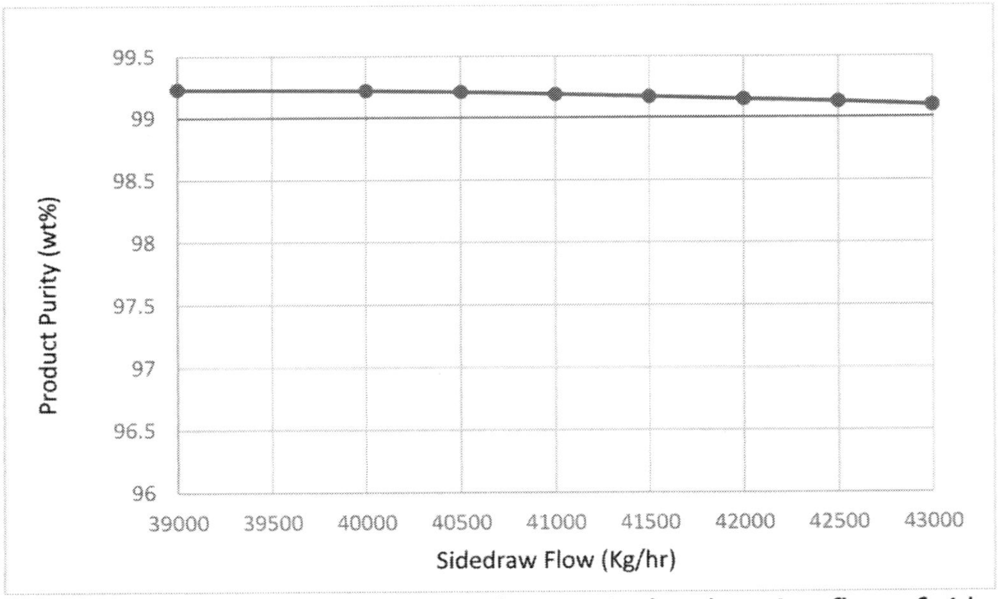

Figure 8. Variation in 1,3-butadiene product purity by changing flow of side draw stream at turndown capacity.[25]

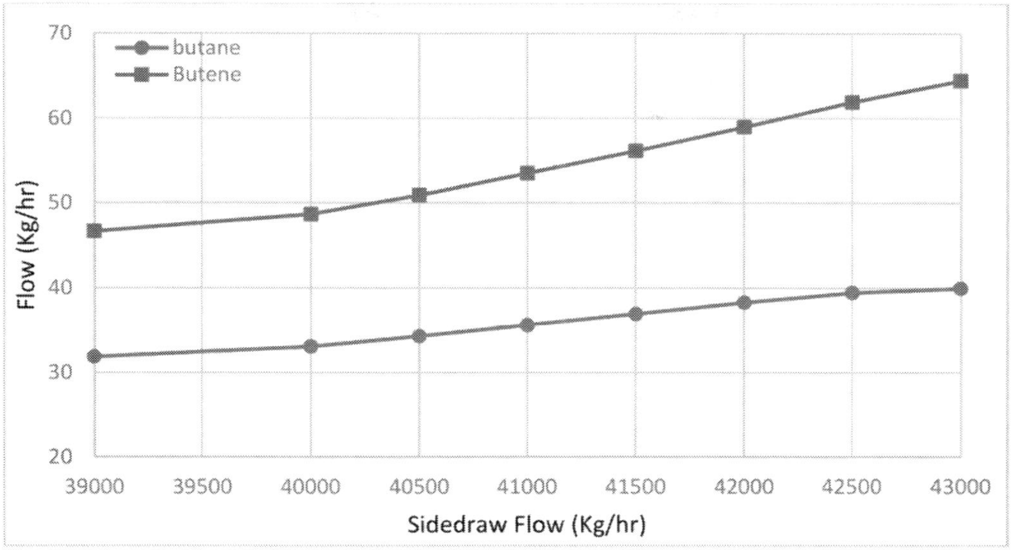

Figure 9. Variation in butanes and butenes flow in product stream by changing flow of side draw stream at turndown capacity.[25]

4.2 Lignin based Processes

ASPEN PLUS simulations were run using the same equipment specifications for lignin flow rates ranging from 480,000 – 720,000 kg/hr of lignin feed. For each case, the water feed was adjusted to the stoichiometric ratio shown in Eq. (1). The total amount of ethylene, propylene and 1,3 butadiene produced in each of the three routes was computed via these simulations. It is observed from Figures 10-12 that as the lignin flow rate increases, a greater amount of ethylene, propylene and 1,3 butadiene is produced in each of the three routes.

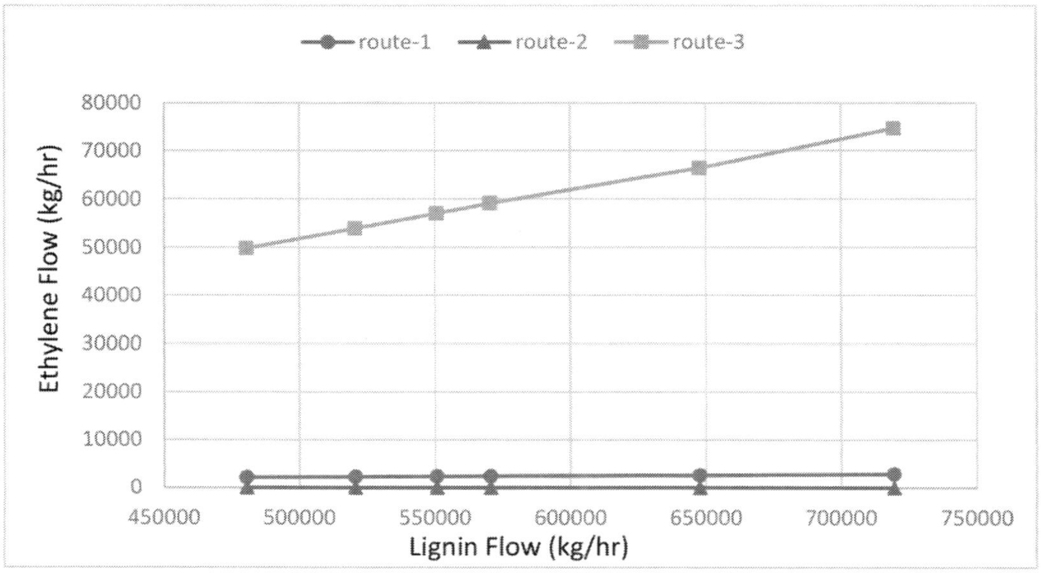

Figure 10. Ethylene productivity under three routes.[26]

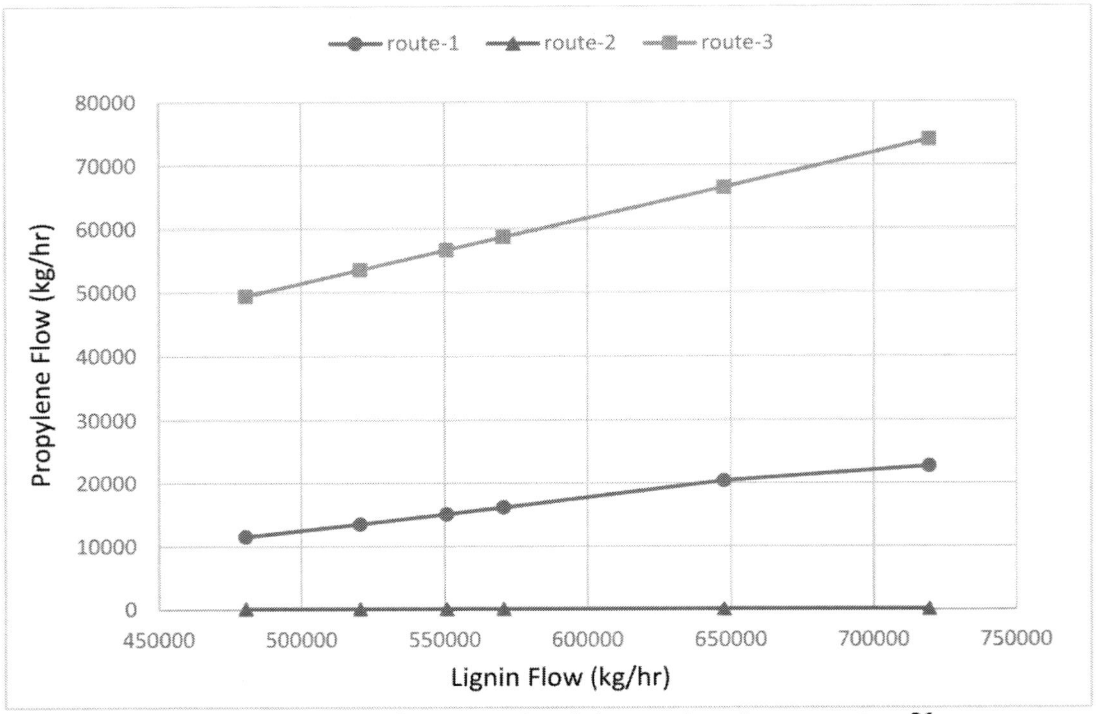

Figure 11. Propylene productivity under three routes.[26]

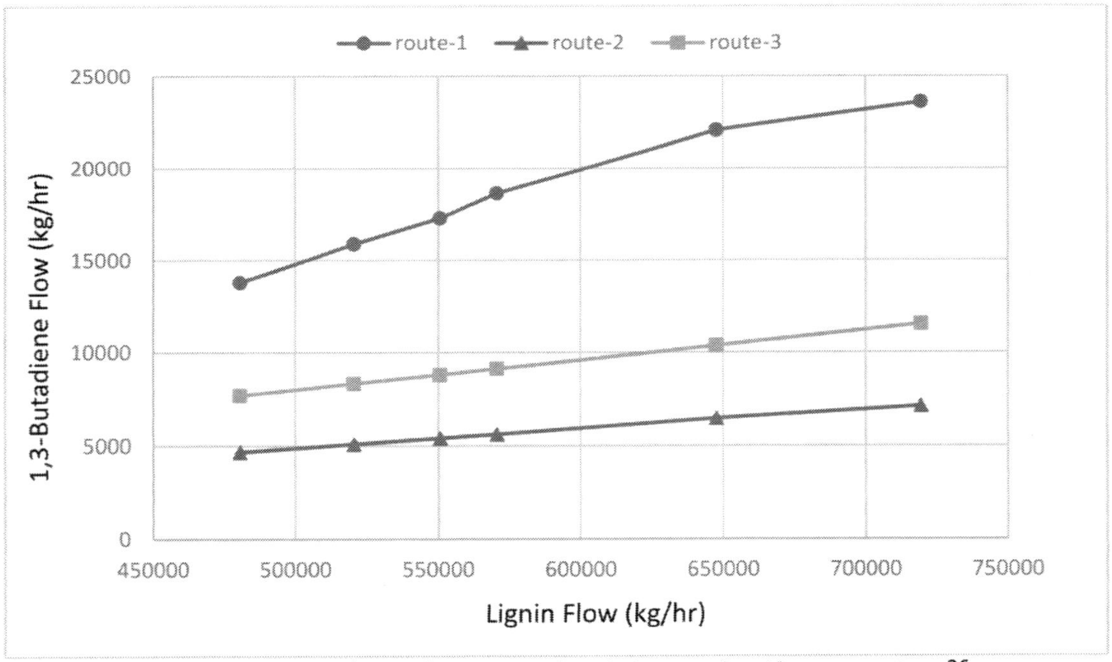

Figure 12. 1,3-butadiene productivity under three routes.[26]

Table 3 shows a comparison of the yields of ethylene, propylene and 1, 3-butadiene in each route in terms of weight of useful product per unit weight of lignin used when the lignin feed is 647,655 kg/hr. It is observed that the yield of 1,3-butadiene is low in all three routes. This is mainly because the commercial catalyst SAPO-34 used in the conversion of syngas products to alkenes in the last reaction step is not tuned for high yields of 1, 3-butadiene29. Route 1 gives the highest yield of 3.4 weight %. For this reason, it is necessary to look for other valuable products, such as propylene and ethylene, produced via the three routes for economic viability of using lignin as a feed source. With this additional consideration, it is observed that route 3, which produces 1.6 weight % 1,3-butadiene in addition to 10.4 weight % ethylene and 10.2 weight % propylene, can prove to be economical in an integrated plant that utilizes not only 1,3-butadiene but also ethylene and propylene for making valuable products.

Table 3. 1,3-butadiene, ethylene and propylene yield results[26]

Method	1,3-Butadiene Yield (wt% Lignin Basis)	Ethylene Yield (wt% Lignin Basis)	Propylene Yield (wt% Lignin Basis)
Route 1	3.4	0.4	3.1
Route 2	1.0	0	0
Route 3	1.6	10.4	10.2

In a previous publication[25] we had considered the production of 1,3-butadiene from a conventional naphtha cracker feed. Normal and turndown throughputs of 1,3-butadiene in the naphtha feed were 25,050 kg/hr and 12,625 kg/hr respectively based on values from BASF[27]. If we assume that such a plant is operating at a turndown throughput of 12,625 kg/hr due to a shortage of naphtha cracker feed and we would like to produce an additional 12,625 kg/hr of 1,3-butadiene from a lignin source. can be utilized to calculate the necessary lignin feed to produce this additional 1,3-butadiene and this result is shown in Table 4. It is observed that we need a lignin feed of 408,064 kg/hr if we use Route 1 and a lignin feed of 790,625 kg/hr if we use Route 3. It can be seen from Table 5 that the composition of a conventional naphtha cracker feed, which is used to recover 1,3-butadiene is substantially different from the product stream of the lignin processes (route 1 and route 3). Routes 1 and 3 have a significant amount of ethylene and propylene that need to be integrated in other parts of the plant to make other high-value products. Furthermore, the separation equipment for these product streams will not be the same as the one used for conventional naphtha cracker streams and is the subject of a future publication.

Table 4. Lignin required to produce required additional 1,3-butadiene in routes 1 and 3[26]

Method	Lignin flow (kg/hr)	1,3-Butadiene flow (kg/hr)
Route 1	408,064	12,625
Route 3	790,625	12,625

Table 5. Comparison of butadiene stream from conventional naphtha cracker and from lignin processes[26]

Component	Conventional Feed (kg/hr) (Turndown capacity)	Feed from route 1 (kg/hr) (@ Lignin flow of 408,064 kg/hr)	Feed from route 3 (kg/hr) (@ Lignin flow of 790,625 kg/hr)
Propene	2.52	4487.78	41594.1
n-Butane	1917.72		
i-Butane	892.08		
1-Butene	1592.64	488.414	29316.3
i-Butene	5896.8		
Trans-2-Butene	824.04	1139.04	
Cis-2-Butene	650.16	687.514	
1,3-Butadiene	12751.2	12650	12650
1-Pentene		497.328	
Hydrogen		381.5	4397.17

5. Conclusions

Based on our previous publications,[25,26] two aspects on emergent 1,3-butadiene production issues have been presented and summarized in this paper. First, both steady-state and dynamic models are employed in this paper to study how operation at turndown ratios of feed rates affects the production of 1,3-butadiene using N-methyl-2-pyrrolidone (NMP) as a solvent in an olefin plant. Dynamic simulations are conducted to study how the product purity changes when the plant is operated at turndown capacity. In particular, the recycle flow of the solvent to both extractive distillation columns are reduced and its effect on the final product purity is studied. It is observed that the butadiene product purity is more significantly affected by solvent flowrate changes to the second extractive column as compared to the first extractive column when the plant is running at turndown capacity. This provides the rationale for tighter monitoring the solvent flowrate in the second extractive column. While a solvent to feed ratio in the range 6-7 is sufficient for normal operations, this ratio has to be increased to a range of 10-11 when the plant is operated under turndown conditions in order to maintain the desired product purity. Next, the flowrate of the side draw stream from the rectifier to the bottom of after washer is increased and its effect on the 1,3-butadiene product purity is studied at normal throughput and turndown capacity conditions.

Second, three different routes that utilize biomass to produce 1,3-butadiene and associated co-products ethylene and propylene from lignin are also presented in this paper. Steady-state simulation models are developed and the yield of useful products from these routes are studied. It is found that the highest yield of 3.4 weight % of 1,3-butadiene is produced in a process that converts lignin to 1,3-butadiene via syn-gas, dimethyl ether and light olefins. However, if we consider the production of ethylene and propylene as additional useful products, the process that converts lignin to 1,3-butadiene via syn-gas and light olefins has a better economic potential. Finally, the amount of lignin needed to produce additional 1,3-butadiene to augment the shortfall from a traditional 1,3-butadiene plant is calculated.

References

1. W.C. White, *Chem.-Biol. Interact.*, **2007**,166 (1-3), 10-4.
2. http://www.grandviewresearch.com/industry-analysis/butadiene-market (Accessed on December 14, 2019).
3. https://www.aspentech.com/v10 (Accessed on December 12, 2019).
4. Y.H. Kim, S. Y. Kim, B. S. Lee, *Korean J. of Chem. Eng.*, **2002**, 29(11), 1493-1499.
5. R. Saffari, F. Abbasi, F. Jalali-Farahani, N. Mostoufi, in *Proc. 2005 Inter. Conf. Sim. Modeling*, Thailand, **2005**.
6. Lv, P.; Yuan, Z.; Wu, C.; Ma, L.; Chen, Y.; Tsubaki, N., Bio-syngas production from biomass catalytic gasification, *Energy Convers. Manage*, **2007**, 48, 1132−1139.
7. Richardson, Y.; Blin, J.; Julbe, A., A short review on purification and conditioning of syngas produced by biomass gasification: Catalytic strategies, process intensification and new concepts, *Prog. Energy Combust. Sci.*, **2012**, 38, 765−781.
8. Couto, N.; Rouboa, A.; Silva, V.; Monteiro, E.; Bouziane, K., Influence of biomass gasification process on the final composition of syngas, *Energy Procedia*, **2013**, 36, 596−606.
9. Yamada, S.; Shimizu, M.; Miyoshi, F. JFE Technical Report, **2004**, No. 3, pp 21−26. http://www.jfe-steel.co.jp/en/research/report/003/pdf/003-05.pdf (Accessed on December 17, 2019).
10. Norbeck, J. M.; Johnson, K., http://www.methanol.org/pdf/Hynolexec.pdf (Accessed on April 22, 2019).
11. Brandberg, Å. R. L.; Ekbom, T., http://www.ecotraffic.se/pdf/isafxii.pdf (Accessed on April 22, 2019).
12. Mitsubishi Heavy Industries. Biomass gasification methanol synthesis system. http://www.mhi.co.jp/power/e_power/techno/biomass/, April, 2006 (Accessed on April 22, 2019).

13. Holdings, J.F.E., Inc. U.S. Patent 2006/0052647, March 9, 2006.
14. Palanki, S.; Balzaretti, D., Design of Forest Products Process Plants for Production of Platform Chemicals, *Journal of Bioprocess Engineering and Biorefinery*, **2013**, 1, 1-4.
15. Hanaoka, T.; Fujimoto, S.; Yoshida, M., Efficiency estimation and improvement of the 1,3-butadiene production process from lignin via syngas through process simulation, *Energy Fuels*, **2017**, 31, 12965−12976.
16. Park, S.; Watanabe, Y.; Nishita, Y.; Fukuoka, T.; Inagaki, S.; Kubota, Y., Catalytic conversion of dimethyl ether into propylene over MCM-68 zeolite, *J. Catal.*, **2014**, 319, 265−273.
17. New Zealand Institute of Chemistry. Chemical processes. http://nzic.org.nz/ChemProcesses/energy/7D.pdf (Accessed on December 17, 2019).
18. Kumita, Y.; Gascon, J.; Stavitski, E.; Moulijn, J. A.; Kapteijn, F., Shape selective methanol to olefins over highly thermostable DDR catalysts, *Appl. Catal., A*, **2011**, 391, 234−243.
19. Schulte, H. J.; Graf, B.; Xia, W.; Muhler, M., Nitrogen and oxygen-functionalized multiwalled carbon nanotubes used as support in iron-catalyzed, high temperature Fischer-Tropsch synthesis, *ChemCatChem*, **2012**, 4, 350−355.
20. Mitsubishi Chemical Corporation. Method for producing conjugated diene, Jpn. Patent No. PCT/JP2010/058842.
21. K. Kindler, H. Puhl, US Patent, 6 337 429 B1, **2002**.
22. H. Renoni, J.M. Prausnitz, *Ind. Eng. Chem. Process Des. Dev.*, **1969**, 8 (3), pp 413–419
23. Wooley, W. R.; Putsche, V. *Development of an ASPEN PLUS physical property database for biofuels components*, NREL/TP-425-20685 • UC Category: 1503 • DE96007902.

24. C. Ruiz, J. Coca, A. Vega, F.v. Diez, *Ind. Eng. Chem. Res.,* **1997**, 36 (11), 4934-4939.

25. N. Tripathi, Q. Xu, and S. Palanki, "Modeling and Simulation of 1,3 Butadiene Extraction Process at Turndown Capacity," *Chemical Engineering and Technology,* **2019**, doi:10.1002/ceat.201900019.

26. N. Tripathi, S. Palanki, Q. Xu and K.D.P. Nigam, "Production of 1,3 Butadiene and Associated Co-Products Ethylene and Propylene from Lignin," *Industrial and Engineering Chemistry Research*, **2019**, 58, 35, 16182-16189.

27. BASF Management's report Segments of investment on Chemical Products, https://report.basf.com/2017/en/managements-report/segments/chemicals/products-and-investments.html (Accessed on December 8, 2019).

AIChE Paper Number: 590265

Turbomachinery Rotordynamics: Modeling, Analysis, and Reporting Overview

Dustin Pavelek, P.E.
Sr. Consultant
Kelm Engineering, LLC

and

Ray Kelm, P.E.
President/Chief Engineer
Kelm Engineering, LLC

Prepared for Presentation at the 2020 Spring National Meeting
Houston, Texas, March 30, 2020

AIChE and EPC shall not be responsible for statements or opinions contained in papers or printed in its publications

Turbomachinery Rotordynamics:

Modeling, Analysis, and Reporting Overview

Dustin Pavelek, P.E.	Ray Kelm, P.E.
Sr. Consultant	President/Chief Engineer
Kelm Engineering, LLC	Kelm Engineering, LLC

Abstract

This paper will provide an overview of the modeling, analysis and reporting process for lateral rotordynamics of turbomachinery. The paper will focus on the application of the newly published API Technical Report 684-1 (API Standard Paragraphs Rotordynamic Tutorial: Lateral Critical Speeds, Unbalance Response, Stability, Train Torsionals, and Rotor Balancing, First Edition, November 2019).

Lateral rotordynamics for the typical equipment used in Ethylene plants such as steam turbines and centrifugal compressors with fluid-film bearings will be examined. Using excerpts from previous rotor studies, both correct and incorrect results that can be produced by failing to follow good engineering practices in rotor model creation, bearing coefficient calculation, unbalance response analysis and lateral stability analysis will be discussed. The Torsional Analysis and Balancing sections of API TR 684-1 will not be discussed in this paper.

The intended audience for this presentation includes engineers that are tasked with evaluating a lateral rotordynamic analysis report produced by an equipment manufacturer or third party consultant during procurement of new machinery or design changes for existing equipment.

Introduction

The American Petroleum Institute (API) Subcommittee on Mechanical Equipment publishes specifications that define the mechanical acceptance criteria for new rotating equipment. The standards are organized by equipment type; such as Axial and Centrifugal Compressors and Expander-Compressors (API 617), Special Purpose Steam Turbines (API 612) and Centrifugal Pumps (API 610). Each of these documents include a chapter on Dynamics that outline the lateral and torsional rotordynamics and balancing criteria designed to ensure equipment mechanical reliability. The specific language in the Dynamics chapter of each standard differs based on the subject equipment but all are based on the Dynamics Section (6.8) of the API Standard Paragraphs. The Standard Paragraphs represent specifications that share commonality between different equipment types and are intended to be a starting point for designing the specification language used by the API Task Force for each equipment type.

The origin of the API Standard Paragraphs on Dynamics dates back to the first publication of the Centrifugal Compressor Standard (API 617) in June of 1958. Subsequent editions of API 617 expanded on the original dynamics requirements by including torsional analysis, balancing criteria, requirements or limits for separation margin, amplification factor, and amplitude [1]. API 684 was first published as a *Tutorial on the API Standard Paragraphs Covering Rotor Dynamics and Balancing* in February 1996. The purpose of this document was to "...describe, discuss, and clarify... the API Standards Paragraphs."[1]

This document was revised in the 2nd Edition published as an API Recommended Practice (RP) in 2005 [3]. Requirements for consideration of active magnetic bearings (AMB) and variable frequency drives (VFD) were added in the 8th Edition of API 617 in September 2014.

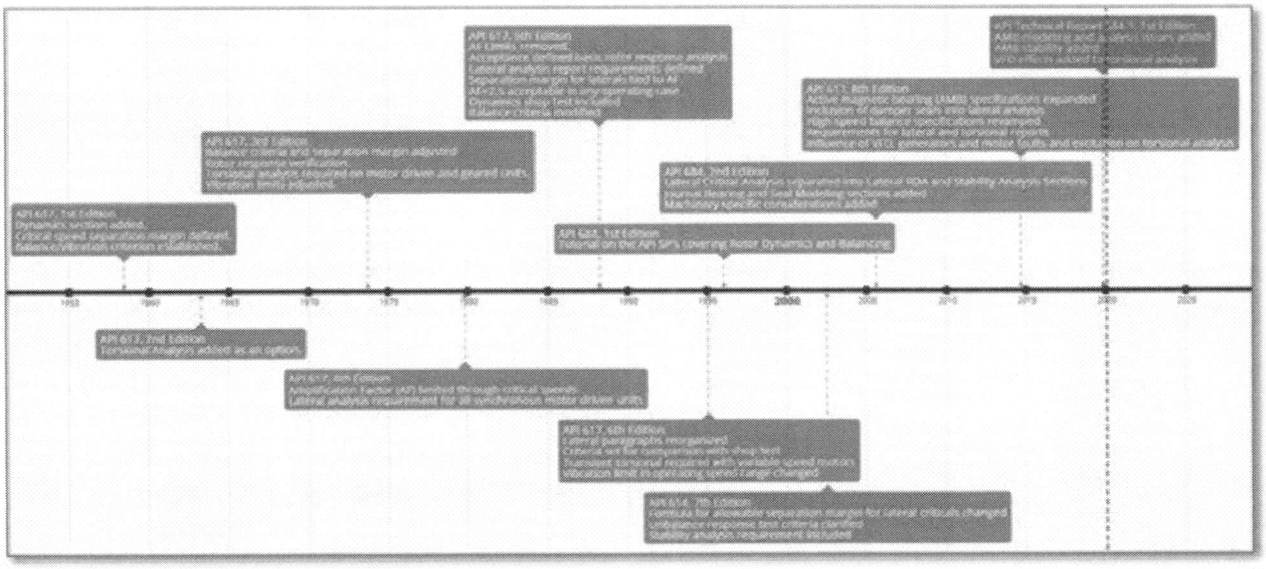

Figure 1 - Historical Timeline of API Standards on Rotordynamics

API Technical Report 684-1, 1st Edition (*API Standards Paragraphs Rotordynamics Tutorial: Lateral Critical Speeds, Unbalance Response, Stability, Train Torsionals, and Rotor Balancing*) [4], was published in November 2019. This document supersedes API RP 684, 2nd Edition and adds sections on modeling, analysis, and stability considerations with AMB's and VFD effects on torsional analysis.

This paper will provide a general overview of lateral rotordynamic analysis as it applies to turbomachinery and will focus on the requirements outlined in the API Standard Paragraphs as described in API TR 684-1. It is not intended to be used as a stand-alone document in place of the applicable equipment standard or in place of API TR 684-1.

Lateral Rotordynamic Analysis

Lateral rotordynamic analysis (RDA) is conducted for new equipment purchases, re-rating of existing equipment that include speed changes or power increases, component upgrades such as bearings, seals, or couplings, and for troubleshooting vibration issues. Lateral RDA described in the API Standard Paragraphs includes:

- Undamped Critical Speed Analysis
- Damped Unbalance Rotor Response Analysis
- Stability Analysis

Successful completion of a lateral RDA requires accurate input data, a well-constructed rotor model, appropriate support models (bearings, seals, substructure), analyst insight into machine dynamics, and careful application of relevant standards.

Several commercial software packages are available to the public for purchase or lease. Since most of these modern software tools use a proven finite element method and have had results validated against test data for thousands of rotors, software selection often comes down to analyst preference, availability of seal and bearing codes, ease of use, and cost. In the words of rotordynamic pioneer, Dr. Dara Childs:

> "Good engineers get good results from good models, leading to sound engineering judgements with a variety of algorithms. Superior algorithms or computer codes will not cure bad models or a lack of engineering judgement."[5]

Model Construction

In general, a lateral rotordynamic model will include the shaft, any lumped masses/inertia such as compressor or pump impellers or turbine wheels, coupling mass, journal bearings, seals, and support structure. The analyst will start with a shaft drawing or listing of mass-elastic data. The shaft will be discretized by defining stations with an outer (and sometimes inner) diameter and length. Shafts are typically broken up with stations starting at any change in shaft diameter, locations corresponding to the center of gravity for lumped masses, bearing midspans, radial vibration probe locations, thrust collars, and seal midspans.

In the finite element analysis world, the stations are the same as nodes. The term station is traditionally used in rotordynamics since the term node is also used when referring to locations on a shaft (or structure) that have zero modal displacement for a given mode shape. The shaft stations are connected by elements. The shaft stations locations are then refined to limit the ratio of the element length to outer diameter (L/D) to less than 1.0. L/D for individual stations between 0.1 and 1.0 are preferred to balance model accuracy and computation time. Longer elements (L/D > 1.0) may not provide enough resolution to accurately calculate higher natural frequencies and mode shapes. Shorter elements (L/D < 0.1) can produce numerical issues with some computational methods when large differences in stiffness exists between adjacent elements and can result in longer run times for large models. Computation time is much less of a concern with modern processing capability unless a model is exceptionally large and computationally expensive transient analyses are to be conducted. Nevertheless, following the guideline of keeping the L/D for elements between 0.1 and 1.0 with a preferred maximum of 0.5 has proven to give adequate results in most cases. Modeling detailed shaft geometry such as small grooves for steam turbine packing sections would result in elements with an L/D much less than the recommended minimum. These small details can typically be omitted without loss of accuracy in the results.

Material properties assigned to elements should reflect the actual materials of construction for the rotor and the operating temperature of the rotor. Steam turbines operating at high temperature or refrigeration compressors operating at low temperature should have the modulus of elasticity adjusted to reflect operating conditions. A plot of the variation of elastic modulus vs. temperature for some common low alloy steels is shown in Figure 2 [4].

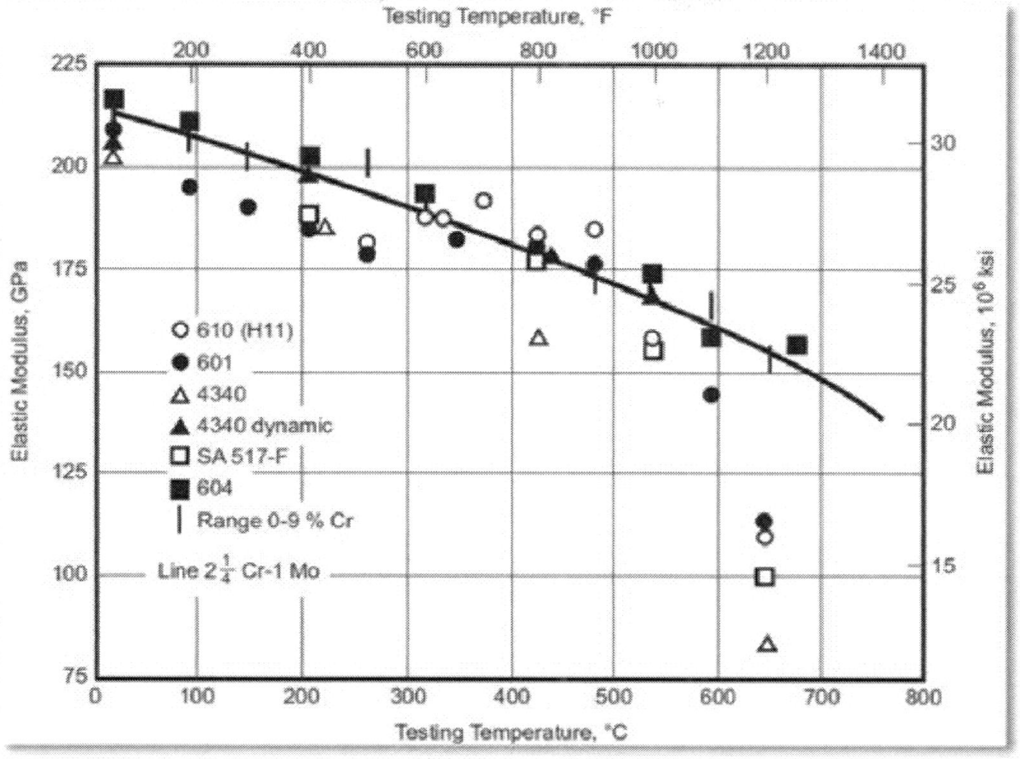

Figure 2 - Elastic Modulus vs. Temperature

Shaft Sleeves and Disks

For rotors that have shrunk on sleeves or disks, the bending stiffness added by the sleeve or disk is often less than what is calculated by assuming the full diameter of the base shaft plus the sleeve. Instead, the sleeve is often assumed to add mass and inertia but no additional bending stiffness. Most RDA software allows the analyst to enter the mass diameter and stiffness diameter separately so that the mass and inertia of sleeves and disks can be accounted for without adding additional stiffness. Based on analyst experience or test data, some compromise between no added stiffness and the full added stiffness for a shrunk on sleeve or disk may be included. This can be accomplished by entering the actual outer diameter (OD) for the shaft, the actual inner diameter (ID) for the sleeve, and an OD for the sleeve that is the average of the inner diameter and outer diameter for the sleeve. Another option is to include the actual sleeve dimensions but defining the sleeve material properties such that the actual sleeve density is used but a reduced elastic modulus is entered.

Disks such as turbine stages or compressor or pump impellers are typically modelled as lumped masses and diametral inertias applied at a shaft station at the axial location of the disk center of gravity. For lateral RDA, the disk mass and diametral inertia are important for accurate results. Disk polar inertia may be entered in the software, though it is not important for lateral RDA. If the lateral model will be used to evaluate torsional natural frequencies, the polar inertia for disks in the model should be accurate. For a disk with a large L/D, the mass and inertia are typically distributed at multiple adjacent stations.

Bearings

Turbomachinery rotors are typically supported on fixed geometry or tilt-pad journal bearings. These require careful consideration for accurate results from a model. Inputs to most commercially available bearing codes include bearing geometry, bearing loading, and lubricant properties.

Bearing geometry refers to the number of pads, load configuration (load on pad or load between pad), pivot offset, set clearance and preload. Preload is calculated from the assembled bearing or set clearance and the pad clearance as shown in Figure 3. For preloaded bearings, the pad clearance will be greater than the set clearance. Typical values for tilt pad bearing preload are 0.0 to 0.6 with a negative preload undesirable. Careful consideration for machining tolerances should be included in the bearing model and is required by API [4].

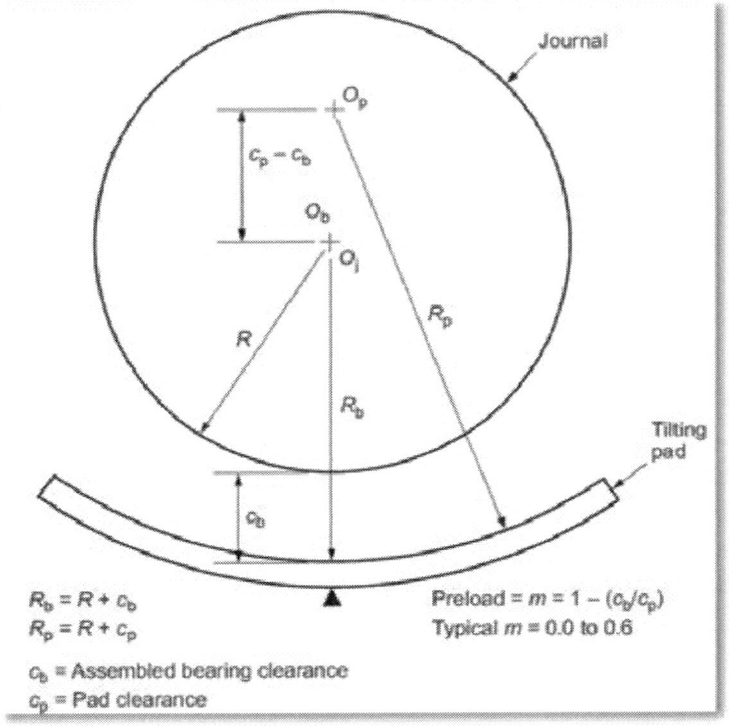

Figure 3 - Tilt Pad Bearing Preload Schematic [4]

Outputs used in a lateral model are the direct and cross-coupled stiffness and damping coefficients. Other bearing parameters that are calculated by most bearing codes that are important for bearing designers are the film and bearing metal temperatures, horsepower loss, minimum film thickness, and maximum film pressure, though these parameters are not directly included in any lateral RDA calculations.

The bearing model should use the actual lubricant properties for the subject machine and account for changes in viscosity with temperature. Additionally, the calculated bearing coefficients are highly dependent on the static load applied to the bearing. This is calculated from the vector sum of the static rotor weight (calculated from the rotor model) and any external loading such as steam admission forces for a turbine or gear load for a gearbox. An example of the difference in bearing static load vectors for a steam turbine with and without consideration of steam admission forces is shown in Figure 4. In this example, the Outboard/Steam End bearing was configured in a load between pad configuration with respect to the static rotor weight but operated as a load on pad bearing with the addition of the steam admission forces.

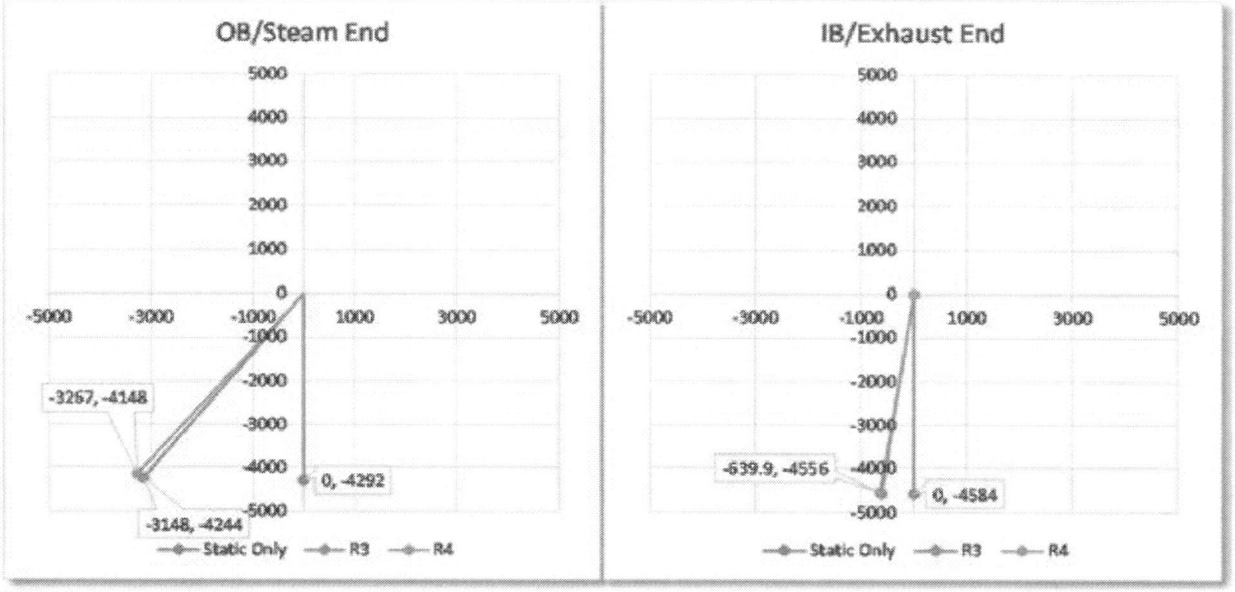

Figure 4 - Steam Turbine Bearing Static Loads with and Without Steam Admission Forces

Bearing Supports

For very heavy rotors, machines with flexible bearing supports ($K_{support}$<3.5*$K_{bearing}$), or machines where a rigid bearing support model does not accurately reflect operating characteristics, a bearing support model should be included. Several methods have been developed to model support stiffness effects. For existing equipment, the bearing support mass elastic properties can be determined from an impact test with an instrumented hammer. For proposed equipment, a 3D finite element model of the support structure can be used to determine support stiffnesses. This is placed in series with the bearing stiffness and damping coefficients. For machines that can be accurately modeled by assuming rigid bearing supports based on similar designs, the bearing support stiffness effects can be omitted from the model.

Undamped Critical Speed Analysis

The Undamped Critical Speed (UCS) Analysis is a preliminary calculation of the rotor-bearing system critical speeds and associated mode shapes. It is considered preliminary because it only considers the influence of the rotor construction and fluid-film bearing direct stiffness on the critical speeds and mode shapes. Any support stiffness and bearing damping effects are neglected.

The primary output of the UCS Analysis is the UCS Map. The UCS Map shows how the critical speeds change as the bearing stiffness varies. The first four critical speeds are plotted as the support stiffness is varied from 0.1x to 10x the expected support stiffness. The expected support stiffness in the horizontal and vertical directions are typically plotted on the UCS Map along with the minimum and maximum operating speeds for the subject rotor.

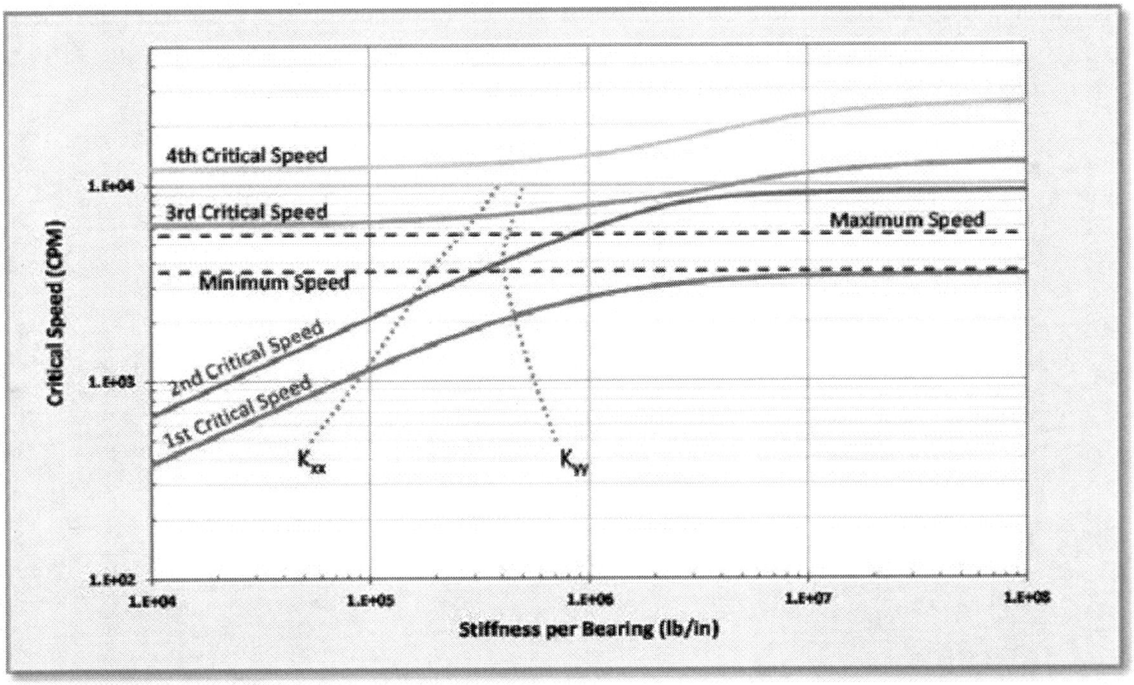

Figure 5 - Typical Undamped Critical Speed Map

Undamped mode shapes are also calculated in the UCS Analysis. These mode shapes are the normalized free (unforced) deflection shapes associated with each critical speed for a given bearing stiffness. At very low stiffness, the first two modes are rigid body modes with a natural frequency of zero. The third mode is the first rotor free-free mode shape. This natural frequency, calculated at zero bearing stiffness, and the associated mode shape can be used to validate the accuracy of a rotor model by comparing the calculated natural frequency and mode shape with the measured natural frequency and mode shape determined by suspending the rotor in slings and conducting a modal test of the rotor. Figure 6 shows a typical UCS Map and the associated undamped mode shapes at the extremes of support stiffness.

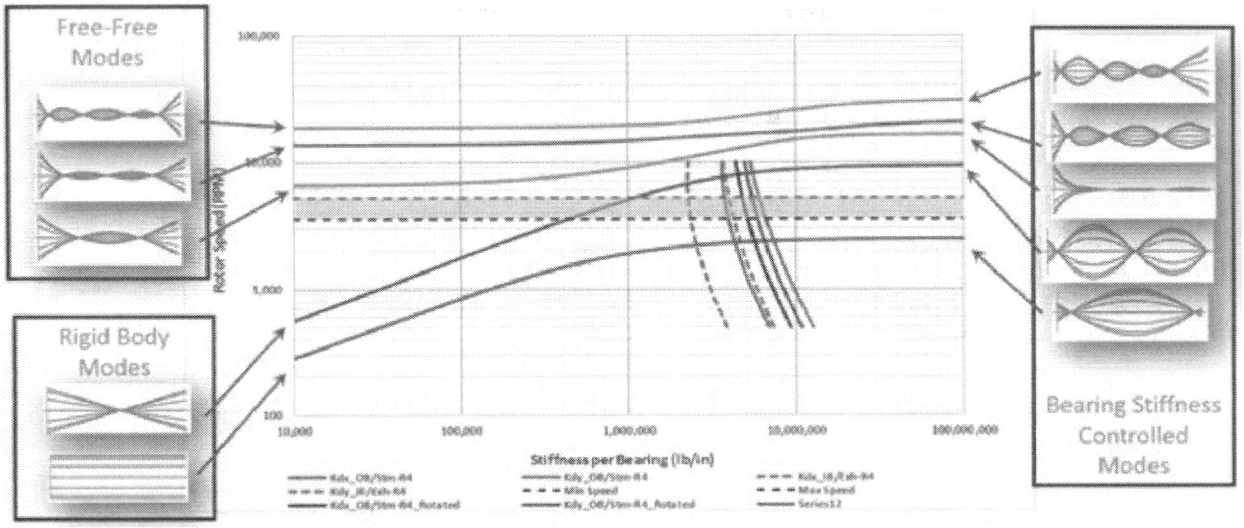

Figure 6 - UCS Map and Associated Mode Shapes

The undamped mode shapes are also used to define the unbalance force location in the Damped Unbalance Rotor Response Analysis. The advantage of using the undamped mode shapes for identification of the unbalance force location is that the analyst can easily determine the proper location to excite the relevant critical speeds based on the critical speeds relative to operating speed. Locations and amplitudes are identified in Figure 7 with *U* representing the unbalance magnitude (oz-in), *W* representing the journal weight (lbf), and *N* representing the maximum continuous operating speed (RPM).

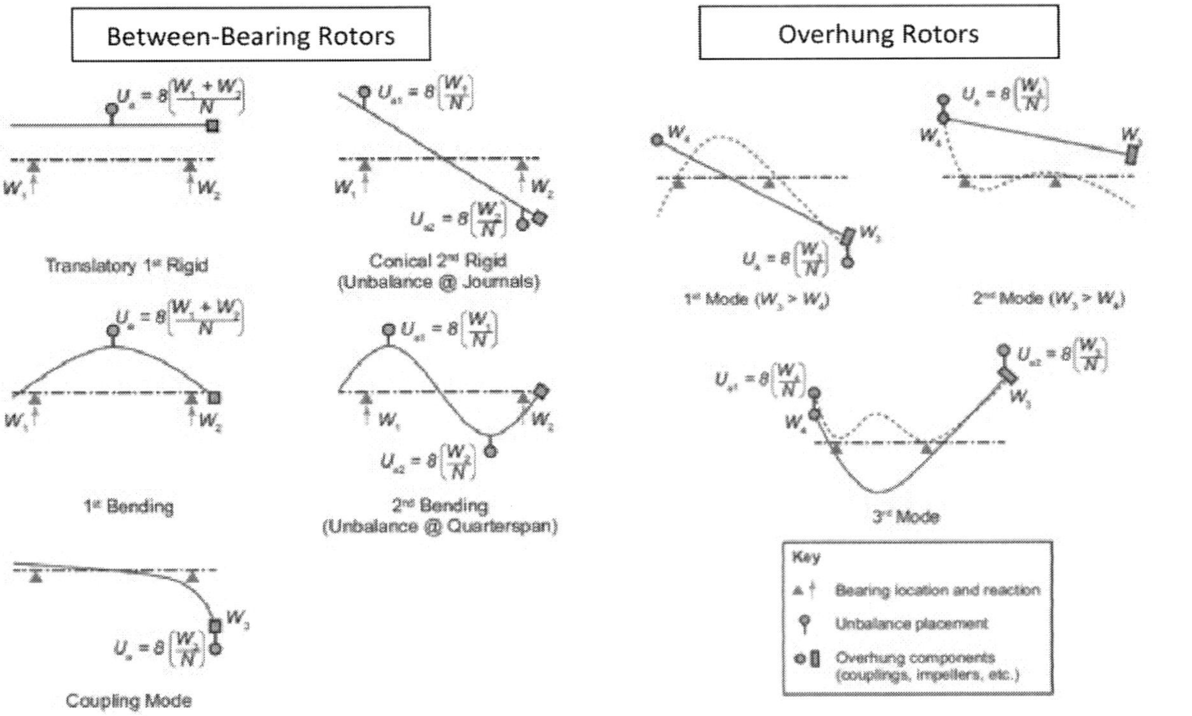

Figure 7 - Unbalance Force Locations from Undamped Mode Shapes [4]

Damped Unbalance Rotor Response Analysis

The damped unbalance response analysis is conducted to determine critical speeds, amplification factor (AF), separation margin (SM), and predicted amplitude at close-clearance locations (bearings and seals) and probe locations. This analysis differs from the UCS Analysis in that the predicted bearing and support stiffness and damping and the direct and cross-coupled stiffness and damping from bearings and seals are included in the analysis.

Imbalance is applied to the rotor model based on the analyst's assessment of the undamped mode shapes. The imbalance location should be selected to excite the modes of interest. This often requires multiple runs with imbalance at multiple shaft stations. The imbalance magnitude is selected per API 684 to be 4*4W/N where W is the journal static load and N is the maximum continuous operating speed (MCOS). The rotor model is analyzed at speeds from 0-125% of the rotor trip speed. The RDA software is used to calculate vibration amplitude and phase at all stations in the rotor model.

Acceptability is determined from evaluation of critical speed location and separation margin based on amplification factor and predicted amplitude at close-clearance locations. The predicted amplitude should be adjusted by multiplying by a correction factor (CF) defined as:

$$CF = \frac{A_l}{A_{4X}}$$

$$A_1 = \sqrt{\frac{12{,}000}{N}}$$

CF = correction factor,

A_l = amplitude limit, calculated using Equation 5 in µm (mils) peak to peak,

A_{4X} = peak-to-peak amplitude at the probe location

API limits the predicted major axis amplitude after correction to 75% of the normal design clearance up to trip speed. This limit does not apply where abradable seals are used.

A generic response plot vs. rotor speed is shown in Figure 8. This plot defines the separation margin and amplification factor (calculated for each critical speed using the half-power method).

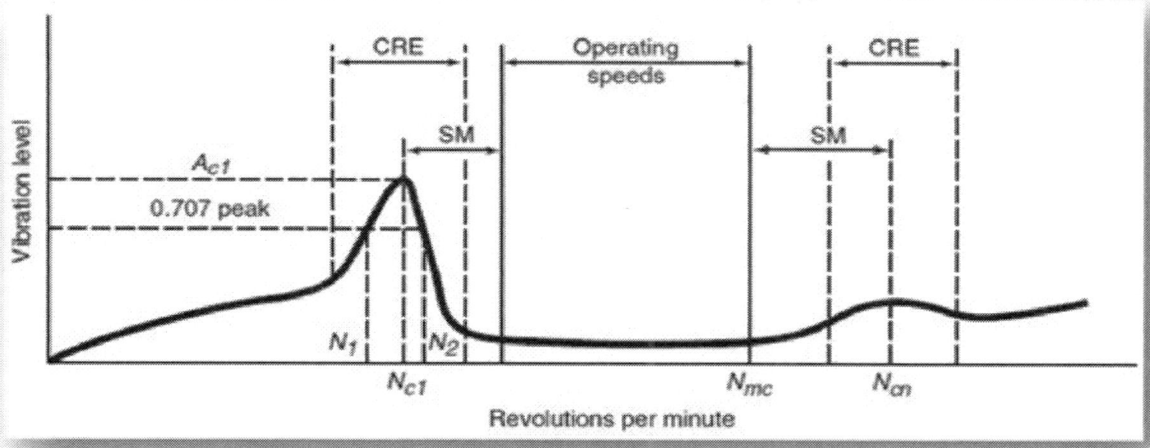

Figure 8 - Example Predicted Vibration Amplitude vs. Rotor Speed [4]

The required separation margin is based on the amplification factor for each critical speed. If the amplification factor is less than 2.5, the critical speed is critically damped and no separation margin from the operating speed range is required.

If the amplification factor is greater than 2.5 and the critical speed is below the operating speed range, the required separation margin is the lesser value of 16% or the value calculated from:

$$SM = 17\left(1 - \frac{1}{AF - 1.5}\right)$$

If the amplification factor is greater than 2.5 and the critical speed is above the operating speed range, the required separation margin is the lesser value of 26% or the value calculated from:

$$SM = 10 + 17\left(1 - \frac{1}{AF - 1.5}\right)$$

The damped unbalance response analysis should include results from predictions of the bearing coefficients at the minimum, maximum, and design bearing clearance as well as the extremes for bearing lube oil supply temperature.

For the unbalance response analysis, the effects of seals, other than the added mass and inertia from any shaft-mounted sleeves, are typically excluded. One exception to this would be lubricated compressor shaft end seals. Pump seals (wear rings, eye seals, breakdown bushings) have a large effect on unbalance response analysis results. Labyrinth seal effects can be omitted.

Stability Analysis

Rotors operating above their first damped natural frequency are subject to stability analysis. Other machine characteristics may make them more susceptible to rotordynamic instability:

- Plain journal bearings
- High pressure compressors
- High pressure turbines (partial arc admission)
- Center (rotor midspan) balance piston seal (back-to-back compressors)
- Large (>10) Slenderness ratio (bearing span / shaft diameter)
- High speeds
- Oil seals for shaft end seals (compressors)
- High mole-weight process (compressors)
- High density process (compressors)

API specifies two levels of stability analysis. Level I is a method of screening a rotor-bearing system using assumptions for destabilizing force magnitude using empirical equations. For Level I, the destabilizing "force" in the form of a lumped cross-coupled stiffness should be applied at a location that would excite the mode of interest based on the mode shape.

Level II is a more detailed analysis incorporating the destabilizing forces from individual components. This is only required where the system fails a Level I analysis.

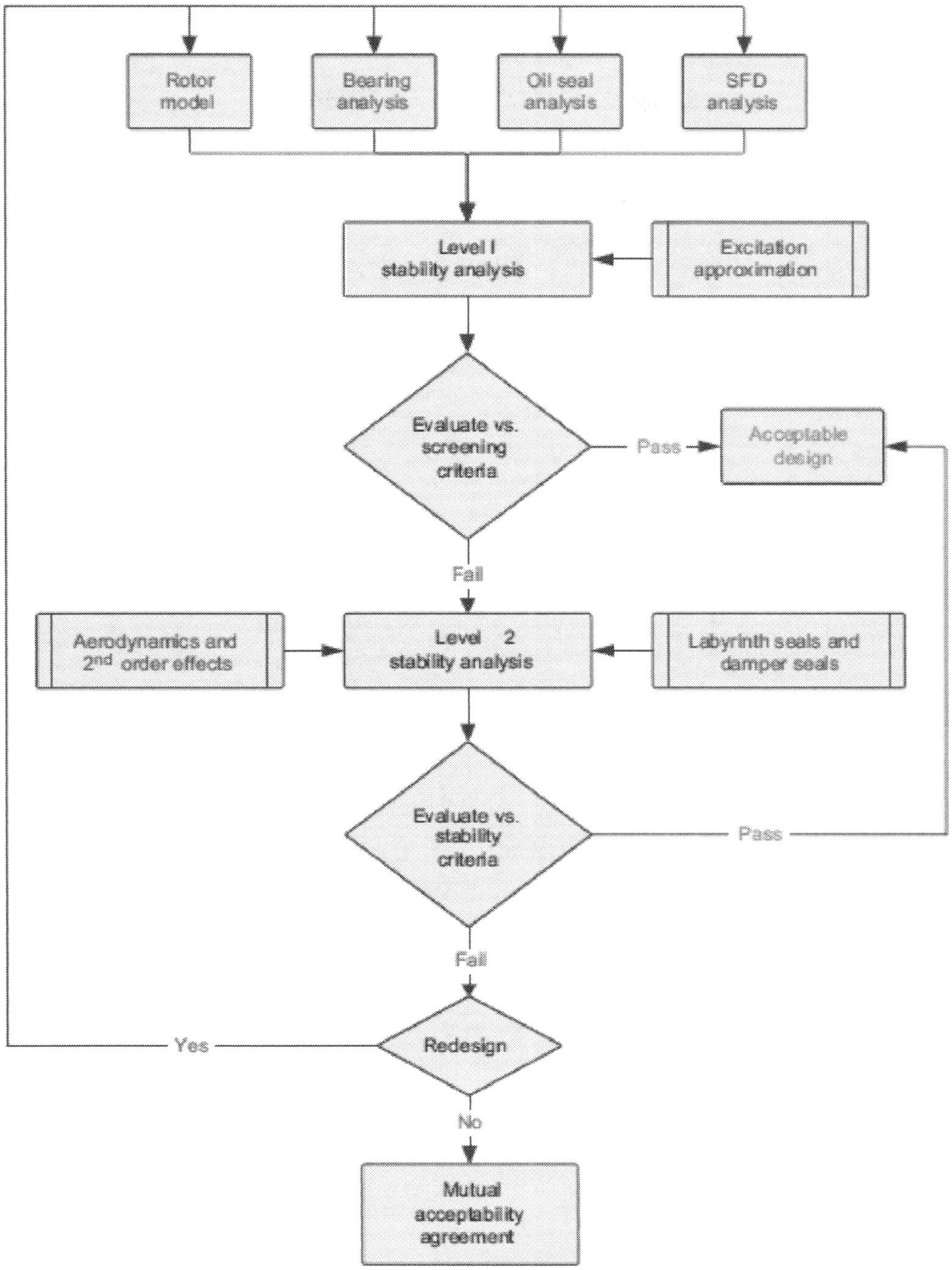

Figure 9 - Stability Analysis Flow Chart [4]

Acceptance criteria for a Level I stability analysis for a steam turbine only requires that the calculated logarithmic decrement (log-dec) is greater than 0.1. For centrifugal compressors, if the calculated log-dec is less than 0.1, the cross-coupling required to make the rotor unstable (log-dec < 0.0) is less than twice the expected cross-coupling, or the cross-coupling required to make the rotor unstable is less than 10 and the critical speed ratio (maximum continuous operating speed / critical speed) falls in Region B of the experience plot published by API [4], a Level II analysis is required.

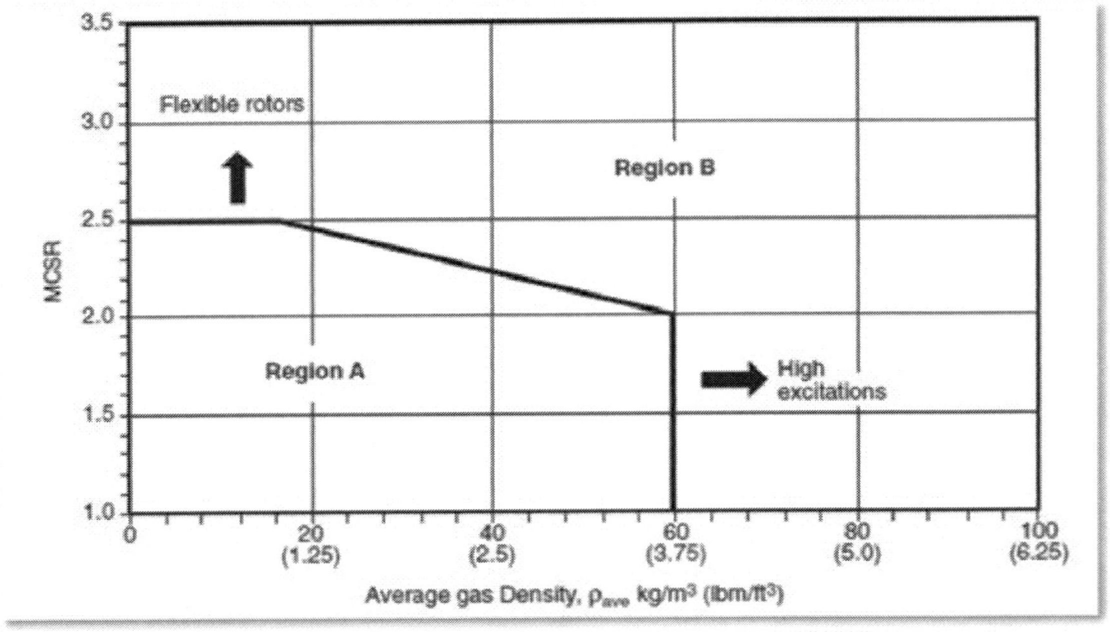

Figure 10 - CSR Experience Plot [4]

A Level II stability analysis is much more time consuming for the analyst since predictions for rotordynamic coefficients for all bearings and seals must be made and included in the model. The acceptance criteria for a Level II stability analysis is that the calculated log-dec is greater than 0.1, regardless of machine type.

If a rotor-bearing system fails a Level II stability analysis, design modification will be required. Changes to the rotor such as decreasing the bearing span or increasing the shaft diameter are not often viable. Modifications to bearings such as changing from fixed geometry to tilt pad, increasing the bearing L/D, increasing the clearance, decreasing the preload, or optimizing pad number and orientation may also be pursued. For high pressure compressors, alternate seal designs may also be viable.

Tips for Evaluating an RDA

The following check list provides some guidance for use when evaluating a rotordynamic analysis report. These items can be checked based on the contents of a report without the need for specialized software or detailed knowledge of rotordynamic analysis. Formal details of the required contents of a report can be found in reference [4].

- Rotor Model
  - Rotor Length
  - Rotor Weight
  - Bearing Span
  - Disk Mass/Inertia and C.G. Locations
  - Element L/D between 0.1 and 0.5
- Bearing Model
  - Correct Geometry (pivot offset, preload, pad orientation)
  - Tolerance Range Evaluated

    - Static Bearing Load Considers All Influences
    - Lubricant Properties are Correct
- Seal Model
    - Shaft End Seals are Included (where applicable)
- UCS Map
    - Bearing Coefficients are Overlaid
    - Not Used as Final Method for Identifying Critical Speeds
- Forced Response
    - Imbalance Location/Phase Correct for Relevant Modes
    - Imbalance Magnitude Calculated Correctly
    - Response Evaluated from 0-125% of the Trip Speed
    - Speed Increments are Reasonable
- Stability Analysis
    - Destabilizing Force Correctly Applied (location)
    - Destabilizing Force Magnitude Calculated Correctly
    - Level I Acceptance Criteria Met or Level II Analysis Conducted

As an example of a failure to adequately apply good engineering judgement and accepted standards, forced response analysis results from an RDA conducted by an aftermarket vendor as part of a compressor re-rate project are shown in the left hand side of Figure 11. The vendor only calculated the forced response analysis at the minimum and maximum operating speed. A third-party analysis identified a critical speed in the operating speed range using a more suitable speed increment (right side of Figure 11).

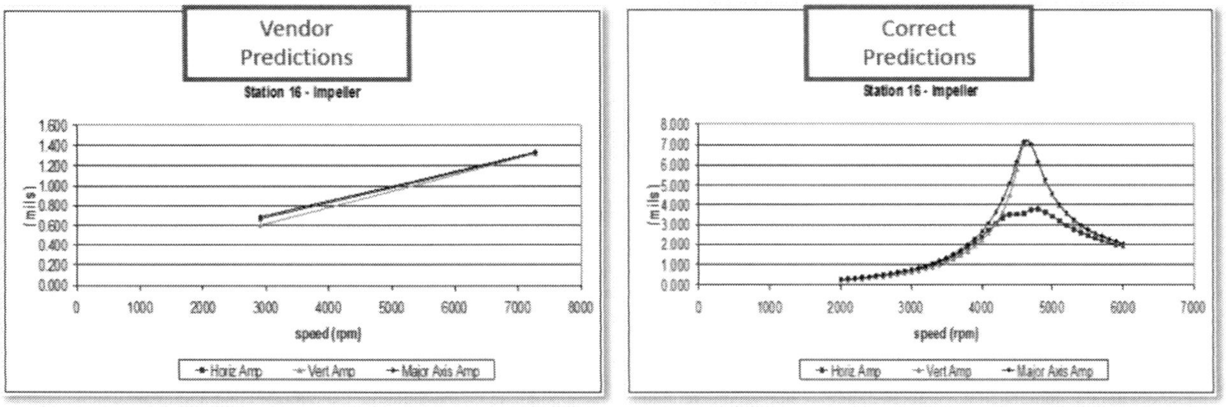

Figure 11 - Improper Speed Increment for Forced Response Predictions

Another example of questionable results were identified in a third-party review of an OEM RDA conducted during a compressor re-rate. A third-party review found a discrepancy in their calculated bearing coefficients when compared to those in the OEM report. At higher rotor speeds, the third-party predictions resulted in much lower direct stiffness and damping compared to the OEM report. After some additional review, it was determined that the OEM analyst used a constant viscosity model for the bearings. The third-party was able to duplicate the OEM results using a constant viscosity model. This method does not consider the change in the lubricant viscosity as the temperature increased. This would result in under-predictions of critical speeds near the upper end of the speed range and over-predicting their amplification factors.

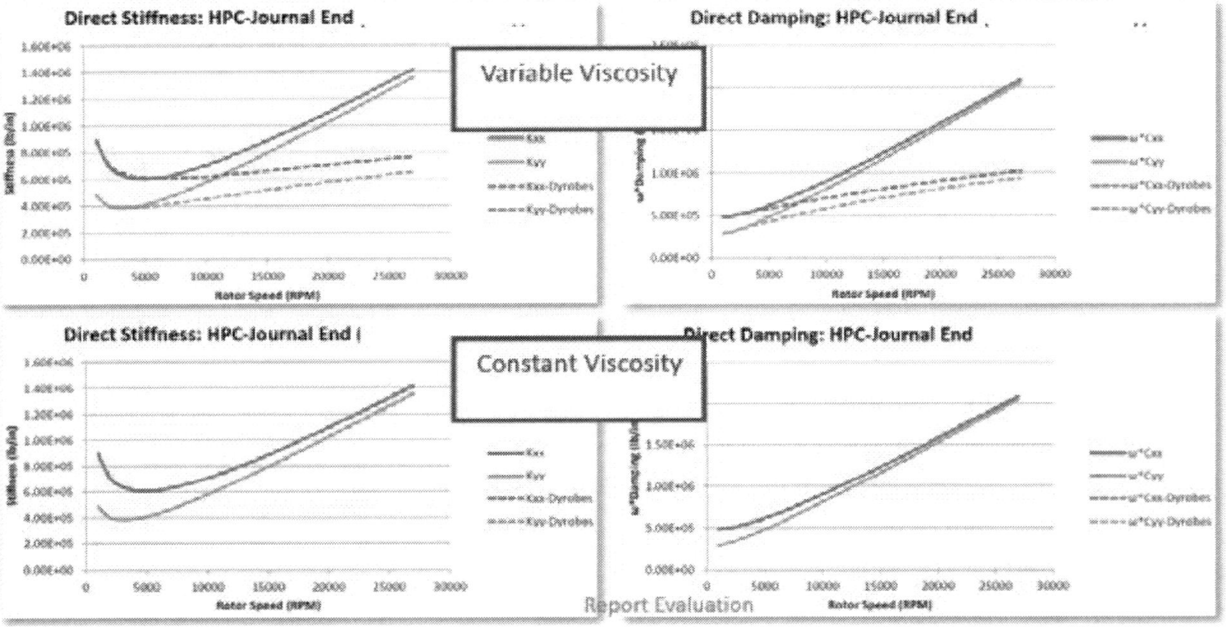

Figure 12 - Calculated Bearing Coefficients using Variable and Constant Viscosity Models

Conclusion

Lateral rotordynamic analysis is a mature field of study. The industry and academia have produced a tremendous amount of resources and data to aid in accurate predictions of rotor response and stability. Modern software tools have streamlined the modeling process and provide beautiful graphics for reports.

The remaining challenge for machinery engineers is to apply sound engineering judgement in the development of a rotordynamic model, application of standards, and assessment of results.

References

[1] Pettinato, B.C., Kocur, J.A. and Swanson, E.E., 2011, "Evolution and Trend of API 617 Compressor Rotordynamic Criteria", Turbomachinery Society of Japan: *Turbomachinery*, 39, pp 292-303.

[2] API, 1996, "Tutorial on the API Standard Paragraphs Covering Rotor Dynamics and Balancing: An Introduction to Lateral Critical and Train Torsional Analysis and Rotor Balancing,", API RP 684, 1st Edition.

[3] API, 2006, "Tutorial on the API Standard Paragraphs Covering Rotor Dynamics & Balancing: An Introduction to Lateral Critical & Train Torsional Analysis & Rotor Balancing," API RP 684, 2nd Edition.

[4] API, 2019, "API Standard Paragraphs Rotordynamic Tutorial, Lateral Critical Speeds, Unbalance Response, Stability, Train Torsionals, and Rotor Balancing," API Technical Report 684-1, 1st Edition.

[5] Childs, D. W., "Turbomachinery Rotordynamics: Phenomena, Modeling, and Analysis," John Wiley and Sons, 1993.

591076 Experiences with Hydrotesting and MIC on Recent Projects

Abstract:

Microbiologically Influenced Corrosion (MIC) is a well understood corrosion mechanism that involves corrosion of metals either caused by or promoted by microorganisms (usually bacteria). One situation that frequently can lead to MIC is hydro testing of new piping and equipment as part of a capital project. This paper will include the following components:

- A high level overview of MIC
- Provide some recent examples of MIC encountered during construction / commissioning of new ethylene units
- Provide some recommended actions which have been successful in avoiding MIC

Copyright © American Institute of Chemical Engineers. All rights reserved.

615071 Popcorn Polymer Fouling in a Depropanizer Tower System

Abstract:

Popcorn polymer formation is a known process safety risk in ethylene processing units with relatively high butadiene concentration (e.g. butadiene purification units).

This presentation will summarize a 2019 loss of containment event involving popcorn polymer plugging a unit with relatively lower bulk butadiene concentration, namely an olefins purification unit in a depropanizer tower system.

As part of the event investigation, it was determined that popcorn polymer can form in units at lower butadiene concentrations when combined with initiators such as oxygen, peroxides or rust, particularly in areas with restricted fluid circulation. A characteristic of popcorn polymer (vs. other forms of butadiene-related polymer) is growth which can result in mechanical damage, unit plugging and possible loss of containment if not mitigated.

This presentation will provide a broad overview of popcorn polymer fundamentals, highlight the risks associated with popcorn polymer via an incident summary and suggest mitigation ideas to address this process safety risk.

Copyright © American Institute of Chemical Engineers. All rights reserved.

589402 Risks and Accumulation Points for Radioactive NORM Foulant in an Ethylene Plant

Abstract:

Naturally occurring radioactive materials (NORM) are found in the earth's crust in ppm levels, and the decay of these unstable radioactive elements produces other radionuclides. During the hydrocarbon recovery process, these subterranean radionuclides become mobile and flow with the oil, gas, and water mixture to surface; eventually finding its way into the ethylene process equipment. NORM materializes in the form of radon gas and deposits of mineral scales from the sparingly-soluble salts. Since radon has a lack of water solubility and boiling point similar to propane, it materializes in the de-ethanizer. The decay of radon gas also forms thin films on interior surfaces of gas processing equipment, piping, and vessels. The mineral scale is typically sulphate scale (BaSO4 & SrSO4) and carbonate (CaCO3). Radium has similarities to barium, strontium and calcium and other alkaline earth metals in group 2. Because of the similarities, radium co-precipitates with the other scale to form radium sulphate, radium carbonate, and in some cases radium silicate. This precipitation leads to accumulation that not only leads to processing problems (reduced flow, increased pressure, poor heat transfer), but also may pose health and safety risks due to the elevated levels of radioactivity.

Copyright © American Institute of Chemical Engineers. All rights reserved.

189d:

SAFETY CONCERNS WITH SELF-HEATING MATERIALS, INCLUDING A PACKED TOWER INDUSTRY EXAMPLE

Tania Almazan
Olefins Recovery Technology Hot Ends Lead Engineer
ExxonMobil Chemical Company

Giselle Larish
Hydroprocessing and Lubes Advisor
Imperial Oil

Selma Kwok
Olefins Recovery Technology Specialist
ExxonMobil Chemical Company

Prepared for Presentation at 2020 Spring National Meeting
Houston, TX, August 17 – August 21, 2020

AIChE and the EPC shall not be responsible for statements or opinions contained in papers or printed in its publications

SAFETY CONCERNS WITH SELF-HEATING MATERIALS, INCLUDING A PACKED TOWER INDUSTRY EXAMPLE

Giselle Larish
Hydroprocessing and Lubes Advisor
Imperial Oil
Tania Almazan
Olefins Recovery Technology Hot Ends Lead Engineer
ExxonMobil Chemical Company
Selma Kwok
Olefins Recovery Technology Specialist
ExxonMobil Chemical Company

Abstract:

Managing materials that have self-heating characteristics is a key process safety concern for many industries, including the ethylene industry. Self-heating materials include materials that, when in contact with air, are liable to self-heat over time, potentially resulting in fire or other damage. Pyrophoric materials are a subset of self-heating materials that ignite rapidly when in contact with air.

This presentation will summarize a 2019 self-heating event involving a section of packing in a tower at a large petrochemical facility. This event will highlight the risks associated with certain vessel internals and self-heating materials. Further, the presentation will provide a broad overview of the risks associated with self-heating materials in an ethylene plant, highlighting areas of specific concern along with potential mitigations.

Self-Heating and Pyrophoric Materials Overview

Background

Managing materials that have self-heating characteristics is a key process safety concern for many industries, including the ethylene industry. Since many polymers, foulants, coke, catalysts, and adsorbents found in olefins facilities are potentially capable of self-heating or pyrophoric activity when exposed to air, ongoing monitoring and mitigation is recommended any time these materials remain inside of open equipment as well as during removal and disposal.

Self-heating materials exhibit a temperature increase when contacted with air, even without an external heat source. Pyrophoric materials are a subset of self-heating materials. They are liquids or solids that can ignite within five minutes after coming into contact with air, even in small quantities and without an external heat source (1). One example of a pyrophoric material is iron sulfide, which is formed when iron reacts chemically with sulfur, typically H_2S.

Since the probability and intensity of self-heating or pyrophoric behavior can vary widely throughout a plant, past experience is not a reliable indicator of future self-heating or pyrophoric activity. Although certain self-heating materials such as popcorn polymer (PCP) are highly reactive and can ignite in minutes, other self-heating materials can take days to react (e.g., very slow heating materials and pyrophoric materials encased in foulant). Moreover, self-heating or pyrophoric behavior can vary with conditions such as humidity, temperature, particle size, and other factors.

Fractionating columns with internal packing and equipment with demister screens are particularly susceptible to self-heating. In general, equipment with a large surface area, narrow flow passages, or regions of low flow can trap iron sulfide scale or other foulants, leading to an increased self-heating/pyrophoric risk.

Where in Olefins Plants are Self-Heating/ Pyrophoric Risks?

Self-heating or pyrophoric materials are common in an olefins facilty. The ethylene industry has experience with these materials in all areas of the plant:

Area	Examples of Self-heating/Pyrophoric Material Observed	Comments
Feed Systems	Iron sulfide, foulant, spent adsorbent	
Furnace Area	Coke, iron sulfide, foulant, deposits	Most common downstream of quench systems
Primary Fractionator/Quench Systems	Coke, iron sulfide, foulant	Primary fractionator, tar filters, tar exchangers, and tar tanks are most vulnerable
Process Gas Compressor Area	Iron sulfide, foulant, spent adsorbent	Acid gas treating units and equipment upstream are most likely to see iron sulfide

Area	Examples of Self-heating/Pyrophoric Material Observed	Comments
Adsorbent/Catalyst	Silver acetylide, copper acetylide Unregenerated and spent methanator catalyst Unregenerated and spent hydrogenation catalyst	Silver acetylide can be formed from acetylene contacting adsorbent/catalyst beds containing silver, and copper acetylide from beds containing copper
Cold Box, Demethanizer, Refrigeration Systems	Foulant	Although kept clean, self-heating has been observed
Deethanizer/Depropanizer	Foulant	Foulant could include popcorn polymer, which is highly pyrophoric (2)
Debutanizer	Popcorn polymer, other foulant	
C5+ Processing	Popcorn polymer, other foulant, nickel sulfide, iron sulfide, chromium sulfide	

Equipment history is not a reliable indicator of future potential self-heating/pyrophoric activity, as demonstrated by the above observations in even "clean" areas of an olefins plant.

Packed Tower Industry Example

Creep Buckling Failure Incident Summary

Even for facilities with long experience favorably managing risk associated with opening equipment containing self-heating or pyrophoric material, self-heating and pyrophoric incidents can still occur. For example, in April 2019, a large fractionation tower failed approximately one hour after opening three manways for maintenance. The tower was located at the back end of a refinery hydrotreating unit. Although no one was injured, this event resulted in an appreciable disruption of plant operations.

The tower was installed in 1968 and had been opened five times with no history of pyrophoric material. The 2019 turnaround was the first time the tower had been opened since 2011. In 2011, nine bottom trays in the tower were replaced with structured packing to improve fractionation efficiency.

This unit produces a light hydrocarbon fuel, and was considered by stakeholders associated with the design, operation, and maintenance of the tower to be a "clean" service. There was no history of self-heating foulant being observed in the tower circuit prior to the event. The non-pyrophoric history of the tower and its use in clean hydrocarbon service resulted in an underestimation of pyrophoric risk.

The sequence of events in this incident started with a standard chemical cleaning and cooldown of the tower. After blanking, Manway 4 was opened and its air mover was turned on. Manway 1 was then opened. When Manway 2 was opened and its air mover turned on, temperatures unexpectedly increased.

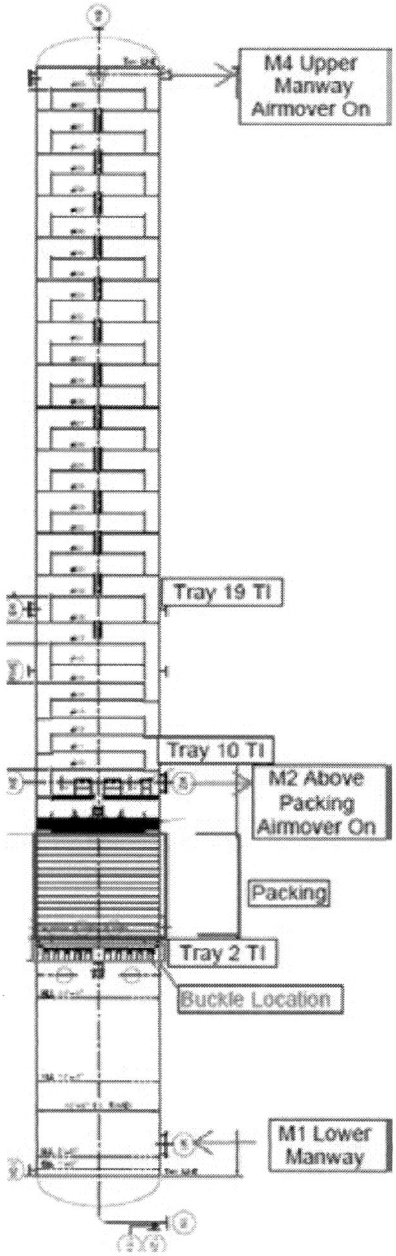

2019 refinery tower drawing with manways labeled

The temperature rise was associated with a pyrophoric reaction in the packed bed. This led to local overheating of the shell, resulting in creep buckling and failure of the tower about one hour after Manway 2 was opened.

Risk Considerations

Carrying out a more rigorous self-heating/pyrophoric risk assessment beforehand may have lessened or avoided the tower damage and operations difficulties that resulted from the event. A rigorous self-heating and pyrophoric risk assessment of plant equipment should include consideration of at least the following risk factors:

- A history of self-heating or pyrophoric behavior in the equipment
- Chemistry that can generate self-heating/pyrophoric material (either in the equipment itself or at a location that is upstream of the equipment)
- Known or suspected iron sulfide accumulation in the equipment
  - Examples of indicators of iron sulfide accumulation are increased pressure drop, visual evidence of iron sulfide in filters, or the presence of sediment in stream samples
- Equipment has changed service since it was last opened
  - Changes in service may include feed property changes, operating severity changes, and metallurgy changes
- Equipment has seen a change in on-stream duration versus the previous cycle
  - A longer runlength provides an opportunity to accumulate a larger quantity of self-heating/pyrophoric material
- Changes in hardware since the equipment was last opened

Features that can increase self-heating/pyrophoric risk include structured or random packing, demisters, fixed bed reactors, filters, and regions of low flow. The larger the quantity of packing, for example, the greater the potential consequence of a self-heating/pyrophoric event. In the case of fixed bed reactors, catalysts are often self-heating, and the fixed bed itself acts as a very efficient collector of foulants. Regions of low flow provide a place where self-heating/pyrophoric materials can settle out and accumulate.

Consider triggers in project and change management work processes to prompt a self-heating/pyrophoric risk assessment.

Potential Mitigations

Self-heating/pyrophoric mitigations can be implemented before and after exposing potentially self-heating/pyrophoric material to oxygen. Before opening equipment, reducing the amount of self-heating/pyrophoric material in the equipment can be done via special procedures for equipment preparation, chemical cleaning (not a sufficient mitigation by itself), and cooling to reduce the vapor pressure of flammable residues (also does not eliminate the risk).

It is important to limit exposure of self-heating/pyrophoric materials to oxygen, e.g. when equipment is opened to the atmosphere. For example, air flow across packing should be minimized before it is wetted. After an initial wetting with water, the material needs to be periodically or continuously re-wetted to avoid dry-out. Consideration should be given to the water supply rate and adequate water distribution to cover the equipment cross-sectional area. Air can be humidified to slow dry-out between periodic wettings.

Another option to limit exposure to oxygen is to water-flood the equipment, but only after assessing the equipment's foundation/support should for handling the additional load.

Provided reasonable measures are taken to prevent asphyxiation risk, inerting the equipment with nitrogen is an alternative method to exclude oxygen. This is a common technique used for spent catalyst removal from fixed-bed reactors.

When equipment is opened to the atmosphere, plans should be in place to quickly detect the onset of self-heating/pyrophoric reaction in the vessel. Because the chemistry can be very fast, an integrated process/mechanical opening and risk mitigation plan should be developed prior to opening the equipment. Special procedures for opening process equipment could include allowing limited initial oxygen exposure to reveal self-heating/pyrophoric potential, paying special attention to manway opening sequence, and allowing natural draft for a period prior to starting air movers. Monitoring parameters include gas monitoring (CO and SO_2) and temperature monitoring.

In the event that monitoring reveals that self-heating/pyrophoric chemistry has initiated, the integrated process/mechanical team should plan for the following emergency response:

- Remote shutdown of air movers to decrease air exposure of the material
- Rapid closure of manways or another means to starve the chemistry of oxygen (i.e. a lightweight fitted plate or fire blanket)
- Further wetting of the vessel or equipment internals
  - Consider the volume and rate needed for internal configuration and potential reaction with the material
- Nitrogen or steam could be used to purge the oxygen from the vessel
  - This method could pose risks to nearby personnel and have limited effectiveness with multiple manways open
- Water flooding of the equipment
  - The vessel and foundation structure's ability to hold the weight would need to be considered before mechanical work begins
- Application of foam
- An emergency response plan with egress plans

Key Messages

As shown by this pyrophoric incident, history is not a reliable indicator of potential self-heating/pyrophoric risk. It is important to ensure that equipment maintenance planning includes an assessment of this risk. Chemical cleaning can be a part of the mitigation plan, provided it is recognized that chemical cleaning by itself is not a sufficient mitigation against self-heating. Finally, pyrophoric chemistry happens very quickly. Ensuring that opening process equipment work processes include self-heating/pyrophoric discussions and development of integrated process/mechanical risk mitigation plans will greatly reduce the risk of another incident like this one. All key personnel involved in the planning and execution of opening process equipment need to understand self-heating/pyrophoric hazards.

References

1. United Nations Manual of Tests and Criteria (2019).
2. Broekaart, Paul and Romagnoli, Massimo (2019), "Recommendations for Preventing Popcorn in Steam Crackers and Butadiene Plants," *European Ethylene Producers Committee.*

588948 An Industry Perspective of Cold Box NOx Safety Practices

Abstract:

Managing NOx safety considerations are a key issue for all ethylene producers. A task group has met intermittently to discuss safety issues associated with NOx compound accumulations in ethylene plant cold boxes over the past 20 years. The group has presented updates and provided tutorials on this subject to increase industry awareness.

This paper summarizes the most recent activity of the task group which is focused on proceedings from a 2019 group meeting. Since the task group last met in September 2012, the meeting was focused on understanding how industry NOx knowledge and operating experience has continued to evolve. Specifically, the paper summarizes NOx mitigation methods and implementation learnings, new areas of chemistry understanding and NOx risk analysis basis setting. In the end, this summary will further enhance industry NOx understanding and application.

Copyright © American Institute of Chemical Engineers. All rights reserved.

AIChE Paper Number: 123a

CAN YOU AFFORD TO NOT UPGRADE YOUR COOLING TOWER?

Ben Sherrod
Civil Engineer
Chevron Phillips Chemical Company LP

Tom Garza
Manager – Fixed Equipment, Quality and Inspection
Chevron Phillips Chemical Company LP

Jim Dykes
Asset Integrity Manager
Chevron Phillips Chemical Company LP

Stuart Bell
Lead Energy and Optimization Engineer
Chevron Phillips Chemical Company LP

Prepared for Presentation at the 2020 Spring National Meeting
Houston, TX, March 29 – April 3

AIChE and the EPC shall not be responsible for statements or opinions contained in papers or printed its publications.

CAN YOU AFFORD TO NOT UPGRADE YOUR COOLING TOWER?

Ben Sherrod
Civil Engineer
Chevron Phillips Chemical Company LP
Tom Garza
Manager – Fixed Equipment, Quality and Inspection
Chevron Phillips Chemical Company LP
Jim Dykes
Asset Integrity Manager
Chevron Phillips Chemical Company LP
Stuart Bell
Lead Energy and Optimization Engineer
Chevron Phillips Chemical Company LP

Abstract:

Chevron Phillips Chemical Company LP (CPChem) operates a 12-cell Marley 600 Series crossflow tower at an ethylene cracker on the Gulf Coast. The cooling tower receives hot process water from the cracker unit and cools it via evaporative cooling. The cells were originally constructed using Douglas fir wood framing beginning in 1976. In 2016, an external visual inspection identified deflection in the 54-inch return header and supporting structure. Further analysis suggested that degradation of the main wooden columns was a likely root cause and could have an impact on their load carrying capabilities. A quick repair was performed to maintain the tower as fit-for-purpose.

The existing tower was then evaluated using structural analysis software to determine the required equivalent fiber reinforced plastic (FRP) member sizes. A thermodynamic model proved there were opportunities to increase the efficiency by installing the latest technology splash-fill. This paper presents the improvements in reliability, maintenance costs, and production after rebuilding the cells with pultruded FRP members and modern splash fill. The other cells are planned to be upgraded at the next scheduled opportunity.

Background

The exterior appurtenances of the cooling tower can be seen in **Figure 1**. Behind the louvers and external partitions is where the main supporting structure is located. The subtle deformation observed in the louvers at the center of the picture was noticed during a routine visit to the field by the operations group. There were no other external indications of the deficiencies within the structure. However, the internal deficiencies were serious enough to lead to a collapse of the structure if not addressed.

Figure 1. Exterior view of cooling tower.

A cross section of the existing transverse end wall bent is shown below in **Figure 2**. There are various wooden beams and columns that are responsible for supporting loads from the fan, gearbox, cooling water header, hot water basin, equipment, fill, drift eliminators, stairs and ladders. The structure must also resist hurricane force winds, ice, and the continual erosion that occurs from the water flowing down and across the wooden members. The member sizes may be familiar to anyone that has been exposed to wood frame construction. There are 4x6s, 4x4s, 2x8s, and 2x10s just to name a few of the sizes used for beams and columns.

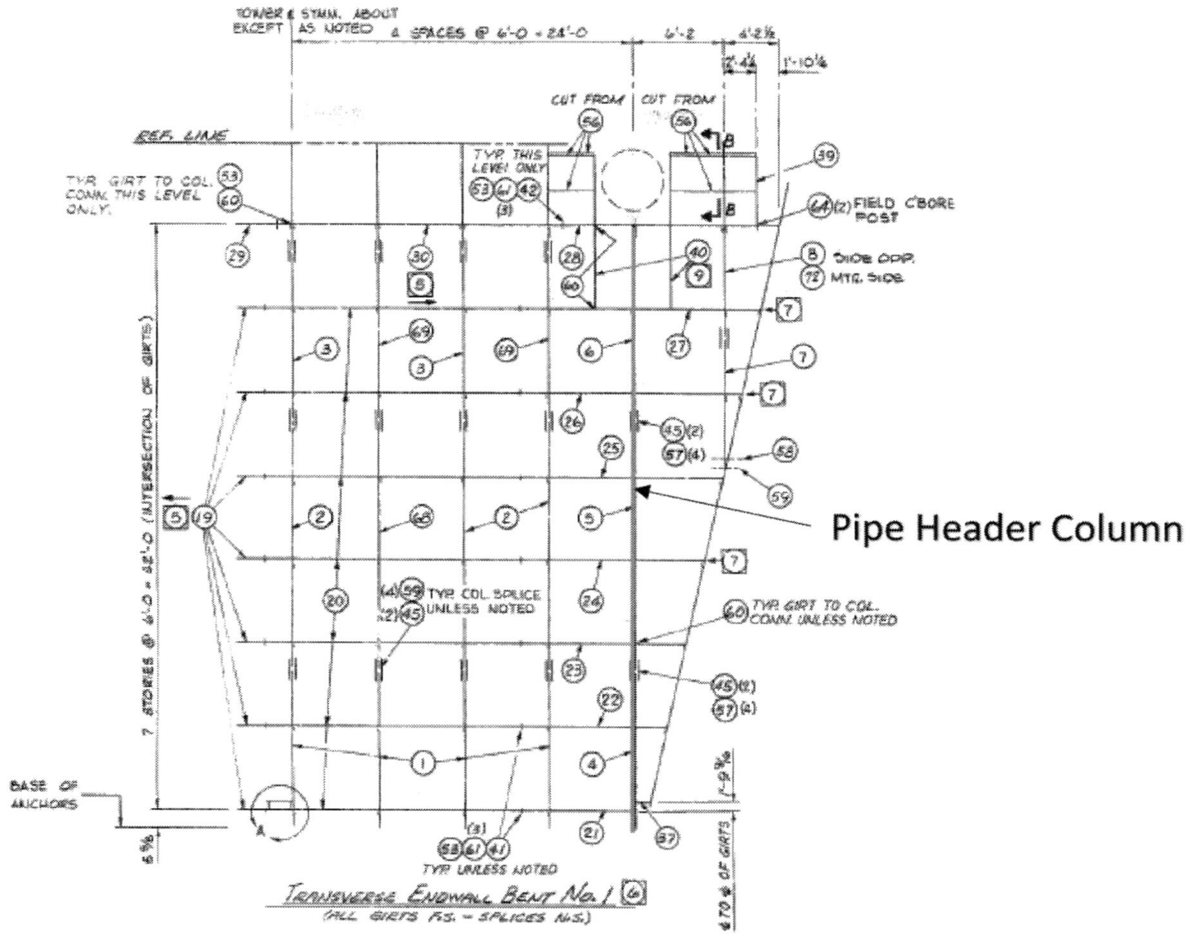

Figure 2. Cross-section of wooden support structure.

The red vertical member labeled "pipe header column" is the most heavily loaded member in this design. It is a series 4x6 timber column spaced every 6 feet and laterally braced every 6 feet about the strong and weak axis. Forty years of water erosion had reduced the net section area of each one of the pipe header columns and over time caused buckling and failure of some columns.

With the columns being subjected to stresses beyond what is allowed by the *National Design Specification for Wood Construction*, it was determined the 54-inch header needed additional support to prevent collapse of the structure. Two strategies were developed. The short-term strategy was to install temporary supports and the long-term strategy was to re-design and rebuild each cell. The remaining sections detail the steps taken for each strategy and summarize the cost and performance benefits of the upgrade.

Structural Assessment

Visual Assessment

Louvers, fill and drift eliminators were removed to reveal deformed and broken pipe header columns in the wet area as shown in **Figure 3**. The pipe saddles were splitting due to long-term creep from the 54-inch pipe bearing on the plywood. Delignification and erosion caused the hot water support beams to fail at the splice locations.

Figure 3. Damaged wooden structural members.

Physical Measurements

The columns and beams were measured to determine as found dimensions and calculate the reduced net section area. The wooden columns were originally 3-1/2" x 5-1/2". Some of the pipe header columns were found to be 2-3/4" x 4-3/4". This represents a reduction of 32% in cross section, which has a large impact on the allowable load each column can support.

Analysis

An analysis was performed using the as-found dimensions of the 4x6 columns to determine the reduced allowable load. The results of this analysis determined the existing header support columns were inadequate and a temporary support system was designed to help support the 54-inch pipe header and redistribute the loads. Fiber reinforced plastic (FRP) columns were chosen based on the long-term reliability compared to the quality of modern fast growth wood members. **Figure 4** shows a comparison between old-growth wood and present-day fast growth wood.

Figure 4. Old growth vs. fast growth wood.

Comparison of Costs for Repair vs. Rebuild

As discussed, the repair strategy involved installing temporary structural support frames constructed from FRP. This required the use of a large crane to support the header while the work was being executed. The goal was to provide adequate support of the 54-inch header until a new structure was designed and analyzed. Time was also needed to secure funding, conduct a bid process and select/award a contract.

The rebuild strategy focused on evaluating the loads and designing a structure that complied with the current ASCE-7 wind code. The goal was to construct new cells on the existing basin that would last for 30 years and require minimal maintenance costs. The need to use fit-for-purpose fill that meets or exceeds the original thermal performance was also a priority.

Cost data was compiled for the rebuild and repair projects. Based on the data shown in **Figure 5**, the rebuild of each cell was 10% less expensive than the repairs that were required to keep the equipment in service. Furthermore, the life expectancy and reliability were greatly improved by performing a complete rebuild with FRP.

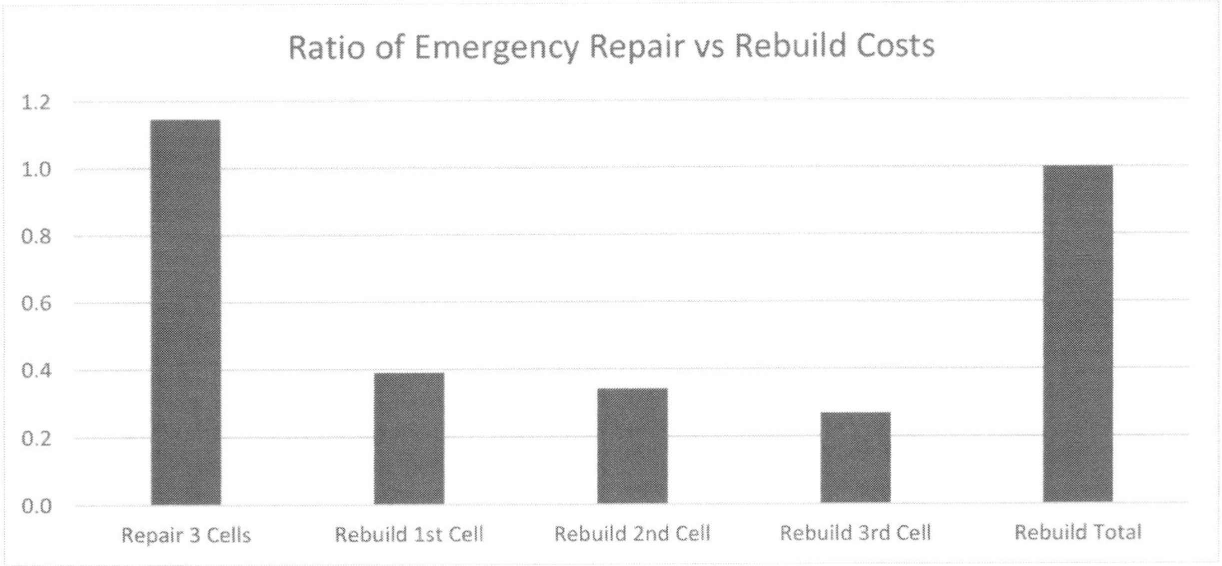

Figure 5. Total installed cost of repairs and rebuild as a ratio.

Thermal Performance

Following the complete rebuild of Cells 1, 2 and 3, a thermal study was conducted to examine the performance of the tower. Dry bulb and wet bulb temperature, liquid temperature in the cold-water basin and return water temperature were measured and recorded for each cell on both sides.

Efficiency for each cell was then calculated using the following formula:

$$\textit{Efficiency} = \frac{T_{Hot} - T_{Cold}}{T_{Hot} - T_{Wet\ Bulb}}\ x\ 100$$

The resulting efficiency for each cell is shown in **Figure 6**. The three new cells with modern fill and drift eliminators are more efficient than the existing cells. The results also suggest there is an opportunity to have the most effect by rebuilding Cells 10, 11 and 12 before addressing the other cells.

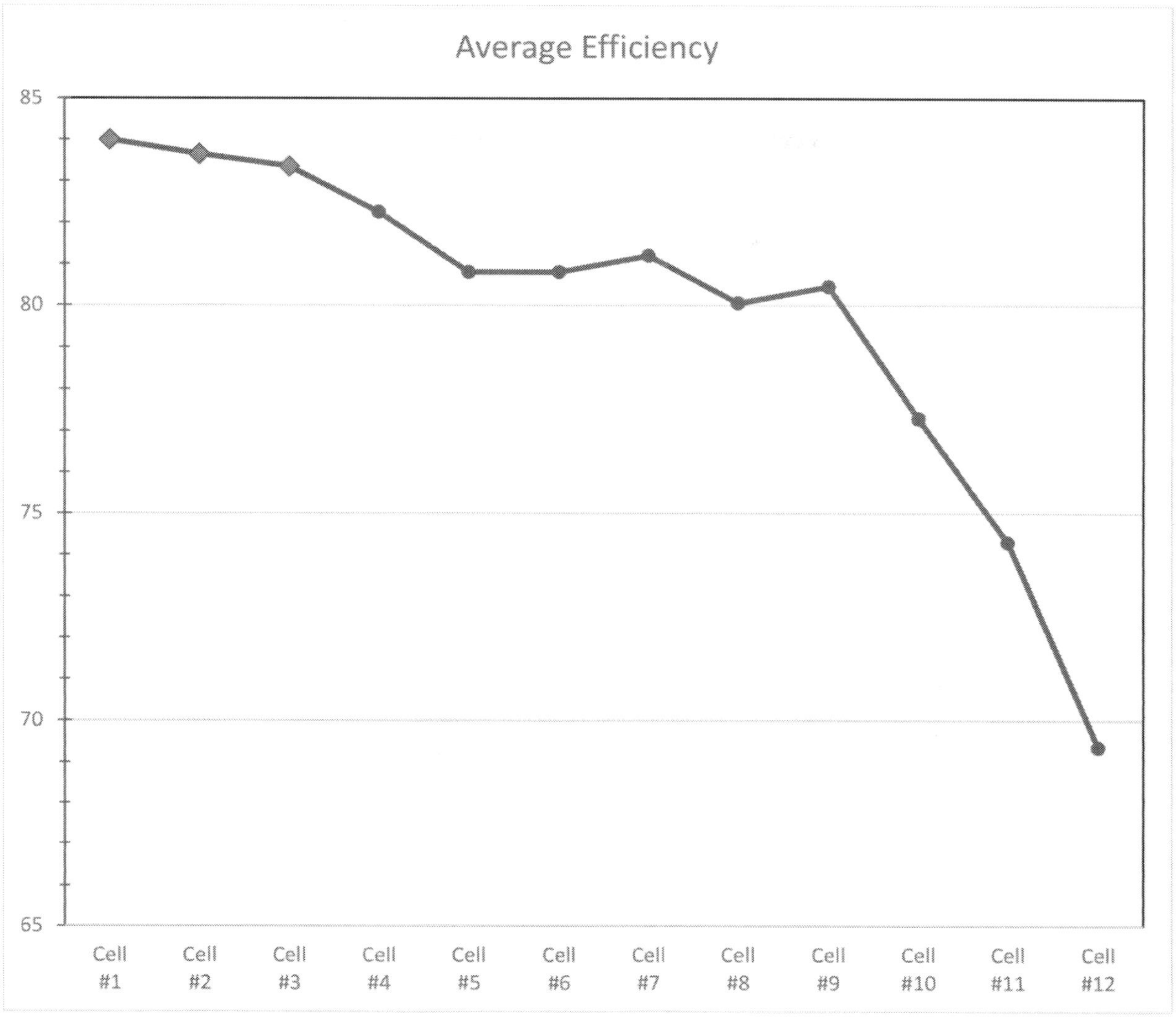

Figure 4. Thermal efficiency of each cell.

Conclusions

Cooling towers should be viewed as critical to the operation of a facility and given the same inspection priority as other fixed equipment to ensure reliability. Wood is no longer the construction material of choice for this application and FRP should be considered in order to meet a 30-year design life without incurring repeated maintenance costs. Inspection and testing programs can identify opportunities for reducing maintenance costs and increasing the performance of these valuable assets.

References

1. NATIONAL DESIGN SPECIFICATION® for Wood

AIChE Paper Number: 123b

ON-LINE MONITORING OF THE EFFECTIVENESS OF CHEMICAL TREATMENT TO ABATE FOULING IN FILM FILL COUNTER-FLOW COOLING TOWERS UTILIZING GAMMA SCANS

Charles Winfield
President
QUANTUM Technical Services, A Nalco Champion Company

Paul Chila
Engineering Consultant
QUANTUM Technical Services, A Nalco Champion Company

James Green
Global SME – Cooling Towers
Dow Chemical

Prepared for Presentation at the 2020 Spring National Meeting
Houston, Texas, March 29 – April 2, 2020

AIChE and EPC shall not be responsible for statements or opinions contained in papers or printed in its publications

ON-LINE MONITORING OF THE EFFECTIVENESS OF CHEMICAL TREATMENT TO ABATE FOULING IN FILM FILL COUNTER-FLOW COOLING TOWERS UTILIZING GAMMA SCANS

Charles Winfield
President
QUANTUM Technical Services, A Nalco Champion Company
Paul Chila
Engineering Consultant
QUANTUM Technical Services, A Nalco Champion Company
James Green
Global SME – Cooling Towers
Dow Chemical

Abstract:

Chemical treatment is widely used to remove or reduce fouling in counter-flow cooling towers using film fill as the cooling medium. Effectiveness monitoring and associated optimization of chemical injection is difficult without shutdown, entry, and visual inspection/verification.

This paper describes how gamma scanning techniques have been adapted to monitor both progression and removal of fouling in these types of cooling towers while the cooling tower remains on-line. Case histories will be presented showing how this application has been utilized to track film fill fouling, monitor the effectiveness of a chemical treatment program, and improve the overall cooling system reliability.

Background

High efficiency cooling tower fill has become common in industry due to its ability to deliver more evaporative cooling capacity per square foot than the splash-type fills historically used. One square foot of film-type fill typically provides the same heat transfer as ten square feet of splash-type fill [1].

High efficiency film fill

Unfortunately, cooling towers with high efficiency film fill are more prone to fouling and can require more diligent water treatment protocols. The fill can easily become fouled with some combination of silt, microbiological growth and mineral scale. In addition to the penalties in heat transfer, continued deposition of foulants increase the loading in the tower causing excessive stress on the fill and the support structure. Left untreated, the cooling tower can eventually collapse resulting in unplanned maintenance costs, lost throughput, and potential safety and environmental concerns. Fouling also impedes air flow, further reducing efficiency. Care must be taken to monitor the cooling tower fill for fouling. Having a robust chemical treatment program in place can also help to minimize the fouling potential.

Gamma Scanning is an online diagnostic tool allowing inspection of cooling towers and other chemical process vessels, such as columns and reactors, while in operation. Using a collimated beam of gamma rays which pass through the vessel walls and measuring the intensity of the radiation exiting the vessel creates a bulk density profile of all materials inside the vessel.

Gamma scanning has proven to be an ideal online tool for diagnosing problems inside operating vessels. Gamma scanning can identify process problems such as damage to trays and packing, level control problems, foaming, vapor and liquid maldistribution, fouling, etc.

Figure 1 shows a typical gamma scanning setup used on a vessel with fill as its medium.

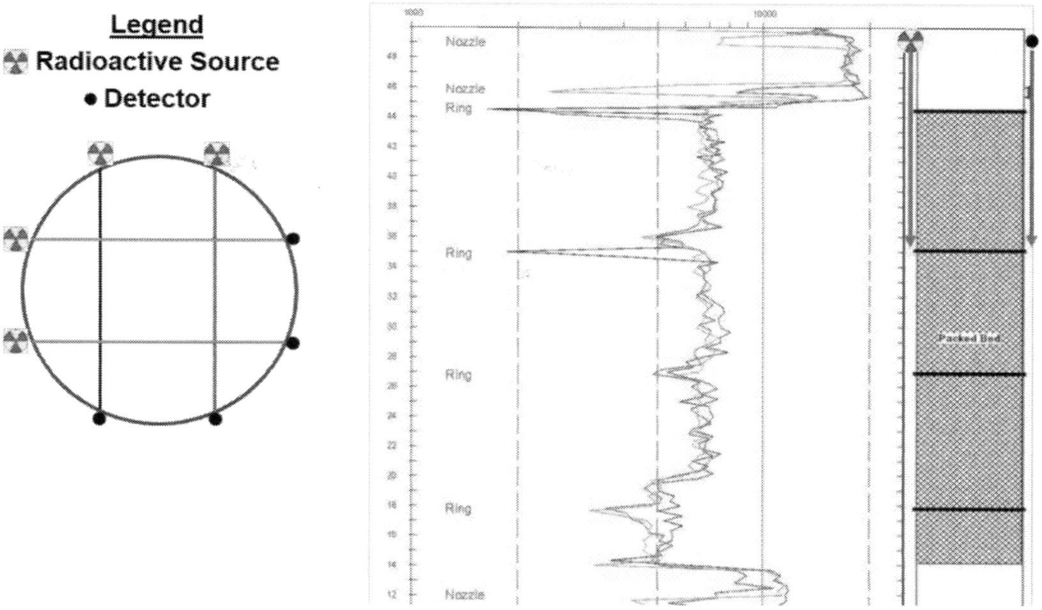

Figure 1. Typical setup to perform gamma scans of a packed tower

Theory

Gamma radiation can be explained by a series of equations including the Beer-Lambert Law [2]. This equation applies to radiation passing through a single medium. The Beer-Lambert Law can be adapted for multi-component systems such as cooling towers [3]. This adaptation results in the ability to calculate a bulk density representing the weight of film fill, water on the film fill, and other materials that may be building up in the film fill of a cooling tower.

Film fill in counter-flow cooling towers is analogous to structured packing in distillation towers. The degree of fouling in film fill can be effectively monitored by applying the principles of gamma scanning while the cooling tower is in operation. This is of great benefit to assess the effectiveness of chemical treatment.

Setup

The equipment setup to scan a cooling tower is very simple. A gamma source and detector are placed on each side of the tower to create a scan line that is roughly tangential to one side of the fan shroud. The source and detector are then lowered incrementally (typically 3 inches per reading) down to the bottom edge of the sidewalls. Once a scan is complete, the source and detector are raised and repositioned to perform another scan tangential to the opposite side of the fan shroud. Figures 2 and 3 show scan direction and paths created during data collection.

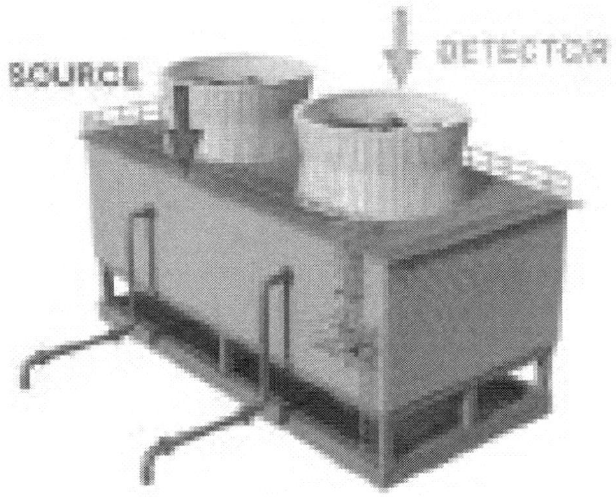

Figure 2. Scan line direction (vertical)

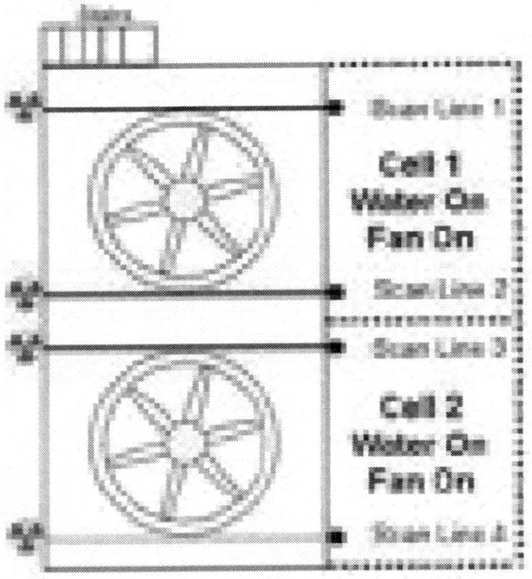

Figure 3. Scan lines (radiation passes horizontally through the cooling tower)

For larger towers a slightly more complex setup may be required. Holes may need to be drilled through the top deck of the cooling tower and PVC piping installed through the film fill to allow for scanning. In cases where the cooling towers are very wide, we adopt a "divide and conquer" strategy to obtain data for analysis. Figures 4 and 5 illustrate how this is accomplished.

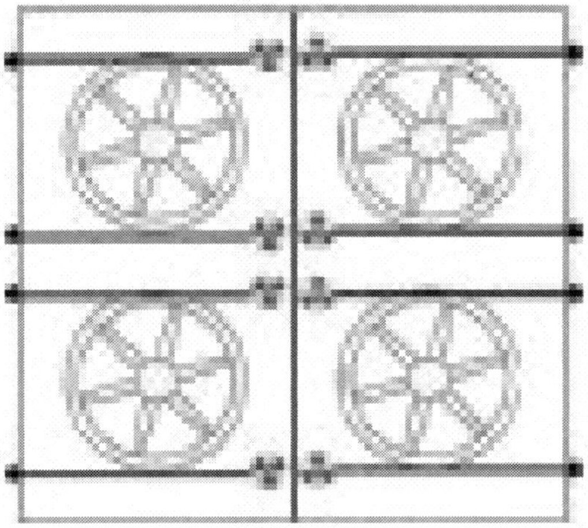

Figure 4. Complex setup

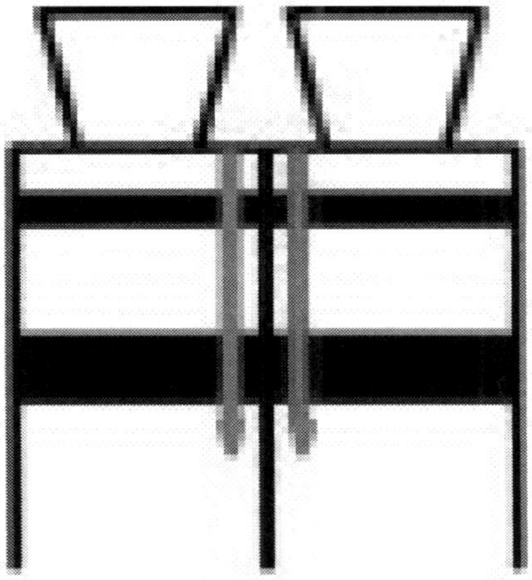

Figure 5. PVC piping installation to accommodate scanning through film fill

Application

A monitoring program utilizing gamma scans which typically includes an initial set of scans and a schedule for follow-up scans based on the findings. Data is presented as an X-Y plot of bulk density versus elevation. The bulk density reported is representative of the combined weight of the film fill, water on the film fill, fouling on the film fill, debris in the film fill, etc. The contribution of tower wall density is considered negligible. Typically reported in units of lb/ft^3, the data is also available for reporting in any units of density requested.

Experience has generated data, allowing guidelines as to what treatment strategy to recommend based on levels of average bulk density found in film fill during operation. Table 1 is an example of an Equipment Reliability Strategy based on fouling rate and weight gain. The equipment owner should seek the specific design value of their tower and establish a risk mitigation strategy such as Table 1.

Table 1: Recommendations Based on Average Film Fill Bulk Density	
2.2 – 6 lb/ft^3	Tower is operating well – no adjustments to treatment needed. Annual monitoring will be adequate.
6 – 10 lb/ft^3	Consider increasing aggressiveness of treatment slightly in the near future. Do first scan 1 – 2 months after adjusting to determine next best course of action for treatment and monitoring program.
10 – 13 lb/ft^3	Consider increasing aggressiveness of treatment in the near future. Do first scan 1 – 2 months after adjusting to determine next best course of action for treatment and monitoring program.
13 – 16 lb/ft^3	Begin increasing aggressiveness of treatment immediately. Do first scan 1 – 2 months after adjusting to determine next best course of action for treatment and monitoring program.
> 16 lb/ft^3	FILL IS AT HIGH RISK OF COLLAPSE - Begin increasing aggressiveness of treatment immediately. Do first scan 1 – 2 months after adjusting to determine next best course of action for treatment and monitoring program.
The values listed in this table come from years of empirical data supplied and verified by clients. They are meant to serve as a general guideline for monitoring film fill performance utilizing gamma scanning technology.	

When a new cooling tower is commissioned, low bulk densities are expected to be seen throughout the height of the film fill. There should also be a strong coincidence of observed bulk densities between the two scan chords. Figure 6 shows a healthy profile of uniformly low bulk densities in a new tower that had been in operation for 1 month. Both scan lines coincide well with each other and the bulk densities seen throughout the height of the film fill are in the expected range.

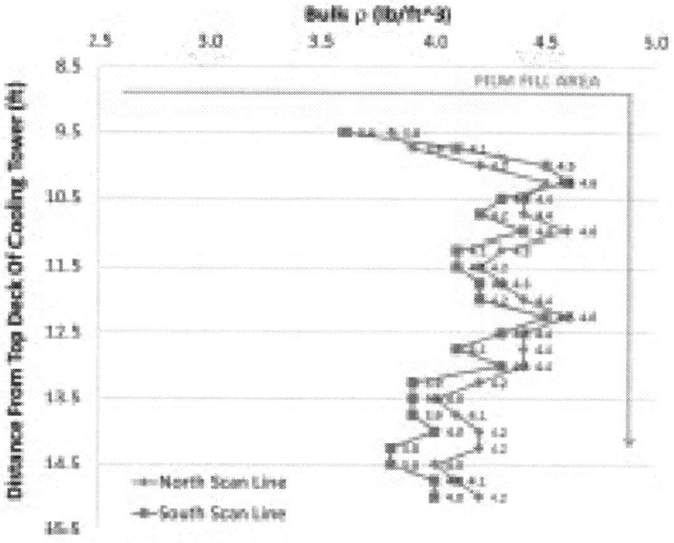

Figure 6. New tower in operation 1 month

Case Study

A Texas refinery with a 2-cell counter flow cooling tower was concerned that fouling was putting their cooling tower fill integrity at risk. Gamma scans were performed, and the initial scans confirmed that the tower was nearing a critical bulk density weight. An increase in the chemical feed rate was recommended. Approximately one month after increasing the chemical treatment dosage, a follow-up set of scans were performed to determine the effectiveness of the change in treatment. The second set of scans indicated a slight improvement but warranted a recommendation to consider an even more aggressive treatment of the film fill. Approximately two months later, a third set of scans were performed to determine the effectiveness of the increased chemical treatment adjustments. The scan results showed that the more aggressive chemical treatment strategy successfully removed the fouling (Figure 7). The lower bulk densities were in alignment with patterns seen in clean new film fill (Figure 6).

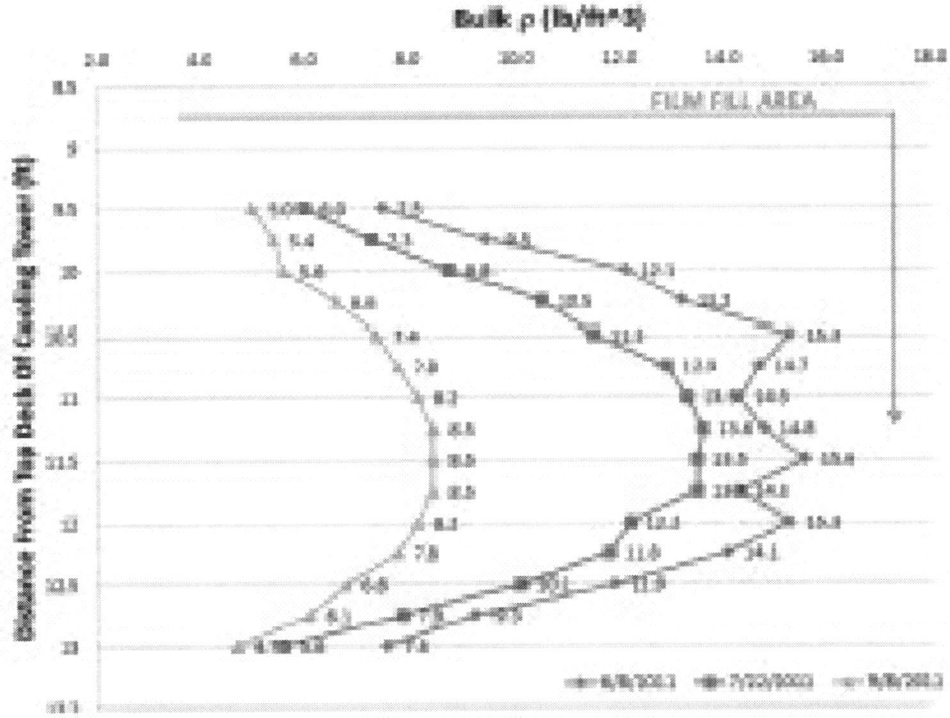

Figure 7. Successful aggressive chemical treatment

The ability to monitor and optimize the chemical treatment program while online prevented the refinery from having to shut down the cooling tower for a very costly inspection and significantly reduced the risk of a catastrophic event such as a tower collapse.

Monitoring on a Maintenance Schedule

Feeding biodispersants in conjunction with biocides can help reduce biofouling on film fill [4]. If a cooling tower becomes fouled, an aggressive chemical treatment strategy consisting of some combination of oxidizing and non-oxidizing biocides, biodispersants and biodetergents can typically restore efficiency to the fill. Figure 8 illustrates an example of where a power plant was able to apply a biodispersant program to measurably decrease the tower fill weight, indicating the removal of foulants [5]. This plant's tower rests on a scale which continuously reports the fill weight. Few plants have this type of fill weight measurement capability and so plant personnel look for other means to monitor the effectiveness of the microbiological treatment program and cleanliness of the tower fill.

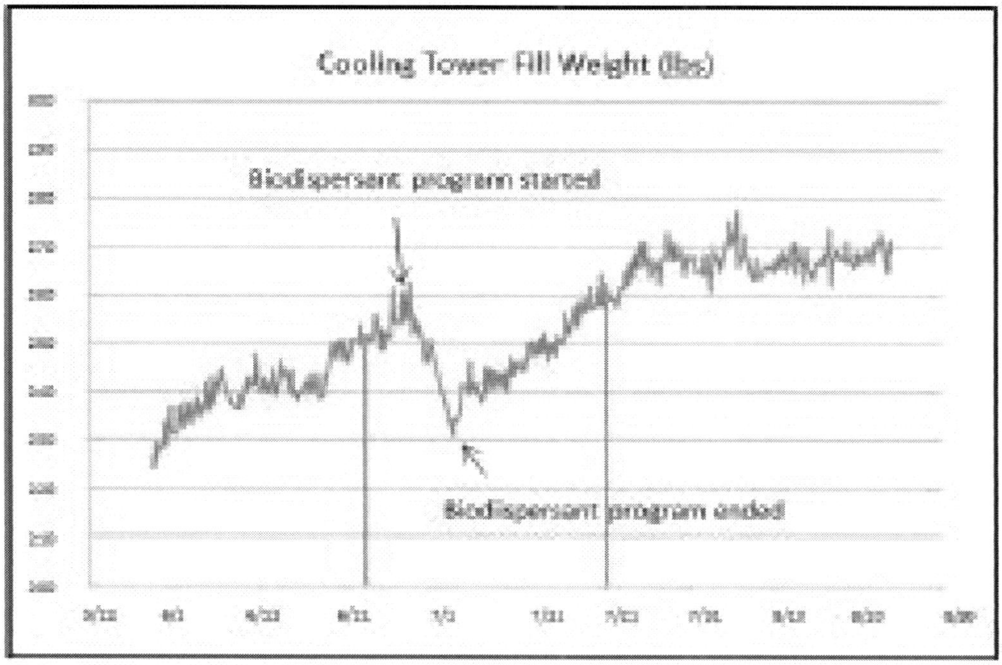

Figure 8. Measurement of fill weight of tower on a scale

Utilizing gamma scanning to monitor the effectiveness of a chemical treatment program is very beneficial. Often plant operations will elect to discontinue the use of a biodispersant after the initial cleanup. However, this can come with consequences if not monitored. Figure 9 shows scan results for a single cell cooling tower performed over a 2-year period.

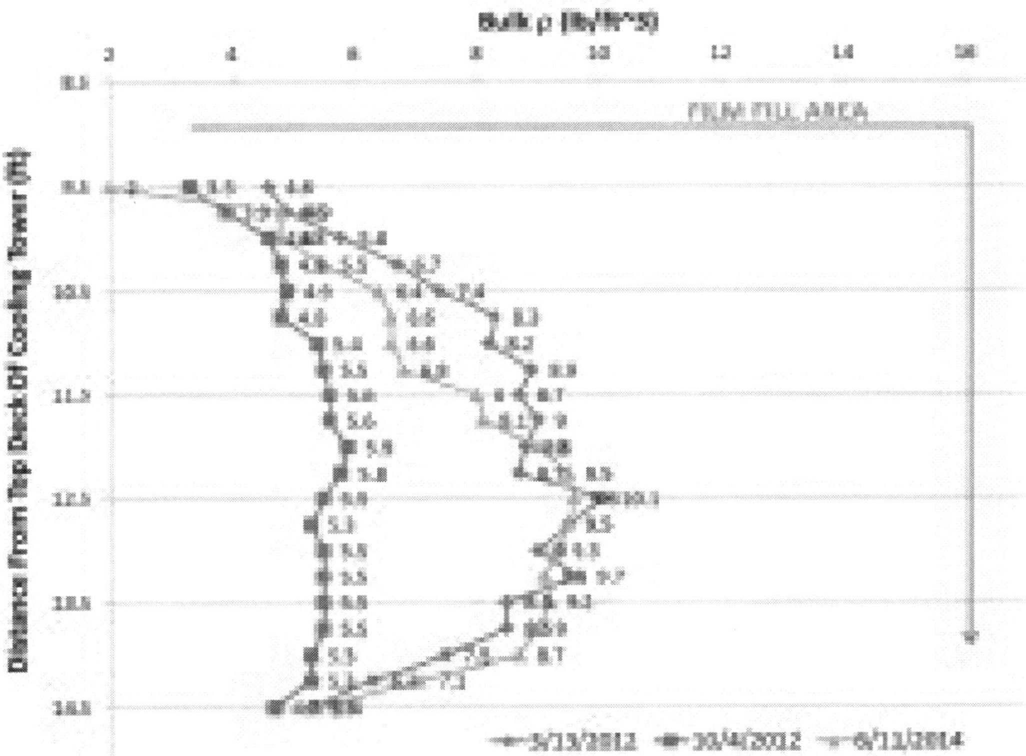

Figure 9. Monitoring the effects of removing biodispersant program

A base scan was performed and indicated that the film fill was fouled, and treatment was recommended. Upon feeding the biodispersant for three months, the film fill appeared to be like new clean fill with observed bulk densities around 5 lb/ft3. Over time, the plant phased the biodispersant out of the chemical treatment strategy. A subsequent set of gamma scans were taken roughly a year later and found that buildup in the bottom of the fill had begun again.

Periodic monitoring of film fill bulk density in a cooling tower is a definitive way for plants to understand the effectiveness of their existing treatment strategies. It can also be utilized to determine the true effectiveness of proposed new treatment chemicals. This patented [6] method of scanning and reporting bulk densities in cooling tower film fill provides operating plants with the data to optimize their chemical treatment strategies while the cooling towers are online and, in many cases, allow extended run times between shutdowns on their cooling towers. Because a history of the film fill performance is kept, plant personnel can be notified when it appears that film fill performance has been severely degraded and alert them to consider aggressive chemical treatment or replacement of film fill well in advance of a catastrophic failure.

Conclusions

The case studies presented demonstrate how gamma scanning can help plants with film fill cooling towers monitor and optimize the effectiveness of their chemical treatment strategy. Knowing and monitoring the bulk density in the fill can assist maintenance and operations personnel in extending run times between tower shutdowns and help prevent a catastrophic tower collapse.

Gamma scanning technology has several advantages over other methods used to monitor fouling in film fill:

- Can be performed while the tower is online or offline and does not require the film fill to be removed
- Results are available immediately
- Provides an actual bulk density of the cell in lb/ft^3
- Profile includes the effect of water being held up by fouling
- Does not rely on inferred results based on water chemistry and other ancillary data
- Subsequent scans provide results that can be directly compared to previous scans
- Identifies critical conditions allowing for immediate proactive management that can be monitored for effectiveness

References

1. Puckorius, P., Kanuth, J., "Cooling Tower Film Fill Water Quality/Operations Guidelines to Successful Utilization," Proceedings, Cooling Technology Institute, 1992 Annual Meeting, Houston, TX

2. Transmitted Intensity and Linear Attenuation Coefficient. NDT Resource Center. www.ndt-ed.org/EducationResources/CommunityCollege/Radiography/Physics/attenuationCoef.htm

3. Chila, P. (2014). Preventive maintenance for cooling towers utilizing gamma scanning technology. Cooling Technology Institute 2014 Annual Conference, Houston, TX

4. Flynn, D. (Ed.). (2009). The Nalco Water Handbook, 3rd ed. (p.21.24). New York, NY: McGraw Hill.

5. Nalco Company. (2014). Nalco Case History CH-1738, Power plant uses Quantum Technical Services to measure cooling tower fouling, assess fill efficiency.

6. US Patent number 8,785,877

AIChE PAPER NUMBER: 123c

BENEFITS OF DIRECT DRIVE MOTORS IN COOLING TOWERS

William (Bill) E. Martin
Senior Development Engineer
ABB Motors and Mechanical Inc.

Robert McElveen
Senior Development Engineer
ABB Motors and Mechanical Inc.

Richard A. Holub
Electrical Engineering Principal Consultant
DuPont Manufacturing Infrastructure Group

Prepared for Presentation at the 2020 Spring National Meeting
Houston, TX
March 29 – April 2, 2020

AIChE and the EPC shall not be responsible for statements or opinions contained in papers or printed in its publications.

BENEFITS OF DIRECT DRIVE MOTORS IN COOLING TOWERS

William (Bill) E. Martin
Senior Development Engineer
ABB Motors and Mechanical Inc.
Robert McElveen
Senior Development Engineer
ABB Motors and Mechanical Inc.
Richard A. Holub
Electrical Engineering Principal Consultant
DuPont Manufacturing Infrastructure Group

Abstract:

Steam Cracking of hydrocarbons to yield olefins produces low temperature waste heat flows; the largest are associated with condensers, pre-coolers, and compressor intercoolers, as well as surface condensers for turbine drivers. The waste heat is typically rejected to cooling water, and the efficiency and reliability of the cooling towers is critical to maximizing throughput and maintaining ethylene plant production capacity in summer. As plants age, cooling tower maintenance costs consume an increasing, but essential part, of the maintenance budget in the ethylene industry.

The most prevalent cause of cooling tower downtime is failure of the right-angle gearbox or associated mechanical components, such as the driveshaft or disc couplings, which are needed to link the slow speed fans to high speed motors. The elimination of these components reduces the losses in the drive system and offers the possibility for improved overall system efficiency. This paper presents a motor topology which allows the high speed motor, driveshaft, couplings, and gearbox to be replaced with a slow speed, direct drive permanent magnet (DDPM) motor, directly coupled to the fan shaft, and operated with an adjustable speed drive (ASD).

A case study is presented. Maintenance and efficiency improvement data is provided. Data on the existing installed base of this motor design is included to establish that the technology has been field tested and proven.

I. INTRODUCTION

A cooling tower is a structure which allows waste heat from a process to be exhausted to the atmosphere. The most common design approach is to collect waste heat with circulating water streams, then allow heated water to fall through a moving air stream created by a fan located at the top of the cooling tower. A small amount of water evaporates, removing a large amount of heat from the process. An open circuit cooling tower distributes heated water over a labyrinth-like packing or "fill". The water is cooled as it descends through the fill. The cooled water is then collected in a cold-water basin below the fill, from which it is pumped back through the process to absorb more heat (1). Towers may consist of multiple "cells", with each cell having its own fan. A typical tower arrangement with two cells is shown in Fig. 1.

Commonly, the size of a tower is described by the diameter of the fan. Fan sizes range from 7 to 42 feet, with the most common applications in the 16 to 28-foot range. The speed of the fan is limited by the mechanical stress on the fan blades. Typical fan speeds are in the range of 125-300 rpm, resulting in blade tip speeds of 10,000-12,000 ft./min. The horsepower of the motor required to drive the fan at the desired speed is calculated based upon the air volume required.

Fig. 1 – Typical Tower Arrangement with Two Cells

Historically, the most common solution for driving the fan utilized an induction motor, driveshaft, disc coupling, and gearbox arrangement, as shown in Fig. 2. Until 2008, few changes to this design approach had been deployed commercially.

The motor employed has traditionally been a NEMA induction motor. In some cases, for reduced energy consumption, two speed motors have been applied for use when full fan speed is not required due to decreased heat load. As the horsepower required to drive the fan varies as the cube of the fan speed, it is advantageous to reduce the fan speed when possible. When the heat load decreases enough, the two-speed drive motor can be run at half speed. This lowers the horsepower required to only 12.5% of the rated value (2). When the heat load requires air flow more than that provided by half speed operation, a two-speed motor must operate at full speed and horsepower as there is no other speed available. Two-speed motors do provide some energy savings, but must be cycled on and off to maintain the desired water temperature. This cycling involves many "across the line", full voltage starts drawing high current and placing unnecessary strain on the mechanical components of the system (3). While providing some flexibility in the tower control logic, two speed motors are not optimal when it comes to maximizing energy savings during times of reduced heat load.

Fig. 2 – Fan Drive Arrangement – traditional solution, NEMA induction motor, driveshaft and 90 degree (or right angle) gearbox

To optimize the effectiveness and reduce the cost of operating the tower, the use of ASDs has become much more commonplace in recent years. ASDs are being installed in most new towers being constructed. Additionally, towers being upgraded or refurbished are also being equipped with ASDs. ASDs have the advantage of a soft mechanical start, no large starting current draw, and the ability to run the fan at any desired speed from zero to the maximum design speed for the application (4). The energy savings realized by using an ASD are well recognized and documented, so no further discussion will be introduced here (5). Two factors that must be considered when applying an ASD are the critical speeds of the mechanical system and the proper lubrication of the gearbox at slow speeds. For practical purposes, the fan of a geared solution tower is generally not run at speeds below 30% of the nominal design speed in order to avoid damaging the gearbox due to improper lubrication.

Historically, the mechanical components of the fan drive system have been the largest maintenance issue for cooling tower installations (6). Gearbox failures, oil leaks, oil contamination, failed drive shafts, misaligned drive shafts and excessive vibration are all significant problems related to this type of fan drive system (7), (8). In 2008, a new, Direct Drive Permanent Magnet (DDPM) fan motor technology was introduced for cooling towers. This solution eliminates the gearbox, driveshaft, disc couplings and other associated mechanical components. An example of a DDPM motor with shaft mounted fan is shown in Fig. 3.

Fig. 3 – Direct Drive Fan Motor – modern solution

In the following sections, it is demonstrated how the direct drive motor solution can be used to improve cooling tower reliability and reduce maintenance costs. The design and installation of a 50 horsepower, 208 RPM DDPM motor for a retrofit application is described in detail. The possibility of improved system efficiency and lower energy consumption with the proposed solution is considered.

There are over 2000 DDPM units installed in over 30 countries. The oldest units have been in operation more than 10 years at the time of this writing.

II. DIRECT DRIVE MOTOR DESIGN

A. Permanent magnet motors

Permanent Magnet (PM) motors have long been recognized as providing higher efficiency and power factor than comparable induction motors. In the past, limitations in terms of motor control, as well as magnet material performance and cost, have severely restricted their use. Due to dramatic improvements in magnetic and thermal properties of PM materials over the past 30 years, as well as significant cost reductions, synchronous PM motors now represent viable alternatives (9).

B. Laminated frame design

Laminated frame motors consist of a stack of laminations permanently riveted under controlled pressure. The cast iron outer frame normally associated with a NEMA motor is eliminated, allowing more room for active (torque producing) magnetic material, as illustrated in Fig. 4. Fig. 5 shows how the stator frame is constructed.

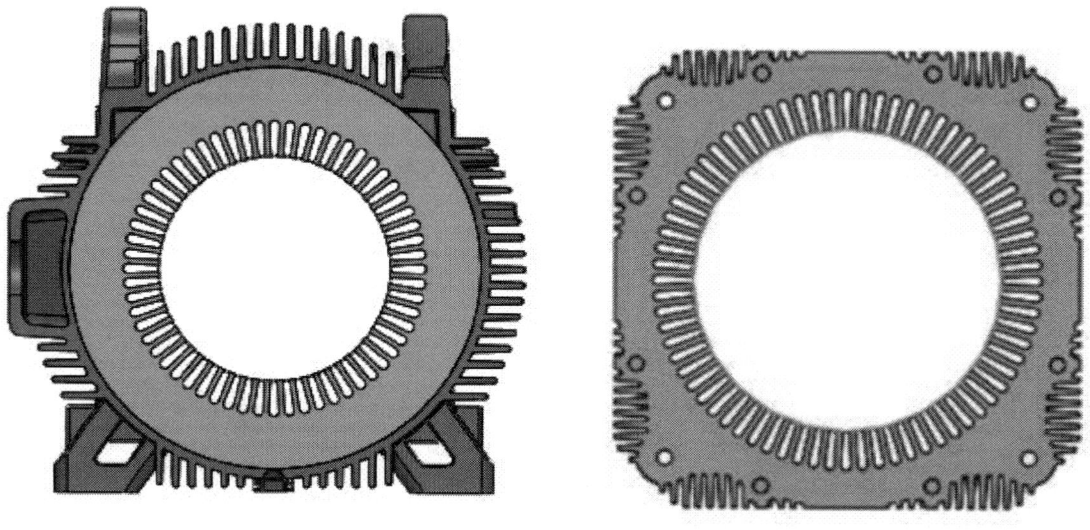

Fig. 4 – Active Material Comparison – Cast Iron (left), Laminated Frame (right)

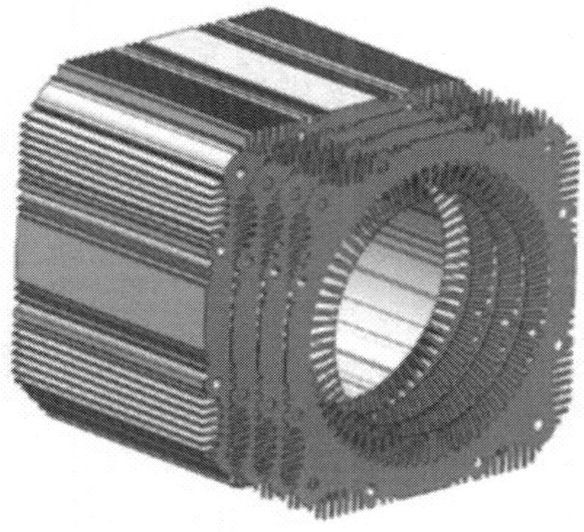

Fig. 5 – Laminated Frame Construction

Another advantage of this construction is that the air used to cool the motor is in direct contact with the stator laminations. The heat transfer mechanism in a cast iron frame motor is highly dependent upon the stator to frame fit. Laminated frame construction eliminates this issue since there is no cast iron frame.

It is this improved cooling method, along with the higher efficiency achieved with the PM technology that allows for increased power density in this type of motor design. This increased power density offers the opportunity to fit a low speed motor in the same space as the right-angle gearbox in cooling tower applications and keep the weight at a reasonable level. For comparison, a study was performed to determine the approximate sizes and weights of various motor types for possible use in this application. The results are shown in TABLE I. The rating is 100 horsepower at 175 rpm.

TABLE I
MOTOR SIZE COMPARISON

Motor Type	Height (in.)	Width (in.)	Wt. (lbs.)
Cast Iron Frame Induction	50	28	9393
DDPM	31	22	3145

C. Improved process control

As mentioned earlier, the addition of the ASD allows the user to more accurately and efficiently control the process. Fig. 6 shows how the motor speed is adjusted automatically with control logic as the heat demand on the system changes with time.

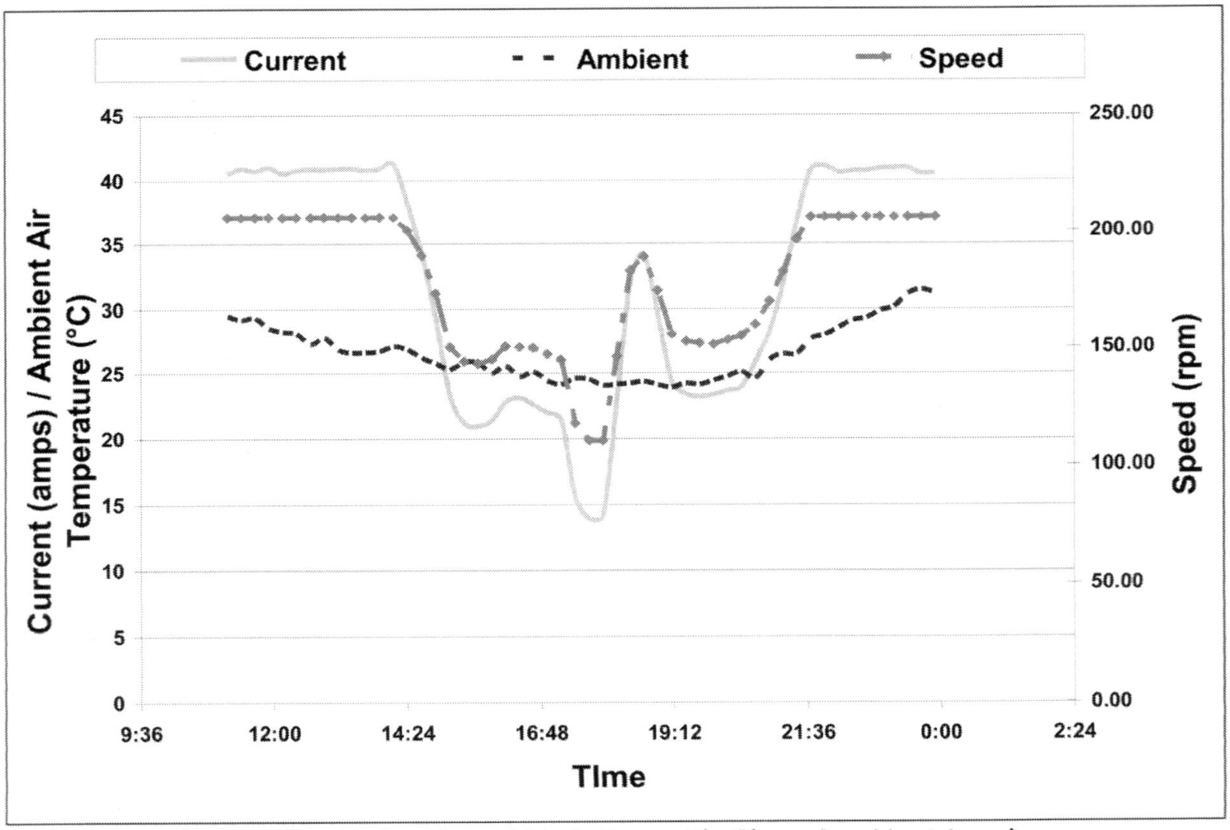

Fig. 6 – Motor Speed Variation with Changing Heat Load

D. Braking and condensation control

The use of an ASD provides the opportunity to offer some additional benefits that across the line systems do not. The ASD may be configured to apply a trickle current to the motor windings to act as a brake during down time. This prevents the fan from wind milling due to nominal winds or adjacent cooling tower turbulence. A mechanical locking mechanism should be employed during any maintenance procedures, as is commonly used with the existing technology. This trickle current acts an internal, built-in space heater by raising the winding temperature above ambient to prevent condensation when the motor is not running.

E. Insulation system

Inside the fan shroud of a cooling tower is an extremely wet environment. The insulation system of the stator windings must be robust and highly moisture resistant. For this reason, an insulation system utilizing an epoxy compound applied via a vacuum pressure impregnation (VPI) system is employed. The VPI system is widely recognized as a superior insulation system for harsh applications such as this.

F. Mounting

Typically, the DDPM motor is mounted with the shaft vertical, using a flange located on the opposite drive end. In many cases, the motor mounting holes can match the bolt hole pattern of the existing gearbox. In other instances, an adapter plate is used to mount the motor to an existing base designed for the gearbox. On new towers, this would not be an issue, however it should be considered for retrofit installations.

One concern expressed by some customers (10) and original equipment manufacturers is the weight of the DDPM motor compared to that of the existing gearbox. Direct drive motors have been installed on concrete, fiber reinforced plastic (FRP), and steel package towers. Individual applications must be evaluated for dead (static) and live (dynamic) loads as well as torque impact (transient).

With direct across the line starting of the geared solution, the fan accelerates from zero to full speed in a matter of just a few seconds, resulting in a large shock load placed on the gearbox and tower structure itself. Torque impact is minimal with the direct drive solution due to the soft start provided by the ASD. Induction motors started across the line do not have this advantage and it is common for an induction motor to provide more than 200% of its rated torque during startup.

Fig. 7 shows the torque vs. time characteristics of an across the line induction motor start and the DDPM motor start on a 50 HP, 208 RPM cooling tower with an 18' diameter fan. With the DDPM solution at no time is torque greater than rated applied. During across the line starting of the induction motor, however, the mechanical system is subjected to torque above 250% of rated. Note the acceleration time of the fan in both instances.

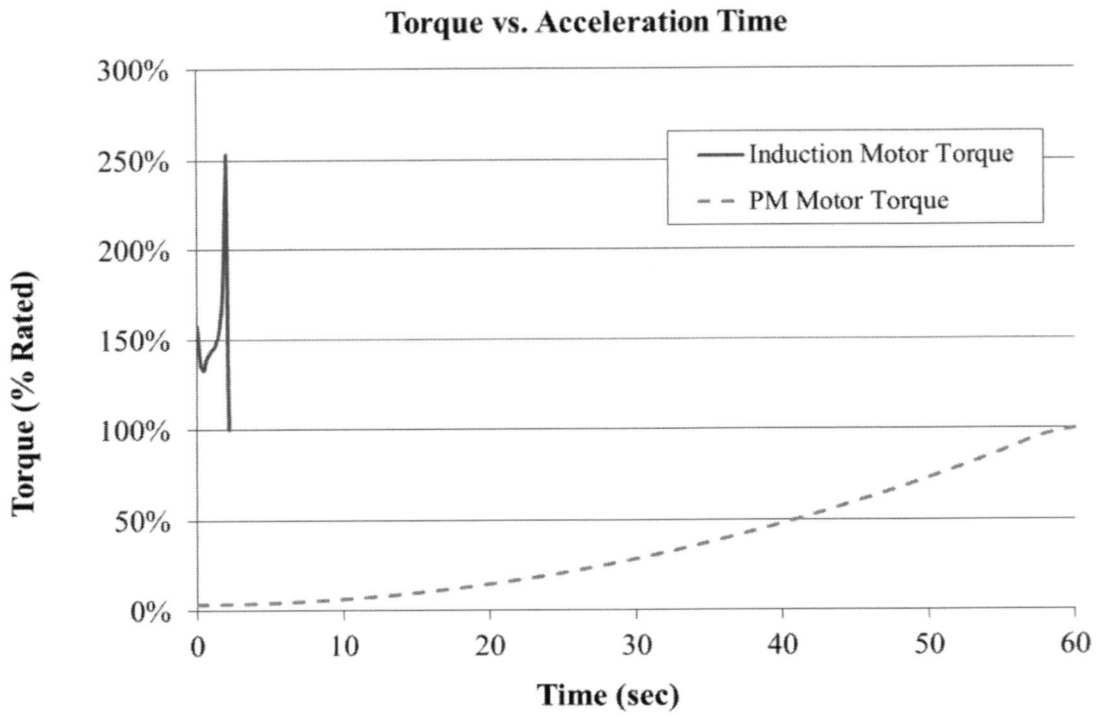

Fig. 7 – Torque vs. Time for Geared and Direct Drive Solutions

Structural issues for retrofits are best addressed by an experienced installer who can analyze the tower structure and apply the applicable structural codes to the retrofit. Existing torque tubes and drive equipment sleds may be reused in a retrofit application as determined by actual tower conditions.

Structural analysis of all known installations to date has indicated that the additional weight is not an issue. However, smaller more compact motors are desirable, especially for retrofit applications where there is limited flexibility as far as shaft height is concerned. Smaller motors also use less material and thus offer reduced initial installation costs for new or retrofit applications.

G. Shaft seal

Due to the harsh environment inherent with a cooling tower application, the motor's drive end is protected by a metallic, non-contacting, non-wearing, permanent compound multi labyrinth shaft seal that incorporates a vapor blocking ring to prevent ingress of moisture. This seal has been proven to exclude all types of bearing contamination and meets the requirements of the IEEE-841 motor specification for severe duty applications. This type of seal has been successfully used in cooling tower gearboxes for many years (11). There are only two ingress points for water into the motor, the shaft and the conduit box. By choosing the right seal and conduit box configuration, an IP56 protection rating can be achieved.

H. Corrosion and Paint

The environment inside the cooling tower is well known to be highly corrosive and is a concern for any equipment installed within. High levels of chlorides and sulfides are not uncommon in cooling tower water and any piece of equipment operating inside the fan stack is subject to a constant mist of water containing these chemicals. For this reason, it is important to address corrosion resistance as it applies to a motor being employed in this application.

The steel portions of the motor, such as the stator laminations, are coated with an epoxy resin. The cast iron parts are e-coated (an electrically applied paint coating), which is a process that provides superior adhesion and corrosion resistance. Finally, a multi-layer epoxy top coat is applied. If corrosion of the motor shaft is a concern, stainless steel shaft material can be used.

I. Maintenance

Another consideration is overall system maintenance. For motor / gearbox combination drives, the lubrication interval is determined by the high speed gear set. The recommended lubrication interval for this type of gear is typically 2500 hours or six months, whichever comes first. In addition, gear manufacturers recommend a *daily* visual inspection for oil leaks, unusual noises, or vibrations. As these units are installed in areas that are not readily accessible or frequented, this is an unreasonable expectation and burden on maintenance personnel. When a gear is to be idle for more than a week, it should be run periodically to keep the internal components lubricated because they are highly susceptible to attacks by rust and corrosion. When being stored for an extended period, it is recommended that the gearboxes be completely filled with oil and then drained to the proper level prior to resuming operation.

Because the high speed input has been eliminated with the slow speed DDPM motor design, the lubrication cycle can now be extended to 12 months. The PM motor does not need to be inspected daily for oil leaks, as the motor contains no oil. When not in operation, the ASD provides a trickle current to heat the stator windings to a temperature slightly above ambient to prevent moisture from forming inside the motor.

J. Vibration

With the elimination of the high speed input to the gearbox, the system dynamics from a vibration standpoint have been simplified. There are no longer any resonance issues with the driveshaft. The maximum rotational excitation is now limited to the rotational speed of the fan. The number of bearings in the drive system has been reduced from six to two for a single reduction gearbox and from eight to two for a double reduction gearbox. This reduces the number of forcing frequencies present in the system. For added protection, a vibration switch can be used as with typical gearbox installations. To this end, a flat pad may be incorporated on the side of the motor, which can be drilled and tapped to accept commonly used vibration sensors.

K. Hazardous location application

There are applications in this industry that require hazardous area motor certification. Some protection concepts rely on maintaining acceptable surface temperatures on both the internal and external surfaces of a motor. For many induction motor designs, the surface temperature of the rotor is the limiting factor. Due to negligible rotor losses seen in many PM designs, it is likely that the rotor will no longer be the hottest surface (12).

L. Motor Cooling

Direct drive motors for cooling tower fans are totally enclosed air over (TEAO) designs. This means that the motor is cooled by air blowing over the fins of the motor frame. This air is provided by the cooling tower fan itself. See Fig. 8 for an illustration.

Fig. 8 – Air Flow Over Cooling Fins

III. CASE STUDY

The case study involves the retrofit of an existing cooling tower constructed in 1986 at a university in the Southeast USA (which has a marginal football team). The tower information is as follows:

Fan Diameter:	18'-0"
Flow Rate:	4,250 gallons per minute (GPM) per cell - 8,500 GPM total
Original Motor Information:	Frame – 326T induction HP – 50/12.5 Speed – 1765/885 rpm
Gearbox:	Size – 155, Ratio – 8.5:1

This tower is comprised of two identical cells. For this study, one cell was retrofitted with a DDPM motor and ASD while the other was left as originally configured. This allows for a direct comparison of the two fan drive solutions. Fig. 9 (left) shows the original configuration, while Fig. 9 (right) shows the PM motor installed in place of the gearbox.

Fig. 9 – Original Installation (left) & DDPM Solution (right)

A power analyzer was used to measure the input power to both solutions. The fans were both running at 208 rpm. Data was taken at both the input and output of the ASD to allow for a direct comparison of the induction motor / gearbox combination to the DDPM motor. 4.5 kW less power consumption (13) was measured on with the DDPM motor installed. The results of the measurements are shown in TABLE II.

TABLE II
POWER CONSUMPTION COMPARISON
TESTING SERVICE DATA

Measurement Location	Volts	Amps	Input kW
Input to Induction	478	54.3	37.9
Input to ASD, DPPM	477	49.8	33.0

The DDPM motor has an advantage over the induction motor / gearbox combination. Using an average price of $0.08/kWh, the annual cost savings for various applications and duty cycles are shown in TABLE III. This table does not account for the additional savings achieved by using an ASD, which offers the ability to run at speeds between 50% and 100% of rated. The largest savings will be related to reduced downtime and maintenance costs as discussed in section V.

TABLE III
ANNUAL ENERGY SAVINGS BASED ON VARIOUS DUTY CYCLES

		Annual Savings (%High Speed/%Low Speed)		
Application	Daily Use	100 / 0	75 / 25	50 / 50
Petro-chem Plant	24 hrs.	$3154	$2488	$1822
Hospital	18 hrs.	$2365	$1866	$1367
University	12 hrs.	$1577	$1244	$911

V. PLANT IMPLICATONS

A. Reduced Maintenance Costs

To quantify the potential benefits of using the direct drive motor solution, two petrochemical facilities with multiple cooling towers were audited over a six-year period (14). There were 16 towers with a total of 102 cells. The average maintenance cost was $5212/cell/year, which includes all maintenance associated with the cell. The cost of unplanned maintenance attributable specifically to the geared drive train (gearbox, drive shaft, couplings, etc.) was $3335/cell/year (61% of total maintenance cost). The data presented here does not include the cost of oil disposal from the gearbox or other "planned maintenance" items that are associated with the geared solution. One example of this is the elimination of the need to manually adjust the fan blade pitch depending on the season. The use of an ASD would eliminate the need for this planned maintenance as well, because the speed of the fan could be increased or decreased to optimize air flow rather than using manual adjustment of the fan blades. Based on a large cooling tower manufacturer's study, which includes these additional planned costs, maintenance costs can be reduced by as much as 90% (15).

B. Serviceability

As with all motors, the ability to maintain and service a PM motor is important to ensure a long and reliable life. Other than the different rotor, the rest of the machine is the same as an induction motor. Typical maintenance procedures used on induction motors can and should be implemented for their PM counterparts (bearing lubrication, for example).

For repair of PM motors, some guidance can be found in IEEE Standard for the Repair and Rewinding of AC Electric Motors in the Petroleum, Chemical, and Process Industries (IEEE 1068-2015) (9). This revision of the standard includes, for the first time, a section discussing PM machines. PM motors used in the petrochemical industry will typically use RE magnets as described earlier. As such, it is not necessary to place the rotor in a "keeper" when disassembled as described in IEEE 1068. If there is any concern as to what magnet material is used, contact the motor manufacturer prior to disassembly.

As stated in IEEE 1068, "the repair work that can be performed on rotors by most service shops is limited..." (9). Unless the motor is subjected to extreme operating conditions (well outside the motor's rated capability – load or temperature) and/or is designed with insufficient operating margin, there is little chance the rotor will need any type of service. When removing the PM rotating assembly from the motor, the magnetic forces acting to hold the rotating assembly in the stator core must be overcome. It is recommended that the motor be disassembled and reassembled in a vertical, drive end shaft up position using a hoist to remove the rotating assembly while securing the frame/stator assembly.

After the rotating assembly is removed, do not allow the rotor to come into contact or near other ferrous materials, especially small metal shavings. The rotor should be protected with a non-ferrous cover (e.g. cloth, Styrofoam, cardboard, wood, etc.) to prevent any metallic contamination from adhering to the rotor outside surface. The rotating assembly should also be kept clean and free from dirt, oil, and grease.

Bearing replacement can be performed in the same manner as for induction motors, using generally accepted installation procedures. IEEE-1068 states "many of the permanent magnet materials used in these applications lose much of their magnetism if heated. Heat shall not be used in any operations of dismantling or reassembly of the machine (9)". Eliminating the use of heat to install a bearing (especially on a large machine) is an impractical and unnecessary restraint. The bearings of a properly designed PM motor can be installed in the same manner as other machines.

PM motor stators utilize a traditional lap or concentric winding with either random or form wound coils. PM machines may be rewound if needed, again using standard practices applicable to induction motors. For specific repair or maintenance instructions, always refer to manufacturer's data or the user's manual.

VII. CONCLUSIONS

Failures of the gearbox, driveshaft, or disc couplings have been the biggest reliability issue facing tower manufacturers and end users. Increasing energy costs have placed a premium on power consumption for all motors and applications. Over the past decade a new technology has emerged which mitigates or eliminates these issues.

Many of the problems associated with cooling tower maintenance and reliability are solved with a direct drive, low speed, PM motor. The relatively high speed (typically 1750 rpm) induction motor has been eliminated. The motor itself has not historically been a problem, but the associated resonances and potential vibration concerns have been an issue. The driveshaft and associated disc couplings have been removed, thus eliminating problems associated with misalignment, improper lubrication, natural frequencies, or delaminating of the driveshaft itself (16). The right-angle spiral-beveled gearbox has been removed. Difficult maintenance issues associated with changing the oil, proper oil fill levels, contamination of the oil, oil leaks, and gearbox failures are no longer a concern.

New motor technology now provides an alternative solution, the direct drive of cooling tower fans (17). PM motor technology combined with a finned, laminated frame design makes possible the construction of low speed, compact motors for use in place of the existing gearbox (18). This solution eliminates the problems associated with the right-angle gearbox and drive shaft design (15). By eliminating the gearbox, which is a significant source of loss in the system, improved system efficiencies can be realized.

VIII. REFERENCES

1. Retrieved April 2, 2009 from http://www.cti.org/whatis/coolingtowerdetail.shtml
2. Benjamin Cohen, Adjustable Frequency Drives: Operation and Application with Evaporative Cooling Equipment, Cooling Technology Institute Paper No. TP07-22, 2007.
3. William F. Immell, Variable Speed Fan Drives for Cooling Towers, Cooling Technology Institute Paper No. TP96-03. 1996.
4. Rick Foree, Cooling Towers and ASDs, Cooling Technology Institute Paper No. TP01-07, 2001.
5. M.P. Cassidy and J.F. Stack, Applying Adjustable Speed AC Drives to Cooling Tower Fans, PPIC, 1988.
6. Jim Horne, How to Address Your Cooling System Woes, PTOnline, 2008.
7. Dave Gallagher, Condition Monitoring of Cooling Tower Fans, Reliability Direct.
8. Philadelphia Gear, The Cooling Tower Gear Drive Dilemma: Why Applying Commodity Products to an Engineered Solution Can Cause Premature Failure.
9. McElveen, Martin, Frank, Jarvinen, "A Motor Primer the Next Generation: PM Machines", Petroleum and Chemical Industry Conference, 2018
10. Robbie McElveen and Bill Martin, "Effect of Air Flow on Size and Weight of Direct Drive Motors", Cooling Tower Technology Conference, 2011.
11. Inpro/Seal Company, An Introduction to Bearing Isolators, March 2005.
12. McElveen, Martin, Massey, "Applying Permanent Magnet Motors in an Ex Environment", Petroleum and Chemical Industry Conference, 2013.
13. Dustin Warrington, Clean Air Engineering Report: Clemson East Chiller Plant, July 2008.
14. McElveen, Martin, Lyles, "A New More Reliable Solution for Cooling Tower Fan Drives", Petroleum and Chemical Industry Conference, 2009.
15. Retrieved January 24, 2020 from http://www.baltimoreaircoil.com/english/video/enduradrive-fan-system
16. Robert Poling, Natural Frequency Characteristics of Drive Shafts, Cooling Technology Institute Paper No. TP05-04, 2005.
17. Robbie McElveen and Bill Martin, Improve Reliability and Reduce Maintenance With Direct Drive for Cooling Tower Fans, Electric Power Research Institute: 2009 Cooling Tower Technology Conference, August 2009.
18. Robbie McElveen, Bill Martin, and Ryan Smith, Recent Developments in Motor Technology Allow Direct Drive of Low Speed Cooling Tower Fans, Cooling Technology Institute Paper No. TP09-18, 2009.

AICHE PAPER NUMBER: 123d

DAMAGE MECHANISM IDENTIFICATION AND CONTROL FOR ETHYLENE UNITS

Kenneth (Ken) Kirkham
Principal Engineer II
E[2]G | The Equity Engineering Group, Inc.

Prepared for presentation at the 2020 Spring National Meeting
Houston, Tx, March 29 - April 2, 2020

AIChE and EPC shall not be responsible for statements or opinions contained in papers or printed in its publications.

DAMAGE MECHANISM IDENTIFICATION AND CONTROL FOR ETHYLENE UNITS

Kenneth (Ken) Kirkham
Principal Engineer II
E[2]G / The Equity Engineering Group, Inc.

Abstract:

In the last 5 years, two API documents for identifying and controlling damage mechanisms have been issued. These documents are likely to be considered recognized and generally accepted good engineering practice (RAGAGEP) by OSHA, and this is anticipated by many chemical and petrochemical companies. An overview of API 584 on integrity operating windows (IOWs) and API 970 on corrosion control documents (CCDs) will be given. The importance of the damage mechanism review to both activities will be described, and the advantages of co-development will be emphasized. The benefits of CCDs & IOWs to mechanical integrity programs will be highlighted, and ethylene unit examples will be shared.

Introduction

API 584 Integrity Operating Windows (IOW) was first issued in 2014. The second edition is in development. API 970 Corrosion Control Documents (CCD) was issued in 2017. Recent industry incidents have shown that the damage mechanisms that affect equipment in process units are not fully understood and appropriate mitigation and monitoring have not been effectively implemented and applied. OSHA citations from audits and incidents have increasingly referred to API documents. Likewise, the Chemical Safety Board (CSB) refers to many API documents when issuing recommendations to API to revise them in response to their (CSB) incident investigations, which effectively identifies them as recognized and generally accepted good engineering practice (RAGAGEP). Developing CCDs and IOWs for process units including ethylene units will improve mechanical integrity performance.

API 970 Overview

The goal of the CCD is to provide a comprehensive document that links damage mechanisms (DM) to the equipment. By fully understanding the DMs, the appropriate mitigation can be applied. The functional sections of the document are listed and described briefly below.

Introduction to Corrosion Control Documents/Systems

This section of the document lays out the typical components of a CCD. There are many possible components that could be detailed in the document. A few of the truly essential components are listed below:

- A brief process description including details relevant to damage mechanisms.
- Drawings based on current process flow diagrams (PFD) that show materials of construction, corrosion loops and damage mechanisms.
- An explanation of how each damage mechanism is prevented, detected, controlled, monitored, or managed.
- A list of recommendations developed as a result of the review process.
- An IOW table.

The primary users of the document tend to be Inspection/Reliability, but other organizations in the plant can benefit from the information. Thoughts are given on how a CCD may support inspection and other mechanical integrity (MI) programs.

CCD Work Process

Much detail is given in API 970 describing a very robust process to develop a CCD. In many cases, streamlining is possible and desirable.

The list starts with having a site procedure that describes program expectations, goals, work process and implementation. There are many potential premises to consider in CCD development, the most common tends to be avoiding process leaks during the turnaround cycle.

A multi-disciplinary team from organizations at the site and perhaps some specialists from outside the site may be needed to execute the project. The minimum required functions are the following:

- Meeting/process facilitator
- Corrosion/Materials expert
- Site Corrosion/Materials Engineer (if not serving as the expert)
- Operations rep
- Process Engineering rep
- Inspector for the unit

Most companies hire outside Corrosion/Materials experts who can facilitate the face-to-face meeting. The responsibilities for each role are fully defined as is the data needed to conduct the on-site meetings.

Additionally, a site coordinator for the program is highly essential. Simply scheduling and running the meetings is not enough to drive the program and reach the desired goal. Expediting and follow up are critically important when involving several plant organizations and pulling together a massive amount of documentation and input.

The damage mechanism review is the heart of the process. Skill and knowledge are needed to obtain the best results in the most efficient manner. A skilled practioner can identify the most likely damage mechanisms and the biggest threats to the unit quickly, which minimizes the time invested by the site personnel in face to face meetings.

Implementing a CCD Program and Integrating CCDs with Other Work Processes

After the rev. 0 document is issued, the work of implementation and integration begins. The implementation and integration steps are defined if there is a site procedure. Most companies set the revalidation cycle consistent with process hazard analyses (PHA) (5 years) or RBI (10 years). Updating should be considered when modifications or significant process changes to the unit are made.

API 584 Overview

Many engineers are familiar with CCDs, but many are not familiar with IOWs. IOWs are process parameters that are monitored and/or controlled to ensure predictable damage rates. In keeping with the definition, IOWs provide a direct link between operation and mechanical integrity.

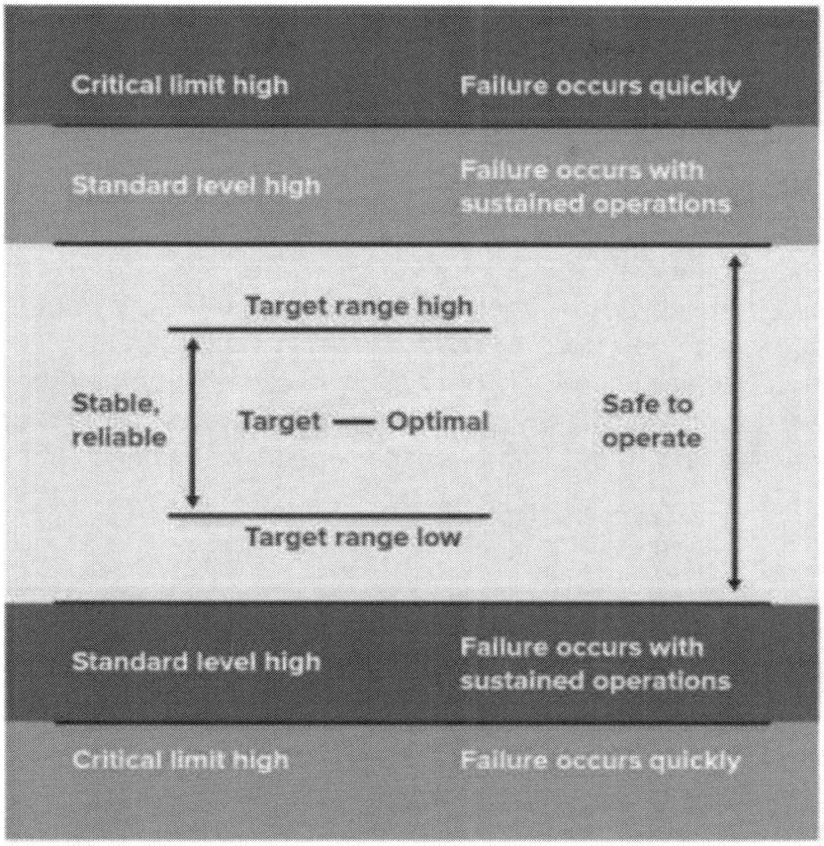

Figure 1: This Graphic Shows the Relationship Between IOW Limits and Potential Damage Rates (After Figure 1 in API 584).

Types and Kinds of IOWs

IOWs can be characterized as physical or chemical. Examples of physical IOWs are as follows:

- Temperature
- Pressure
- Flow rates

Examples of chemical IOWs are as follows:

- pH
- Crude sulfur content
- Crude TAN
- Caustic strength

Examples that incorporate both chemical and physical parameters are as follows:

- Hydrogen partial pressure
- H2S partial pressure
- Water dew point

IOWs can be further defined by type as follows:

- Critical
- Standard
- Informational

Critical IOWs tend to be the few parameters that can lead to very rapid damage or very severe consequences. Generally, the recommended reaction times for correcting exceedance of critical IOWs tend to be immediate to as quickly as possible.

Standard IOWs tend to be the majority of the controllable parameters that will cause unacceptable damage rates if left uncorrected for weeks or months.

Informational IOWs are important to monitor, because they may indicate process creep or trends for increased damage rates. For example, some water samples are analyzed for pH and iron. Control is based on pH, but iron provides supporting evidence of damage.

What Is an IOW?

What is an IOW (as opposed to what is not)? There are 2 concepts to consider:

- Time to react and consider options
- Limits better covered by other programs

Temperature and pressure limits such as the name plate values given on an ASME pressure vessel or pipe spec P & T ratings are not IOWs. It is understood that these limits will be followed; so, they are not included as IOWs. Minimum pressurizing temperatures to avoid brittle fracture for pressure vessels should be covered in maintenance or operating procedures. Where failures can occur rapidly from loss of cooling, interlocks are used; there is no time to discuss options. Process safety hazards such as ethylene decomposition would not be covered by IOWs.

IOW Work Process

The work process has many similarities with CCDs regarding information gathering and identifying DMs. Identifying the equipment most vulnerable to damage is a key part of the process to ensure that the most effort is expended to protect it. The effort requires a multi-discipline team. Involvement by Operations and Process Engineering is paramount, because those organizations must agree to the IOW limits and criticality ratings.

After the DMs are identified the IOW work process differs significantly from CCDs. It focuses on operating limits that will protect the equipment from unpredictable damage rates. One challenge relates to the lack of predictability for many corrosion issues. For instance, calculators exist for sulfidation (sulfidic) corrosion. Temperature or sulfur composition limits can be chosen based on a corrosion rate. Where calculators do not exist, limits that can be calculated and controlled are sought. For example, in concentrated sulfuric acid, flow velocities are limited to 3 feet per second in CS piping. Lack of predictability for many corrosion mechanisms often leads to the selection of conservative limits.

The 2014 Edition of API 584 contains many refining examples of possible IOW parameters. These examples are not of much use for ethylene units, but the next edition which is in publishing, will contain a table of possible IOWs for ethylene units. Not much guidance is provided on establishing limits; so, it is left to the corrosion SME to select them. Risk Ranking is discussed. Risk ranking becomes more important as the number of IOWs identified grows to a large number and when sample stations or instrumentation need to be added for an IOW to be monitored.

In petrochemical units, the composition of feedstocks tends to be less variable than say crude supply for a refinery. Consequently, fewer IOWs are expected in petrochemical units.

In situations where a site is not ready to implement IOWs, suggested IOWs have been provided.

Documenting IOWs

The minimum documentation is the IOW table. See the figure below for a sample table. If only a table is provided, the basis for the limits should be clearly stated. For IOW-only projects, some text detailing key process information is desirable. The best approach is to develop IOWs as a part of the CCD. Then the damage review is well documented.

Table 1: One Row of a Sample IOW Table

MOE #	Parameter (Units)	Sample Location	Variable Type Inst/Lab	Limits Max/min / target	Recom. Sampling/ Monitoring Interval	IOW Type C/S/I	Possible Consequence of Deviation	Recommended Operations Response
1	Niter Content of fresh acid	NA	COA	50/-/0	Every Delivery	S	Increased Corrosion rates are possible in all acid handling piping and equipment.	-Include requirements in the manufacturing spec. -Reject cargoes that don't meet the limit. -Notify Inspection/ Reliability if high niter acid is used.

Site Follow-up and Integration Activities

Training of operators and IOW owners is critical to the success of implementation. Projects may need to be initiated to add alarms, analyzers or sample stations for new IOWs. Notification strategies need to be established. All IOWs are reviewed regularly by at least the responsible engineer. More folks and functions are typically notified for critical and standard IOW exceedances; therefore, a communication strategy has to be defined. In some organizations, critical IOW exceedances are considered incidents that must be investigated. Some organizations review critical IOWs and exceedances during PHAs. Revalidation is more likely to be done on the PHA cycle (5 years) rather than the RBI cycle (10 years).

Damage Review Comparison

Identifying the DMs is at the very core of both work processes. The DMs must be identified and understood to control or otherwise mitigate them. A large amount of mechanical, process and operational data has to be compiled and analyzed to execute the analysis for both.
The CCD has a goal of identifying all DMs to assess the need for mitigation. The CCD can recommend mitigation steps that include increased inspection, changes to or development of operating procedures, upgrading materials and IOWs. In effect, IOWs are a subset of a complete CCD.

In the case of IOWs some DMs are not amenable to process control. For instance, many shutdown related DMs are avoided by maintenance or operating procedures. Consequently, IOWs don't address all DMs.

Benefits of CCDs & IOWs to Plant Organizations

The benefits of CCDs and IOWs to plant organizations are listed below in Table 2.

Table 2: The Benefits of CCDs and IOWs to Plant Organizations

Reliability/Inspection	Operations	Education & Training
• Where to inspect • How to inspect (what type of damage to look for) • Supports RBI • Supports root cause analysis for FE failures • Supports new designs	<u>IOWS</u> • Set limits • Establish integrity monitoring • Define recovery actions & steps when outside limits <u>SOPs</u> Operations Practices to avoid/minimize damage during: • Shutdowns • Downtimes • Startups	• Process Engineers, Reliability Engineers, Operators, Inspectors • Understanding how a process change or IOW impacts damage mechanisms • Understanding impacts when operating outside of desired limits • Enhances & reinforces API 571 certification

Ethylene Unit Examples

Drawings/Damage Mechanisms Descriptions/Mitigation/Recommendations

Several of the key components of a CCD are reproduced below to show the functionality of a CCD. The figures and attachments should be considered stylized examples for educational purposes (as opposed to actual plant examples).

Drawings are a key component of a CCD. A corrosion materials diagram (CMD) is frequently used to show materials of construction and damage mechanisms for a process unit.

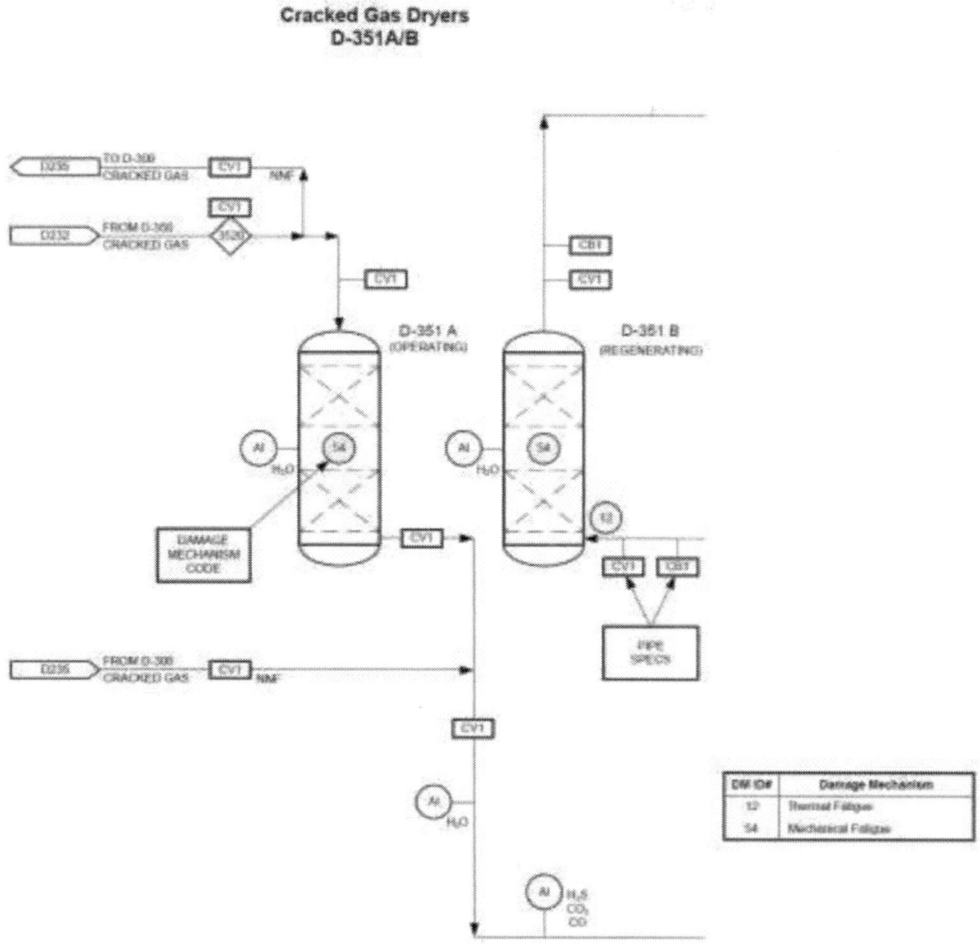

Figure 2: A Corrosion Materials Diagram for Cracked Gas Dryers at an Ethylene Unit.

In this CCD, the mitigation options were contained in the DM Descriptions. Here is an excerpt from the Mechanical Fatigue description:

Fatigue cracking is a mechanical form of degradation that occurs when a component is exposed to cyclical stresses for an extended period, often resulting in sudden, unexpected failure. These stresses can arise from mechanical loading and are typically below the yield strength of the material. There are two types of fatigue issues in the unit. The first is frequent pressure cycles, and the second is alternating stress on fixed equipment caused by nearby machinery.

There are several dryers that experience frequent regenerations. The difference between the operating and regenerating pressures were recognized as being so different that these vessels were analyzed for fatigue during the design.

It would be prudent to log the cycles and validate the assumptions in the fatigue analysis after a year of operation. The fatigue analysis identifies the highest stress locations that should be inspected for surface breaking cracks. API 510 requires an inspection for fatigue at the half-life.

Recommendations:

- *Log the cycles for the vessels designed for fatigue.*
- *Review the fatigue analysis and confirm the highest stress locations.*
- *Compare the operating conditions and number of cycles to validate the design assumptions. The rate of heating on the Regen Gas inlet nozzle of the driers should be measured after start-up. Confirm that the fatigue analysis covered both thermal and pressure cycling. Inspect the life limiting locations at the half life as required by API 510.*

This DM stems from a normal plant operation that cannot be avoided; it is not amenable to control with an IOW.

Here is an excerpt from the DM description for Thermal Fatigue:

Fatigue is the result of cyclic stresses caused by variations in temperature. Damage is in the form of cracking that may occur anywhere in a metallic component where relative movement or differential expansion is constrained, particularly under repeated thermal cycling.

Thermal fatigue needs to be considered if heating to 400°F is rapid. The regen gas inlet nozzle would be most at risk. The dryer bed will act as "thermal insulation" for the rest of the vessel.

Recommendations:

- *Measure the heating rate at the Regen Gas inlet nozzle of the driers.*
- *Confirm that the fatigue analysis covered both thermal and pressure cycling.*
- *If so, inspect the life limiting locations at the half life as required by API 510.*
- *If not, inspect for cracking at the ID and OD surfaces of the nozzle at the first full turnaround.*
- *Consider an IOW to restrict the heating rate.*

In the example shown above, the link between the equipment, DM and DM mitigation are clearly shown.

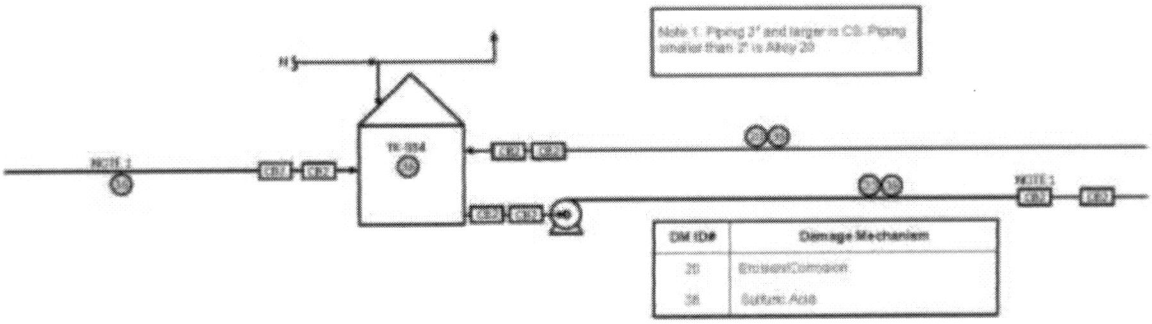

Figure 3: Sulfuric Acid Tank and Piping

Here is an excerpt from the Sulfuric Acid DM description:

Concentrated sulfuric acid has been handled in CS equipment and piping for years. The site will be using 93% acid which eliminates the need for heat tracing. Unfortunately, 93% acid is more corrosive to CS than 98% acid. At 75F, the corrosion rate can be as high as 20 mpy without any velocity effects. CS is useful in concentrated sulfuric acid because a slushy ferrous sulfate scale forms and slows down further corrosion. If the scale is disrupted, rapid corrosion begins again, hence, the requirement for low flow velocities (2-3 fps) in CS concentrated sulfuric acid piping. Higher flow velocities produce enough turbulence to remove the protective scale.

The corrosion reaction produces hydrogen, which can also disrupt the protective scale. The consequences of hydrogen grooving on tanks' shells creates a significant mechanical integrity concern. See NACE SP0294 for more information regarding the design, fabrication, and inspection of sulfuric acid tanks. Consider a design review of TK-884. The tank is painted white which helps keep it cooler in the sun. When the resistance of CS is inadequate for piping, upgrading to Alloy 20 is done selectively. Columbium (Cb) has been added to avoid sensitization of Alloy 20 wrought products. Any maintenance repairs to cast Alloy 20 require a solution-annealing heat treatment to avoid subsequent intergranular corrosion.

In 93% acid, 316L is marginally resistant and 304L should not be used.

There is a corrosion concern in regenerated acid called niter. Niter is nitrosylsulfuric acid (NSA). At concentrations above about 50 ppm, it accelerates corrosion of CS and Alloy 20. Typical fresh acid is reported to contain 25 to 40 ppm.

Recommendations:

- *Consider a design review of TK-884 per SP0294. It may help determine inspection frequencies and future design upgrades.*
- *Calculate the flow velocities in the concentrated CS sulfuric acid piping for various flow rates. Provide a table for operations.*
- *Consider size or alloy upgrades if velocity limits cannot be met.*
- *Consider adding a requirement to the sulfuric acid specification for 50 ppm maximum NSA/NOx and an IOW.*
- *Inspection is typically focused on welds and bends.*

IOWs

The suggested IOW for heating rate from the previous Thermal Fatigue section is modeled in Table 3.

Table 3: Example Heating Rate IOW

MOE #	Parameter (Units)	Sample Location	Variable Type Inst/Lab/ Calculation	Limits Max/min/ target	Recom. Sampling/ Monitoring Interval	IOW Type C/S/I	Possible Consequence of Deviation	Recommended Operations Response
1	Heating Rate during Regen (°F/hr)	Regen Inlet Nozzle to Dryers D351A/B	Calculation	200/-/0	Every Regeneration	S	Increased fatigue damage accumulation at the inlet nozzle.	-Reduce heating rate if it is exceeded. -Notify inspection if rate is exceeded.

The suggested IOW for Niter content and flow velocity of fresh concentrated sulfuric acid is given in Table 4. (The Niter IOW was used as an example earlier in the IOW sections.)

Table 4: Example of IOW Limits for Concentrated Sulfuric Acid

MOE #	Parameter (Units)	Sample Location	Kind of Variable Inst/Lab	Limits Max/min/ target	Recom. Sampling/ Monitoring Interval	IOW Type C/S/I	Possible Consequence of Deviation	Recommended Operations Response
1	Niter Content of fresh acid	NA	COA	50/-/0	Every Delivery	S	Increased Corrosion rates are possible in all acid handling piping and equipment.	-Include requirements in the manufacturing spec. -Reject cargoes that don't meet the limit. -Notify Inspection/ Reliability if high niter acid is used.
2	Flow Velocity (ft/sec)	All CS conc. H2SO4 lines	Calculated	3/--/2	Calculate flow velocity for all CS at max. rates and when capacity increases are planned.	S	Rapid corrosion of CS piping.	-Reduce flow rates if the velocity limit is exceeded. -Inform Inspection if flow velocity limits will be exceeded for a week or more.

References:
1. API 571 Damage Mechanisms Affecting Fixed Equipment in the Refining Industry, SECOND EDITION, APRIL 2011
2. API 584 Integrity Operating Windows, FIRST EDITION, MAY 2014
3. API 970 Corrosion Control Documents, FIRST EDITION, DECEMBER 2017
4. NACE SP0294-2006 Design, Fabrication, and Inspection of Storage Tank Systems for Concentrated Fresh and Process Sulfuric Acid and Oleum at Ambient Temperatures

587491 Ethylene Risk and Technology Review What's the Outlook?

Abstract:

Abstract:

Per the court-mandated timeline, the EPA is currently working through the statutorily-required risk and technology review (RTR) for the ethylene NESHAP (40 CFR, Part 63, Subpart XX and Subpart YY) process. The proposed rule, was published in the Federal Register October 9, 2019, and will impact current ethylene operations. The proposed rule mirrors certain aspects of the refinery rule, such as the flare, heat exchange systems and PRD refinery language, decoke emissions (both decoke to the fire box and decoke to an atmospheric pot) and waste water emissions.

This paper will discuss the proposed rule and the potential impact on the ethylene facilities.

Copyright © American Institute of Chemical Engineers. All rights reserved.

590057 *Flaring Enforcement into Next Generation Flare Rule Making: Flare Management Plans and Beyond*

Abstract:

AIChE

2020 Spring Meeting and 16th Global Congress on Process Safety

The 32nd Ethylene Producers' Conference

Title: *"Flaring Enforcement into Next Generation Flare Rule Making: Flare Management Plans and Beyond"*

Topic: Flaring, Flare Management Plans, Sustainability,

Authors:

Linda Bartlett, Principal Consultant and Brian Woodbury, Senior Consultant

Environmental Resources Management (ERM), 3838 N. Causeway Blvd. Suite 3000, Metairie LA 70002

(504) 831-6700; linda.bartlett@erm.com, brian.woodbury@erm.com

Abstract:

Flaring has been a focus for EPA enforcement for several years. Regulations are now catching up with the Consent Decrees and the National Enforcement Initiative to cut Hazardous Air Pollutants (HAPs). Clean Air Act 114 information requests are now extending into other industries beyond refining including chemicals operations. These non-refinery operations can learn from the refinery journey and potentially stay a step ahead of coming regulations.

An early step in this next generation of flaring regulation is focused on refineries under Title 40 Code of Federal Regulations Part 60, New Source Performance Standards Subpart Ja (NSPS Ja) and Part 63, National Emission Standards for Hazardous Air Pollutants Subpart CC (NESHAP CC). Refineries are required to prepare and submit a lengthy, in-depth flare management plan (FMP) which prescribes how the facility will minimize flaring. ERM has been working with multiple refiners and chemical manufacturers on their flare management plans. In this process, ERM has identified key lessons which will be summarized in this presentation. ERM's experience will provide historical context and "lessons learned" to the ethylene industry as they prepare for adapting to new proposed standards similar to refinery NSPS Ja and NESHAP CC. Examples of the lessons learned include:

- how to actively engage facility stakeholders to ensure understanding, particularly for developing minimization procedures;
- implications with multiple process units on a single flare header for minimization and alternate baseline flow rates,
- why and how summarized startup and shutdown procedures can be used; and
- important options to consider for monitoring such as sample conditions (wet or dry analysis).

Furthermore, with continued EPA and state enforcement pressure around flaring, ERM will present the following ideas on how to maintain compliance after the flare management plans have been submitted:

- programs and procedures needed to maintain an evergreen plan in a continuously changing facility environment,
- potential cost savings around plan structure and management,
- incorporation of minimization procedures into current operating procedures, and
- potential "red flags" associated with the content submitted in the FMP such as baseline flow compared against the production capacity.

Chemical manufacturers may want to consider a proactive review of flaring operations with respect to the journey that refineries have traversed. Actions taken now can position manufacturers not yet subject to such rigorous flaring regulations for coming capital and operating needs, and potentially reduce operating costs by enacting good flaring practices.

Bio:

Ms. Linda Bartlett is a chemical engineer from Tulane University in New Orleans, Louisiana. Prior to joining ERM in 2014, she supported a large refining and petrochemical complex in Louisiana in various positions including Environmental Manager.

She has over thirty years of experience in facility environmental support, process and project engineering, operations, management, and environmental consulting.

Her experience includes air, water, and waste compliance and permitting. Her expertise lies in air quality including Title V and PSD permitting, compliance management, and environmental personnel training. She has responded to numerous Clean Air Act 114 requests involving complex flare systems and recently permitted two large green field LNG facilities with supporting power plants and concrete batch operations.

Copyright © American Institute of Chemical Engineers. All rights reserved.

589593 Electronic Reporting Requirements for the Proposed Ethylene Production Maximum Achievable Control Technology Standard (EMACT)

Abstract:

On October 9, 2019, EPA proposed Maximum Achievable Control Technology (MACT) Standards for Ethylene Production (National Emission Standards for Hazardous Air Pollutants: Generic Maximum Achievable Control Technology Standards Residual Risk and Technology Review for Ethylene Production). An important requirement of the proposed rule is electronic reporting using EPA's Compliance and Emissions Data Reporting Interface (CEDRI). Specifically, the proposed rule requires that Notices of Compliance Status (NOCS), Flare Management Plans (FMPs), and any emission tests performed using test methods that are supported by EPA's Electronic Reporting Tool (ERT) be submitted through CEDRI. CEDRI is accessible from EPA's Central Data Exchange (CDX).

CDX supports the requirements of EPA's Cross-Media Electronic Reporting Rule (CROMERR). The CROMERR program is intended to ensure the enforceability of regulatory information collected electronically by EPA and EPA's state, tribal and local government partners. Accordingly, understanding CEDRI reporting requirements is fundamental to compliance should electronic reporting remain in the final rule.

This presentation will provide an overview of CROMMER and CDX, review the specific CEDRI reporting requirements of the proposed rule, and outline the registration and reporting process for CEDRI.

Copyright © American Institute of Chemical Engineers. All rights reserved.

590620 Down the Road: The Future of Ground Flares and Amels

Abstract:

Facilities using pressure-assisted multi-point ground flares have requested Alternative Means of Emissions Limitations (AMELs) from the U.S. EPA to operate above the limits on exit velocity found in the General Provisions and Refinery Sector Rule. On October 9, 2019, the U.S. EPA proposed amendments to the Ethylene Maximum Achievable Control Technology (MACT) standards. These proposed changes include a new set of compliance requirements for ground flares without the need to request an AMEL. This presentation will review the new requirements for ground flares found in the Ethylene MACT draft rule. The presentation will also discuss how these proposed changes could affect ground flares in other sectors and the future of the AMEL process.

Copyright © American Institute of Chemical Engineers. All rights reserved.

590043 Increasing Sustainability of the Steam Cracking Industry through Drop-in Renewable Feedstock

Abstract:

It is a known fact that the demand for polymers and chemicals is increasing at a high pace, with studies predicting that demand for polymers will roughly double within the coming 15 years. In order to meet this demand, several investments in new capacity for the production of olefins and subsequent products are being made at a global scale. Consequently, ever-increasing amounts of fossil resources are being utilized, which puts additional strain on the environment.

Steam cracking is already a mature process, and revolutionary technologies are rather scarce in the field. In this context, with the aim of improving the environmental footprint of the process and its products, sustainable feedstock usage is one of the most straightforward, beneficial and scalable solutions readily available for cracker operators.

In this paper, technical aspects of Neste's renewable hydrocarbons are discussed, focused on their suitability as a drop-in steam cracker feedstock, as well as their sustainability record and environmental impact, which is typically evaluated by means of a life-cycle assessment (LCA). It has been determined that the environmental benefit offered by bio-based Polypropylene (PP), produced using steam-cracker-based propylene is substantial [1]. Calculations indicate up to 80% lower CO_2 eq. compared to conventional fossil PP [2].

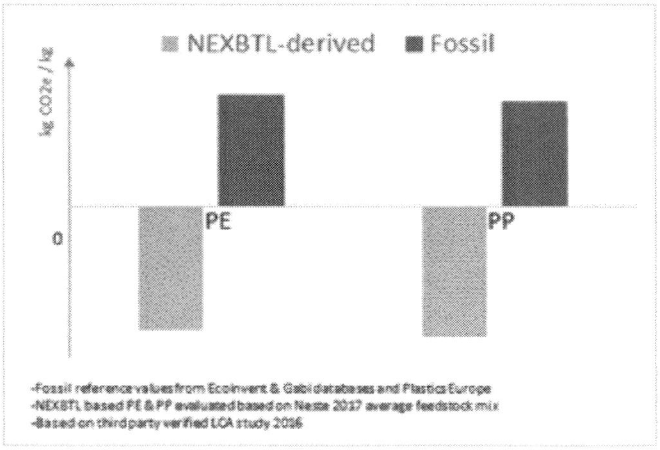

References

[1] Environmental impact assessments of innovative bio-based products, Task 1 of "Study on Support to R&I Policy in the Area of Bio-based Products and Services ", COWI A/S and Utrecht University EN December, 2018

[2] Neste's internal and 3rd party verified Life Cycle Assessment, 2016.

Copyright © American Institute of Chemical Engineers. All rights reserved.

AIChE Paper Number: 123e

CRACKING FURNACE REFRACTORY LINING CHOICES TO IMPROVE PERFORMANCE AND RELIABILITY

Riccardo Cantone
Global CPI Market Director
Morgan Advanced Materials

Steve Chernack
Global Technical Manager
Morgan Advanced Materials

Prepared for Presentation at the 2020 AIChE Spring National Meeting
Houston, TX, March 29 – April 1, 2020

AIChE and the EPC shall not be responsible for statements or opinions contained in papers or printed in its publications.

CRACKING FURNACE REFRACTORY LINING CHOICES TO IMPROVE PERFORMANCE AND RELIABILITY

Riccardo Cantone
Global CPI Market Director
Morgan Advanced Materials

Steve Chernack
Global Technical Manager
Morgan Advanced Materials

Abstract:

Refractory linings are a relatively small percentage of the total cracking furnace cost. For this reason, the importance of the refractory lining can easily be overlooked. Whether the lining is new or approaching the end of service life, any issues with refractory performance or reliability can easily result in large potential liabilities and painful profit loss. Small potential upgrades and attention to refractory product quality, lining design and installation details are extremely important to address, especially when considering a retrofit. This paper, outlines real case histories where alternative refractory systems have been proposed, accepted and installed for improved performance and reliability.

Introduction

Refractories are a relatively small percentage of the total cracking furnace cost. For this reason, the importance of the refractory lining can easily be overlooked. However, the importance of the refractory lining to the reliable operation of these furnaces is undeniable. If an unforeseen problem occurs in your lining, the refractories then become worth much greater than their initial cost! If anyone reading this has faced refractory issues resulting in loss of production/profit, it is easy to understand the above.

Refractory linings in cracking furnaces have evolved through the years for many valid industry wide safety and production requirements such as;

- Worker safety resulting in need for lower casing temperatures.
- Rising operating temperatures relative to radiant tube metal advances (HK-40 to newer super alloys).
- Constant pressure on operators to push furnace production for increased profits
- Need for improved furnace reliability resulting in less down time and higher profits.

Regarding past refractory issues, it is important to understand the root cause. Failure of refractory linings can occur for various reasons, so it is important to consider all the following before deciding the true root cause;

- Poor product selection/quality
- Poor refractory lining design
- Poor installation of the refractory lining and anchors
- Operational issues / upsets
- Original refractory lining designs are no longer suitable to handle today's operating conditions.

The most notable refractory lining change over time, is older style cracking furnaces had much thinner refractory linings than current designs. In the past, it was not uncommon to have radiant walls and roof linings 8-9" (200-225mm) thick. Today, standard wall and roof linings can be 12-14" (305–350mm) thick. Floors have gone from 10-12" (250–305mm) thick to 12-14" (305-350mm) thick. While it is always desirable to design to have the highest efficiency newest technology grass roots furnaces, it is still very cost, time and process effective to revamp using today's technologies to increase production and profitability.

This paper will address ways to upgrade old style furnace refractory linings using today's well proven refractory value propositions. Lining propositions will start from radiant floor up and through the convection section of the cracking furnace to address potential options. It is important to be thinking just a little outside the box versus merely copying from past linings. The results are some unique but well proven refractory linings designs that can well exceed the original design, without the need to increase lining thickness.

When discussing best options in revamping a refractory lining, make sure to consult experienced refractory people or refractory companies. Ideally, it is best to find folks who have experience with;

- The original OEM lining designs
- Witnessed typical lining issues with multiple furnaces during the life of the refractory
- Designed/supplied next generation refractory linings proven to meet today's needs

Please note, all the examples presented are actual examples already proven globally in multiple units of the key cracking furnace OEM's.

Discussion

Floor Linings

It is quite traditional to have multi-layer castable systems on floors. Normally a dense, strong material is used on the hot face for impact and load resistance. From there your typical back-up lining consists of progressively lower temperature, lighter weight and better insulating options to meet your desired cold face.

Example of a 12" (305mm) thick lining (old style) would be (Figure 1);
3" (75mm) Dense Castable (2500F-120pcf / 1371C-1920kg/m3) +
3" (75mm) Lightweight Insulating Castable (2300F–60pcf / 1260C-960kg/m3) +
6" (150mm) Ultra-Lightweight Insulating Castable (1800-2000F–30pcf /982C-480kg/m3)

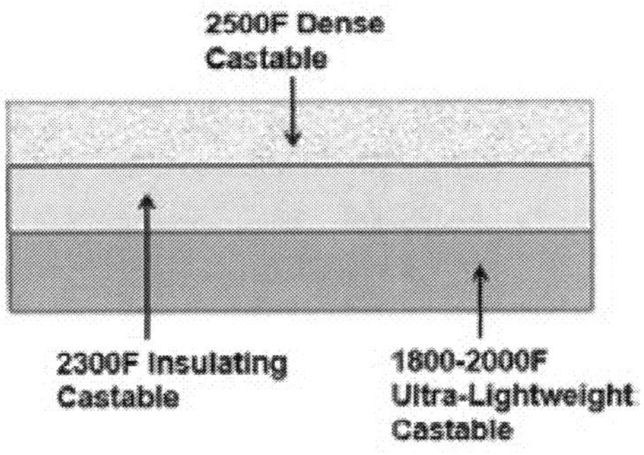

Figure 1

Running the base case heat flow simulation (ASTM C680 method), using an operating temperature of 2200F (1260C), 80F (27C) ambient and no wind, results in a projected cold face temperature of 235F (113C).

Looking closer there are other lining options to enhance performance (Figure 2). Castable materials remain a good option, but require water to mix and install, plus removal of water during dry-out. If this dry-out process moves too quickly, it can result in explosive spalling and lining damage. An interesting and proven value upgrade is a

Dry Floor using no castables, thus minimizing potential placement and dry-out issues due to water removal. A lining design of this style consists of the following;
2.5" (63mm) of High Duty Firebrick +
6" (150mm) 2300F (1260C) Insulating Firebrick +
2.5" (63mm) 2000F (1100C) Block Insulation

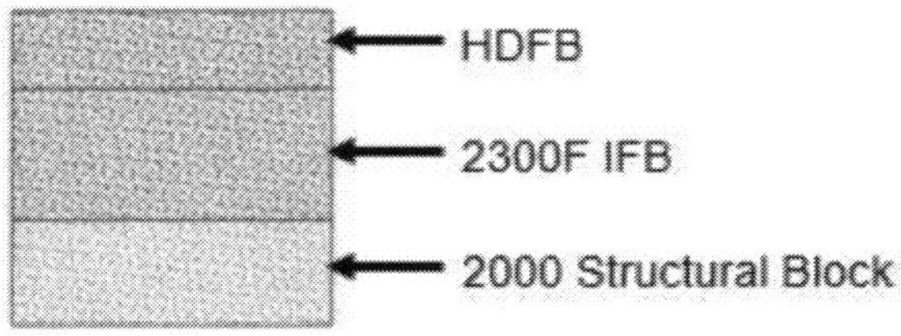

Figure 2

Running the heat flow simulation for this option at the same operating conditions yields a cold face of 208F (98C). Advantages of this system include;

1. Lower casing temperature by 27F (15C) – worker safety
2. Improved heat savings - less fuel consumed, and less CO2 produced
3. No water - lower weight, no dry-out, no controlled temperatures for placement and curing
4. Time savings / money savings – Other workers can simultaneously work inside the radiant box during installation versus castables installation that requires separate casting and curing. In addition, no dry-out is needed which in aggregate will save several days' time and allow additional furnace production time.
5. Lower heat storage – 30% savings – more heat to the process than stored in the lining during operation and during start-up and shut-down.

This type of option is now used or proposed on nearly every new cracking furnace lining. While there may not be an initial product cost savings for new construction, EPCs like the idea of better product quality control using less water and no dry-out, while End Users like how user friendly, efficient and reliable this floor lining can be.

Lower Radiant Wall Lining

In floor fired units it is quite typical due to flame impingement to utilize a high temperature insulating firebrick (IFB) lining system. For a 12" (305mm) thick lining, a typical lining design (Figure 3) is comprised of;

9" (230mm) 2800F (1540C) IFB +
3" (75mm) of some type of 1900-2000F (1037-1100C) structural/insulating block

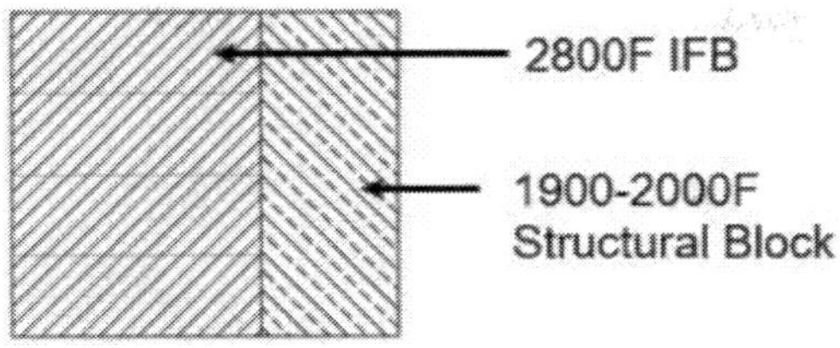

Figure 3

Since this area sees direct flame impingement, hot face conditions would typically exceed 2400F (1316C). Using this design temperature, the resulting cold face temperature would be 204F (96C). But additional value options are available to improve the casing temperature, the total refractory lining weight and at the same time lowering the cost of the lining.

Lining option 1 – Typically IFB tieback anchors are located every 4th row of brick. So, change out the IFB between tieback levels to (Figure 4);
4.5" (114mm) 2800F (1540C) IFB +
4.5" (114mm) 2300F (1260C) IFB +
3" (75m) 1900-2000F (1037-1100C) Structural Block

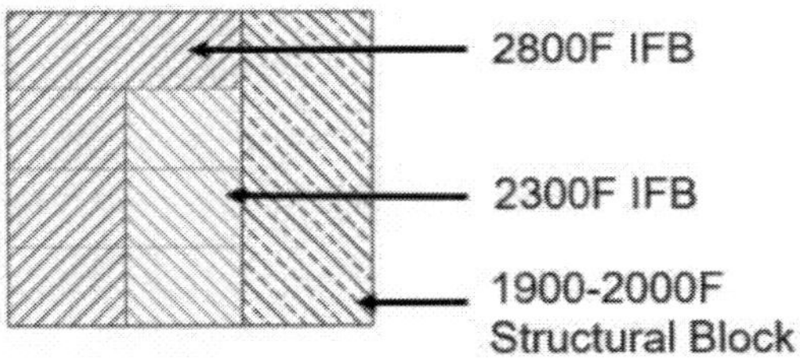

Figure 4

Due to the lighter weight and better insulating properties of the 2300F (1260C) IFB, the resulting cold face temperature is lowered to 186F. In addition, both heat storage and cost of the lining goes down as you replace volume of higher cost 28 class IFB with 23 class IFB.

Lining option 2 – For the ultimate in lower energy costs, replace only ½" (12mm) of the back-up Structural Block with Microporous Board (Figure 5). While Microporous Board may be more expensive than traditional options, there is no denying the tremendous benefit to heat loss and lowering of casing temperatures.

The 3" (75mm) thick back-up lining behind the IFB is now comprised of;

½" (12mm) Microporous Board +
2.5" (63mm) 1900-2000F (1037-1100C) Structural Block

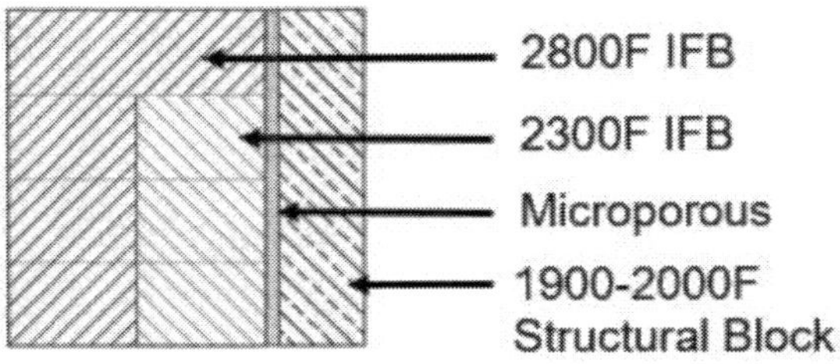

Figure 5

The resulting heat flow yields;
Cold face temperature of 168F (76C) – lower by 36F (20C)
Cutting heat loss by 35% and heat storage by 20%.

Two words of caution when retrofitting this lining;

The casing steel over time will likely warp. It is also important for long term reliability to have a flush hot face meaning all stresses run parallel to the casing. In order to achieve a "flush" hot face surface, you may need to place a layer (or two) of flexible refractory ceramic fiber (RCF) or alkaline earth silicate (AES) fiber blanket directly against the casing plus design your IFB tieback anchor system with enough "play" (longer) to account for the warpage.

It is also important to run heat flows to check your anchor metallurgy. Improving back-up insulation results in driving higher interface temperatures between the boards and the IFB, thus raising tieback anchor temperatures as well.

Upper Walls / Arch

In floor fired units it is quite normal to have fiber modules located above the flame height in all walls and the arch. As mentioned earlier, older style units had linings 8-9" (200-230mm) thick where today most are 12" (305mm) thick. But if the lining thickness cannot be changed, what can be done to improve the properties?

Fiber systems since they are open pore / open porosity, will improve in thermal conductivity as density increases. This phenomenon maxes out at roughly 25pcf (400kg/m3). So, it stands to reason, increasing product density will improve the thermal properties of your lining. Its true most folded modules can reach densities of around 12pcf (192kg/m3) before fibers become crushed and lose resiliency (spring-back). However monolithic modules can be made in nominal densities of up to 15pcf (240kg/m3) before compression so can more easily achieve this density increase and conductivity decrease.

Comparing heat flows, a typical older style lining would be designed using a 9" (230mm) folded module at 9.3pcf (150kg/m3). Running the thermal analysis at 2200F (1200C) yields a cold face of approximately 195F (91C).

Taking this to a more efficient option (Figure 6), use 8" (200mm) 15pcf monolithic module backed by 1" (25mm) Blanket resulting in 172F (78C) cold face. Results of this option lowered the casing temperature 23F (13C) and saved 25% on the heat loss! All this just by raising the module density and designing in a back-up blanket/safety lining. Higher density modules also have benefits of being stronger and more resistant to velocities and other mechanical abuse.

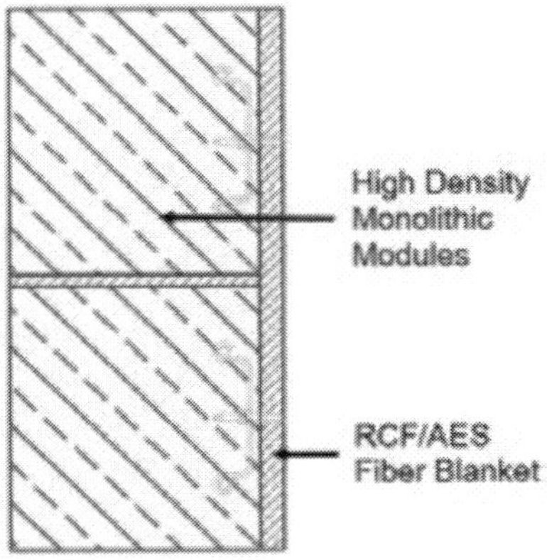

Figure 6

Going to an even higher efficient option, a layer of Microporous Board can be introduced. Taking the example above, reduce the blanket to ½" (12mm) but include a layer of ½" (12mm) Microporous. This results in a casing temperature of 160F (71C). Comparing to the original base lining results in a 35F (19C) casing temperature drop and heat loss improvement of 37%.

Important design note when using Microporous Board;
When using the Microporous, design it into the lining layer where most efficient. Since the highest thermal conductivity differences between Microporous and Block or Fiber occur somewhere in the 1000F (538C) range, it is important to place the Microporous Board in this temperature zone. This results in the most efficient and cost-effective use of the product, yielding the greatest benefits to the end user.

One point of caution with these designs;
Improving the cold face will again raise anchor temperatures within the modules. Care should be taken to ensure the module anchor system is designed for long term and reliable operation. This may include a higher metallurgy and/or more robust parts.

Peep Door Surrounds

One of the most problematic areas for refractory lining are in the peep door surrounds. Many units suffer from hot spots directly behind this surround area. Reviewing older style refractory lining design practices, it is typical in IFB linings to use a pre-formed castable or vacuum formed ceramic fiber peep door surround. While in fiber linings, it is more typical to have some type of vacuum formed peep door surround.

The difficulty with the above designs in not in the surround design itself, but the proper sealing against heat leaks between dissimilar refractory lining materials. For this reason, a better solution is to match the peep door surround materials with the adjacent refractory lining. This means using cut IFB shapes in the IFB lining (Figure 7A) and cut modules in a fiber module lining system (Figure 7B).

Below are pictures, showing both an IFB and module lining peep door surround designs. Key point here is to pay attention to the refractory design details to ensure proper mating so the seal is not lost.

Figure 7A

Figure 7B

Tube Bay / Shadow Box Linings

Any entrance or exit to the heater is a potential heat leak or influx of cold air which lowers efficiency. These areas, like other refractory areas will degrade over time to the point you may not realize you've lost your seal. Material and design improvements are available to mitigate potential heat leaks and cold air seals. Included in these mitigations; modules designed to ensure a tight fit, not affected by tube movement and to ensure there is no potential trapping of heat within steel structures that may compromise reliability of those critical areas.

The ideal design is normally a combination of sealing modules close to or against radiant tubes inside the radiant box (Figure 8A) and high temperature textile tube seals (Figure 8B) extending outside the radiant box. The modules can be pre-designed or are easy to cut and fit around the radiant tubes. It's also possible to design this shadow box area for potential removal of modules from the top if required to access the radiant tubes. This eliminates the need for scaffolding, shortening downtime. The tube seal is a pre-designed and prefabricated unit, consisting of high temperature textiles with some fiber blanket inside for additional insulation value. However, in both cases it is important to have a proper refractory design considering the tube movement, maintenance practices and to avoid trapping undue heat on structural members.

Figure 8A

Figure 8B

Convection Sections

Convection sections have traditionally been lined with castable (insulating concrete) materials. A typical older configuration in the lower shield section (Figure 9) consists of;

4"-2500F (100mm-1371C) insulating castable +
4" (100mm) of an 1800F (982C) ultra-lightweight insulating castable.

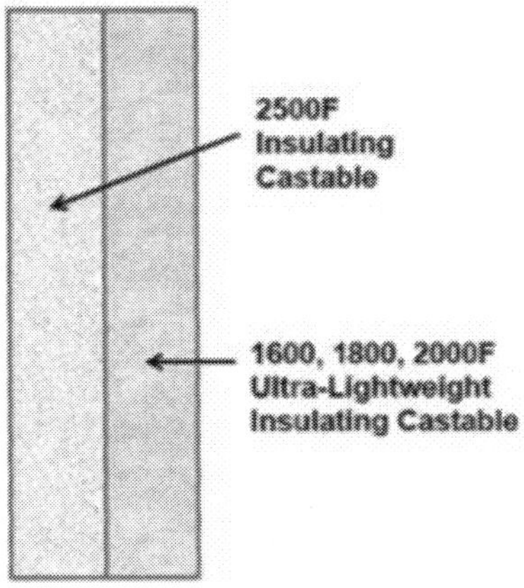

Figure 9

Running a heat flow at 2200F (1204C) results in a cold face of 216F (102C).

When reviewing this lining we find the typical Ultra-Lightweight Castable requires roughly 150% water by weight and is quite weak. Earlier we discussed that castables require expertise in placement with addition of water added to place, time to cure and a controlled dry-out to remove the water.

It stands to reason at the least, we can eliminate the high amount of water contained in the back-up castable by using Structural block to replace the Ultra-Lightweight Castable (Figure 10). If we run a heat flow simulation on this lining, we see a cold face temperature of 210F (99C). While this may be only a modest improvement, additional benefits can be had since there's no water to dry-out in the back-up, we eliminate potential cost and time to dry-out resulting in added production time. Another benefit is strength of the Structural Block is at least equal to and typically greater than the Ultra-Lightweight Castable.

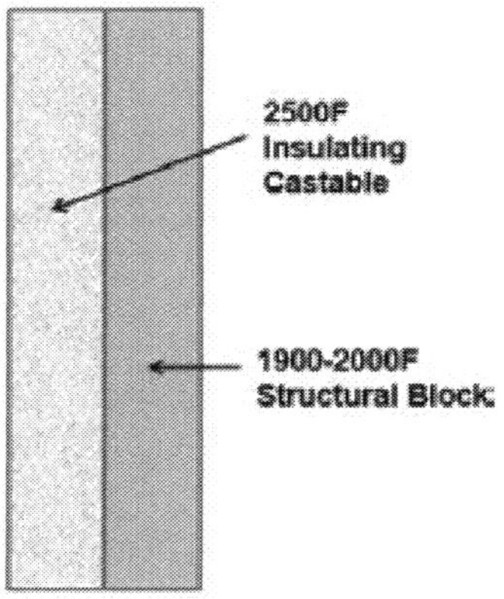

Figure 10

Taking this again one step further, replace ½" (12mm) of the structural block with ½" (12mm) of the Microporous Board (Figure 11).

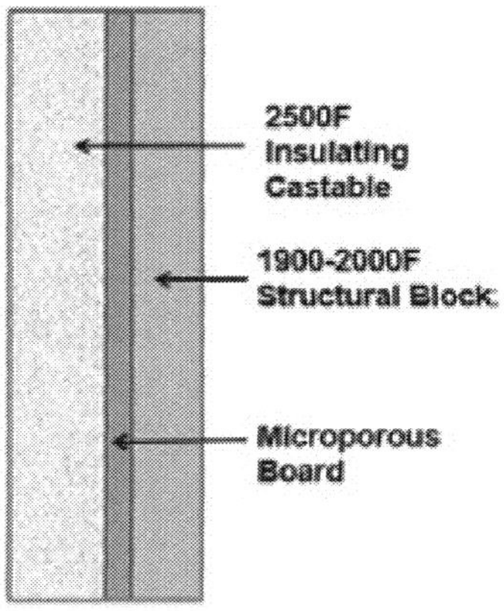

Figure 11

Resulting cold face temperature drops dramatically to 179F (82C). Results of this option yields improvements of 37F (21C) on the cold face and 33% lower heat loss! This in addition to the time savings by lessening the dry-out time and potential weight savings if shipping from a fab shop to the field!

One last and very interesting option is to use Fiber Modules in the convection section (Figure 12). First question is always, what do you mean, use fiber in this tough and normally castables application? YES! High density monolithic fiber modules can withstand velocities up to 150ft/s (46m/s) without damage. If we also apply a proper rigidizer, this resistance can be enhanced another 30+ %.

Running a heat loss calculation using a 15pcf (240kg/m3) monolithic fiber, this results in a cold face of 204F (96C). Saving over the original design as 11F (6C) and 10% heat loss. But the other benefit is there is absolutely NO water. So, you save weight for shipment and eliminate any time for dry-out.

Figure 12

Conclusion

In summary, old style cracking furnace refractory linings differed from today's design as they were typically much thinner. Any potential revamp and upgrade of these older units begs for safer and higher efficiency refractory linings to meet today's tougher requirements. Efficiencies gains can be measured by the fuel savings and less CO_2 produced. Safety improvement comes from lower casing temperatures and more reliable linings. But on top of this are potential time savings resulting in additional production time and profitability.

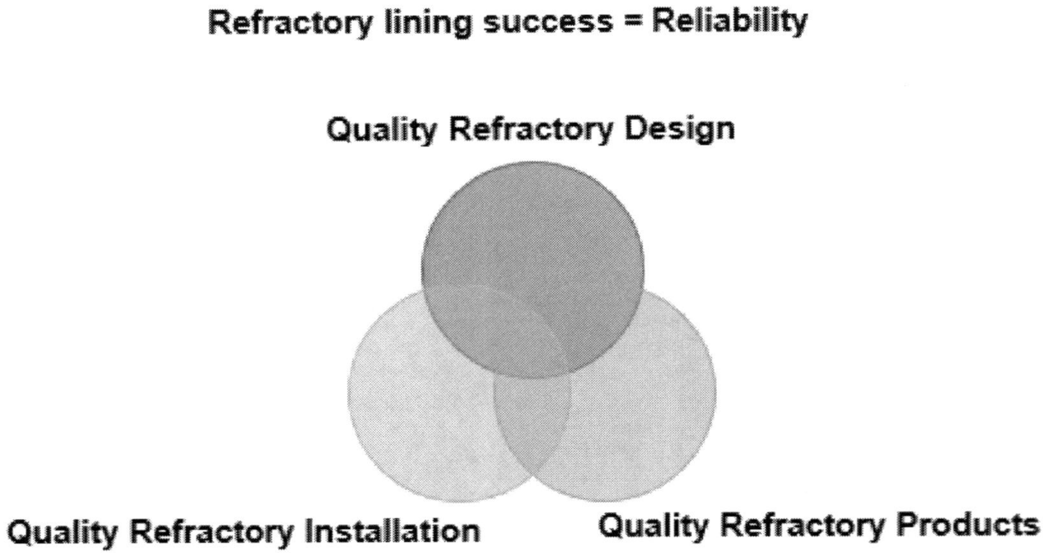

Figure 13

Overall it makes great sense to pay strong attention to your refractory lining in revamp situations. While some companies have refractory experts, it is also very easy and cost effective to have a reputable refractory company come in to do an IR Camera study on your casing and lining health assessment. After assessment, a simple paper study can explore potential options. But most importantly, if you decide to make refractory upgrades, choose high quality products from top notch companies, ensure your refractory design is sound, and employ reputable installers with oversight from the refractory company advisor, to insure the proper installation of your refractory lining (Figure 13). Once complete, it is easy to see the results of your due diligence and refractory value upgrades with improved furnace efficiencies and a refractory lining that will reliably and safely perform over the next 20+ years.

References

1. ***ASTM C680*** - *Standard Practice for Estimate of the Heat Gain or Loss and the Surface Temperatures of Insulated Flat, Cylindrical, and Spherical Systems by Use of Computer Programs*

AICHE PAPER NUMBER: 123f

CONTINUOUS ONLINE PH MONITORING AND CONTROL FOR DILUTION STEAM SYSTEM RELIABILITY

Lindsay Hernandez
Global Marketing Manager - Ethylene
NALCO Water, An Ecolab Company

Deepak Sharma
Industry Technical Consultant – MEA
NALCO Water, an Ecolab Company

Sultan Saud Saif-Al
Sr. Operations Engineering Manager
SABIC

Mustafaa Waheed
Account Manager
NALCO Water, an Ecolab Company

Prepared for Presentation at the 2020 Spring National Meeting
Houston, Texas, March 29 – April 1, 2020

AIChE and EPC shall not be responsible for statements or opinions contained in papers or printed in its publications

Continuous online pH Monitoring and Control for Dilution Steam System Reliability

Lindsay Hernandez
Global Marketing Manager - Ethylene
NALCO Water, an Ecolab Company
Deepak Sharma
Industry Technical Consultant – MEA
NALCO Water, an Ecolab Company
Sultan Saud Saif-Al
Sr. Operations Engineering Manager
SABIC
Mustafaa Waheed
Account Manager
NALCO Water, an Ecolab Company

Abstract:

Corrosion and fouling in the quench water and dilution steam systems (DSS) of an ethylene plant are an ongoing problem within the petrochemical industry. Constant attention is required to maintain the mechanical integrity of affected heat exchanger equipment to achieve safe, reliable, and profitable operation. Traditional means of managing these systems is based on periodic sampling and analysis to determine the corrosivity and fouling likelihood throughout the quench water and dilution steam systems. In most cases, 90% of the corrosion occurs during 10% of the total operating time due to events like furnace decoking or corrosion inhibitor pumping failure. The advent of continuous online monitoring and unique online control of neutralizing chemistry adds another layer of protection for increased system reliability.

The challenges of quality monitoring and control in a dilution steam system (DSS) include: (i) potential for exposure and extensive manpower required for capturing samples and tracking system changes, (ii) high variability of impurities causing emulsions, fouling, and corrosion, (iii) increasing corrosion rates due to frequent pH swings, and (iv) low visibility to system updates due to infrequent sampling and lack of data points. Automation technology to receive continuous pH and water turbidity data, as well as allow active control of neutralizer, has been applied to reduce overall corrosion rates and extend run lengths of the quench and DSS for the optimization of steam cracker operations.

One ethylene plant was able to improve turbidity control due to operational changes and introduce an improved dosing philosophy for neutralizer to minimize pH swings, resulting in a more robust system that is better able to adapt to furnace swapping and operational upsets. By maintaining the target pH, another plant reduced the corrosion rate and emulsions in the system, thereby decreasing the amount of Total Organic Carbon (TOC) carryover and fouling in high heat flux areas, while also improving dilution steam generator run lengths.

Introduction

Ethylene producers typically utilize dilution steam systems to supply the steam to their furnaces, and these process water systems are often plagued by corrosion, fouling, and emulsions. Constant attention is required to maintain the mechanical integrity of affected towers and exchangers to achieve safe, reliable, and profitable operation. Additionally, this portion of the plant is often home for many errant recycle streams which can upset stable operating conditions, and it is common for significant problems to be detected too late for correction. Based on consensus from Ethylene industry expertise, in the majority of crackers 90% of the corrosion occurs during 10% of the total operating time. Without real-time monitoring many of these corrosive events can be missed and cause irreversible damage. Multiple case studies utilizing new automation technology to accurately and precisely measure online pH, turbidity, conductivity, and corrosion rates will be presented.

Common Problems in Dilution Steam System

Ethylene Cracker Dilution Steam Systems (DSS) are typically monitored on an inconsistent basis, with process water conditions assessed between once a shift to once a day, or even once a week spot sampling. Procedures and intervals can differ greatly between complexes, without a common methodology. DSS incorporates several individual functions including process water recovery, filtration, hydrocarbon stripping, and dilution steam generation. Each function is closely linked to changes in plant operation; i.e., cracking severity, feedstock, and imported recycle or waste streams.

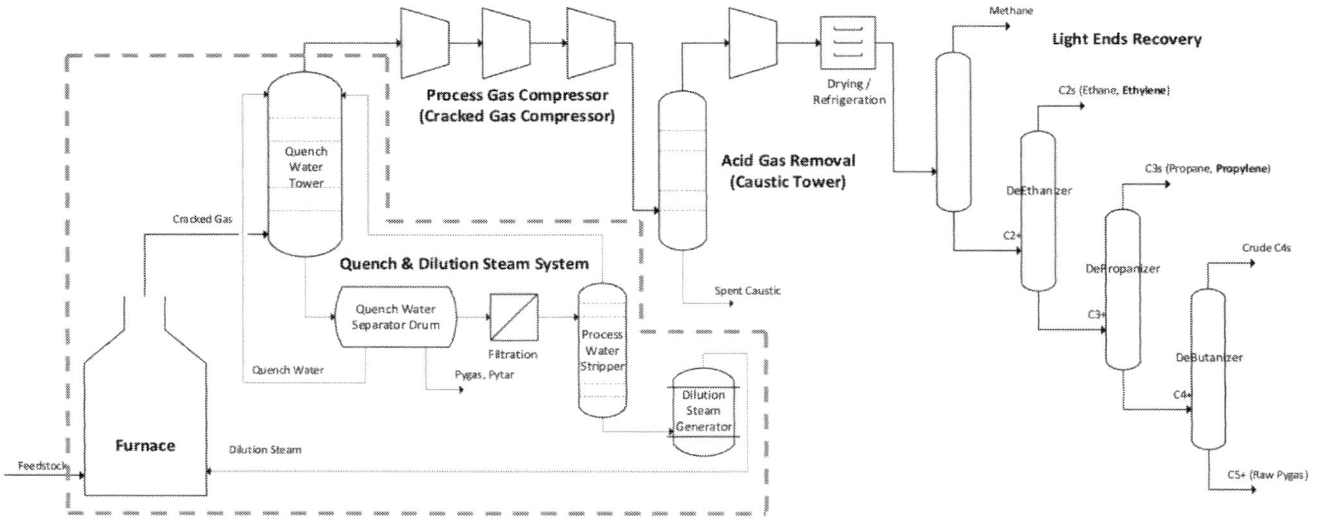

Figure 1 Schematic of a typical Ethylene unit, highlighting the dilution steam system

Managing these systems is based on periodic sampling and analysis to determine the corrosivity in the quench water circulation, process water stripper, Dilution Steam Generator (DSG) blowdown and the dilution steam pH and iron content. For fouling and emulsion concerns, a broad range of variables must be monitored, including Total Organic Carbon (TOC), oil and grease, and turbidity in the process water indicating a hydrocarbon emulsion in the water that could lead to fouling in high heat flux areas. This intermittent approach is reasonable during periods of stable operation, but this approach does not account for changes to operation or periods of unstable operation. Today's industry best practices impact 90% of typical operation but fail to address the 10% of time during upsets when maximum damage can occur. Such events include furnace decoking or neutralizer/corrosion inhibitor pump dosing failure. The traditional methods of routine grab sample monitoring can easily miss these problems or may only detect issues after significant damage has already occurred.

Table 1 Typical Process Water Quality Measurements

TEST	QW/Separated Water PWS Feed			PWS Bottom			DSG Blow-Down			Dilution Steam		
	Min	Aver	Max	Min	Aver	Max	Min	Aver	Max	Min	Aver	Max
PH	3.7	7.85	10.6	5	7.8	10	4.2	8.3	11.8	4	8.3	9.8
Suspended Solids (ppm)	0	22	110	0	19	140	0	21	80			
Total Iron (ppm)	0.05	1.2	2.8	0.01	1.6	21	0.001	0.6	6.8	0.001	2.35	22
TOC (ppm)	65	280	380	92	240	1060	4	476	3900			
Total Hardness (ppm)	0.3	1.08	5.1	0.3	1.46	8.6	0.3	3.8	38			
Magnesium (ppm)	<0.1	0.85	1.6	<0.02	0.22	1.4	0.01	0.21	1.9		0.4	
Calcium (ppm)	0.1	0.22	5.9	0.08	1.4	8.6	0.01	2.7	38		1.9	
Aluminum (ppm)		0.1			0.2			0.17			0.1	
Silicon (ppm)	0.1	0.67	1	0.1	0.9	1.4	0.1	1.8	3.8			
Total Acid (mmol/l)	<0.1	0.3	1.6	<0.1	0.32	2.4	0.05	0.32	1.8	0.1	0.28	2.1
Chlorides (ppm)	<1	13	74	<1	11.2	39	<1	19	160			
Total Sulfur (ppm)	0.1	20.5	110	0.8	20.4	120	0.1	25	63			
Sulfites (ppm)	0.1	2.2	6.4		1.5		<1	5	14			
Sulfates (ppm)	1	6.2	15	1	10	23	4.8	56	120			
Thiosulfates (ppm)	1	9.8	20	1	11	26	0.05	16	40			
Sulfides (ppm)	0.1	0.4	0.8		0.2		0.05	0.3	0.8			
Phenols (ppm)	1.8	60	98	1.5	48.7	95	0.01	26	77			
Phosphate (ppm)	<0.1	0.1	0.1	<0.1	3.7	53	0.02	25	440			
Styrene (ppm)	0.005	18	90	0.005	0.4	3.8	15	1186	25000			
Total Alkalinity (ppm)	0	106	1086	24	144	1052	12	308	1100			
Carbonate (ppm)				0	38	328		80				
Bicarbonate (ppm)				12	81	724		215.5				
Ammonia (ppm)	2	17	89	<0.5	6.5	40	<0.5	6.5	40	1.1	10	32
Sodium (ppm)	0.09	18	280	<0.1	31.5	280	<0.09	121.5	320	<0.09	4.3	43
Potassium (ppm)				0.07	0.32	1.6	0.07	0.43	2.4	<0.1	0.88	32
BTX (ppm)	101	101	158		<5			<10				

Fouling

The process water from the Quench Water Separator (QWS) contains dissolved, suspended and emulsified hydrocarbons, together with small amounts of free (entrained) oil, and for a gas cracker, suspended solids (mainly coke fines), polymer oils, tar, and heavier hydrocarbon components with densities close to that of water.

In order to minimize organic fouling in the Dilution Steam Generation (DSG) system and Process Water Stripper (PWS) and thereby increase plant reliability, it is desired to remove these suspended solids, free oil, and any suspended or emulsified hydrocarbon upstream of the DSG

System. Monitoring these characteristics enables the operator to take corrective actions, otherwise at the operating temperatures of the PWS and DSG precursors such as dienes, styrene, and indenes, can lead to organic fouling in the quench loop, process water stripper, or the DSG reboiler.

Consequences of this fouling include reduced throughput, increased differential pressure, decreased heat transfer, and occasionally blockages. Fouling can also create an environment for under deposit corrosion, even when the bulk pH is properly controlled.

Corrosion and Emulsion

In the quench water circuit corrosion can occur due to low bulk pH. Acids in the cracked gas can condense in the quench water causing low pH conditions. Typical acidic species include organic acids (predominantly acetic and formic acid) and carbonic acid.

The primary best practice guideline for corrosion inhibition in DSS performance is to have good pH control. Corrosion rate increases with the acid concentration, temperature, and oxygen level. Effects of poor pH control can be seen in Figure 2, where pitting and metal loss are visible when examining failed exchanger tubes.

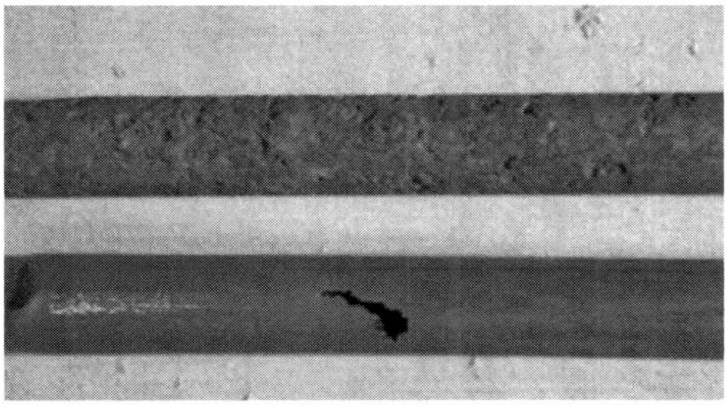

Figure 2 Thin-edged holes along tube OD surface and areas with comparatively shallow but irregular OD metal loss.

Settler pH is also a critical parameter affecting not only corrosion, but also emulsions. Poor pH control in the quench water tower and separator drum(s) may result in fouling problems downstream at the PWS and DSG. This occurs because of inefficient hydrocarbon and water separation, resulting in high TOC carry over. The ease of breaking an emulsion is related to not only the stream composition and relative densities, but also the pH. In many cases an emulsion breaker chemistry is needed to assist in the hydrocarbon and water separation.

In all aspects of pH control, to avoid corrosion and emulsions, gathering accurate and reliable data for monitoring is the most important aspect of any corrosion inhibition treatment.

The Monitoring Challenge

There are four key roadblocks in gathering accurate and reliable data for monitoring:

i. Potential for exposure and extensive manpower required for capturing samples, analyzing, and tracking any system changes
ii. High variability of impurities causing emulsions, fouling, and corrosion
iii. Corrosion rates increase with frequent pH swings
iv. System upsets are often missed

Considering these factors, several ethylene producers have sought to install a complete performance management system, intended to maximize safety, reliability, and minimize operating costs associated with dilution steam systems. The desired equipment would continuously measure key system parameters and, when upsets occur, take immediate corrective action, preventing operational problems and maximizing performance. The automation equipment should also have the ability to communicate with system users via a live alert system, including text messages or email, keeping the system owners informed of changes or upsets. This innovation would allow the plant to monitor and accurately capture significant changes in the dilution steam system in near real-time to make any changes in operational and chemical dosing parameters before any major upset or damage could occur. More importantly, the increased frequency of data gathering could provide an in-depth understanding of the impact of process changes and upsets on the system, in turn allowing operations to be more proactive to avoid consequences in the DSS.

This paper will discuss the implementation of this innovative automation and how this technology has been used to successfully reduce the overall corrosion rates and extend the run lengths of the quench and dilution steam system for the overall optimization of cracker operations, thereby reducing the total cost of operation (TCO).

Advanced Monitoring and Control

Incomplete and Missing Information

Quench Water pH plays an important role for both corrosion and emulsion stability, which in turn leads to hydrocarbon/water separation efficiency and finally to the fouling at the maximum heat flux areas. Controlling and maintaining of pH in the right pH window is really important for the DSS performance as this system is so dynamic with streams from all over the plant adding up to many independent variables to control. Any lack of data monitoring would not only impact the corrosion but simultaneously the fouling in the system.

In most areas of the ethylene plant, the samples are usually taken once per shift or once per day based on the severity of the concerns in the plant. As discussed earlier, to be successful treating a dilution steam system to prevent corrosion, emulsions, and fouling, one must obtain accurate pH data, whether it comes from instrumentation or lab analysis. Once a shift samples equate to 2-3 samples per day or up to 1,100 analysis data points per year. pH deviation is an infrequent event, which usually takes place during unstable and infrequent operational practices. These unstable events include anything from furnace decoking to feedstock changes leading to corrosive conditions. In addition, caustic slippages from the Acid Gas Removal Unit (AGRU) to the quench system can lead to high pH, in turn stabilizing an emulsion and then leading to fouling in the DSS, and one step further, if ammonia is freed from the caustic before these slippages are noticed, it will continue to cycle up in the system without removal possibilities. To get the system back to normalcy after a spot sample is collected in the field, analyzed in the lab, and then corrective action taken with regards to chemical treatment changes could take about 4 to 6 hours at a minimum. During this period corrosion and/or fouling would have already taken place. This feedback loop takes too long to provide adequate asset protection.

Ideally a system could achieve the following:

1. 1,100 annual data points replaced by a single analyzer which provides over 52,000 annual data points (based on 1 data point every 10 minutes)
2. Constant monitoring with real-time pH, conductivity, turbidity, and corrosion analyzers
3. Consistent and trustworthy pH monitoring that reliably matches with handheld pH spot measurements
4. Secure network communication for remote monitoring, alarming, and data management

As demonstrated in the case studies that follow, this can be achieved using automation technology.

Automation Technology Solution

The automation technology should feature real time monitoring and control of the chemical treatment program, data visibility and alarming capabilities. Benefits realized include consistent and reliable online monitoring of key parameters throughout the dilution steam system and quick upset response leading to improved total cost of operations.

The automation technology applied in the case studies can be installed and setup for two different modes: monitoring only or automatically controlled optimization mode. Monitoring mode provides online measurements, but all dosage adjustments are done manually. Data is then transferred to a database using wireless technology. Data can also be transferred to the DCS system and all adjustments will be controlled by operations. The automatically controlled optimization mode has all the features of monitoring mode and adds PID functions to control pump strokes of the neutralizer chemical pumps to adjust to a target pH set-point. Set points are selected based on the plant's optimization range and readjusted at least quarterly based on plant economics. The key parameters to be monitored are pH, conductivity, turbidity, and corrosion rate.

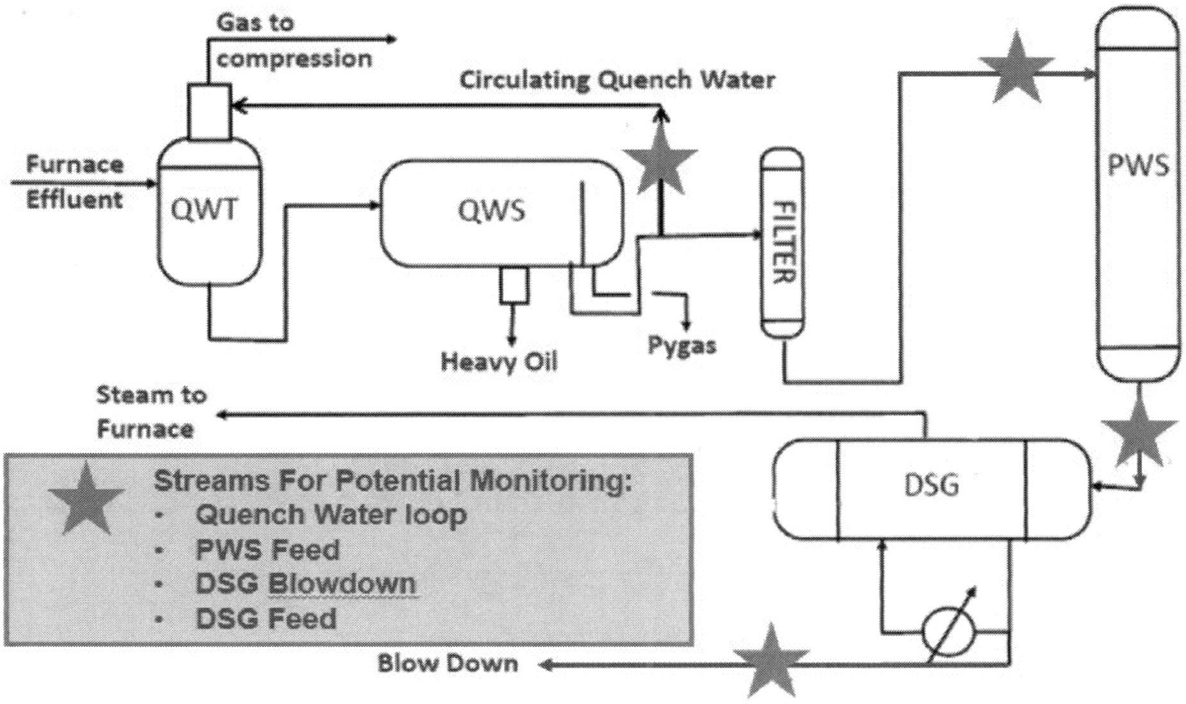

Figure 3 Quench and Dilution Steam System Diagram with Possible Monitoring Locations (denoted by stars)

The pH sensors are specifically designed to resist fouling from the stream's water/hydrocarbon mixture. For turbidity, there is an industrial turbidimeter with a range of 0-8000 NTU measuring in the IR range and eliminating any potential interference from color. The conductivity and corrosion probes are established technology, already widely accepted independently throughout the industry.

The controller receives signals from the sensors (turbidity and pH) and can communicate with the plants DCS system. The collected data is also sent by an onboard cellular modem to the company's servers where users can download the raw data, examine data trends, and create automated reports. The cellular communication is also utilized to alert users of alarm conditions identified by the controller, such as high effluent turbidity. Several additional features were developed to address technology reliability, accuracy, and sensor fouling. An automated ultrasonic cleaning methodology was incorporated during development to reduce organic fouling and increase sensor reliability.

This new automation technology is delivering value across the globe to ethylene producers looking to improve their reliability and reduce maintenance on their Quench and Dilution Steam System. Below two case studies from different stages of commercialization will display how continuous online monitoring not only validated the periodic spot sampling analysis, but captured previously unseen events, and led to quicker response times – consequently improving the reliability and maintenance of the dilution steam system.

Case Study 1: Early Stages – Beta Unit

A gas cracking ethylene producer in the Middle East was an early adopter of this equipment to monitor and eventually control all aspects of their Quench water and Dilution Steam System. Trialing the equipment with real plant operating conditions would test the hypothesis that online monitoring could improve asset life and reliability, while also reducing required maintenance and system upsets.

Trial Scope and Goals

The goal of this trial was to measure how accurately online analysis of pH, conductivity, oxidation and reduction potential (ORP), and turbidity could be carried out, while minimizing human errors that often occur during periodic sampling and lab analysis. This work should result in equipment that will provide the following benefits for the operator:

- Validated data from the equipment, by comparison with periodic lab data
- Less manpower required for analysis and controlling the system
- Feedback control of pH by adjusting amine pump (based on pH values)
- Stabilized pH trend and subsequent corrosion control
- Corrosion prevention, due to optimized neutralizer dosage
- Feedback control of emulsion by adjusting emulsion breaker injection pump (based on turbidity measurements)
- Optimized oil and water separation to reduce overall system fouling

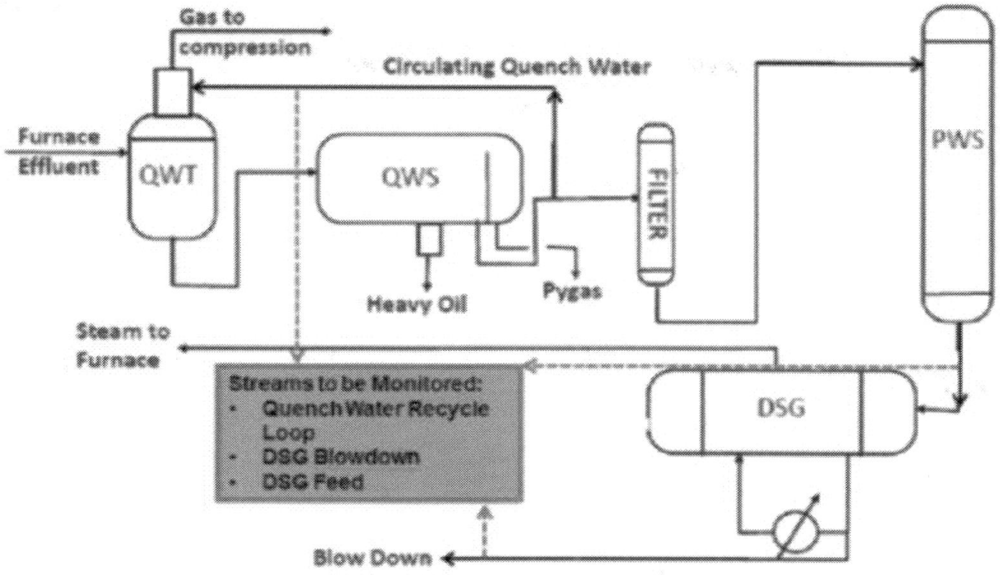

Figure 4 Integrated Automation Monitoring Strategy Showing Stream to be Monitored

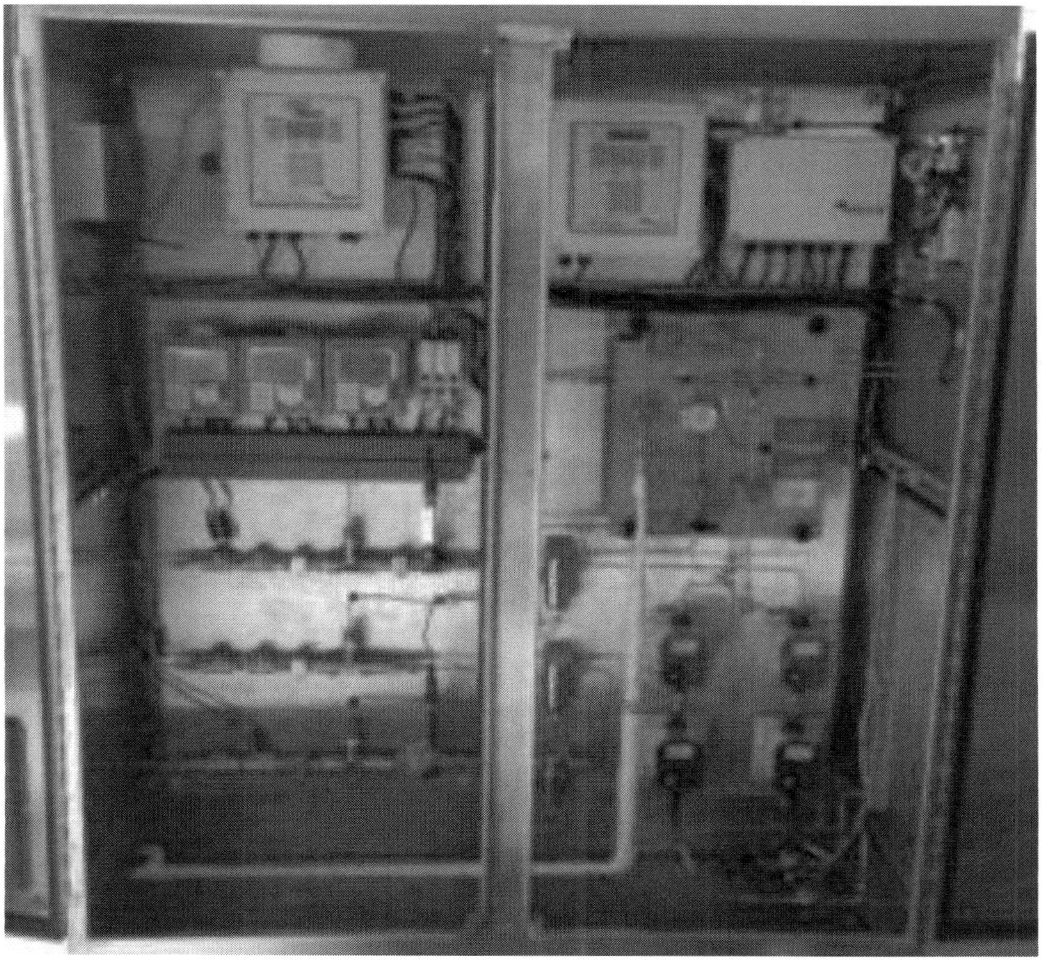

Figure 5 Internal View of Beta Unit

With agreement from the producer, all monitoring tools were combined into one piece of equipment (shown in Figures 4 and 5) to analyze multiple streams simultaneously. This helped to understand the performance and the repeatability with the issues related to the live process measurement in dirty water. The ultimate goal of the first trials was to determine what, when, and how to online analyze specific dilution steam samples with the intent to automate the chemical feed to optimize overall system performance.

The final vision was a compact single stream analyzer package that could be installed in prime locations for monitoring and chemical feed control to optimize the entire dilution steam system. The key performance indicators (KPIs) that were measured during this trial were: pH, turbidity, and Fe loss.

Online Monitoring Data

Once the equipment was placed in the plant, data was received for pH, turbidity, and corrosion rates for the installed streams. The major criterion was to validate the probe data with the lab data for pH and turbidity values. Samples were collected from the units for conducting lab analysis, and average data for the pH and turbidity were collected for 6 data points to have lower sampling error. Close to 70 lab analyses were carried out in a span of 6 months.

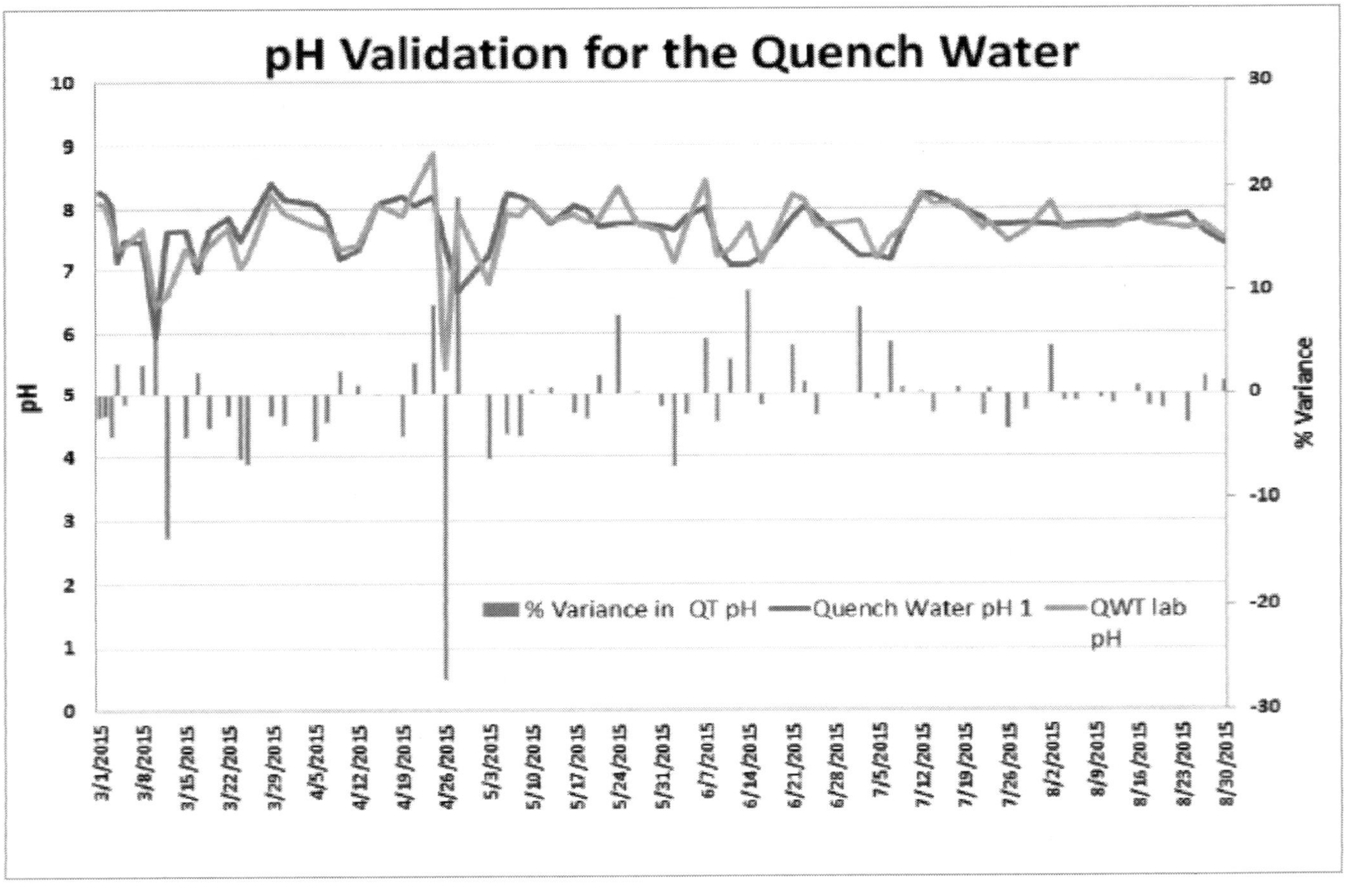

Figure 6 pH Trend Comparison for Quench Water

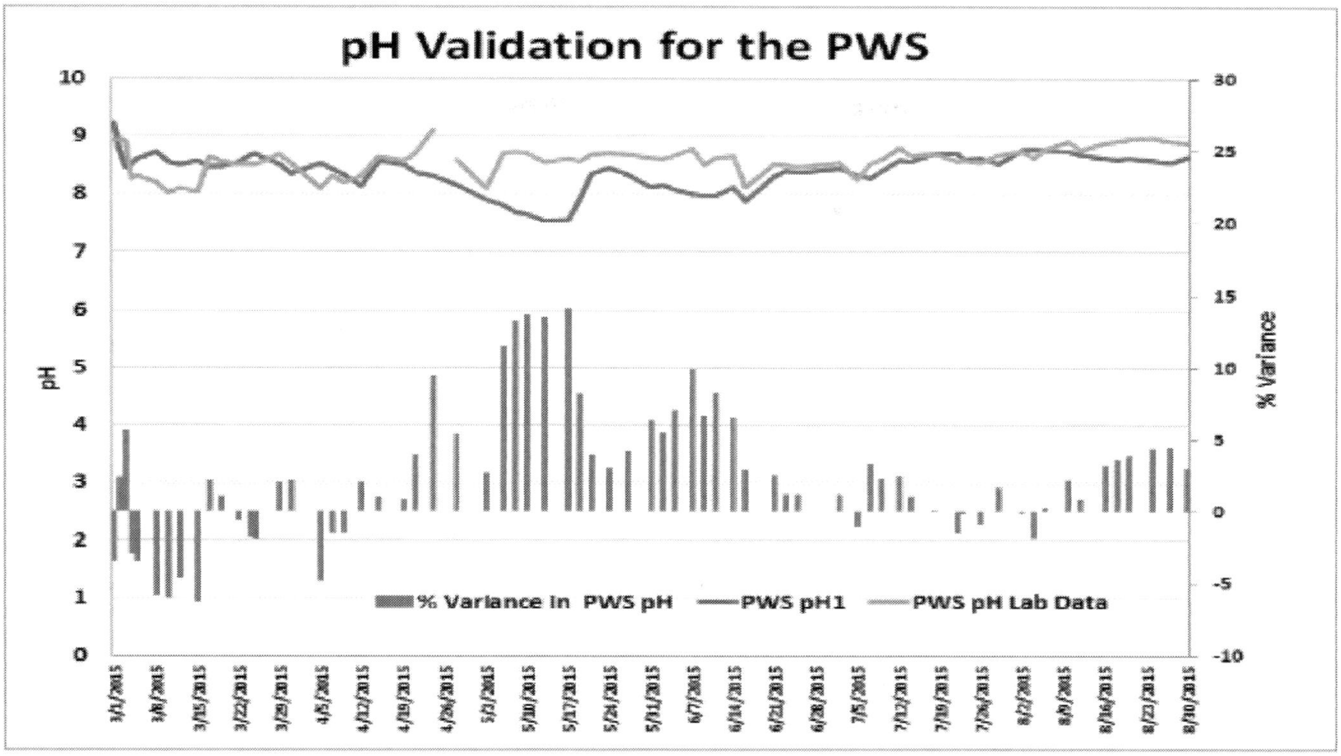

Figure 7 pH Trend Comparison for Process Water

It was clear from the quench water results that the maximum deviation of the probe data from the lab analysis was <10% , and 90% of the time the deviation was <3%. The same was done for the process water stripper, and the maximum deviation of the pH probe data from the lab analysis was <15%, and 90% of the time the deviation was <5%. This indicated both precision and accuracy, and further validated the selection of this particular pH probe for reliable measurement.

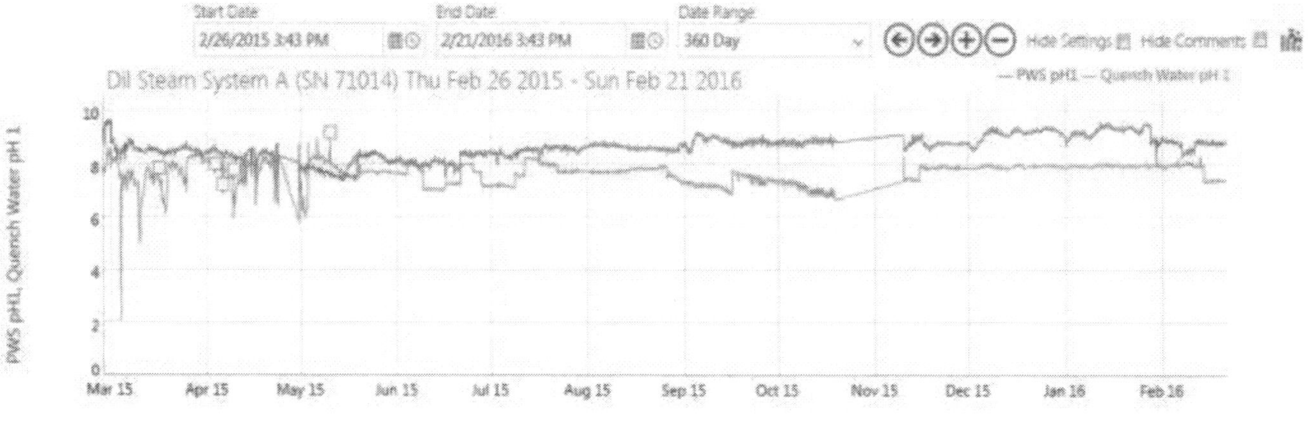

Figure 8 pH Measurements from online monitoring

Similar field sampling for lab to online probe comparison was done for the turbidity. This testing was not as straightforward as the turbidity probes were more prone to foul due to system upsets caused by various independent variables allowing the process water to turn dirty, leading to bad data. Once the probe was coated with the foulant material, as indicated from the cell fouling reading in the system, it was supposed to be manually cleaned and replaced in the system. The frequency varied between 5 days and 2 weeks, and cleaning was cumbersome and time consuming.

It was also observed that once the cell fouling readings exceeded 50%, the turbidity reading variances were large enough to be discarded as bad data points. In order to avoid this cumbersome manual process, a patented[1] ultrasonic cleaner was installed for self –cleaning.

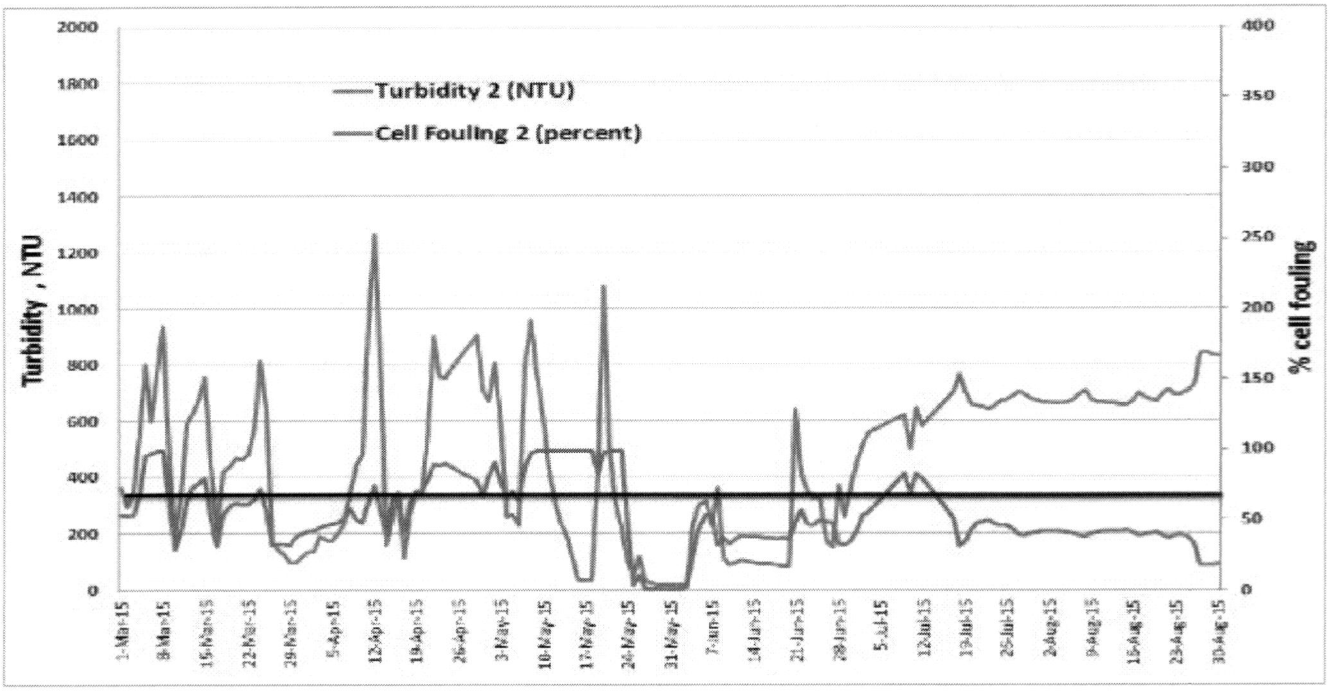

Figure 9 Turbidity and Cell Fouling Measurements for the Quench Water Tower

After this addition the turbidity trends more closely matched the lab analysis numbers. Even after the overall turbidity increased in the system the probes were not getting fouled to the point of giving false data, as witnessed in the past. Despite increased propane cracking resulting in increased turbidity levels, the frequency of manual cleaning of the probe was extended from once per week to about once per to 2 months.

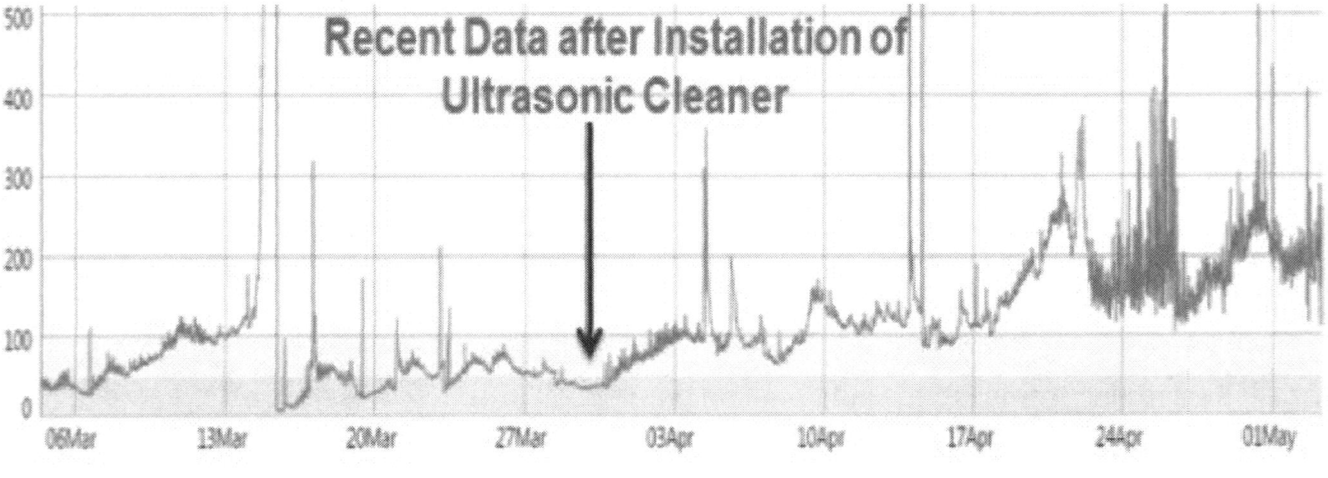

Figure 10 Turbidity trend showing introduction of self-cleaner

Outcome of Technical Trial

- pH readings across the system showed good results with a deviation of <5% from the lab data, which was +/- 0.1 pH, for the majority of the time (>90% of operation)
- Installation of an ultrasonic cleaner was effective in keeping the probe clean
- Turbidity data is within 20% deviation compared to the lab data, when cell fouling <50%

Resulting Operational Gains

The ethylene producer was able to control the turbidity much better around events like furnace swapping and operational changes, as it was clear that the turbidity was impacted due to the pH fluctuations in the quench system. The dosing philosophy for neutralizer was changed and a new standard operating procedure (SOP) was generated for proactively increased neutralizer addition to make sure that the pH would remain in the operating band. Corrosion rates and Fe counts in the DSG and process water stripper were maintained as per the KPIs.

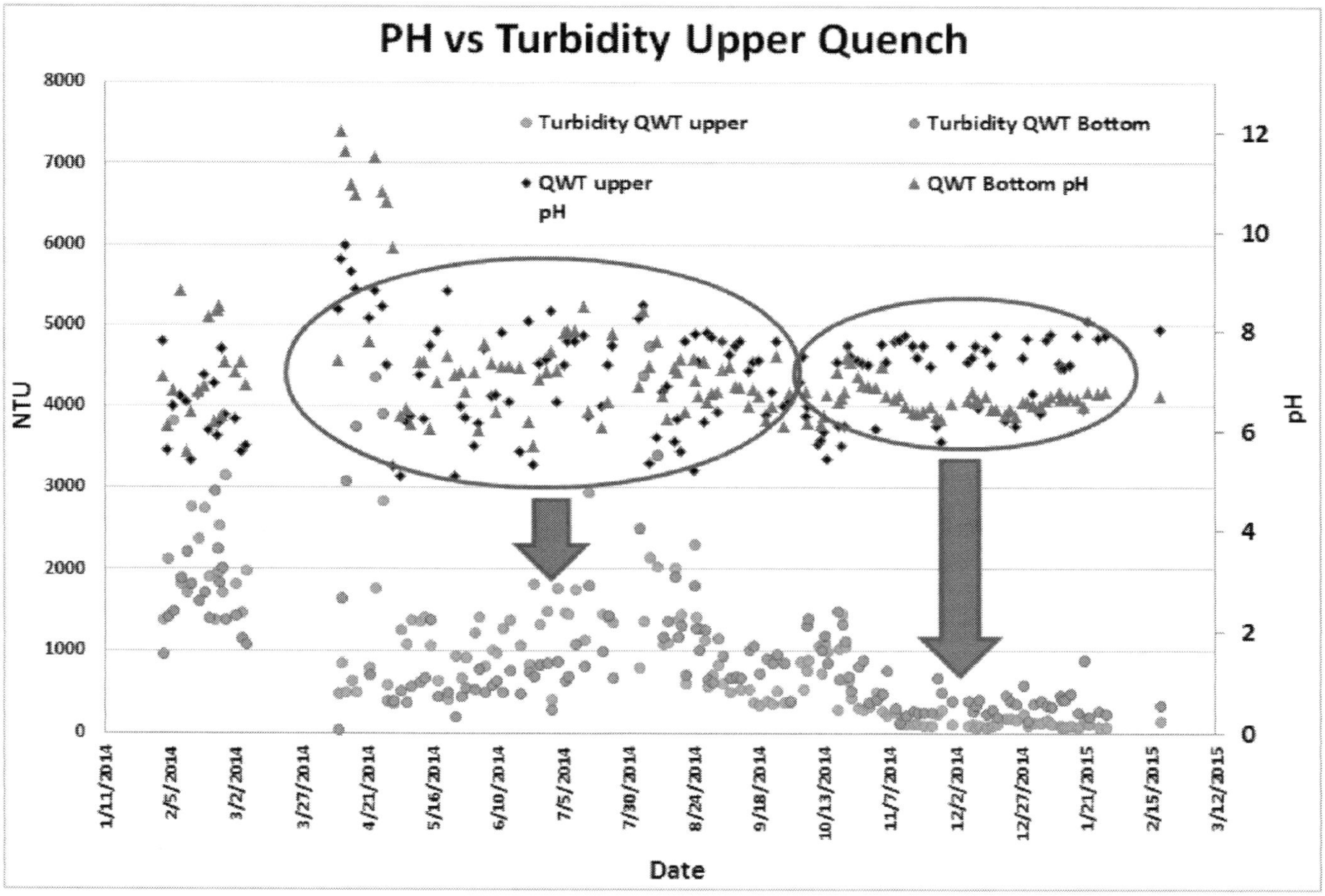

Figure 11 pH and Turbidity measurements before and after automation installation

The data above in Figure 11 shows that with more consistent control of the pH the overall emulsions in the quench water system were reduced. Keeping the pH under control and more closely aligned with the targets (by adjusting the neutralizer in the system) also eliminated the frequent pH swings below 6, reducing the likelihood of increased corrosion attacks.

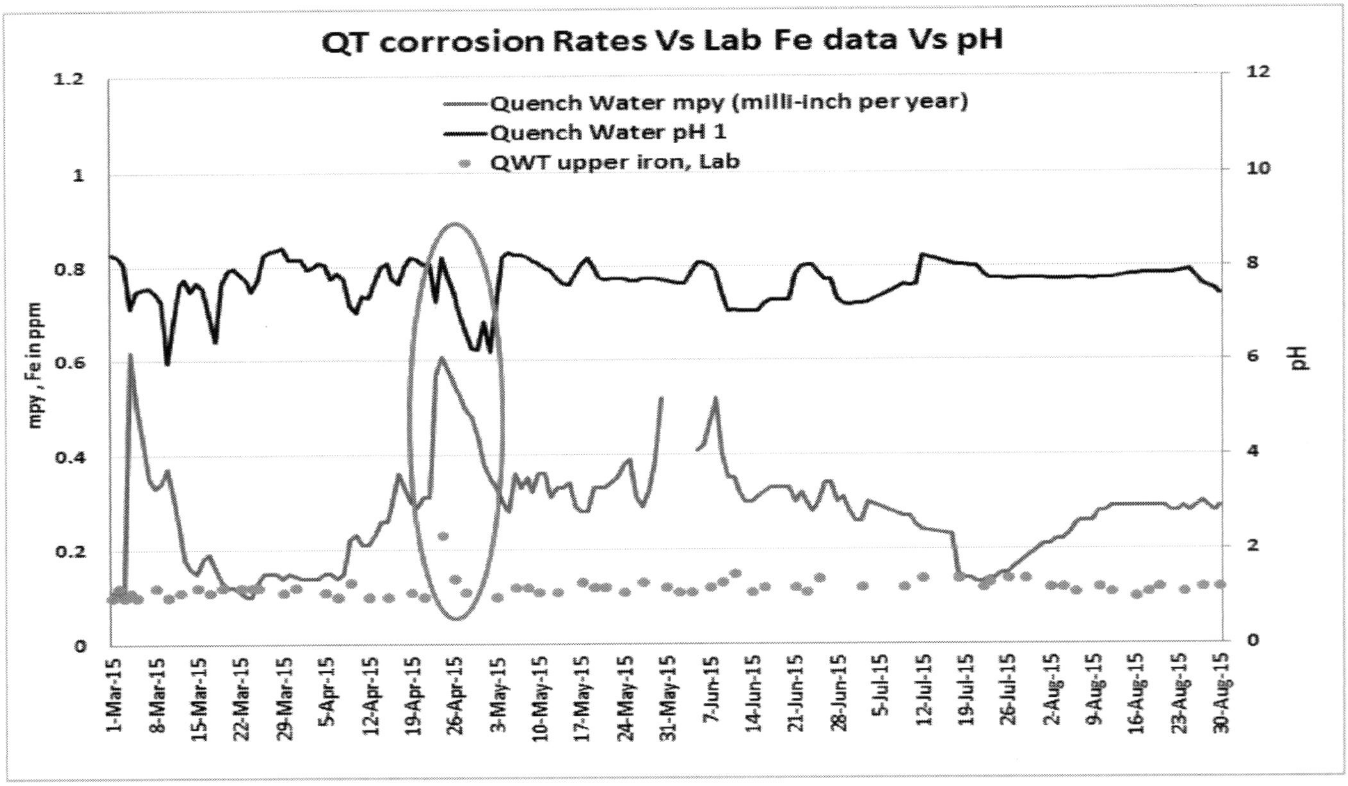

Figure 12 Corrosion Measurements in QWT

The quench tower corrosion rate was also analyzed and is shown in Figure 12, showing that a seemingly short (as measured by spot check analysis) drop in pH can drastically increase the corrosion rate and corresponding iron losses.

This data shows the benefit of continuous online monitoring to provide more information on the system and advise faster decisions, but the automation of reactive chemical dosage adjustments is what truly helped to shorten the feedback loop and provide the long term value to the assets.

Case Study 2: Commercial Automation Unit Success

Background

Based upon the results of the technical trials, an improved version of the dilution steam system automation technology was commercialized. Another commercial user of this technology is a different Middle Eastern ethylene unit that utilizes ethane only as the feedstock. With the limitation on the dosing points for neutralizers in the system, the corrosion previously found at the process water stripper was being controlled by neutralizer addition upstream in the quench system. While this was not optimal, the ability to correct course more quickly could improve the operation with this configuration. Increased fouling in the DSG reboilers was seen due to huge pH swings upstream in the Quench Tower. These pH swings in the quench system led to an increased disturbance in the water quality downstream, which impacted the DSG reboiler run lengths. Based on the various discussions with the unit engineers, it was agreed to have more focused monitoring on the system.

This focused monitoring required more manpower for sample collection and an increased workload on the lab for the analysis, along with increased potential for exposure. The Key Business Driver for the ethylene producer was the improvement in productivity, and aligning with Safety, Healthy, and Environmental standards.

Initial Trial Period

Initially, the automation technology was granted a 6-week trial to prove the efficacy on the quench water system. Handheld pH measurements were used during the trial run to compare with the automation technology unit readings. The 6-week reliability trial criteria and results are captured in Table 2. This innovative technology not only provided a single solution for multiple parameters but also reduced the manual work involved in collecting the samples, controlling the pH, and tracking the system changes thereby significantly reducing the risk of any kind of exposure. Over the run, it also helped to understand the system operations and helped in taking proactive actions to avoid deterioration of the system.

Table 2 Trial Criteria and Results Summary

Trial Criteria	Delivered
+/- 0.5 pH	+/- 0.3 pH (all data points)
90% of the time	91% (all data points)
Statistical Data after Handheld Outliers Removed	
Time within +/- 05=.5 pH	100%
Percent Error	2%
pH Deviation	+/- 0.1 pH

The trial run proved that the technology for dilution steam can reliably monitor pH and match with handheld pH measurements within +/- 0.5 pH units at least 90% of the time when compared to the actual lab data. This was similar to the results of the trial conducted with the beta unit. The trial period was extended based on these results, as the ethylene producer began to utilize this data to optimize their operations. A snapshot of this data is shown in Figure 13.

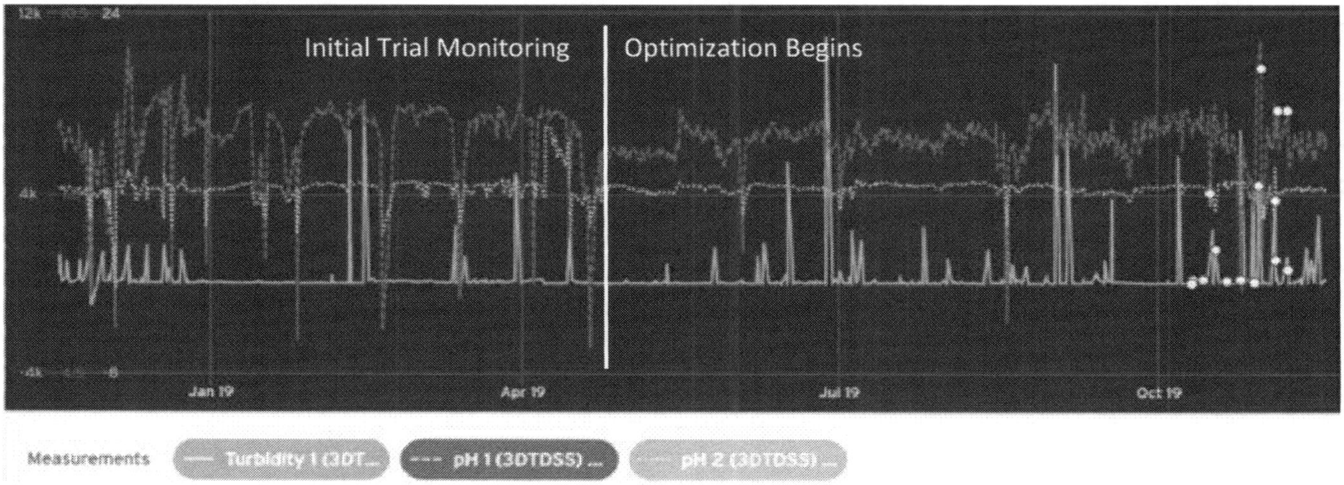

Figure 13 Online Monitoring pH and Turbidity Data

Automation Integration & Benefits

Consistent pH with Turbidity data over one year period proves that the probes can withstand the fluctuations of the process water due to operational changes, and that with minimal maintenance and cleaning of the probes reliable data can be captured every 10 minutes over long periods of time to increase the data and subsequent system information available to any ethylene producer with concerns about the reliability and maintenance of their DSS.

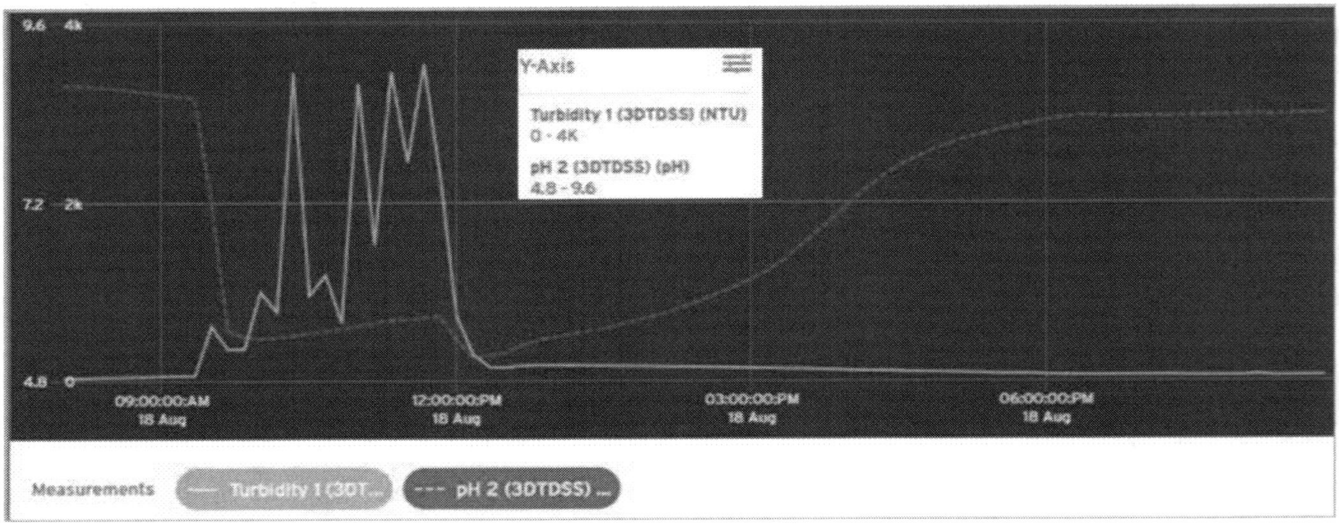

Figure 14 pH and Turbidity Data from an Operational Upset

Apart from regular monitoring, the unit also helped in other proactive ways to make sure that the system parameters were kept in check during variabilities in the unit operations. One such event where the pH rapidly dropped and the process water turbidity increased to 150 NTU was captured at the early stages, if this had been prolonged the low pH could have caused more corrosion damage and the increase in turbidity could have turned into a larger upset condition (Figure 14). However, neutralizer addition was increased and the system was brought back under contol in less than 24 hours, with the turbidity recovering even more quickly in only 3 hours as the pH began to rise to a safer range.

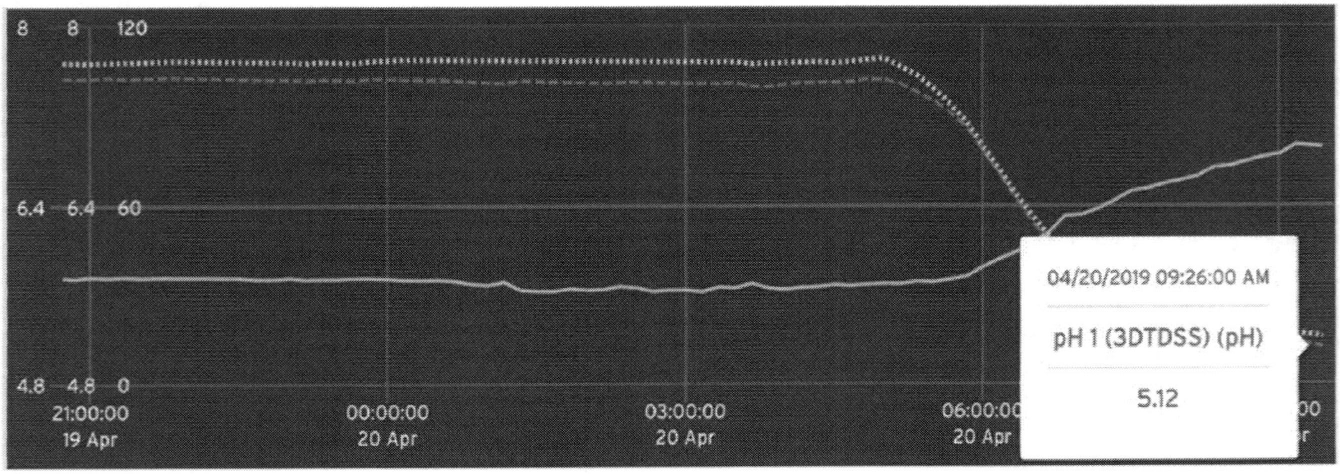

Figure 15 pH and Turbidity Data from another Operational Upset

With a repeated similar event (Figure 15), it became clear that during furnace pre-sulfiding with higher DMDS injection rates, there would be a corresponding dip in pH in the quench and process water units as well, leading to these excursions from stable operation and impacting corrosion, emulsions, and downstream fouling. As a corrective measure, an increase in the neutralizer addition at the quench system was put as a standard operating procedure to to make sure that the pH remains within the recommended band throughout furnace changeovers and thereafter. This action, albeit still manually reliant, helped the operators to be more proactive to preempt these pH swings in the future – beginning the jounrey to turn the lagging feedback control into more leading and predictive control measures.

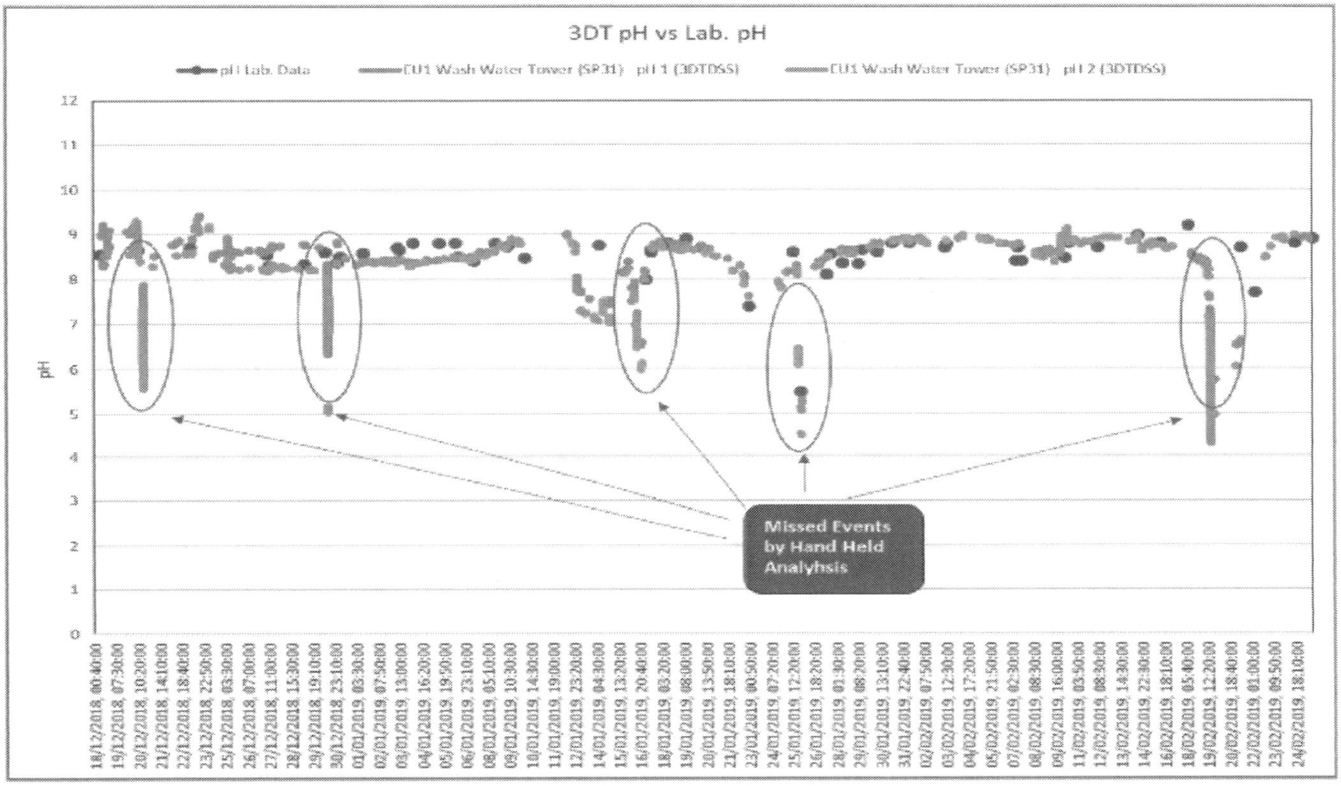

Figure 16 pH Excursions Identified by Online Analysis

Additionally, a comparison of the periodic handheld pH readings with the online monitored pH shows that there were several pH excursions that would have been missed entirely or not corrected as quickly without the automation technology, Figure 16. These excursions were caught by the monitoring system, engineers were notified by email of the alarm, they were able to take corrective action, and then able to quickly restore the pH in the system, usually before the next shift began. This was done before the unit was operating in control mode, so the feedback cycle would have been shortened even more if the automation unit had been controlling the neutralizer addition injection pumps. Benefits like these are why this ethylene producer now relies heavily on the automation technology to monitor and help them maintain the integrity of their Quench and Dilution Steam System.

Conclusion

Additional information is always helpful in troubleshooting: whether a spike in conductivity indicates more free caustic in the system, a pH dip corresponds to the addition of a new recycle stream, or an increase in turbidity ties to reduced heat transfer in downstream exchangers. These are all parameters that this unique compact automation technology can measure: pH, conductivity, turbidity, and corrosion rate.

When 90% of the corrosion happens in only 10% of the operating time, and is easily missed with spot monitoring, by adding continuous reliable online monitoring to the equation, there is visibility into that troublesome 10% and the opportunity to correct it before corrosion and fouling can impact the system. To effectively manage the entirety of the quench and dilution steam system and avoid corrosion, fouling, and emulsions to increase the reliability and decrease the maintenance of the unit, it is crucial to have a comprehensive and reliable monitoring program. , This innovative automation technology, shown in Figure 17, has been demonstrated to provide that capability.

Figure 17 Views of automation technology, showing array of measurement probes and sensors

References:

1. Bradley, Michael E., Michael J. Murcia, Daniel E. Schwarz, Mita Chattoraj (2015), "Fouling Reduction Device and Method," *United States Patent Office*, US9032792 B2.
2. Arnst, Dr. Theodore C., Debby Rossana (2019), "Tutorial: The Fundamentals of Dilution Steam Generator / Quench Water System Fouling," *AIChE Spring Meeting, Topical: The 31st Ethylene Producers*.

AIChE Paper Number: 123g

Creep Strength Enhanced Ferritic Steels: Why They're Great and What Can Go Wrong

Wendy Weiss
Associate
Structural Integrity Associates, Inc.

Terry Totemeier, PhD
Associate
Structural Integrity Associates, Inc.

Ben Ruchte
Consultant
Structural Integrity Associates, Inc.

Prepared for Presentation at the 2020 Spring National Meeting
Houston, Texas March 29-April 1, 2020

AIChE and EPC shall not be responsible for statements or opinions contained in papers or printed in its publications

Creep Strength Enhanced Ferritic Steels: Why They're Great and What Can Go Wrong

Wendy Weiss
Associate
Structural Integrity Associates, Inc.
Terry Totemeier, PhD
Associate
Structural Integrity Associates, Inc.
Ben Ruchte
Consultant
Structural Integrity Associates, Inc.

Abstract:

The steam operating conditions in both petrochemical plants and power generation facilities have increased through the years. Recent ethylene plant projects have been designed and built with steam pressures as high as 1700 psig/11.7 MPa and superheat steam temperatures of 960°F/515°C. The power industry has long since moved past ethylene plant steam pressure and temperature levels to supercritical steam systems that operate at pressures above 4400 psig/30.3 MPa and temperatures above 1100°F/600°C. For these systems, creep strength enhanced ferritic steels (CSEFs) and austenitic steels are used at the higher-temperature locations. However, CSEF steels such as Grade 91 have also been increasingly selected as materials for steam piping in new construction projects for petrochemical plants, despite the lower temperatures and pressures where improved creep resistance is not necessarily required.

These steels are based on the conventional chrome-moly low alloy steels (e.g., 2¼Cr-1Mo and 9Cr-1Mo) but utilize small additions of carbo-nitride formers such as vanadium and niobium to provide enhanced creep strength. Due to their normalized-and-tempered heat treatment condition, CSEF steels have considerably higher strength than conventional low alloy steels at all temperatures. This allows less material to be used (e.g., thinner walls, less weld metal, fewer supports, etc.), often making CSEF steels more cost-effective than low alloy steels even outside the creep range.

The power generation industry has routinely been using CSEFs since the 1980s, and several issues have been encountered in the transfer of these laboratory-developed steels to the real world of steam piping and headers. Unlike carbon and low alloy steels, the improved properties of Grade 91 and related CSEF steels rely on achieving and maintaining a specific martensitic microstructure. The martensitic microstructure is attained through very specific and controlled temperatures and cooling rates during the steel processing. Virtually any incident during manufacture, construction, or operation that disrupts this microstructure can compromise the material's integrity and prevent it from achieving its designated properties.

Drawing on experience from the power generation industry, including work done with and by the Electric Power Research Institute (EPRI), this paper will present some of the history and background of these steels and examples of current applications and issues.

Introduction

Creep-strength enhanced ferritic steels (CSEF) are alloys that develop high strength through heat treatment - a rapid cooling or quenching to form martensite or bainite, followed by a temper to form the proper microstructure and improve ductility. As a result, these alloys have a much higher allowable stress than conventional low-alloy steels, which means that thinner sections provide adequate strength for high-temperature service. Grade 91 steel with a nominal 9% chromium – 1% molybdenum composition, including ASME SA-335 Grade P91 pipe and SA-213 Grade T91 tube, is the most commonly used CSEF steel in the power and chemical processing industries, and it is the focus of this paper.

The history of Grade 91 steel began in the late 1970s when 9 to 12 percent chromium alloys were being studied by Combustion Engineering and Oak Ridge National Laboratory to develop improved steels for the liquid metal fast breeder reactor program (1,2). The reactor program did not come to fruition in the US, which shifted the attention for the alloy's use to boilers and pressure vessels in the fossil power generation and petrochemical areas. Grade 91, which is a modification of the 9Cr-1Mo alloy identified as Grade 9 in traditional ASME specifications, emerged from the reactor program. The changes to the standard 9Cr-1Mo alloy that resulted in the development of Grade 91 steel included, among other things, controlled additions of vanadium, niobium, and nitrogen. With the addition of these elements, the steel not only had a lower thermal expansion, higher thermal conductivity, and improved oxidation resistance compared to lower chromium alloys, it also had a substantial increase in creep-rupture strength.

Composition and Microstructure

The properties of CSEF steels are critically dependent on strict control of composition, steelmaking, processing, and heat treatment all resulting in *one specific and unique* microstructural condition. Lack of control of these factors can lead to compromised strength, ductility, or both. In the case of Grade 91, a tempered martensitic microstructure is a requirement, but the simple achievement of this structure is not sufficient to ensure the superior creep-rupture properties of the alloy. The formation of a stabilizing network of optimally sized and distributed, sub-microscopic, temper-resistant precipitates during original fabrication is the foundation for the enhanced material properties. The additions of alloying elements such as vanadium, niobium, and nitrogen result in the precipitate formation. In the absence of these precipitates, the material will not achieve the expected level of improved properties even with an initial tempered martensitic microstructure. Figure 1 shows a proper Grade 91 microstructure based on optical microscopy and transmission electron microscopy (TEM); the stabilizing precipitates are only visible at the very high TEM magnification.

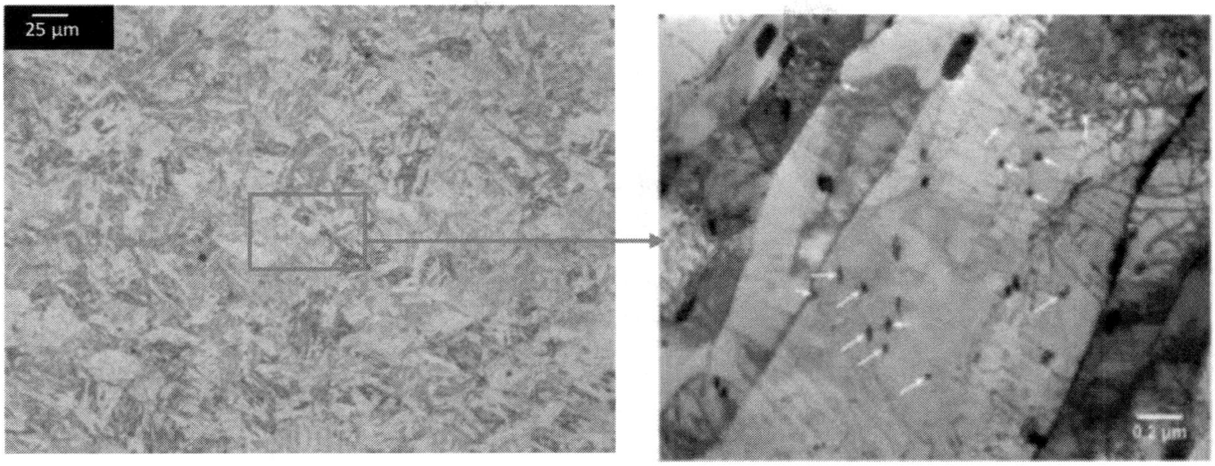

Figure 1. Optical metallographic image showing normal martensitic microstructure and TEM image showing precipitates in the sub-structure (right image from Ref. 3).

In addition to the improved elevated temperature properties, strict controls on composition have been shown to improve other areas of the material performance. For example, there is evidence indicating that reducing the amount of nickel specified for Grade 91 steels from 0.40 to 0.20 weight percent provides steel that is more resistant to excessive tempering. Current research has also shown that controlling the material composition to the levels specified in Class 2 (ASTM A335 P91 Type 2) requirements improves cavitation resistance, which is of particular concern for components operating at higher temperatures. (4)

Processing, Fabrication, and Heat Treatment

The formation of the necessary microstructure is dependent on composition and heat treatment. Proper fabrication of Grade 91 components requires normalizing at temperatures between 1040 and 1080°C (1900 - 1975°F) and tempering between 730 and 780°C (1345 - 1435°F). Furthermore, specific cooling rates are required after normalizing to ensure the formation of a martensitic microstructure and stabilizing precipitates. Welding and the associated post weld heat treatment temperatures must also be properly controlled, as shown in Figure 2. Ultimately, properly manufactured Grade 91 material should have hardness values between 190 and 250 Brinell for base metal.

Because Grade 91 requires strict adherence to the prescribed heating and cooling requirements to form the proper structure, any action that either prevents the initial formation of the stabilized sub-structure or that disrupts that sub-structure after its initial formation will adversely affect the material's long-term performance at elevated temperatures. The essential stabilized sub-structure either can fail to form properly or can be disrupted after its formation in several ways.

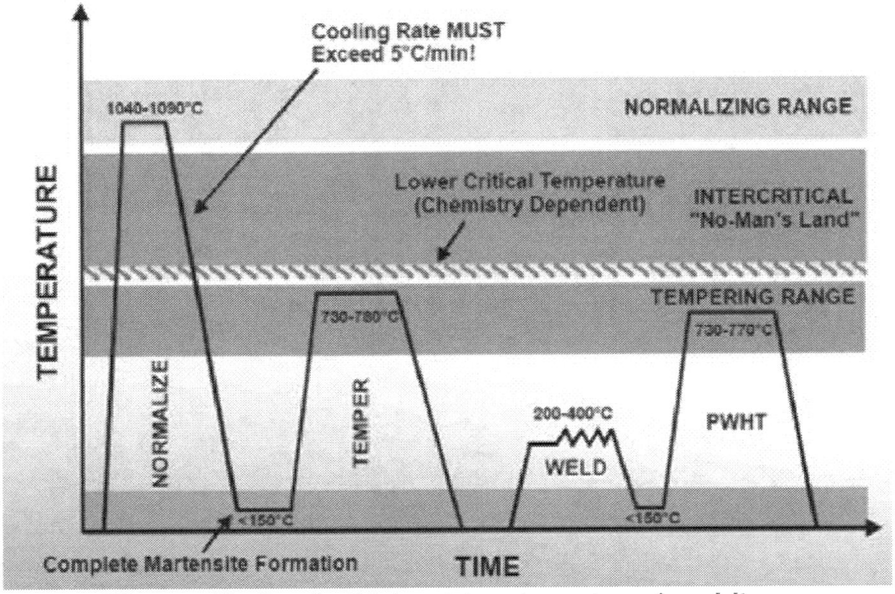

Figure 2. Grade 91 heat treatment and welding.

First, a heat of Grade 91 material with an insufficient amount of active nitrogen, due for example to an unfavorable ratio of nitrogen to aluminum, will not form the desired level of stabilizing precipitates; it can still develop a "normal" tempered martensitic structure and hardness level, but the long-term creep strength of the material will be adversely affected due to the shortfall in precipitates.

For a similar reason, if the original heat treatment for a heat of material is above the upper critical transformation temperature (5) but below an effective solutioning temperature for the precipitate-forming elements, sufficient amounts of vanadium or niobium will not be available in the solution to form the required number of precipitates during the tempering phase of the heat treatment. Again, while a "normal" tempered martensitic structure and hardness level will form under these conditions, the material will not exhibit the improved properties due to lack of the stabilizing precipitates.

Finally, and probably most importantly with regard to the condition of installed Grade 91 components, if the material is heated above the lower critical transformation temperature (5) at any time after the initial normalizing and tempering heat treatment, the critical sub-structure may become disrupted and the enhanced properties deteriorated. Exposure to temperatures above the lower critical temperature may not result in evidence of the overheating in either hardness values or in the microstructure at an optical microscopic level of magnification.

Translation to Performance

When Grade 91 components are properly manufactured and installed, they offer several benefits including higher allowable stress for a given temperature compared to traditional low alloy steels, lower coefficient of thermal expansion than austenitic stainless steels, and the potential for lowered thermal-fatigue cracking susceptibility because of thinner walls. Figure 3 shows the substantial savings in wall thickness afforded by using Grade 91 compared to Grade 22 for a given set of design conditions.

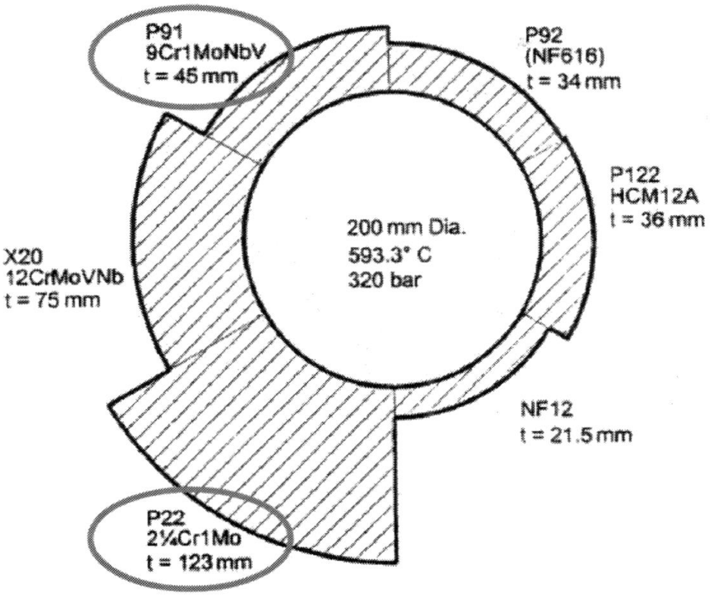

Figure 3. Required component thickness for Grade 91 compared to Grade 22 and other low-alloy and CSEF steels (from Ref. 6).

The ability to use thinner components allows for additional material savings, such as the use of less weld metal during fabrication and installation and the need for fewer supports in piping systems, which in turn, lead to lower costs. Cost savings are particularly notable when service temperatures exceed 540°C (1000°F) and the improved high temperature properties become most important, as shown in Figure 4.

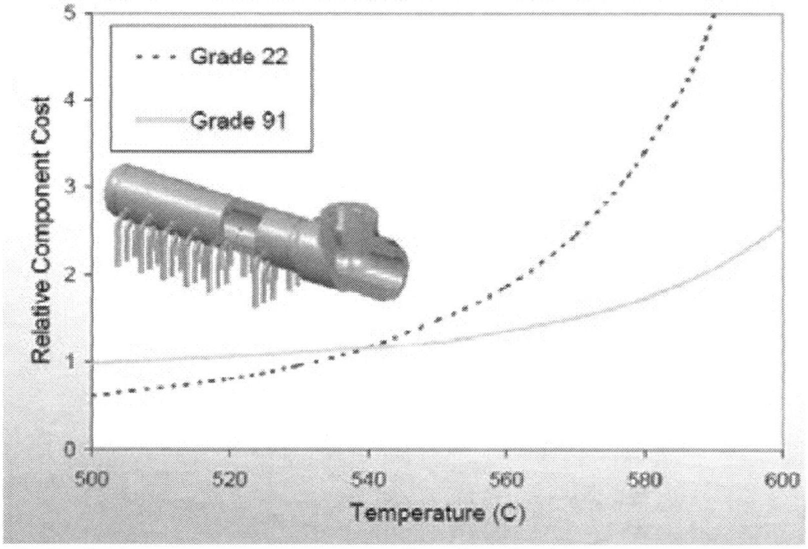

Figure 4. Relative component cost for Grade 91 compared to Grade 22.

What Can Go Wrong

Based on the previous discussion, it should come as no surprise that Grade 91 components can experience a number of difficulties while in service. Problems with the material can form during almost any step of the manufacturing and installation processes. The following "industry issues" and case studies are relevant to the ethylene industry.

Inadequate Cooling Rate in Normalizing – Soft Material

Improper heat treatment during manufacturing can result in Grade 91 material essentially having properties similar to those of Grade 22 with the optically observable microstructure being ferritic rather than martensitic and the hardness values being below the specified minimum. The swollen P91 main steam pipe section bend shown in Figure 5 was found to be in the "soft" condition, resulting in a steam leak after only five years of operation. Field examination of the component revealed that the pipe spools upstream and downstream of the bend and the associated field welds were in satisfactory condition. Laboratory examination of the component identified an average hardness of 160 HB in the swelled area compared to 200 HB outside of the swelling. The microstructure consisted patches of alloy carbides in a ferrite matrix within the swollen area and tempered martensite outside of the swelling (Figure 6). Both the low hardness and ferritic structure are characteristics of Grade 91 in the "soft" condition, which exhibits reduced high temperature properties.

The fact that the low hardness values were confined to the bent spool piece and that the field welds and pipe spools on either side of the bend were in good condition indicated that improper processing led to low hardness and improper microstructure. When a large portion of a component is affected, a situation where an inadequate cooling rate was used as part of the normalizing heat treatment is often the culprit. This can occur if the parts are furnace cooled, if parts are stacked to cool, or if forced air cooling is not used for thick components. Hardness testing beyond that specified in fabrication codes may be necessary to identify areas of soft material.

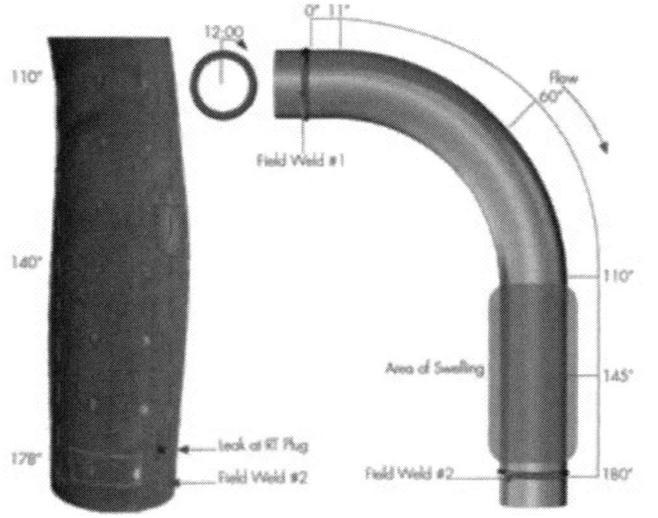

Figure 5. The swollen portion of the main steam pipe bend (left) and a schematic depicting the overall bend.

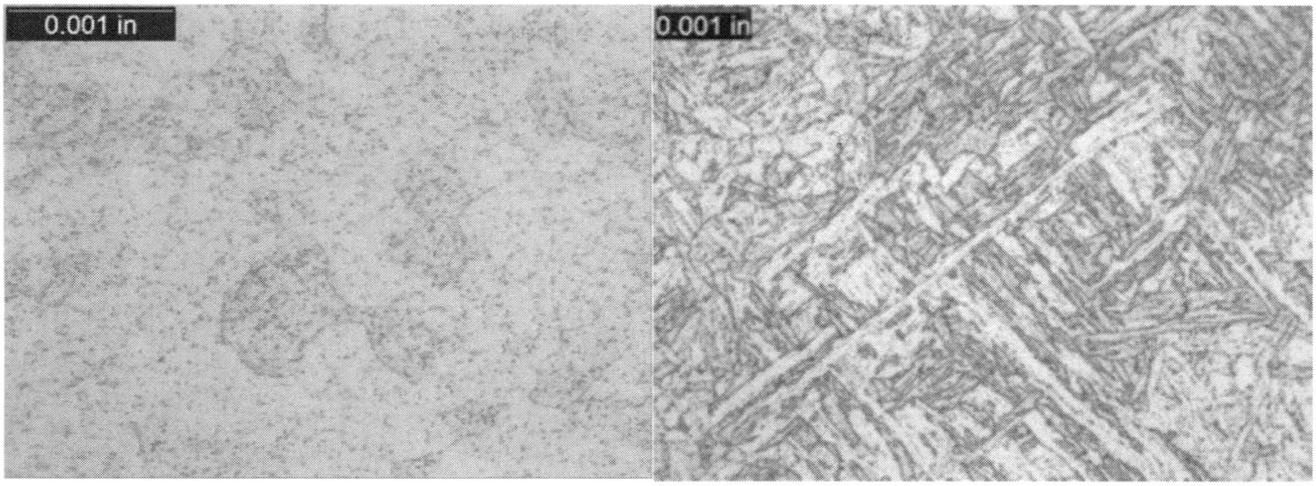

Figure 6. Comparison of the microstructure observed within the swollen region (left) with the microstructure observed on the opposite end of the bend.

Material Left in the Hardened Condition

CSEF steels, including Grade 91, are susceptible to stress-corrosion cracking (SCC) in what would otherwise be considered "benign" environments when left in the untempered or under-tempered condition. SCC generally requires three factors to be present to occur including a susceptible material, sufficient stress, and a specific environment. Grade 91 is susceptible to SCC when left in a hardened condition (greater than ~350 Brinell) and exposed to moisture, so components should be tempered directly after normalizing and undergo post weld heat treatment following welding. If components must be stored in a hardened condition, they should be maintained in a dry environment such that even condensation cannot occur.

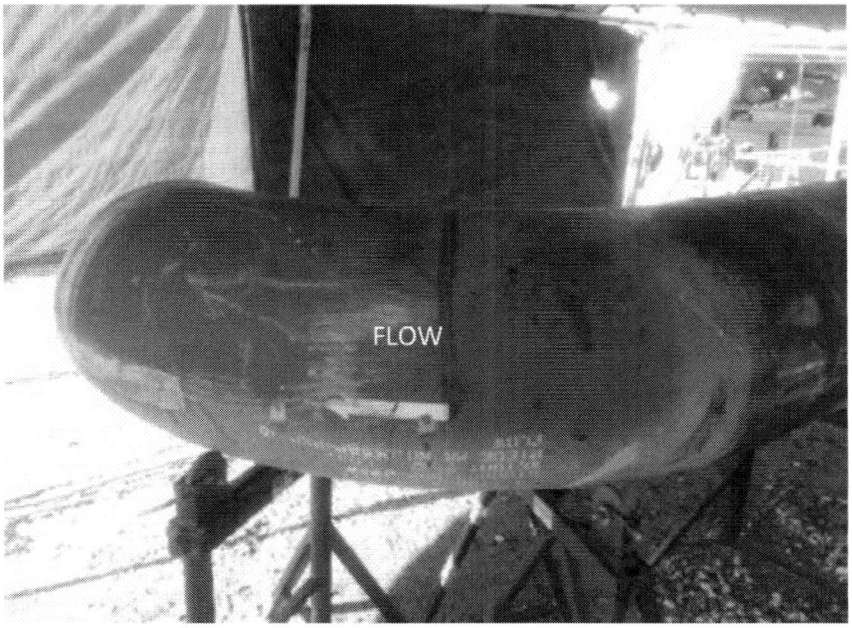

Figure 7. Elbow prior to installation but after field welding.

The elbow shown in Figure 7 had not yet been installed when cracking was identified at both the field-welded and open ends. Field evaluation of the cracks revealed they were intergranular and oxide-filled (Figure 8). The hardness values in the HAZ/base metal on the elbow ranged from 209 to 239 HB. Even though the hardness values were within the normal range for Grade 91, the cracking present in the elbow was characteristic of SCC. The presence of oxide within the cracks indicated the cracking was present when the material was heated to a high enough temperature to form the oxide, such as during tempering or PWHT. Since cracks were present at both the welded and unwelded ends of the elbow, the cracking most likely formed prior to the field welding. The oxide filled cracking at both ends of the elbow and the hardness values within the normal range indicated that the components were heat treated after the cracks formed.

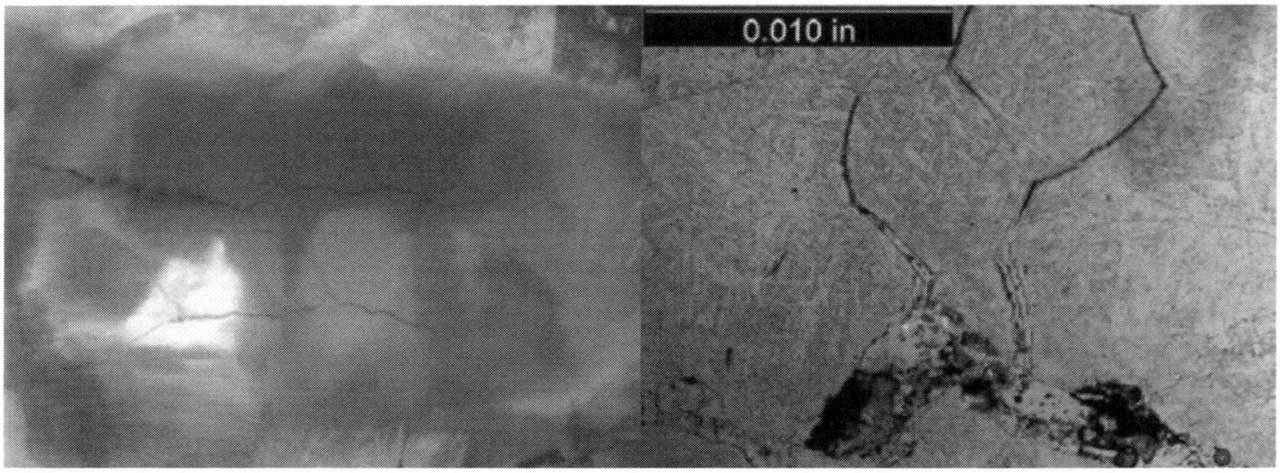

Figure 8. Cracking typical of that observed at both ends of the component shown after preparation for replication (left) and from a replica (right).

Dissimilar Metal Welds

Dissimilar metal welds (DMWs) between Grade 91 and ferritic or stainless steels are common, and service experience has revealed some unique characteristics and failure mechanisms in both types of DMWs.

There are several potential issues that must be considered in the design and fabrication of DMWs between Grade 91 to other ferritic steels such as Grade 11 or Grade 22. First, the diffusion of carbon atoms from the material with lower chromium content to the Grade 91 material will occur during any post-weld heat treatment or service at temperatures above approximately 490°C (900°F). This diffusion creates a carbon-enhanced zone in the Grade 91 material and a weak decarburized layer in the lower chromium material. The decarburized zone can result in premature cracking and failure (Figure 9).

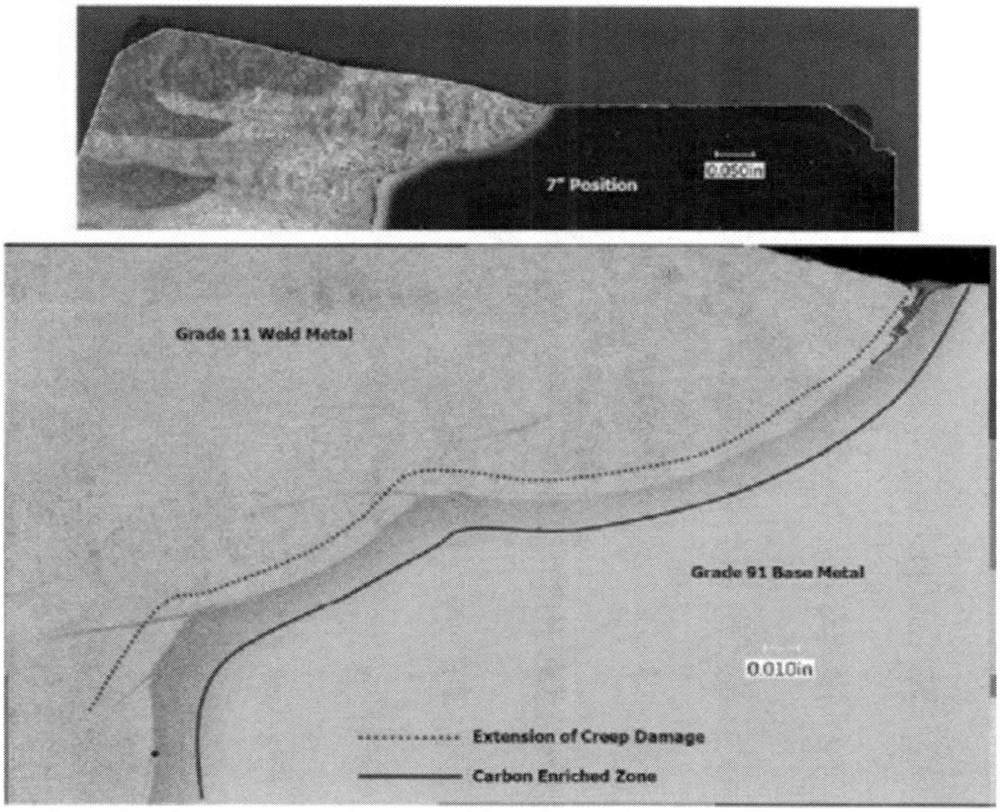

Figure 9. A Grade 11 to Grade 91 weld with cracking in a carbon depleted zone.

A second issue is the significant difference in strength between the low alloy steel and Grade 91 materials, which can result in strain localization in the weaker material if the joint is not designed correctly. The strain localization would be aggravated by any decarburization at the joint. Lastly, selecting a PWHT temperature is not straightforward since the code-required ranges for the different materials do not always overlap.

DMWs between Grade 91 and austenitic (e.g. Type 304 stainless steel) are present in both thin and thick sections. Cracking has been found in thinner sections with fillet welds, including tube attachments, thermowell fittings, and steam sampling line fittings. Cracking is typically along the Grade 91 to nickel-base filler metal fusion line; Figure 10 shows an example. Thermowells are of particular concern because of the significant projectile safety hazard should a through-thickness crack form.

A number of premature failures have occurred in thick section DMWs, and the metallurgical root cause is not fully understood. It is not currently possible to make life predictions for those welds based on their nominal service loading, as early failures have occurred in welds operating at very low service stresses. An early example of a premature thick section DMW failure was a complete separation at a weld where Grade 91 pipe was welded to a stainless steel in-line flow meter (7). The cracking in this failure was creep related with cavities present along the fusion line ahead of the crack tip (Figure 11). While the use of nickel-base filler metals and "buttering" of the Grade 91 material with a nickel-base filler can help minimize carbon migration issues, metallurgical changes still occur over time.

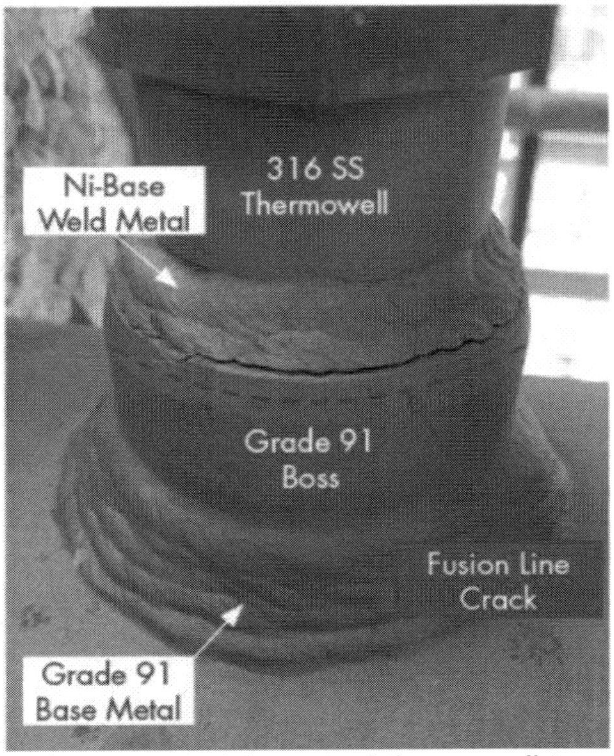

Figure 10. Fusion line cracking at a stainless steel thermowell in a Grade 91 piping system.

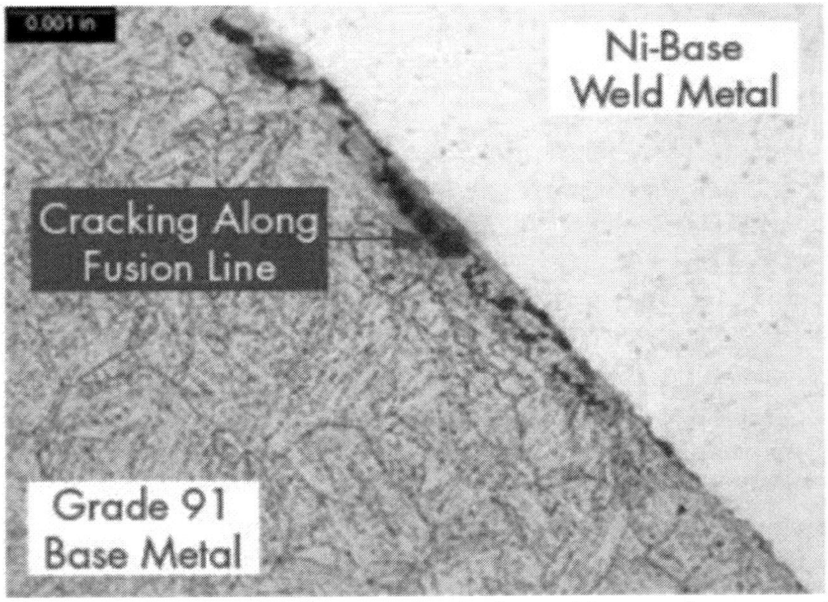

Figure 11. Optical photomicrograph showing typical cracking and creep voids along a Grade 91 to nickel-base weld metal fusion line in a thick-section DMW.

Ultimately, any Grade 91 to stainless steel DMWs should be identified and replaced. If replacement is not an option, then frequent monitoring and inspection should be performed.

Fabricated Laterals, O-lets, Wyes, and Tees

Branch connections between large bore pipes (including headers and manifolds) are often fabricated with a reinforced branch commonly in the form of a "catalogue" (standard size) fitting, such as an 'o-let'. These fittings are typically thicker than the pipes in which they are installed to provide compensating reinforcement for the piping run penetration. Full reinforcement is often not achieved because insufficient weld metal was deposited, the fitting was incorrectly heat treated, or the fitting was simply undersized. Insufficient reinforcement results in higher local stresses and, particularly in the case of CSEF steels, creep cracking in the weld HAZ (Type IV cracking). Damage is therefore most likely to occur in fabricated branches that operate in the creep range.

Damage is primarily in the form of creep cracking at the toe of the weld on the main run side of the connection, as shown in Figure 12. The susceptibility to damage early in life (in some cases before 50,000 hours of service) has been widely reported. As early as 2008 (8) a warning was issued by an architect engineering company to advise of the known problems. Despite that warning, use of these fittings with their associated inadequacies remains prevalent.

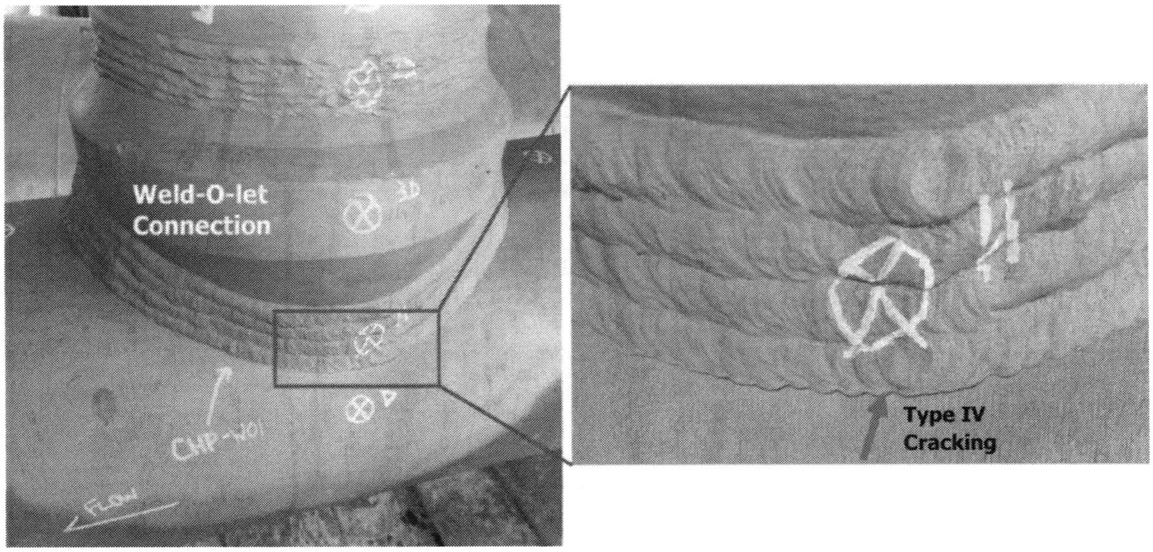

Figure 12. Example of cracking along a weld-o-let connection.

Several steps can be considered to mitigate damage in these types of joints. Weld build-up at the saddle, and in some cases the crown, can be applied to improve the strength of the connection. Finite element analysis can be used to estimate the amount of weld build-up required to appropriately decrease stresses; however, the amount of weld buildup necessary is very often impractically large. Re-normalizing and tempering the component after fabrication can minimize the detrimental effects of the HAZ and reduce the likelihood of Type IV cracking. Replacing (or specifying) fabricated joints with forged fittings, which eliminates welding at the branch connection and provides a more balanced reinforcement, is the best method of dealing with these components.

Longitudinal Seam Welds

The majority of Grade 91 components are fabricated to seamless specifications (ASME SA-335 for pipe, for example) however, in some cases components are fabricated from SA-387 plate, or to a seam-welded pipe specification such as SA-691. SA-234 allows for seam welded Grade 91 fittings. Seam welds in Grade 91 components should receive a normalize and temper heat treatment after welding, which minimizes the detrimental effects in the heat affected zone and substantially decreases the risk of failure. Even if the specification requires heat treatment after welding, e.g., in SA-691, it is not always done. If the component is fabricated to SA-387 or SA-234 specifications, then it may not be clear if a normalize and temper heat treatment was applied after welding.

Because long seam welds are very susceptible to Type IV cracking in their HAZs and because creep damage cannot be detected by ultrasonic inspection methods before microcracking has formed in Grade 91 welds, verification of the normalize and temper heat treatment should be done by replication on all seam welded components operating at temperatures where creep is a possible damage mechanism. Figure 13 shows a Grade 91 flow meter with cones constructed from SA-387 plate; each cone had two seam welds for a total of four seam welds in the component. None of the seam welds received the normalize and temper heat treatment, which is clear in Figure 14.

Figure 13. Seam welded flow meter.

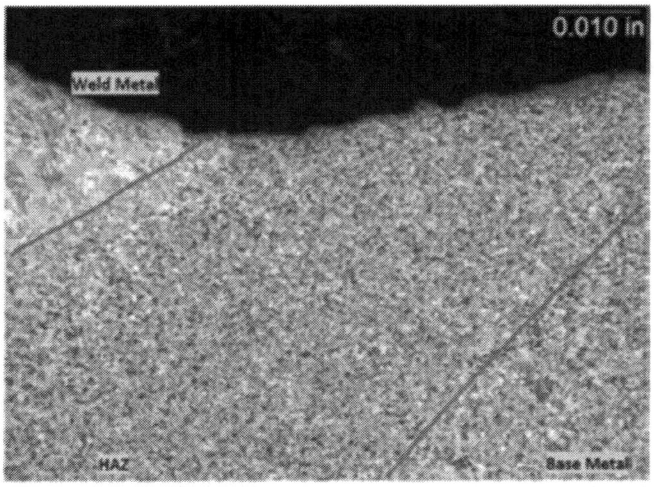

Figure 14. Optical metallographic cross-section through seam weld showing a HAZ, indicating that the weld had not been re-normalized and tempered after welding.

Conclusions

Grade 91 and other CSEF steels are important and useful materials for plant construction. They offer improved properties, especially at higher temperatures, compared to the more traditional low alloy steels. Because of the improvement in properties, less material can be used in the fabrication and installation of components, resulting in lower costs. However, the steelmaking, manufacturing, fabrication, and installation processes must be carefully controlled to ensure proper performance of the material. Careful quality control and detailed verification of these processes should be performed during the various steps of fabrication and erection. Guidance on ordering Grade 91 to ensure proper processing and quality control is provided in Reference 5.

References

1. *Summary of Modified 9Cr-1Mo Steel Development Program: 1975-1985.* Oak Ridge National Laboratory Report ORNL-6303, 1986.
2. V.K. Sikka, C.T. Ward, and K.C. Thomas, "Modified 9Cr-1Mo Steel – an Improved Alloy for Steam Generator Application," in *Ferritic Steels for High-Temperature Applications*, A.K. Khare, ed., Metals Park, OH: ASM, 1983, pp. 65-84.
3. Panait, Clara, Zielinska-Lipiec, Anna, Koziel, Tomasz, Czyrska-Filemonowicz, Aleksandra, Gourgues-Lorenzon, Anne-Françoise, and Bendick, Walter, Evolution of Dislocation Density, Size of Subgrains and MX-Type Precipitates In a P91 Steel During Creep and During Thermal Ageing At 600 °C For More Than 100,000 h. *Materials Science and Engineering: A*, Elsevier, 2010, 527, pp.4062-4069.
4. *The Benefits of Improved Control of Composition of Creep-Strength Enhanced Ferritic Steel Grade 91*, EPRI, Palo Alto, CA, 2014, 3002003472.
5. *Guidelines and Specifications for High-Reliability Fossil Power Plants, 2nd Edition: Best Practice Guideline for Manufacturing and Construction of Grade 91 Steel Components*, EPRI, Palo Alto, CA, 2015, 3002006390.
6. Fleming, A., Maskell, R.V. Buchanan L.W. and Wilson, T., "Material Developments for Supercritical Boilers and Pipework", in *Materials for High Temperature Power Generation and Process Plant Applications*, A. Strang, Editor. 2000, IOM Communications: London. p. 33-77.
7. Matherne, Carl, DeRouen, Dane, and Baxter, Wallace, (2010) "Catastrophic Failure of a Dissimilar Metal Weld in a High Pressure Steam Venturi," presented at the 2010 AIChE Spring National Meeting, San Antonio, TX March 21 – 25, 2010.
8. Accelerated Creep Cracking of Integrally Reinforced Weld-on Lateral Fittings Installed in the Main Steam and/or Hot Reheat Systems, B&V Energy TECHNICAL ADVISORY 2008 – 2.

AICHE PAPER NUMBER: 159b

NATURALLY OCCURRING RADIOACTIVE MATERIALS IN ETHYLENE PLANTS REVISITED

Dr. Debby M. Rossana
Senior Technical Lead
Nalco Water

Dr Phillip James
Technology Director
Evonik Performance Materials GmbH

Justin Harris
Radiation Safety Services Manager
Quantum Technical Services

Dr David Slim
Technology Advisor
Evonik Corporation

Prepared for Presentation at the 2020 Spring National Meeting
Houston, TX (March 29 - April 2, 2020)

AIChE and EPC shall not be responsible for statements or opinions contained in papers or printed in its publications

NATURALLY OCCURRING RADIOACTIVE MATERIALS IN ETHYLENE PLANTS REVISITED

Dr Debby M. Rossana
Senior Technical Lead
Nalco Water

Dr Phillip James
Technology Director
Evonik Performance Materials GmbH

Justin Harris
Radiation Safety Services Manager
Quantum Technical Services

Dr David Slim
Technology Advisor
Evonik Corporation

Abstract

Naturally occurring radioactive materials (NORM) are an occupational health concern in the ethylene plant. NORM (as radon) have been known to be a contaminant of natural gas for nearly 100 years but it was only in 1971 that radon was found to concentrate in the lighter natural gas liquids during processing (ref 1). The presence of radioactivity has not been widely reported in the ethylene industry. A perusal through the EPC archives shows only 6 papers directly related to NORM. A survey carried out in 1994 by Hart, Wolf and Graham (ref 2) looked at 17 US ethylene plants and 53% of the respondents had identified the presence of NORM in their plants. A 2020 survey that looked at 42 ethylene plants shows the number of ethylene plants with identified NORM has increased to 87%. Is this increase due to better monitoring and awareness or could the increase in shale fracking be the root cause?

1 Background

Uranium and thorium and the products of their decay, radium 226 and radium 228, can be found in rock formations where oil and gas are produced (ref 3). Radon gas, which comes from radium, can end up in the natural gas. Therefore, the primary radioactive material that concerns the ethylene plant comes from radon, which can be found in gas feeds such as propane and to a lesser extent ethane (ref 4). The amount of radon in the propane and ethane the ethylene plant receives depends on where the gas originated and how long since it was extracted from the ground. It should be noted that Taylor and Cleghorn (ref 5) measured higher than expected radon in their feeds due to a concentrating effect.

Radon boils between ethane and propane (ref 6) see table 1 and therefore NORM tend to collect in the C2 and C3 areas of the ethylene plant (ref 7), although NORM have been reported in feed pumps, compressors and caustic towers.

Table 1 – key boiling points

	Boiling point °C	Boiling point °F
Ethylene	-103.7	-154.7
Ethane	-88.5	-127.4
Radon	-61.7	-79.1
Propylene	-47.6	-53.7
Propane	-42.2	-43.8

Radon is a colorless, odorless gas, with a half-life of 3.8 days. It decays via alpha particles to polonium to bismuth to lead (ref 8 Hyperphysics site). While radon is likely to be in the gas feeds, because of the short half life of radon, it is usually solid polonium, bismuth or lead that is measured as NORM in ethylene plant equipment. **Figure 1** (ref 9) shows the degradation products of radioactive materials found in rock formations. Radon and its progenies have short half lives, except for lead 210, which has a half life of 22.3 years.

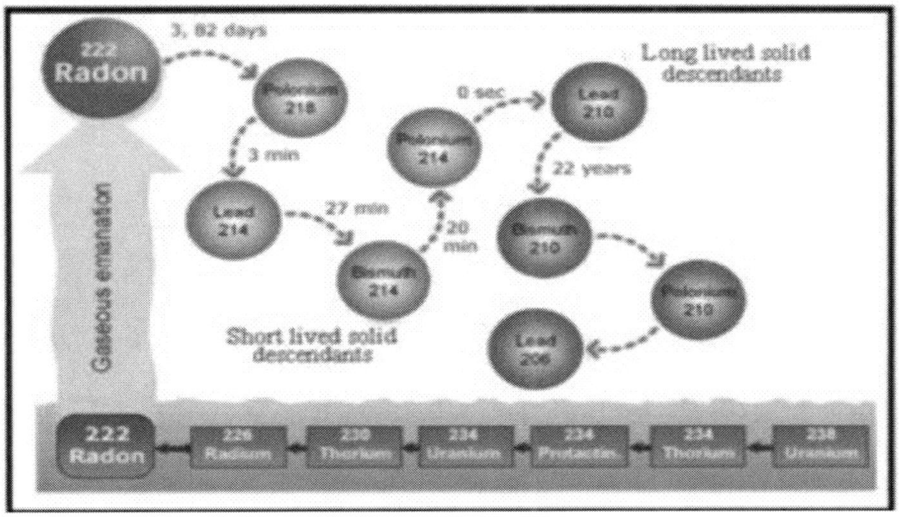

Figure 1: Half-lives of Radon and its progenies

Note: half-life is the average time taken for half the weight of an isotope to disappear: e.g. 1 gram of U-238 becomes 0.5 grams in 4.5 billion years whereas Po-214 disappears almost instantly. **Figure 2** (ref 10) shows the overall time it takes for radon 222 to decay to polonium 218.

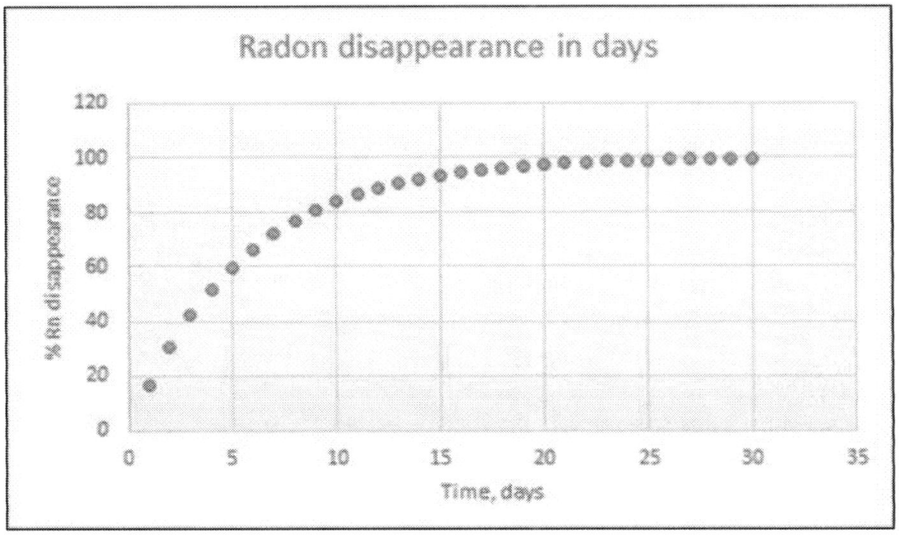

Figure 2: Time for Radon 222 to decay to Polonium 218

As radon is the only gas in the decay pathway, it can escape rock formations with the hydrocarbon gas and depending on how the gas is processed may follow either the ethane or propane. If it follows the ethane, it will enter the steam cracker. If it follows the propane, there are more potential outlets. Propane can be a steam cracker feedstock, go to a propane dehydrogenation unit, or end up in fuel use. For the purposes of this paper we will focus on steam cracking.

Because radon decays so quickly, radioactive contamination of steam crackers is most likely only an issue for plants with a feedstock supply chain from well to plant of less than 1 month. Consequently, it seems unlikely that US shale ethane which is shipped overseas to Europe and Asia will still contain radon by the time it reaches its destination. Instead it will most likely have reverted to solid form, lead 210, as the immediate progenies polonium 218, lead 214, bismuth 214 & polonium 214 are so short-lived. It is expected that the lead 210 will form a solid deposit at the base of the shipping vessel. The risk posed by this material needs further evaluation. Radioactive scales and sludges are discussed in reference 6.

2 NORM in the ethylene plant

As described above, NORM poses the highest risk to steam crackers if it is present in the form of radon gas. Radon is chemically inert, being a noble gas, so it is not expected to react with anything else during its passage through the cracker. The only expected reaction is radioactive decay. The general consensus is that the high temperature to which it is exposed in the cracking furnaces does not affect the rate of decay. After leaving the furnace, the radon will follow ethane and propane through the ensuing units, until reaching the light ends separation unit of the plant. Based on its boiling point (see table 1) it could either end up in the C2 or C3 splitter. In the case of the C2 splitter it will follow ethane and will be recycled to the furnace to complete the loop again. In the case of the C3 splitter, it is likely to follow the propylene and will end up in the propylene product.

Depending on where the radon entering the plant is in its radioactive decay timeframe, will dictate whether it remains as radon during its passage through the cracker. Since it has a half life of just 3.83 days, it is quite likely that at some point it will decay into its immediate progenies. As these are metallic solids, they are much less likely than radon to continue their passage in the gas phase, and therefore will start to form deposits inside the plant. Of these progenies, polonium 218, lead 214, bismuth 214 & polonium 214 are so short-lived that within a few hours, the only remaining species will be lead 210.

If the radon is caught in the ethane recycle loop it will inevitably end up as lead 210 before it can leave the cracker. If it escapes with the propylene, its fate will be decided by how the propylene is handled. If it directly enters a downstream process such as polypropylene, acrylonitrile, acrylic acid or phenol-acetone production, it could also contaminate that process with radiation. If it firstly enters a storage vessel, then the decay may happen at that location, leading to contamination of that vessel.

If the radon in the cracker feedstock has already decayed by the time it reaches the steam cracker, the decay products – primarily lead 210 – may be present as solid deposits at the entry point to the plant.

3 Shale Fracking and NORM

Rock formations contain radioactive materials and radon gas. As fracking increases, there appears to be a likelihood of an increase in NORM in the ethane and propane that is received by the US ethylene plants. Several papers have been published suggesting homes near shale fracking areas have higher than normal levels of radon (ref 11 and 12). In particular, the Marcellus shale play is considered to have higher levels of radioactive materials vs other plays and has been measured at 20 times higher. According to a paper by Frackcident Injury Law (ref 13), the radioactivity is so high that radiation detectors can be used to find deposits.

Figure 3 (ref 14) shows the radon levels measured across the US. The geometric figures superimposed on the map show the approximate locations of shale gas plays. The Bakkan (circle), Niobrara (triangle) and Marcellus (rectangle in the Ohio/Pennsylvania area) are areas of high radon. The odds are greater that the ethylene plants who receive feeds from these areas will receive NORM containing ethane and propane.

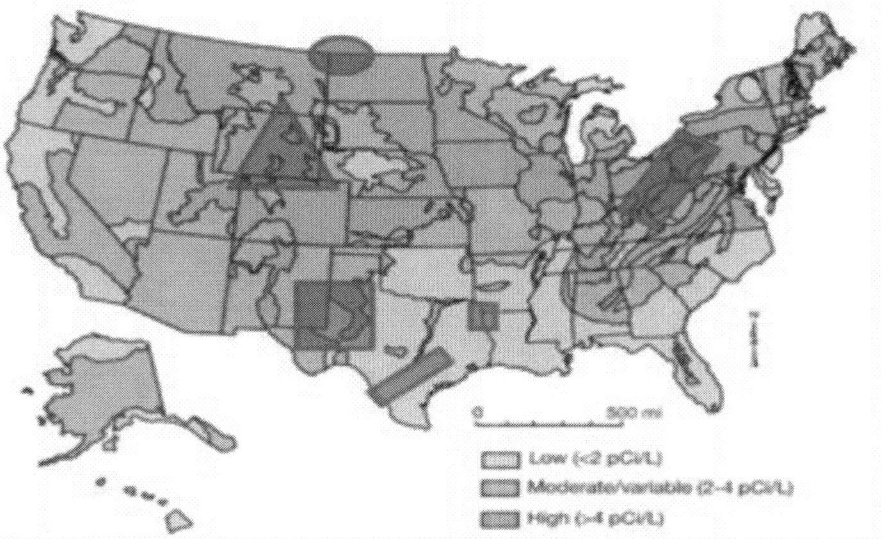

Figure 3: Major shale fracking areas in geometric shapes superimposed on high Radon areas in the US

4 Survey Results

As mentioned earlier, a survey was performed by the authors to determine if the presence of NORM has increased in recent years. All North American ethylene plants were contacted to participate in the 2020 survey. A response rate of 69% was achieved. Of the responses, 6 were new ethylene plants. 2 new plants have already detected NORM in deposits and/or equipment and 4 have not tested yet. In addition, 3 other established ethylene plants responded that they do not test for NORM.

Of the 69% who responded to the survey, 17% do not test for NORM and 83% do test, as shown in **Figure 4**.

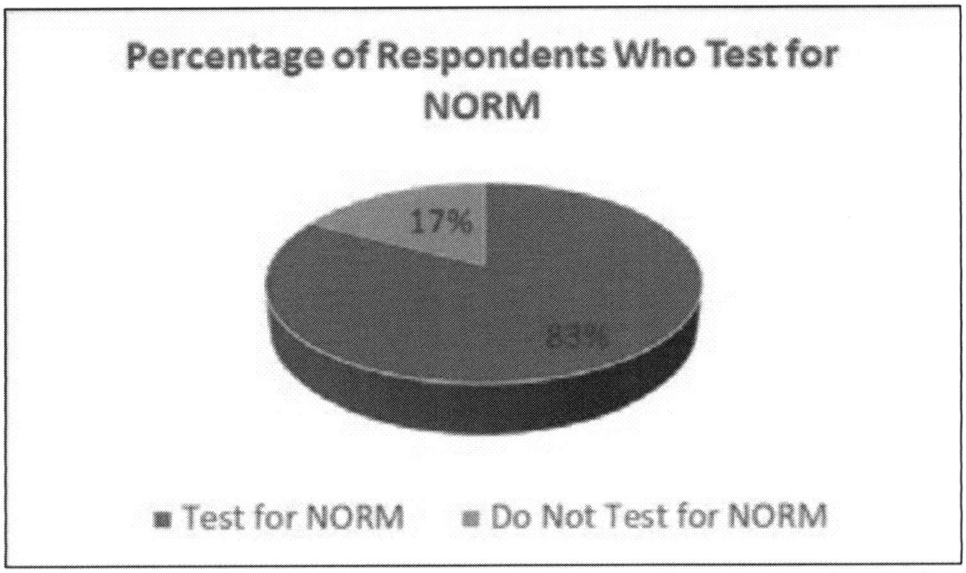

Figure 4: Percentage of respondents who test for NORM in their plants

Of those respondents who do test for NORM, 86% have found NORM in their plant while 14% have not. This is up from 53% detection in 1994, which represents an increase of 160% (**Figure 5**).

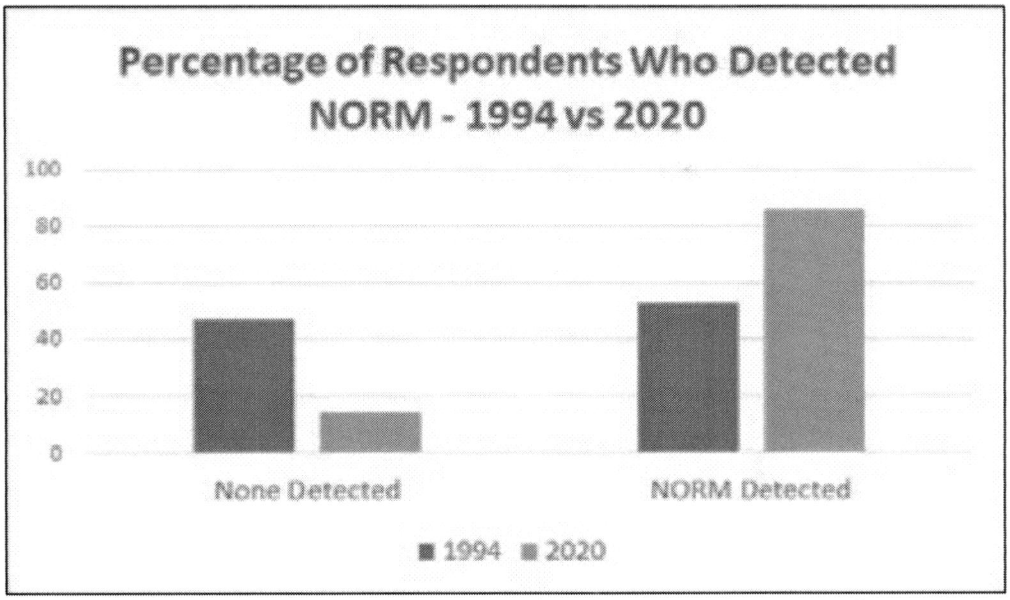

Figure 5: Percentage of respondents who detected NORM in a 1994 survey vs a 2020 survey

Survey data shows that both gas and mixed crackers have detected NORM in their plants (**Figure 6**).

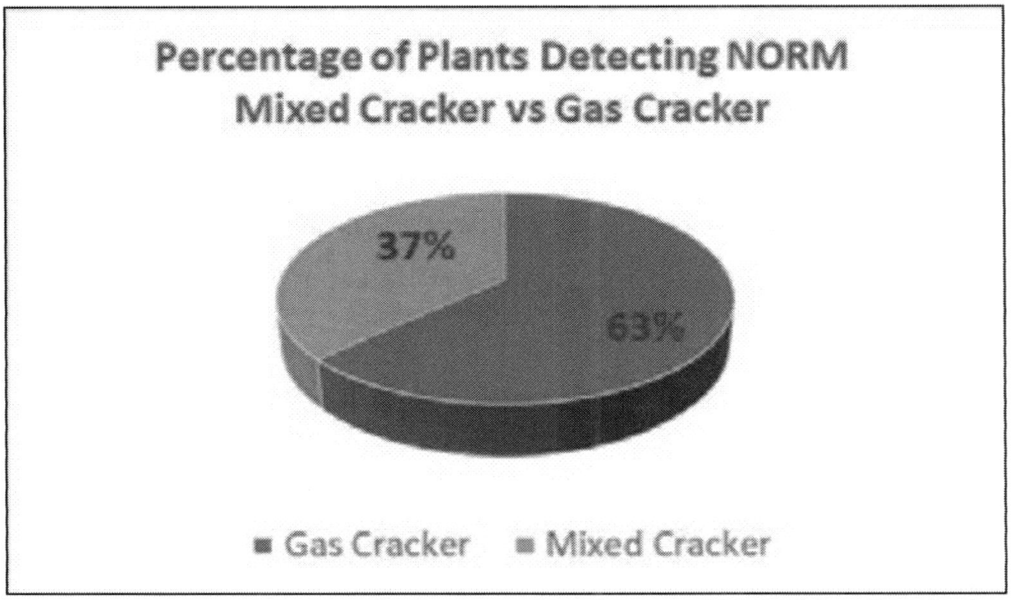

Figure 6: Percentage of plants who detect NORM by feed slate

The survey asked respondents if they felt this increase was due to gas feeds from shale fracking or due to more awareness and better detection. Overwhelmingly, the respondents felt the higher NORM presence was due to better awareness and measurement and not necessarily from shale fracking. A paper by Hlibrich (ref 15), however, suggests this increase is, at least in part, due to shale gas feeds.

Of those who detected NORM in their plant, all found it in the light ends (deethanizer, depropanizer, C3 splitter). Most of the respondents only measure for NORM during maintenance but a number of plants also check their feeds and/or solid waste and some have found NORM in these areas. For some plants NORM have been found in nearly every area of their plant including compressors and caustic towers while other plants see NORM primarily in the light ends towers.

Although most respondents did not know which radioisotopes were detected in the NORM, 2 have identified various amounts of the radon progenies in their plants. Most are only detecting alpha and beta radiation as expected but a few have found gamma radiation as well (ref 15).

The most common measurement tool used by the respondents is the Ludlum produced handheld Geiger counter similar to **Figure 8b**.

The largest variation seen among the surveyed plants was their handling of NORM. Several plants have very detailed procedures with mapping of contaminated areas such as Taylor and Cleghorn discussed in their 1996 EPC paper (ref 5) and others have decontamination stations for workers as they leave NORM areas. Most, if not all, plants use air movers as part of their NORM procedure to blow the NORM out of towers before entering with or without a minimum waiting time before workers can enter. Most require a minimum of a disposable TYVEC™ suit and respirator as shown in **Figure 7** (ref 16).

Figure 7: Proper PPE for NORM exposure

Radioactive waste removal is typically done by an outside company with the proper training to handle these wastes.

5 Handling NORM

During normal operation, there is little concern around exposure to radiation. However, NORM poses a problem to workers particularly during maintenance, waste transport and processing, and decommissioning. However, there may be instances around certain types of equipment (i.e. pumps, filters) where external gamma readings reach a level that need to be posted to maintain compliance with general public dose limits.

Although NORM contamination will be general throughout an ethylene plant, the contamination will usually be greatest in areas of high turbulence, such as in pumps and valves. Thus, the effects of radon progenies are mainly seen in compressors and light ends fractionators. When workers open equipment and vessels, precautions should be taken to prevent contamination. It is recommended that maintenance procedures include the use of respirators, wet grinding, and good hygiene to prevent inhalation of radioactive dust and other materials.

Occasionally, a plant processing ethane and propane is taken out of service and the facility sold or dismantled. Any equipment with internal surface deposits of NORM must receive special consideration when scrapped, sold, transferred or disposed of. In most cases, analysis for lead-210 will be required to verify the extent of contamination and to determine if special handling is required. Care must be used to prevent employee exposure to NORM contamination.

6 Measuring NORM in the Ethylene Plant

External and internal measurements can be performed to measure NORM levels in the ethylene plant (ref 17). External surveys can be done while equipment is running. The primary radioactive component measured externally is the gamma peak from bismuth-214. Internal surveys are done when the equipment is opened for maintenance and inspection. It should be noted that not finding any NORM on the outside of equipment does not mean there are no radioactive components inside the equipment. However, the converse is not true; readings outside of equipment does indicate internal NORM presence.

A means of detecting radioactivity is an electrical device called a Geiger counter. Some examples are shown in **Figures 8a and 8b**. The Geiger counter contains a gas-filled chamber with a thin membrane on one end that allows radiation emitted from radioactive nuclei to enter the chamber and knock electrons off atoms of gas (usually argon). The presence of electrons and positively charged ions causes a small current, which is detected by the Geiger counter and converted to a signal on a meter or, commonly, an audio circuit to produce an audible "click." Detection of alpha and beta radiation is very sensitive to the sample environment, a layer of hydrocarbon on the sample is likely to adsorb the radiation and not be detected by the Geiger counter

Figure 8a – Gamma Scintillator

Figure 8b – Pancake Probe

Internal NORM measurements are done at turnaround or when equipment is opened. During pre-opening purging, some of the radon present will be removed. The solid decay products of radon such as lead-210 and polonium-210 will remain during purging. A pancake probe (**Figure 8b**) is used for checking workers for contamination and for measuring radiation in equipment.

Types of Radiation (refs 18 and 19)

Alpha particles: these are fast moving helium atoms. They have high energy, typically in the MeV (meg electron volt) range, but due to their large mass, they are easily stopped, even by a few inches of air. Alpha particles cannot penetrate most matter. A piece of paper or the outer layers of skin is sufficient to stop alpha particles. Radioactive material that emits alpha particles (alpha emitters) can be very harmful when inhaled, swallowed, or absorbed into the blood stream through wounds.

Beta particles: these are fast moving electrons. They typically have energies in the range of a few hundred keV (kilo electron volts) to several MeV. Since electrons are lighter than helium atoms, they can penetrate further, through several yards of air. Beta particles can be stopped by a layer of clothing or by a few millimeters of a substance such as aluminum. Beta particles can penetrate the skin causing radiation damage, such as skin burns. As with alpha emitters, beta emitters are most hazardous when they are inhaled or swallowed or absorbed into the blood stream through wounds.

Gamma rays and X-rays: these are photons, but have high energy, typically from several keV to several MeV. Depending on their energy, they are capable of greater penetration than other rays. Several feet of concrete or a few inches of lead are required to stop them. Gamma rays and X-rays are a radiation hazard for the entire body. While gamma rays and X-rays can easily pass completely through the human body, some fraction of the energy will always be absorbed by body tissue.

Units of Measurement for Radioactivity (ref 3)

The original unit for measuring the amount of radioactivity was the curie (Ci) which was first defined to correspond to one gram of radium-226 and more recently defined as 3.7×10^{10} radioactive decays per second. In the SI system the curie has been replaced by the becquerel (Bq), which equals 1 radioactive decay per second or 2.703×10^{-11} Ci.

The absorbed dose describes the amount of radiation absorbed by an object or person (that is, the amount of energy that radioactive sources deposit in materials through which they pass). The units for absorbed dose are the radiation absorbed dose (rad), which is defined as the energy of 100 ergs per gram of irradiated material; and gray (Gy) where 1 Gy = 100 rads.

Equal doses of all types of ionizing radiation are not equally harmful. Alpha particles produce greater harm than do beta particles, gamma rays and x rays for a given absorbed dose. To account for this difference, radiation dose is expressed as equivalent dose in units of Sievert (Sv). The dose in Sv is equal to "absorbed dose" multiplied by a radiation weighting factor (1 for beta and gamma and 20 for alpha). 1 Sievert is equivalent to 100 Roentgen equivalent man (REM).

The effects of being exposed to large doses of radiation at one time (acute exposure) vary with the dose are shown below in **Table 2**.

Table 2 - effects of radiation

Dose (sievert)	Effect
10	Risk of death within days or weeks
1	Risk of cancer later in life (5 in 100)
0.1	Risk of cancer later in life (5 in 1000)
0.05	TLV for annual dose for radiation workers in any one year
0.02	TLV for annual average dose, averaged over five years

7 Regulatory Limits (ref 17)

Currently there are only 8 states in the U.S. that have NORM specific regulations. Since regulations vary slightly per state, Texas will be used as an example to put this into perspective. Regardless of the source of the ionizing radiation, exposures to the general public must be kept below 100 mR/yr and not to exceed 2 mR in any one hour. To be in compliance, facilities must maintain survey data to show these limits are not being exceeded. There are also limits for equipment release, contamination in soil, contamination of media (filters), and inhalation/ingestion to employees. **Table 3** illustrates these limits:

Table 3 – regulatory limits

Limit	Description	Reference
>50 μR/hr at any accessible point	Materials and equipment in recycling process & pipe tubulars and other downhole or surface equipment used in oil production	25 TAC 289.259(d)(2) & 25 TAC 289.259 (d)(3)
5,000 dpm/100cm^2	Contamination that cannot be readily removed from surfaces by casual contact, wiping, brushing or washing	25 TAC 289.259(w)
1,000 dpm/100cm^2	Radioactive material that can be removed from surfaces by non-destructive means, such as casual contact, wiping or brushing	25 TAC 289.259(w)
30 pCi/gram of Ra-226 or Ra-228 (soil/other media)	Oil and gas NORM waste	25 TAC 289.259 (d)(1)(A)(i)
150 pCi/gram any other radionuclide (soil/other media)	Oil and gas NORM waste	25 TAC 289.259 (d)(1)(A)(i)(II)
Derived Air Concentration (DAC) Pb-210 ($4x10^{-1}$ μCi/ml) Po-210 ($3x10^{-10}$ μCi/ml)	Concentration of radioactive material in air that will result in an annual limit of intake (based on 2,000 hours)	25 TAC 289.202 (ggg)(2)(F)
Annual Limit on Intake (ALI) Pb-210 ($1x10^{0}$ μCi) Po-210 ($6x10^{-1}$ μCi)	Derived limit for the amount of radioactive material taken into the body by inhalation/ingestion in a year.	25 TAC 289.202 (ggg)(2)(F)

Note: dpm = disintegrations per minute. 60 dpm = 1 Becquerel. 50 microroentgens/hr = 0.5 microSievert/hr

8 Health Concerns from Radon Gas and NORM

Radiation is a mutagen, which eventually can lead to cancer. Radiation can either kill cells or damage the DNA within them, which damages their ability to reproduce and can eventually lead to cancer. When radiation is present, high energy particles pass through the body. These can collide with atoms in the body and disrupt atomic structure and damage DNA. This can lead to uncontrolled cell division which is the definition of cancer.

Radon gas accounts for 55% of a typical person's radiation exposure (ref 20) as shown in **Figure 9**. Exposure to radon gas is the leading cause of lung cancer for non-smokers (ref 21).

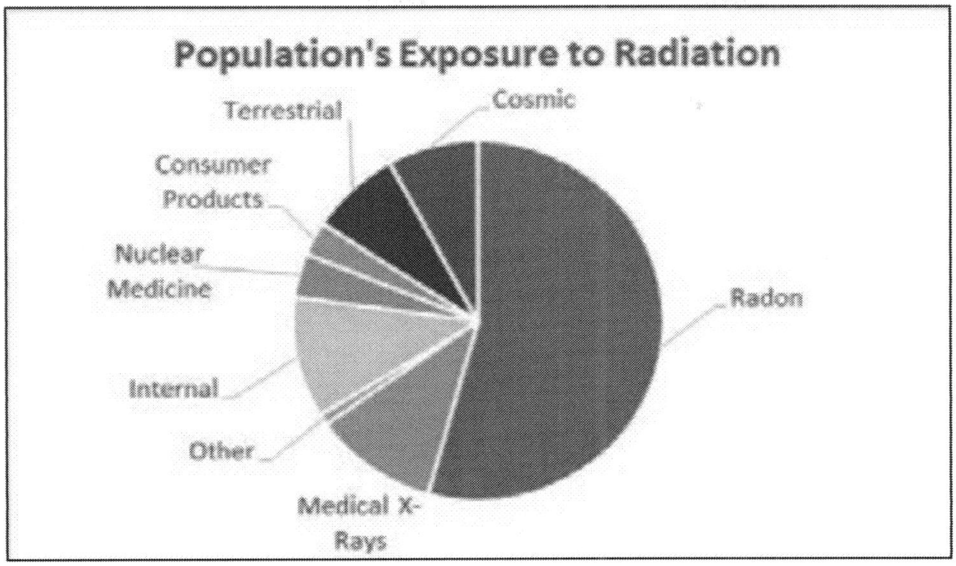

Figure 9 – Break down of typical exposure to radiation

9 Protection of Workers and the Environment

Each location should have a process for managing radioactivity. Considerations include

- A Radiation Safety Officer (RSO)
- Detection equipment (see above)
- Detailed procedures
- PPE (The IAEA has a practical radiation technical manual relating to PPE)
- The IATA held a conference in 2004 describing a methodology for disposal of low active radioactive waste

Any facility containing NORM should have a designated Radiation Safety Officer that is responsible for making decisions relating to NORM maintenance. The main goal of any radiation protection program is keeping all exposures to ionizing radiation **A**s **L**ow **A**s **R**easonably **A**chievable (ALARA). It is impossible to pre-plan for every scenario that may arise. Occasionally, there may be activities that aren't directly outlined in a written procedure. During these times, the RSO will review the work scope and make a determination on the best way to approach the task.

The following safety precautions should be followed when working with NORM contaminated items: (ref 22):

1.) Direct skin contact with radioactive scale, solids, and liquids should be avoided to the maximum extent possible.
2.) Eating, drinking, smoking and chewing shall not be allowed in the NORM work area
3.) Personnel shall thoroughly decontaminate after working with NORM containing equipment, before eating, drinking, or smoking.
4.) Work activities involving NORM should be conducted in areas to which access has been restricted. The number of personnel in the area should be kept to a minimum.
5.) Openings on contaminated equipment shall be sealed or wrapped in plastic or other suitable materials.
6.) NORM containing materials shall be kept wet to limit the possibility of NORM dust becoming airborne.
7.) Plastic ground covers should be utilized to the extent possible to contain NORM and facilitate cleanup.
8.) Workers who may be exposed to NORM must be provided with and wear proper Personal Protective Equipment. (PPE) to reduce the risk of exposure. The most effective PPE for working around NORM includes:
    - Protective eyewear, specifically goggles in high dust areas
    - Disposable or non-porous gloves (two layers of gloves are recommended)
    - Hooded disposable coveralls (i.e., TYVEK™ suits) or non-porous rain suits
    - Rubber or other non-porous protective footwear
    - Respirators with appropriate NIOSH-approved HEPA filters
9.) When respiratory protection is required, the selection, use, and care of any respiratory protective equipment must be in compliance with all company procedures.
10.) Gloves, respirators, coveralls, plastic and rags that have come in contact with NORM should be decontaminated or placed in containers, drums, sealed and held for disposal. Contaminated PPE should not be removed for the restricted work area until it has been decontaminated or placed in approved waste receptacles.
11.) NORM containing material shall be sealed in DOT 17-H steel drums or other suitable containers. Containers of NORM shall be labeled with a durable, clearly visible label bearing the radiation symbol and the words, "Caution Naturally Occurring Radioactive Material (NORM). The label shall also provide applicable information such as the radionuclides present, radiation levels, the quantity of radioactivity (in ρCi/g) if identified, the date for which the activity is estimated, and the type of material. This will assist individuals handling containers or

working near the containers to take precautions to avoid or minimize exposures.

10 Maintenance & Off-Site Repairs

When dealing with NORM, it is imperative that a plan be in place that outlines and controls the maintenance activities of equipment containing NORM. The plan should include the monitoring of work that is performed inside the facility as well as any items being shipped off-site for repair. NORM should be taken into serious consideration while planning for an outage or turnaround. Poor planning could result in a longer downtime as well as a potential liability. NORM contaminated equipment must be decontaminated to levels below the regulatory threshold before being shipped off-site. NORM contaminated equipment exceeding the surface contamination levels can be shipped to maintenance facilities providing they hold a specific license to accept this material. A monitoring program should be in place to identify all NORM contaminated equipment that is being opened or removed during a turnaround. This will ensure that the contaminated equipment does not inadvertently get shipped off-site to an unlicensed company. Having a good understanding of suspect NORM equipment involved in a turnaround results in an improved way of managing it ahead of time. Pre-planning for off-site maintenance will ensure that the facility as well as the employees are not exposed to NORM during repair/maintenance. It is very important to ensure that all decontamination activities are performed by a company holding a specific license to perform that service. The RSO should maintain a current radioactive materials license for any contractor performing decontamination activities at their site.

Summary

A survey carried out in preparation for this paper found that Naturally Occurring Radioactive Materials are present in nearly all ethylene plants. The root cause is known to be radon gas which exists in the rock formations where hydrocarbon feedstocks are often found. Radon is a gas with a short half-life, which poses a challenge to understanding how individual plants will be affected by NORM. In some cases the radon could pass through the plant without leaving a trace, while in others it could decay to non-volatile progenies which are retained in the plant. The known presence of NORM has increased by 160% over the past 25 yrs. This is certainly due to better awareness of the dangers of radioactive materials in our plants. However, the data also suggests shale fracking may be a contributor to higher NORM. How to address the presence of NORM varies plant to plant. It is our recommendation that this topic should be regularly revisited by industry bodies including the US Ethylene Producers Committee, and that guidelines should be developed and shared by the community.

References

1. P.R. Gray. Journal of Petroleum Technology, Vol 45, issue 01, January 1993
2. R.A. Hart, R. Wolf, M. Graham, "Fluoride and Radon Survey Results", EPC article 24B, 1994
3. Texas RRC. NORM (Naturally Occurring Radioactive Material)
4. E. Scarnichia, A. Etchepareborda, M. Arrivere. "Radon in Propylene, Unexpected Influence of NORM in a Chemical Plant.", 2008; SAR; Buenos Aires (Argentina); IRPA 12: 12. International congress of the International Radiation Protection Association (IRPA): Strengthening radiation protection worldwide; Buenos Aires (Argentina);
5. G. Taylor, M. Cleghorn, "Naturally Occurring Radioactive Materials: Contamination and Control in a World Scale. EPC paper 21D, 1996
6. P.R. Gray, "NORM Contamination in the Petroleum Industry." JPT, Jan 1993, pgs 12-16
7. J.G. Barret, "Naturally Occurring Radioactive Material (NORM) Contamination Encountered during Propylene Fractionator Modifications." EPC paper 34B, 1997
8. R. Nave. Georgia State University, HyperPhysics, 2016
9. www.radioactivity.eu.com. Radon Properties.
10. D. Slim, Evonik Industries, private communication
11. "Fracking linked to higher radon levels in Ohio homes." Science Daily, University of Toledo. June 18, 2019.
12. J. Grant. "Fracking is on the rise in Pennsylvania. So are radon levels. Are the two connected?" Living on Earth. August 18,2018
13. Frackcident™ Injury Law. Blue Bell, PA
14. Apex Inspection Service
15. J. Hilbrich, "Experience with NORM", EPC Paper 2020.
16. Picture courtesy of Sheri Snider, Nalco Water
17. J. Harris, "Naturally Occurring Radioactive Material (NORM): Detection and Protection in the Petrochemical Industry", EPC paper 96F, 2012.
18. Howstuffworks.com
19. D. Slim, Radioactivity in ethylene plants, prepared for the Hazardous Chemicals sub-committee of the EEPC, 2019
20. St. Louis Test and Mitigation
21. D. Satterfield. AGU Advancing Earth and Space Science, SEPT 2009 "How Much Radiation Does it Take to Kill You?"
22. BP GoM Region Naturally Occurring Radioactive Material (NORM Manual)

AIChE PAPER NUMBER: 159c

PLASTIC WASTE TO CRACKING FEEDSTOCKS AND THE CIRCULAR ECONOMY

Andre Bernard

NOVA Chemicals

M. David Smith

NALCO Water, an Ecolab Company

Thomas Pickett

Technip Stone & Webster Process Technology

Joe Lally

Evonik Corporation

Prepared for presentation at the 2020 Spring National Meeting

Houston, TX, March 30 – April 4, 2020

AIChE and the EPC shall not be responsible for statements or opinions contained in papers or printing in its publications.

PLASTIC WASTE TO CRACKING FEEDSTOCKS AND THE CIRCULAR ECONOMY

Andre Bernard

NOVA Chemicals

M. David Smith

NALCO Water, an Ecolab Company

Joe Lally

Evonik Corporation

Thomas Pickett

Technip Stone & Webster Process Technology

Abstract:

Processing plastic waste into oil that can be use as feedstock to steam cracking is an alternative that is gaining attention. The prospect of this process, although at an embryonic stage, is believed to provide a significant boost in the overall waste plastic recycle yield when compared to existing methodologies. However, this potential new feedstock to steam cracker comes with a number of contaminants, which would likely impact the reliability and performance of ethylene plants. This paper is a high-level discussion on the potential of using waste plastic as feedstock and the contaminants that can be expected.

Introduction

The negative impact of plastic waste is currently shadowing the benefits of plastic products to society. Public perceptions and pressures on governments to act resulted in 60 countries around the globe banning or being in the process of banning single use plastics.[1] Literature reports that only 12 percent of plastic waste is reused or recycled.[2] This low recycling yield is attributed to the current methodology. Processing plastic waste to cracking feedstock is an emerging technology that has the potential to boost the recycling yield and meet the vision of a plastic circular economy model.

Ethylene Value Chain

Figure 1 illustrates the predominant value chain of ethylene crackers leading to different types of plastic materials. The production of polyethylene represents a large fraction of ethylene consumption, where it accounts for about 36 percent of the end products illustrated on Figure 1. The production of ethylene oxide follows and is used to make ethylene glycol. Monoethylene glycol is then used to manufacture polyester fibers and polyethylene terephthalate (PET). PET accounts for 8 percent of the plastic in the value chain. Ethylene dichloride made by chlorination of ethylene is then cracked into vinyl chloride monomer (VCM) for producing polyvinyl chloride (PVC), which represents 18 percent of the plastics fraction of the value chain. Polystyrene contributes to 8 percent of the plastic pool and uses ethylene as feedstock by reacting it with benzene to form ethyl benzene that is further processed with styrene into polystyrene. Polypropylene which constitutes 26 percent of the plastic pool is made from propylene. Although not shown on Figure 1, butadiene recovered from the mixed C_4s stream from ethylene crackers is used for manufacturing rubber, which contributes for about 3 percent of the plastic pool illustrated on Figure 1.

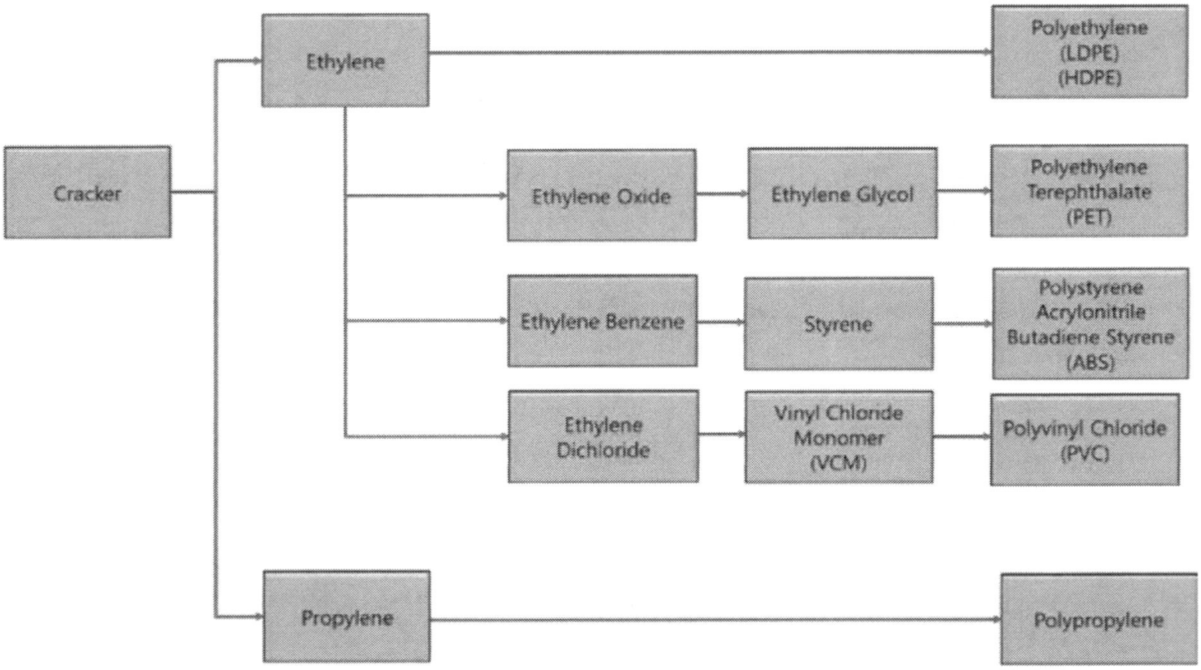

Figure 1: Ethylene Cracker Value Chain

Plastic Production and Usage

Global plastic production is estimated at about 390 million metric tons per year.[3] Although used earlier, the boom of plastic usage started after WWII and started growing exponentially from the 1950s to date. The production of plastic is expected to quadruple by 2050. Figure 2 illustrates the distribution of the different types of plastics produced worldwide. Those categorized as others includes sealants, coatings, polycarbonate, polyamide and bioplastics.

Thirty percent of the plastic produced is use for packaging, which includes a large variety of products. Some are single use such as plastic bags, food wrapping, plastic bottles (water and soft drinks). Others are products intended to be reused like plastic buckets, boxes and multipurpose bottles. About 50 percent of polyethylene production is dedicated to the plastic film market, which is typically single use and rarely accepted by recycling programs as it tends to get stuck in the sorting machinery.

Building and construction consumes about 17 percent of the plastic production through products such as pipes and fittings, siding, flooring and others. Those materials are installed and have a useful life expectancy of many years.

Plastic comprises about 15 percent of the total mass of cars. The use of carbon fiber reinforced plastic is use in the aviation industries to reduce the weight of airplanes allowing gains in fuel efficiencies. Plastic consumption from the transportation sector accounts for about 14 percent.

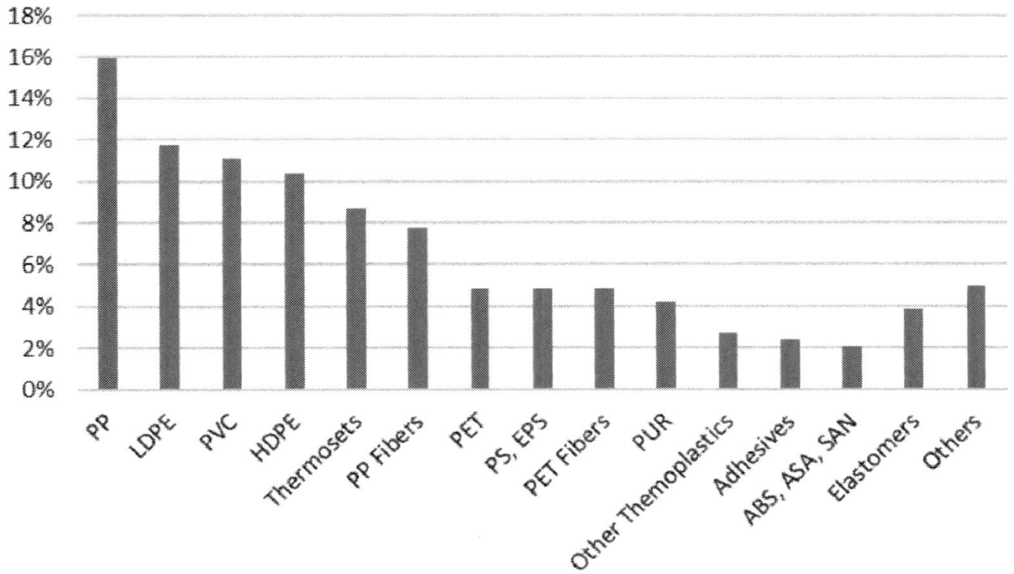

Figure 2: Plastic Production Distribution by Type

Many other products such as toys, healthcare/medical, kitchenware and many other products represent about 11 percent of the total plastic consumption. Textiles and the electric/electronics sector consume 10 and 4 percent respectively of the total plastic production.

Negative Impacts of Plastic Waste

Plastics have largely contributed to the betterment of society. However, their disposal after use creates an environmental problem. The life of plastic material is currently more in alignment with a linear economic model from generation to disposal. The value of the material is lost as waste after single use. A lot of plastic waste finds its way in the ocean. The UN estimated that the equivalent of 1 garbage truck per minute is released to the ocean.[4] It is reported that by 2050 there will be more plastic than fish in the ocean. An increase in synthetic fiber contamination in freshwater systems and terrestrial habitats has also been reported.

Forty years after the launch of the international recycle symbols, only 9 to 15 percent, depending on region,[5,6] of plastic is being recycled. Of that volume, about only 5 percent will be recovered for further use. Ninety percent of plastic ends up in landfills; as such, the plastic content in municipal waste has increased from 1 percent in 1960 to over 10 percent.[7]

The plastic industry is trailing behind paper, metal and glass with regards to recycling. Data from 1990 to 2015 suggest that the recycling rate in the US for paper has increased from 27 to 66 percent, 24 to 34 percent for metal and 20 to 26 percent for glass.[6] The plastic recycle rate has only increased from 3 to 9 percent. The variety of plastics composition and handleability makes it challenging for sorting the different types of plastic. For instance, as mentioned earlier, plastic bags are rarely recycled as they get stuck in the sorting machinery. Recycling facilities focus more on PET, HDPE and PP. Recycling is also limited by the capability of existing methodologies.

Regulatory Response

Many countries around the globe have implemented or are in the process of implementing bans on single use plastic. The European Union voted to ban single use plastic by 2021, targeting products such as straws, plates and cups.[10] In the footstep of the European Union, Canada is proposing a ban on single use plastic to come into effect by 2021. Some states in the US are considering or already have bans on a single use plastic. California and Hawaii already have bans in place, while New York state lawmakers have approved a statewide plastic bag ban to come into effect in March 2020.

Those bans have a large impact on plastic films, which represent a significant fraction of the value chain. The industry must therefore catch up on recycling strategies and methodologies.

Traditional Recycling (Mechanical)

Mechanical recycling is currently the dominating recycling methodology. It applies to polyethylene terephthalate (PET), high density polyethylene (HDPE) and polypropylene (PP). Plastic must first be sorted and separated from other waste (metal, glass, paper). Magnets are used for removal of ferrous metals; an optical sorter is used for removing carton. Eddy current is used for removing non-ferrous metal (mostly aluminum). Plastic is then separated based on its molecular signature using near infrared (NIR)[8] and an air nozzle, which blows the targeted plastic off the conveyor into a recovery bin. The recovered plastic parts are then washed and ground, followed by compounding and pelletizing. Polymer grade quality, however, degrades after multiple cycles due to accumulation of contaminants, primarily color pigments. Higher selectivity relative to pigmentation increases the cycles but at the expense of recovery.

Chemical Recycling and The Circular Economy

In response to the regulatory actions taken against plastics, particularly single-use plastics, and the challenges inherent in mechanical recycling, a new paradigm was developed: the "Circular Economy" for plastics. Rather than mechanically recycling plastic materials to use in increasingly lower-value applications with degrading product quality, the Circular Economy paradigm utilizes technological innovations to chemically break down the polymeric chains in plastic into either fuel, monomer, or feedstock to a steam cracker – a process known as Chemical Recycling.[11]

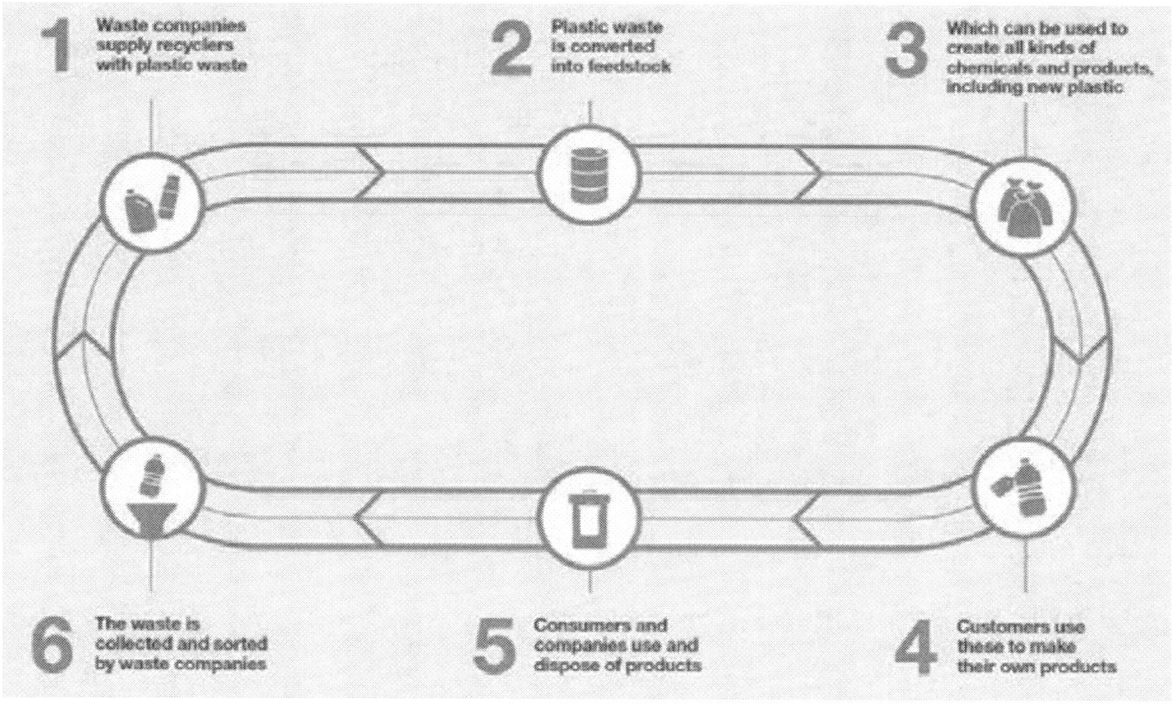

Figure 3: The Circular Economy for Plastics[28]

As shown in Figure 3, there are several steps in the full cycle of chemical recycling. Waste collection facilities provide recyclers with plastic waste. The waste must be sorted carefully, because certain types of plastic have components that are considered feedstock contaminants for petrochemical manufacturers; these will be discussed in more detail below. Next, one of several methods is used to break the polymer chains in the plastic down to smaller molecules that can be used for fuel, or in chemical plants and refineries to produce other consumer products.

Methods of Chemical Recycling

As shown in Figure 4, there are two main subsets of chemical recycling of plastics: decomposition, also known as monomer recycling, and conversion, also known as plastic-to-fuel (PTF) recycling. Each of these subsets utilizes various technologies to accomplish the breakdown of plastic into usable materials.

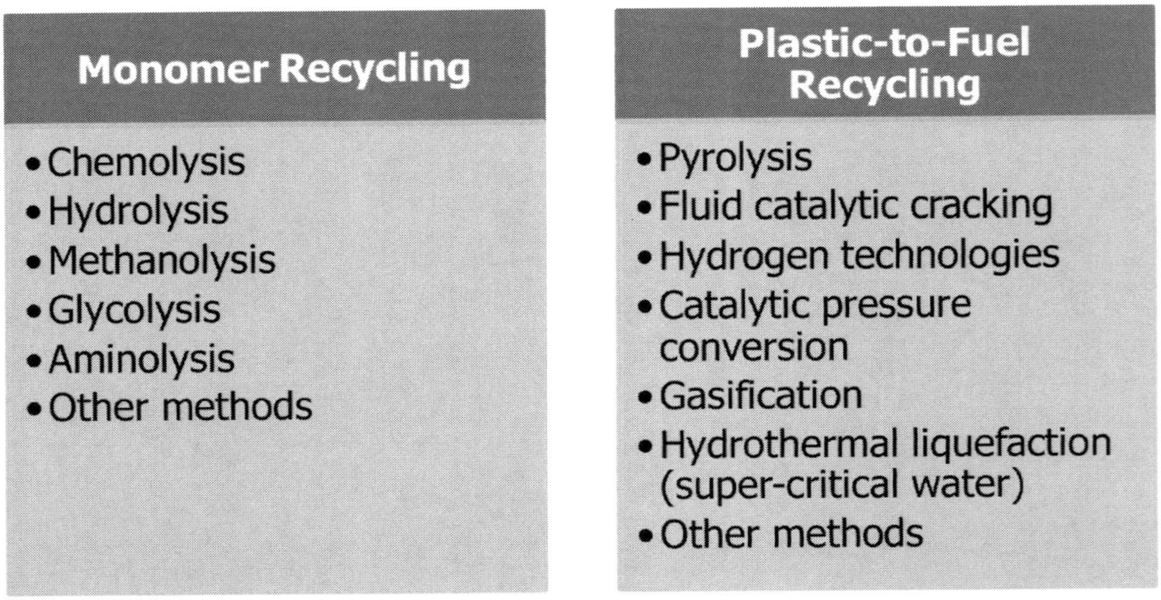

Figure 4: Methods of Chemical Recycling

Chemical Recycling Initiatives in the Industry

Interest in chemical recycling within the industry has been high. An organization known as the Alliance to End Plastic Waste is a consortium of companies that manufacture, use, and recycle plastics. Comprised of such companies as BASF, ExxonMobil Dow, NOVA Chemicals, Sabic, Shell, and others, the Alliance has committed to investing $1.5 billion in promoting chemical recycling efforts over a 5-year period. Their goal is to dramatically reduce the estimated $120 billion in plastics they say enter the environment each year. Another industry group called Styrenics Circular Solutions includes such companies as INEOS Styrolution, Repsol, Total, Trinseo, and Versalis and is focused on recycling polystyrene (PS) to styrene monomer (SM).[12]

BASF's Ludwigshafen steam cracker was among the pioneers of recycled plastic-to-feedstock in ethylene production. They announced a successful run of processing pyrolysis oil supplied by the German chemical recycling firm Recenso, which uses a process called catalytic tribochemical conversion that can be applied to both plastics and agricultural biomass waste. The BASF program, which they call "Chemcycling", utilizes both pyrolysis oil and syngas for petrochemicals production.[13]

One of the first reported polymer-to-monomer processes occurred in the United States. American Styrenics, a joint venture of Trinseo and ChevronPhillips, announced in March 2019 that they had received recycled styrene monomer for processing at their St. James, Louisiana site. The supplier of recycled monomer, Agilyx, has been recycling plastic waste to synthetic crude oil for over a decade, processing 8 million pounds of mixed plastics into 800 thousand gallons of crude in 2014. They commissioned their first polystyrene-to-styrene monomer facility in 2018. Agilyx has also signed a joint discovery agreement with INEOS Styrolution and announced a plan to build a processing facility for polystyrene-to-styrene in Europe.[14,15,16,17]

Another successful industrial application of chemical recycling was announced in November 2019 by Shell Chemical. Nexus Fuels supplied Shell's Norco, Louisiana steam cracker facility with liquid feedstock produced from a pyrolysis process that uses high- and low-density polyethylene (HDPE, LDPE), polypropylene (PP), and polystyrene (PS). Shell has a stated goal of using 1 million tons of recycled plastic as feedstock by 2025.[18,19]

Other companies are involved in partnerships as well. In 2018, Sabic and Plastic Energy signed a Memorandum of Understanding (MoU) to build a commercial facility to produce "TACOIL", a feedstock produced via a process known as thermal anaerobic conversion.[20]Plastic Energy currently operates two units in Spain, with the intention of putting more TACOIL units throughout the world, close to centers of plastic supply. Plastic Energy has also signed an MoU with Petronas Chemicals.[21] Another company, UPM Biofuels, has announced agreements with INEOS and Dow Chemical to supply wood-based feedstocks for plastic production.[22,23]

Among the other companies at the forefront of chemical recycling include the following:

- Recycling Technologies, which uses a thermal cracking process that is similar to a refinery fluidized catalytic cracker to produce a variety of "Plaxx" oils and has signed an alliance with chemical manufacturer Total;[24]
- ReOIL, which recycles used tires into pyrolytic oil and carbon black;[25] and
- ReNew ELP, which uses super-critical water in a catalytic hydrothermal reactor to produce original monomer and oils.[26]

Overall, there are a variety of processes for chemical recycling, mostly based around some version of pyrolysis, and many of the largest petrochemicals producers are developing partnerships with chemical recycling companies to develop and implement an effective Circular Economy strategy.

Challenges to Chemical Recycling: Feedstock Contaminants

Contaminants in petrochemical feedstocks can cause a variety of problems in a steam cracker. To provide information on various contaminants and their potential impacts, the Ethylene Producers Committee (EPC) of the American Institute of Chemical Engineers (AIChE) established a Feedstock Contaminants Database that indexes various contaminants, the locations in an ethylene plant or downstream processing units these contaminants might impact, and what the negative consequences could result from the presence of these contaminants.[27]

Because of the deleterious impacts of certain contaminants, plastics being collected for chemical recycling as petrochemical feedstock must be separated carefully. In particular, two types of plastic must be removed prior to processing chemically for use as feedstock: polyethylene terephthalate (PET) and polyvinyl chloride (PVC).[28]

PET is produced by polymerizing ethylene glycol and terephthalic acid and is commonly used for soda bottles and other food and pharmaceutical packaging.[31] As shown in Figure 5, PET contains a considerable amount of oxygen. Oxygenates in petrochemical feed provide a number of negative consequences in an ethylene plant, including increased acetic acid production leading to corrosion, increased carbonyl concentration that can lead to aldol polymer fouling, and concentration of carbonyls in the crude C_4 product.[27]

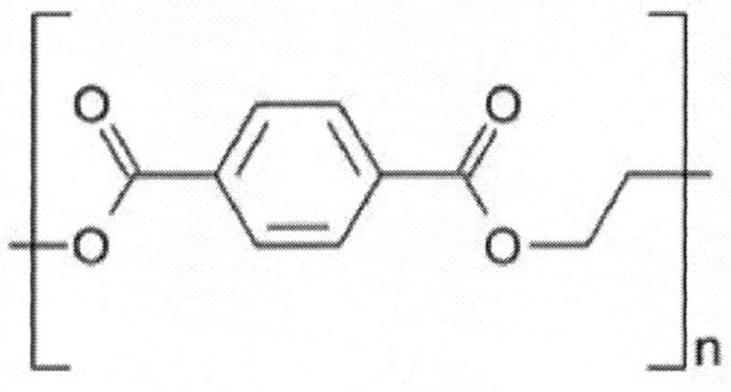

Figure 5: Polyethylene Terephthalate

PVC is produced by the polymerization of vinyl chloride monomer and is used in piping and plumbing, automobiles, raincoats, and a variety of other applications.[32,33] Chlorides in the feed can produce hydrochloric acid (HCl). HCl is highly corrosive, particularly at dew point conditions that can occur in petrochemical plants. Because HCl is a stronger acid than the organic acids typically observed in ethylene plants, it can be especially corrosive.[34]

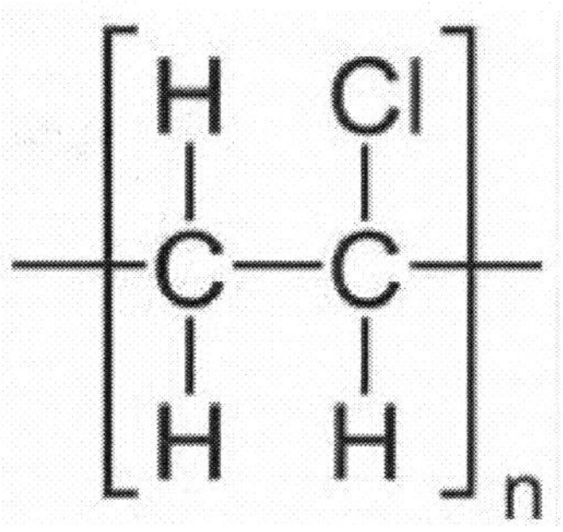

Figure 6: Polyvinyl Chloride

Other potential contaminants could exist as well, such as nitrogen, which can crack to form ammonia, and phosphorous, which can poison the gasoline hydrotreater catalyst. Following the more widespread use of these recycled feedstocks, a more exhaustive survey of observed contaminants and their impacts can be catalogued.[27]

Other Challenges to Chemical Recycling

There are several additional challenges that exist to implementing a widespread program of cracking recycled feedstocks. These include the following:

- Sorting Efficiency. As mentioned above, effective sorting is important to prevent feedstock contaminants. Sorting is currently an inefficient, expensive, and time-consuming operation.
- Geography. In addition to the nature of the sorting itself, sorting centers are still few are far between, requiring transport from collection to the sorting, potentially from the sorting center to the reprocessing facility, and then from the processing facility to the manufacturer.
- Scale. Currently, there are not enough collection, sorting, and processing facilities to provide feedstocks that meet the Circular Economy goals of governments, regulatory agencies, industry groups, and manufacturers.
- Economics. Because of the cost incurred by collecting, sorting, processing, and transporting the plastics and the recycled feedstock, it is still not meritorious from a purely economic standpoint.

Summary

Plastics have become nearly ubiquitous, touching nearly every aspect of our lives. However, the current practice of use-and-dispose has created problems of pollution and waste that are considered unsustainable many consumers, government agencies, and environmentally conscious individuals and groups. As a result, strict limits are being put in place that make recycling of plastics a necessity. In addition to the legacy methods of physical recycling, which recycles products into increasingly less valuable end uses due to product deterioration, chemical recycling promises to provide a method of truly having a circular economy, where new plastics are produced from feedstocks derived from previous goods, without a degradation of product quality. These feedstocks provide challenges that must be overcome for chemical recycling to become a world-scale, economically efficient endeavor.

References:

1. P. Hodges, "Tide will turn against single-use plastic", eChem, January 14th, 2019
2. T. Hundertmark, M. Mayer, C. Mcnally, T.J. Simons, C. Witte, "How plastics waste recycling could transform the chemical industry", Hydrocarbon Processing, April 2019.
3. M.W. Ryberg, A. Laurent, M. Hauscgild, "Mapping of global plastics value chain and plastics losses to the environment", United Nations Environment Programme, 2018
4. L.L. Alexander, "Repackaging the Planet", Chemical Engineering Progress (CEP), May 2018
5. "The new plastics economy. Rethinking the future of plastics", World Economic Forum, Ellen MacArthur Foundation and McKinsey & Company, 2016
6. A. H. Tullo, "Plastic has a problem; is chemical recycling the solution?", Chemical & Engineering News, October 6, 2019
7. R. Geyer, J.R. Jambeck, K.Lavender Law, "Production, use, and fate of all plastics ever made", Science Advances, July 19, 2017
8. "Near Infrared (NIR) Sorting in the Plastics Recycling Process", The Association of Plastic Recyclers, Document number: RES-SORT-2, July 20, 2018
9. K. Ragaert, L. Delva, K. van Geem, "Mechanical and Chemical Recycling of Solid Waste", Waste Management, August 2017.
10. R. Andrews, "The European Union Just Voted Ban on Single-Use Plastics by 2021", Forbes, October 25, 2018.
11. Rubel, Holger, Udo Jung, Clint Follette, Alexander Meyer zum Felde, and Santosh Appathurai (2019), "A Circular Solution to Plastic Waste", Boston Consulting Group.
12. Alliance to End Plastic Waste (2019), "We Have A Clear Vision to Eliminate Plastic Waste: First Anniversary Report, www.endplasticwaste.org.
13. BASF (2018), "BASF for the first time makes products with chemically recycled plastics", Press Release, www.basf.com
14. Recenso (2020), "Plastic Conversion", Press Release, www.recenso.eu
15. Styrenics Circular Solutions (2020), "Teaming Up for Change", www.styrenics-circular-solutions.com/circular-economy
16. American Styrenics (2019), "Agilyx Begins Delivery of Recycled Styrene Monomer to AmSty", Press Release, www.amstyrenics.com
17. Agilyx (2019), "SCS members Trinseo and INEOS Styrolution plan commercial scale polystyrene chemical recycling plant with Agilyx in Europe", Press Release, www.newswire.ca
18. Moore, Stephen (2020), "Shell uses plastic waste to produce resin feedstocks, *Plastics Today,* January, www.plasticstoday.com
19. Shell Chemical (2019), "Shell uses plastic waste to produce chemicals", Press Release, www.shell.com
20. Plastic Energy (2018), "Sabic Signs MoU with PLASTIC ENERGY", Press Release, www.plasticenergy.com

21. Duckett, Alan (2019), "Petronas Chemicals looks to convert waste plastic into feedstock", *The Chemical Engineer*, July, www.thechemicalengineer.com
22. UPM Biofuels (2019), "Dow and UPM partner to produce plastics made with renewable feedstock", Press Release, www.upmbiofuels.com
23. UPM Biofuels (2020), "INEOS and UPM Biofuels announce supply agreement for renewable raw materials to make plastic", Press Release, www.upmbiofuels.com
24. Recycling Technologies (2019), "Citeo, Total, Recycling Technologies, Mars, and Nestle Join Forces to Develop Chemical Recycling of Plastics in France", Press Release, www.recyclingtechnologies.co.uk
25. Reoil (2020), "Tire Recycling by Continuous Pyrolysis", Press Release, www.reoil.us
26. Renew (2020), "We achieve in just 20 minutes what takes nature 200 million years: turning waste plastic into readily usable hydrocarbon feedstock", Press Release, www.renewelp.co.uk
27. Bernard, Andre (2019), "2019 Feedstock Contaminants Database Presentation", presented at the AIChE Spring National Meeting, Ethylene Producers Conference, Session 35.
28. Chitalia, Ashish (2019), "Global Polyolefins: Are the upside risks over and done with?", Wood Mackenzie Americas Olefins Conference, Polymer Focus Session.
29. Editors of Encyclopaedia Britannica, "Polyethylene terephthalate", www.britannica.com
30. Hood, David and Steve Coleman (2007), "Minor Components, Major Issues: An Overview of Contaminants in Ethylene Plants", AIChE Spring National Meeting, Ethylene Producers Conference, Paper 51c.
31. "Polyethylene Terephthalate", www.wikipedia.com
32. Editors of Encyclopaedia Britannica, "Polyvinyl chloride", www.britannica.com
33. "Polyvinyl Chloride", www.wikipedia.com
34. Elder, Sherri, Nicolas Genty, and Wendy Young (2006), "Neutralizer Selection for Control of Acid Corrosion in Ethylene Plants", AIChE Spring National Meeting, Ethylene Producers Conference, Session 218.

Abstract:

In the “Tail End” hydrogenation, the light ends are removed first and hydrogenation of acetylene is on the overhead of the De-ethanizer. The feed stream is a pure C2 fraction and hydrogen is added upstream from the reactors under H_2/Ac molar ratio control. The low hydrogen concentration in reactor feed limits the hydrogenation reaction and thermal runway rarely occurs in “Tail End” operations. Nevertheless, if there is uncontrolled hydrogen addition, the hydrogenation reaction will overheat the catalyst bed and can initiate the exothermic reaction of ethylene polymerization and possibly its decomposition, resulting in a thermal runway event.

There still are plenty of “Tail End” operating units, operation knowledge has improved over the years and bring the catalyst users to the same level of skills is important. In this session we will review the fundamentals of “Tail End” acetylene hydrogenation, weigh potential causes of thermal runway in such units, share experience of a runway event and present best practices of reactor interlock.

CLARIANT

Table of Contents

- C2 Acetylene Overview and Operation Hazards
- Temperature Runaway in Normal Operation
- Temperature Excursion at Start-up, Regeneration and Stand-by
- Takeaways

Hi, my name is Carlos Vianna, I am Product Manager for Selective Hydrogenation Catalysts at Clariant and today I am going talk about different ways a temperature runaway can happen in backend acetylene converters.

Here is how I am going to do it:

I will start with operation and hazards of backend reactors. As we discuss about it, I am going to share real cases of temperature runaway in normal operation, start up, regeneration and stand-by. I will be happy if at the end of this presentation you learn the different ways backend acetylene reactors can have temperature runaway.

If you are listening to this presentation, you probably already know that it is very difficult for a temperature runaway to take place in backend converters. However, I will show you it is more frequent than you expect. And I will tell you more: if you run a backend acetylene reactor, you have possibly experienced a temperature runaway before.

C2 Acetylene Overview and Operation Hazards

CLARIANT

what is precious to you?

CLARIANT

Overview of back-end acetylene hydrogenation

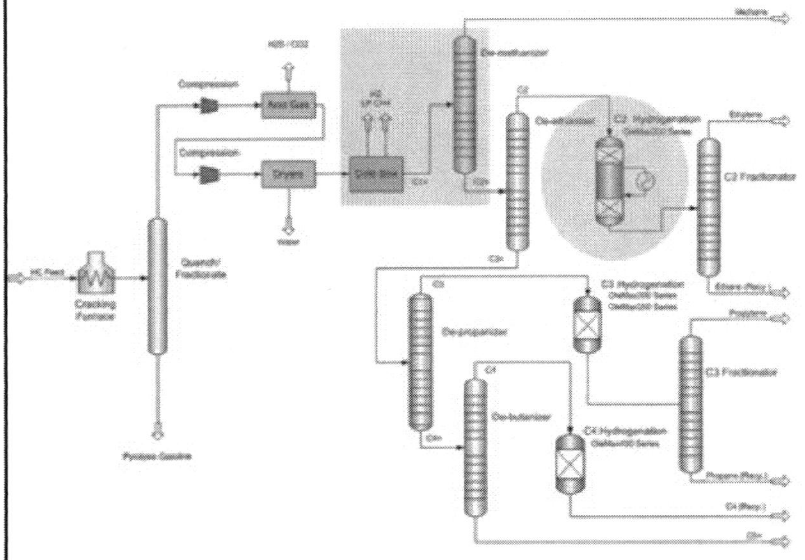

- Demethanizer first sequence
- C2 stream
- Operation at 20-25 barg (285-315 psig) and <120°C (250°F)
- Gas phase
- H_2 added at controlled and stoichiometric quantities
- Process control: inlet T, H_2 and CO (optional)
- Cycle length: vary from few months to several years (green oil formation)
- Requires in situ regeneration facilities

Some of you may not be familiar with back-end process separation thus I will briefly describe this process scheme.

The picture shows a flow diagram of a backend process configuration for ethylene production. Ethylene and by-products recovery begin with removal of methane and hydrogen in the cold box and demethanizer as indicated by the blue area in the flow diagram. So this is why this process scheme is also called demethanizer first sequence. The bottom of the demethanizer is sent to de-ethanizer where acetylene, ethylene and ethane are separated as overhead product. Thus, the backend acetylene hydrogenation is performed in a C2 pure stream.

Acetylene Converters operate around 20-25 barg and temperature in 30 to 120°C range. Because of this set of T and P, the acetylene converter operates in gas phase.

As it is a C2 pure stream and hydrogen is added in small quantities hence limiting the amount of hydrogen for the hydrogenation reactions.

As the catalyst ages, inlet temperature and hydrogen can be adjusted to improve catalyst performance. In addition, the backend configuration allows

CO addition to moderate catalyst activity. Then, temperature, hydrogen and CO addition are the ways of reactor control.

The drawback of low hydrogen content in reactor feed is the higher green oil formation, which will limit the cycle length. The cycle length changes from plant to plant due to different feed composition, operating conditions as well as different catalysts. So, cycle length can be either as short as couple of months or as long as several years.

Due to short cycle length, backend systems are designed with spare reactor and regeneration facilities for in situ-regeneration.

Due to the characteristic of operation with stoichiometric addition of hydrogen, the magnitude of uncontrolled exothermic reaction hazard is significantly reduced, and this process configuration can be called as "inherent" safer process than other ethylene plant process schemes.

So, what are the hazards of backend acetylene reactors?

CLARIANT

Acetylene reactor operation hazards

- Loss of containment – Can produce a flammable vapor cloud (less than liquid phase)
- Uncontrolled exothermic reaction – Greatest hazard
- Excessive hydrogenation - Most common hazard
- Unlimited hydrogenation - Initiates ethylene polymerization (possibly decomposition) or runaway
- Temperature excursions - Initiated by non-hydrogenation exothermic reactions
- Physical integrity of reactors/vessels – Compromised as a result of over temperature

One hazard is loss of containment that can produce vapor cloud. As the reaction is in vapor phase, converter inventory may be considered low compared to liquid phase reactors, where a loss of containment can produce large vapor cloud potentializing the risk of explosion. For this reason, this risk is considered low.

Therefore, the greatest hazard in operation of the Backend Acetylene Converters is uncontrolled exothermic reaction, typically caused by addition of hydrogen in large quantities.

If the exothermic reaction is left uncontrolled, localized overheat can initiate ethylene polymerization and possibly its decomposition.

Moreover, temperature excursion can be initiated by non-hydrogenation exothermic reactions as well, and not only when the reactor is in operation.

If the additional heat released by exothermic reactions increases over the capacity of the system to remove heat, it may result in temperature runaway. If it reaches severe high temperature, it can threaten the physical integrity of the converter.

CLARIANT

There are several exothermic reactions in back-end acetylene converters

Type of reaction	Reaction
Selective hydrogenation	Acetylene + $H_2 \rightarrow$ Ethylene
Non-selective hydrogenation	Ethylene + $H_2 \rightarrow$ Ethane Acetylene + $H_2 \rightarrow$ 1,3 Butadiene
Non-hydrogenation	Oligomerization Ethylene decomposition Ethylene polymerization
Regeneration (carbon burn-off)	Hydrocarbon + $O_2 \rightarrow CO_2$ + water
Regeneration (catalyst activation)	$PdO + H_2 \rightarrow Pd^0$ + water (heat release: not noticeably)

These are the chemical reactions of backend acetylene converters. The desired reaction is the selective hydrogenation of acetylene and a plenty of side reactions can take place along the catalyst cycle length.

Let's see when those reactions take place.

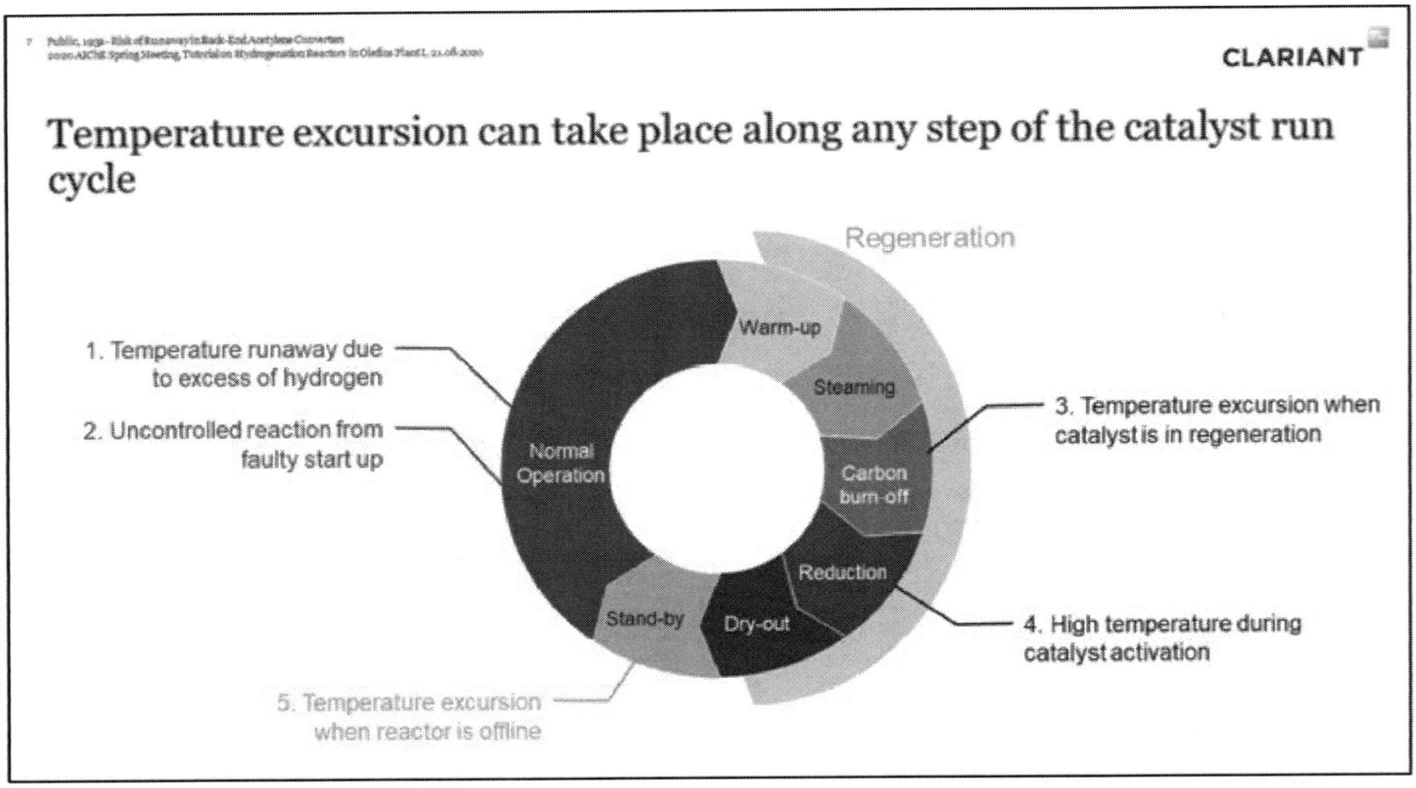

This picture describes the cycle length of backend reactors. Let's split one catalyst cycle as normal operation, regeneration and stand-by.

Once the reactor is on-stream, or normal operation, there are 2 scenarios that temperature runaway may happen: 1) Excess of hydrogen; 2) Uncontrolled reaction from faulty start up.

When the reactor is out of operation for regeneration, it has to go through 5 steps before being in stand-by and to be back in operation. The first 3 steps are the regeneration itself. The other 2 steps are reduction and dry out and they are performed for catalyst activation and conditioning.

A temperature excursion can happen during regeneration and palladium reduction (3 and 4). And here is when you most probably have seen a temperature runaway before.

A remote case of temperature runaway is when the reactor is in stand-by (5). If you have not faced one of those temperature excursions you may be asking yourself: how is it possible?

The numbering is the sequence of cases that we will discuss in the next slides and you will see they are possible and more frequent than you expect.

Temperature Runaway in Normal Operation

CLARIANT

what is precious to you?

CLARIANT

Case # 1 – Runaway temperature due to excess of hydrogen

Main Cause:

- Hydrogen added into acetylene reactor after plant trip

Description:

- Failure of reactor feed inlet control valve
- Feed flow to acetylene converter cut to zero
- No low feed flow interlock:
  - Hydrogen valves left open
  - Excess hydrogen
  - No flow through = no heat removal → temperature runaway
  - Over hydrogenation initiated ethylene polymerization and/or decomposition
- TI's indicated >500°C (930°F) at bottom of the reactor

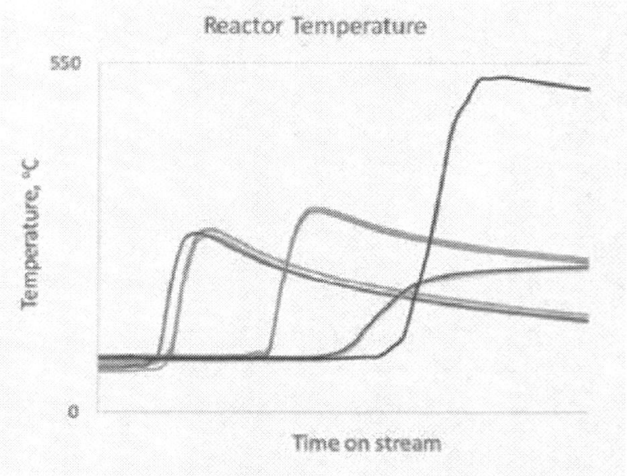

All the cases we will discuss now really happened in backend converters.
Case #1: Hydrogen addition after plant trip.

Loss of signal caused reactor feed control system failure. The control valve fail safety position is closed thus the reactor feed flow was cut to zero shutting down the plant.

However, there was no interlock to close hydrogen control and shut off valves because of low feed flow condition. As a result, hydrogen was continuously added into the reactor causing over hydrogenation reaction and localized high temperature, that triggered ethylene polymerization or potentially its decomposition.

As there was no flow across the catalyst bed, heat losses through reactor vessel wall was not enough to remove the heat released by the reaction chain and the temperature runaway.

The picture demonstrates the heat wave across the catalyst bed. TI's indicated temperature above 500°C in the bed.

CLARIANT

Case # 1 – Best practices to prevent acetylene reactor overheating

Eliminate source of heat	• Isolate hydrogen feed • Isolate ethylene feed • Shut off steam flow to feed pre-heater
Take reactants out	• Block reactor effluent line (avoid to move reacting mass across catalyst bed) • Depressurize to flare • Purge with inert gas (Nitrogen)
Automation	• High temperature and low flow interlocks
Rigorous HazOp Studies	• Assessment of alarms, flow shut off, depressurization • Rank low, medium and high excursion events and set up

Looking at this scenario, what is recommended as best practices to prevent catalyst overheating?

First: eliminate the source of heat by closing reactants inlet valves. Be sure to keep valves well maintained as valves may leak and the leak will not stop reactants to go into the catalyst bed. I will not discuss this as it is hydrogen addition related. Nevertheless, it can happen. Believe me!

Another action is to shut off steam flow and remove the source of feed preheating.

As there is neither hydrogen nor olefins flowing into the reactors anymore, remove them from the reactor by purging with nitrogen. As the bottom of the catalyst bed operates at higher temperatures, flow direction is important. Thus, it is recommended up flow purge to avoid ethylene and hydrogen to move down the bed.

Normally acetylene backend converters are provided with arrays of thermocouples in several bed depths, redundancies and so on. However, temperature interlock sometimes is not enough. So, use the automation to set up the interlock properly and do rigorous HazOp studies to deep dive into failure scenarios and corrective actions. Share experience and lessons learned and implement the best practices to improve your operations.

CLARIANT

Temperature Excursion at Start-up, Regeneration and Stand-by

what is precious to you?

CLARIANT

Case #2 – Uncontrolled reaction in the course of reactor start-up

- **Main Cause:**
  - Feed introduction with reactor containing hot spots
- **Description:**
  - Most of the bed TI's were < 45°C (<115°F)
  - Only one TI indicated T > 100°C (>212°F)
  - Feed was introduced to pressurize the reactor
  - Temperature runaway during reactor pressurization ~16 barg (235 psig)
  - No flow → no heat removed
  - Potential damage to vessel integrity (internal inspection required; early catalyst replacement)

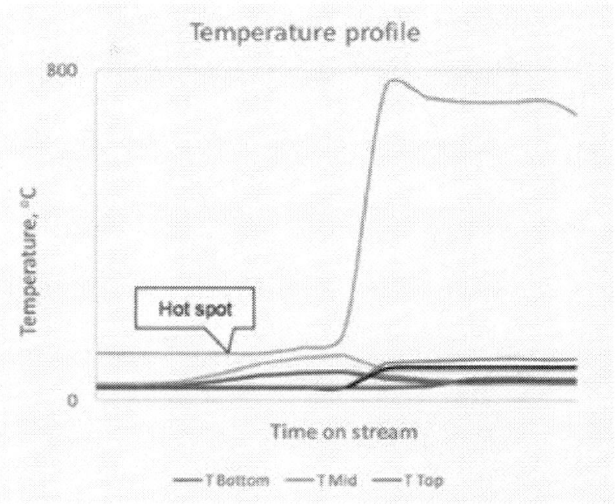

Next, I will discuss temperature excursion when the reactor is not in operation. I mean at converter start-up, regeneration and believe it or not, when the reactor is in stand-by.

Uncontrolled exothermic reaction can also result from faulty start-up. Let's check how it can be:

After plant trip, operation decided to re-start the plant with bed containing one thermocouple indicating temperature above 100°C. While pressurizing the converter, temperature runaway.

In this case, feed was introduced into reactor containing hot spots and initiated ethylene decomposition, which might be triggered by reactor pressurization as well.

It is important to know that a TI indicate temperature of small volume fraction of catalyst bed around the position of the thermocouple in the bed. There is no information about catalyst temperature in the surroundings, where there is no thermocouple. If so, it is impossible to know what the hot spot temperature is and if the hottest temperature is really that one indicated by the thermocouple.

As you can see in the chart, thermocouple indicated above 700°C and that was the end of scale of the instrument, so it is not known what the maximum temperature achieved.

As a safety action, the operator decided to put the reactor off line and unloaded the catalyst for vessel internal inspection.

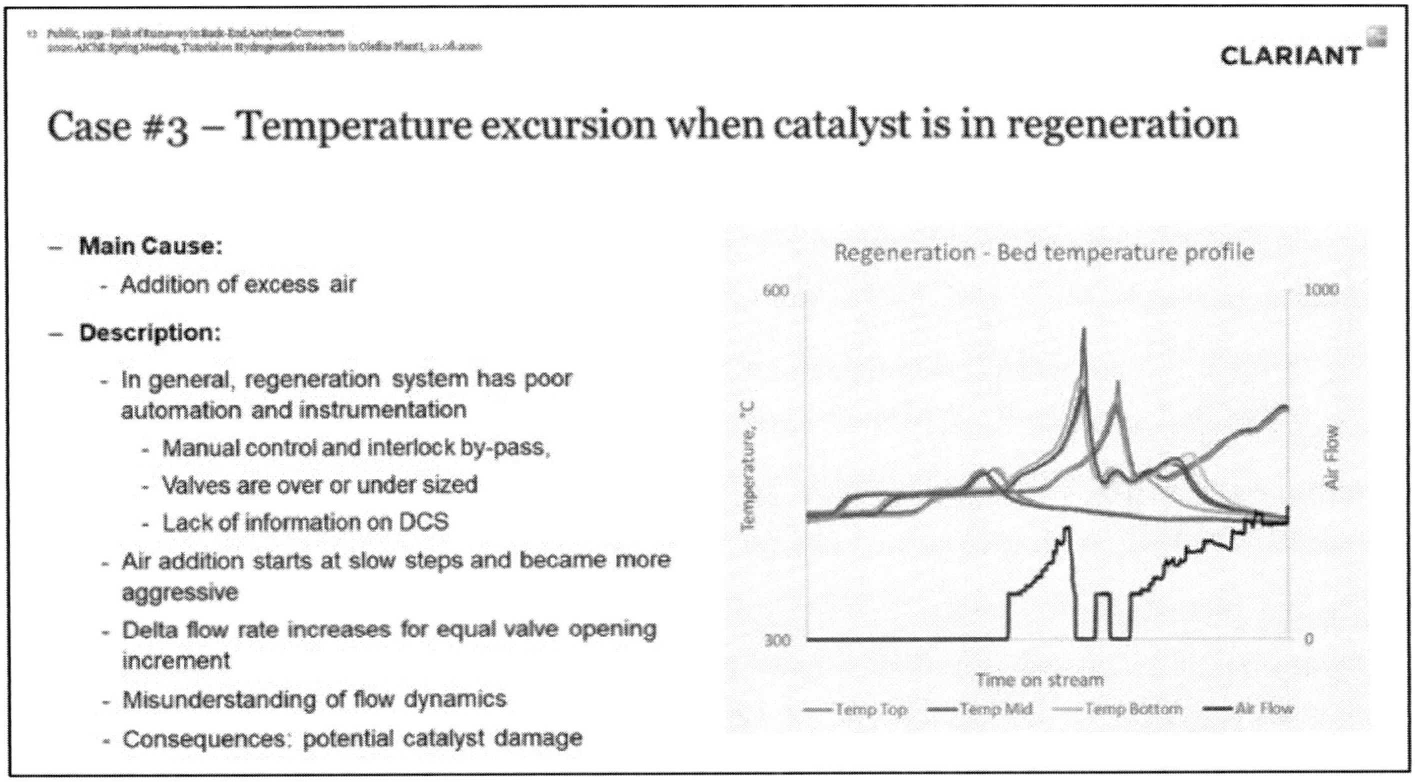

Have you ever seen a similar regeneration chart like this before? This is a temperature runaway caused by addition of air in excess during catalyst regeneration.

Regardless the excessive hydrogenation is the most common hazard of backend acetylene converters, the temperature runaway in catalyst regeneration happens frequently.

If so, I would like to discuss details of how air is added in excess so you can do it better next time.

One of the main reasons is poor instrumentation and automation of regeneration facilities.

In the chart, green, red and blue line represent the TI's in the bottom, mid and top section of catalyst bed. The black line is air flow.

First evidence of poor instrumentation is the bed delta T in the left side of the chart. That is because of oxidation of HC in presence of oxygen. However, the flow rate was not detected by the air flowmeter yet.

Another reason for air addition in large amounts is the pace.

Normally carbon burn-off starts with air flow valve opening in small steps like 1 or 2% at time. As the time pass by, the small valve opening doesn't start air and very little reaction is detected in catalyst bed still. Thus, it is decided to increase the valve travel opening. The biggest mistake!

Due to control valve characteristic, performance is not linear and equal changes in valve position cause equal percentage of changes in flow. So, the higher the flow, the more air is added for the same valve travel.

And what is the result of that? More air (or oxygen) than needed is added and the "Temperature runaway".

In most of the cases, when the acetylene converter is offline for regeneration, the automated safety interlock is by-passed, and the protection layer is upon operators to detect a temperature excursion and take the corrective actions to avoid uncontrolled reactions.

This is how temperature runaway occurs in catalyst regeneration and there is more to discuss about catalyst regeneration. We can further review your regeneration procedure and discuss how to improve it and perform catalyst regeneration better and better.

CLARIANT

Case #4 – High temperature during catalyst activation

- **Main Cause:**
  - Olefins sent to hydrogen during catalyst reduction
- **Description**
  - Operation tried to reduce the catalyst with hydrogen as feed to the PSA
  - Ethylene sent to hydrogen due to fluctuations in the Hydrogen-Methane Separation
  - Temperature excursions up to ~250°C and limited by olefins in gas mixture
  - Consequences: potential for coking and impact on catalyst performance

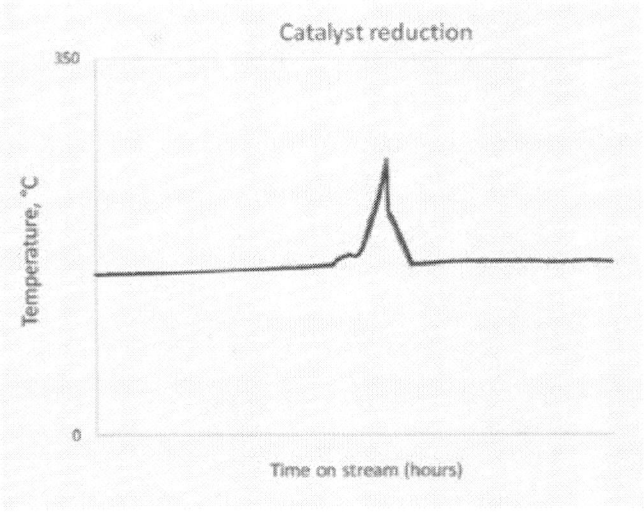

Temperature excursion can also happen during the catalyst reduction. It is not caused by the Pd reduction itself, but for either acetylene or olefins in the reduction gas.

Process fluctuations may result in higher temperatures in hydrogen-methane separation train and consequently changing the vapor-liquid equilibrium. Eventually ethylene concentration builds up in reduction gas and reacts with hydrogen on the catalyst.

Fortunately, this reaction is limited by the concentration of olefins in the reduction gas, but depending on the temperature achieved, it can cause coking and impact catalyst performance next time it is on stream.

A minimum delta T is expected in catalyst reduction. If you notice temperature development, probably there is something wrong.

CLARIANT

Case #5 – Localized high temperature when reactor is offline

- **Main cause:**
  - Line up error resulted in process gas flowing into reactor (hydrogen rich stream)
- **Description:**
  - Reactor was offline for regeneration
  - Hydrocarbon flow established in spare reactor via reactor bottom valve left open through to flare via safety valve bypass
  - Temperatures increased through the catalyst bed
  - The highest temperature reached was over 350°C (660°F)

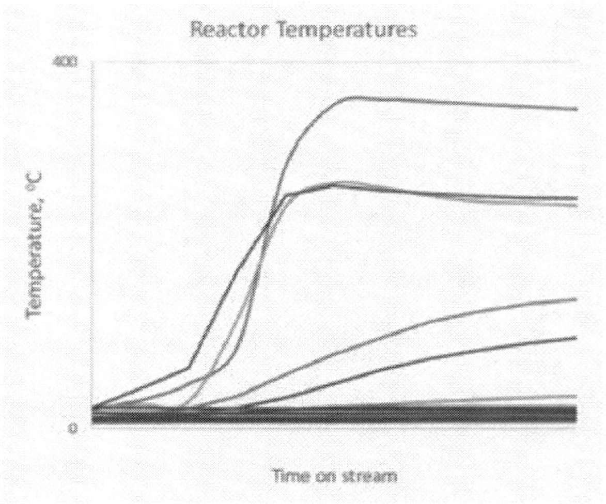

Last but not least, temperature runaway can also happen when the reactor is off-line.

You may be asking to yourself: why in the world that will happen when the reactor is off-line?

The answer is: human error and it can happen with any of us.

A line up error and H2 rich stream was sent into stand-by reactor.

This incident happened after a regeneration. Reactor bottom valve was left open and flow was established through the catalyst bed via safety valve by-pass.

I can imagine this event as a front-end acetylene converter operating with backend catalyst and at low flow condition. There is only one thing to expect from this: Can you see in the chart? Yes, thermal runaway.

Luckily it was detected before extreme high temperature could be achieved in the bed.

Summary

CLARIANT

what is precious to you?

CLARIANT

Takeaways

- Temperature runaway rarely occurs in normal operation due to low hydrogen content in reactor feed
- It does happen in back-end reactors → after a plant trip or when reactor is offline
- Temperature runaway is caused by:
  - Ethylene polymerization and possibly its decomposition, both initiated by:
    - Excessive hydrogenation
    - Introducing feed into a reactor filled with hydrogen or containing hot spots
  - Regeneration: air in excess
  - Catalyst reduction: olefins in hydrogen → limited to olefins content in hydrogen
  - Line up error → reactants flowing through active catalyst
- Potential impact: production losses, catalyst damage, threaten physical integrity of reactor vessel

Lessons learned and rigorous risk analysis will give granular look at potential hazards that may otherwise be overlooked

Those were the cases when temperature runaway can happen in backend acetylene converters that I would like to share with you.
And what is to be learned from this presentation?

Thermal runaway does happen in backend acetylene reactors.

It is difficult to occur in normal operation due to low hydrogen concentration in feed gas. However, hydrogen can be added in excess in plant trips.

High temperature is achieved and triggers either ethylene polymerization or its decomposition.

Both reactions are exotherm and increase the heat release. The result is catalyst overheating and potentially a temperature runaway.

So, temperature runaway is caused by hydrogenation and non-hydrogenation reactions.

Reactor start up with catalyst hot spot is another cause of temperature excursion. We learnt here that high temperatures can result from reintroducing ethylene feed into a converter which hasn't been cooled enough and along the pressurizing step. This hazard is controlled by adherence to start-up procedures.

The most frequent temperature runaway incident is caused by excess of air along the regeneration. So, improve your regeneration facilities and controls to minimize such type of incident.

Attention to reduction gas composition and delta T development during catalyst activation. Palladium reduction exotherm reaction is very low and difficult to be detected in this application.

Protect reactors from other systems by isolation using blinds. That will easily avoid line-up errors and non-desired reaction taking place in spare catalyst bed.

Depending on the extension of a temperature excursion events, it can cause plant shut and production losses, catalyst damage and in extreme case damages to the reactor vessels.

Share your experience. Share with the people. Clearly state them in operation instructions. When you are doing risk analysis, be rigorous. Don't overlook potential hazards that can be mitigated because them can significantly change the outcome of plant trip or faulty operation instructions.

I am sure these real cases will help you somehow!

Thank you!

If you have further questions after the end of this presentation, reach out the authors by e-mail in any of the contact details below:

- Carlos Vianna: carlos.vianna@clariant.com (Product Manager – Selective Hydrogenation)
- Nennette Richard-Jordan: nennette.richardjordan@clariant.com (Head of Sales Americas – Petrochemicals)
- Darren Adams: darren.adams@clariant.com (Global Head of Market Segment – Olefins Purification)

591708 Overview of Acetylene Hydrogenation in Frontend and Backend Acetylene Converters

Abstract:

This session will provide an overview on C2 selective hydrogenation operations in ethylene plants. The presentation will be on selective hydrogenation of acetylene and provide background information for the succeeding papers in this session.

We will share information on the acetylene and MAPD amounts formed as a function of the furnace feed and operating conditions. The paper will cover acetylene hydrogenation in vapor phase, in both adiabatic and isothermal fixed bed reactors. We will highlight the options for locating selective hydrogenation units within ethylene plant sequences and the pros/cons for each location.

In discussing the operating strategy, we will cover topics such as, safety in operations, the mechanism of runaway reactions in front-end acetylene reactors (please note: not covering Tail End systems), the impact of a runaway on catalyst life, green oil formation and mechanisms to protect from runaway reaction.

During the presentation commercial data of different units will be shared which will show safe operations and the risks of temperature runaways.

Copyright © American Institute of Chemical Engineers. All rights reserved.

189d:

SAFETY CONCERNS WITH SELF-HEATING MATERIALS, INCLUDING A PACKED TOWER INDUSTRY EXAMPLE

Tania Almazan
Olefins Recovery Technology Hot Ends Lead Engineer
ExxonMobil Chemical Company

Giselle Larish
Hydroprocessing and Lubes Advisor
Imperial Oil

Selma Kwok
Olefins Recovery Technology Specialist
ExxonMobil Chemical Company

Prepared for Presentation at 2020 Spring National Meeting
Houston, TX, August 17 – August 21, 2020

AIChE and the EPC shall not be responsible for statements or opinions contained in papers or printed in its publications

SAFETY CONCERNS WITH SELF-HEATING MATERIALS, INCLUDING A PACKED TOWER INDUSTRY EXAMPLE

Giselle Larish
Hydroprocessing and Lubes Advisor
Imperial Oil
Tania Almazan
Olefins Recovery Technology Hot Ends Lead Engineer
ExxonMobil Chemical Company
Selma Kwok
Olefins Recovery Technology Specialist
ExxonMobil Chemical Company

Abstract:

Managing materials that have self-heating characteristics is a key process safety concern for many industries, including the ethylene industry. Self-heating materials include materials that, when in contact with air, are liable to self-heat over time, potentially resulting in fire or other damage. Pyrophoric materials are a subset of self-heating materials that ignite rapidly when in contact with air.

This presentation will summarize a 2019 self-heating event involving a section of packing in a tower at a large petrochemical facility. This event will highlight the risks associated with certain vessel internals and self-heating materials. Further, the presentation will provide a broad overview of the risks associated with self-heating materials in an ethylene plant, highlighting areas of specific concern along with potential mitigations.

Self-Heating and Pyrophoric Materials Overview

Background

Managing materials that have self-heating characteristics is a key process safety concern for many industries, including the ethylene industry. Since many polymers, foulants, coke, catalysts, and adsorbents found in olefins facilities are potentially capable of self-heating or pyrophoric activity when exposed to air, ongoing monitoring and mitigation is recommended any time these materials remain inside of open equipment as well as during removal and disposal.

Self-heating materials exhibit a temperature increase when contacted with air, even without an external heat source. Pyrophoric materials are a subset of self-heating materials. They are liquids or solids that can ignite within five minutes after coming into contact with air, even in small quantities and without an external heat source (1). One example of a pyrophoric material is iron sulfide, which is formed when iron reacts chemically with sulfur, typically H_2S.

Since the probability and intensity of self-heating or pyrophoric behavior can vary widely throughout a plant, past experience is not a reliable indicator of future self-heating or pyrophoric activity. Although certain self-heating materials such as popcorn polymer (PCP) are highly reactive and can ignite in minutes, other self-heating materials can take days to react (e.g., very slow heating materials and pyrophoric materials encased in foulant). Moreover, self-heating or pyrophoric behavior can vary with conditions such as humidity, temperature, particle size, and other factors.

Fractionating columns with internal packing and equipment with demister screens are particularly susceptible to self-heating. In general, equipment with a large surface area, narrow flow passages, or regions of low flow can trap iron sulfide scale or other foulants, leading to an increased self-heating/pyrophoric risk.

Where in Olefins Plants are Self-Heating/ Pyrophoric Risks?

Self-heating or pyrophoric materials are common in an olefins facilty. The ethylene industry has experience with these materials in all areas of the plant:

Area	Examples of Self-heating/Pyrophoric Material Observed	Comments
Feed Systems	Iron sulfide, foulant, spent adsorbent	
Furnace Area	Coke, iron sulfide, foulant, deposits	Most common downstream of quench systems
Primary Fractionator/Quench Systems	Coke, iron sulfide, foulant	Primary fractionator, tar filters, tar exchangers, and tar tanks are most vulnerable
Process Gas Compressor Area	Iron sulfide, foulant, spent adsorbent	Acid gas treating units and equipment upstream are most likely to see iron sulfide

Area	Examples of Self-heating/Pyrophoric Material Observed	Comments
Adsorbent/Catalyst	Silver acetylide, copper acetylide Unregenerated and spent methanator catalyst Unregenerated and spent hydrogenation catalyst	Silver acetylide can be formed from acetylene contacting adsorbent/catalyst beds containing silver, and copper acetylide from beds containing copper
Cold Box, Demethanizer, Refrigeration Systems	Foulant	Although kept clean, self-heating has been observed
Deethanizer/Depropanizer	Foulant	Foulant could include popcorn polymer, which is highly pyrophoric (2)
Debutanizer	Popcorn polymer, other foulant	
C5+ Processing	Popcorn polymer, other foulant, nickel sulfide, iron sulfide, chromium sulfide	

Equipment history is not a reliable indicator of future potential self-heating/pyrophoric activity, as demonstrated by the above observations in even "clean" areas of an olefins plant.

Packed Tower Industry Example

Creep Buckling Failure Incident Summary

Even for facilities with long experience favorably managing risk associated with opening equipment containing self-heating or pyrophoric material, self-heating and pyrophoric incidents can still occur. For example, in April 2019, a large fractionation tower failed approximately one hour after opening three manways for maintenance. The tower was located at the back end of a refinery hydrotreating unit. Although no one was injured, this event resulted in an appreciable disruption of plant operations.

The tower was installed in 1968 and had been opened five times with no history of pyrophoric material. The 2019 turnaround was the first time the tower had been opened since 2011. In 2011, nine bottom trays in the tower were replaced with structured packing to improve fractionation efficiency.

This unit produces a light hydrocarbon fuel, and was considered by stakeholders associated with the design, operation, and maintenance of the tower to be a "clean" service. There was no history of self-heating foulant being observed in the tower circuit prior to the event. The non-pyrophoric history of the tower and its use in clean hydrocarbon service resulted in an underestimation of pyrophoric risk.

The sequence of events in this incident started with a standard chemical cleaning and cooldown of the tower. After blanking, Manway 4 was opened and its air mover was turned on. Manway 1 was then opened. When Manway 2 was opened and its air mover turned on, temperatures unexpectedly increased.

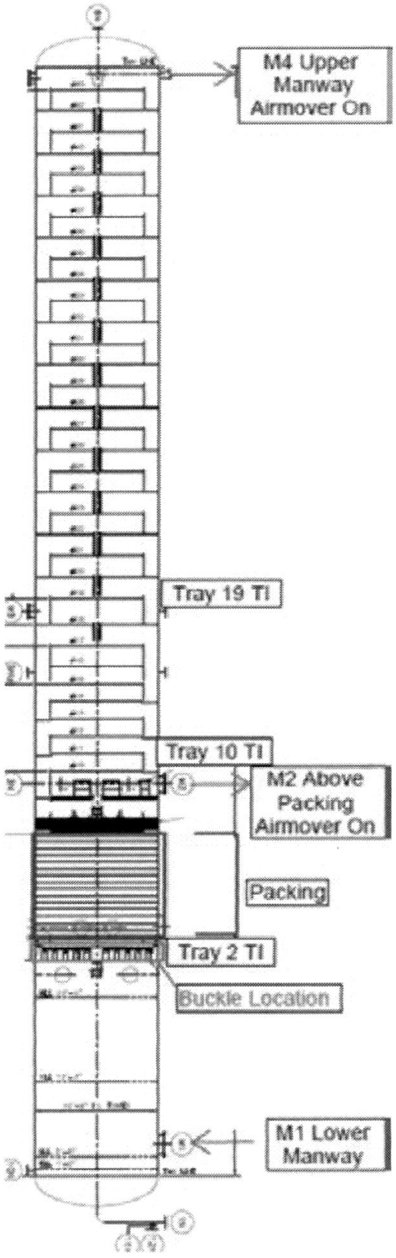

2019 refinery tower drawing with manways labeled

The temperature rise was associated with a pyrophoric reaction in the packed bed. This led to local overheating of the shell, resulting in creep buckling and failure of the tower about one hour after Manway 2 was opened.

Risk Considerations

Carrying out a more rigorous self-heating/pyrophoric risk assessment beforehand may have lessened or avoided the tower damage and operations difficulties that resulted from the event. A rigorous self-heating and pyrophoric risk assessment of plant equipment should include consideration of at least the following risk factors:

- A history of self-heating or pyrophoric behavior in the equipment
- Chemistry that can generate self-heating/pyrophoric material (either in the equipment itself or at a location that is upstream of the equipment)
- Known or suspected iron sulfide accumulation in the equipment
    - Examples of indicators of iron sulfide accumulation are increased pressure drop, visual evidence of iron sulfide in filters, or the presence of sediment in stream samples
- Equipment has changed service since it was last opened
    - Changes in service may include feed property changes, operating severity changes, and metallurgy changes
- Equipment has seen a change in on-stream duration versus the previous cycle
    - A longer runlength provides an opportunity to accumulate a larger quantity of self-heating/pyrophoric material
- Changes in hardware since the equipment was last opened

Features that can increase self-heating/pyrophoric risk include structured or random packing, demisters, fixed bed reactors, filters, and regions of low flow. The larger the quantity of packing, for example, the greater the potential consequence of a self-heating/pyrophoric event. In the case of fixed bed reactors, catalysts are often self-heating, and the fixed bed itself acts as a very efficient collector of foulants. Regions of low flow provide a place where self-heating/pyrophoric materials can settle out and accumulate.

Consider triggers in project and change management work processes to prompt a self-heating/pyrophoric risk assessment.

Potential Mitigations

Self-heating/pyrophoric mitigations can be implemented before and after exposing potentially self-heating/pyrophoric material to oxygen. Before opening equipment, reducing the amount of self-heating/pyrophoric material in the equipment can be done via special procedures for equipment preparation, chemical cleaning (not a sufficient mitigation by itself), and cooling to reduce the vapor pressure of flammable residues (also does not eliminate the risk).

It is important to limit exposure of self-heating/pyrophoric materials to oxygen, e.g. when equipment is opened to the atmosphere. For example, air flow across packing should be minimized before it is wetted. After an initial wetting with water, the material needs to be periodically or continuously re-wetted to avoid dry-out. Consideration should be given to the water supply rate and adequate water distribution to cover the equipment cross-sectional area. Air can be humidified to slow dry-out between periodic wettings.

Another option to limit exposure to oxygen is to water-flood the equipment, but only after assessing the equipment's foundation/support should for handling the additional load.

Provided reasonable measures are taken to prevent asphyxiation risk, inerting the equipment with nitrogen is an alternative method to exclude oxygen. This is a common technique used for spent catalyst removal from fixed-bed reactors.

When equipment is opened to the atmosphere, plans should be in place to quickly detect the onset of self-heating/pyrophoric reaction in the vessel. Because the chemistry can be very fast, an integrated process/mechanical opening and risk mitigation plan should be developed prior to opening the equipment. Special procedures for opening process equipment could include allowing limited initial oxygen exposure to reveal self-heating/pyrophoric potential, paying special attention to manway opening sequence, and allowing natural draft for a period prior to starting air movers. Monitoring parameters include gas monitoring (CO and SO_2) and temperature monitoring.

In the event that monitoring reveals that self-heating/pyrophoric chemistry has initiated, the integrated process/mechanical team should plan for the following emergency response:

- Remote shutdown of air movers to decrease air exposure of the material
- Rapid closure of manways or another means to starve the chemistry of oxygen (i.e. a lightweight fitted plate or fire blanket)
- Further wetting of the vessel or equipment internals
    - Consider the volume and rate needed for internal configuration and potential reaction with the material
- Nitrogen or steam could be used to purge the oxygen from the vessel
    - This method could pose risks to nearby personnel and have limited effectiveness with multiple manways open
- Water flooding of the equipment
    - The vessel and foundation structure's ability to hold the weight would need to be considered before mechanical work begins
- Application of foam
- An emergency response plan with egress plans

Key Messages

As shown by this pyrophoric incident, history is not a reliable indicator of potential self-heating/pyrophoric risk. It is important to ensure that equipment maintenance planning includes an assessment of this risk. Chemical cleaning can be a part of the mitigation plan, provided it is recognized that chemical cleaning by itself is not a sufficient mitigation against self-heating. Finally, pyrophoric chemistry happens very quickly. Ensuring that opening process equipment work processes include self-heating/pyrophoric discussions and development of integrated process/mechanical risk mitigation plans will greatly reduce the risk of another incident like this one. All key personnel involved in the planning and execution of opening process equipment need to understand self-heating/pyrophoric hazards.

References

1. United Nations Manual of Tests and Criteria (2019).
2. Broekaart, Paul and Romagnoli, Massimo (2019), "Recommendations for Preventing Popcorn in Steam Crackers and Butadiene Plants," *European Ethylene Producers Committee.*

589971 Pixel-Based Approach in Combination with GC Ã— GC for Fouling Tendency Prediction in Steam Crackers

Abstract:

Pixel-based approach in combination with GC × GC for fouling tendency prediction in steam crackers

Marko R. Djokic[1,*], Victor Abrahamsson[2], Nenad Ristic[1], Kevin Van Geem[1]

[1] Ghent University, Laboratory for Chemical Technology, Technologiepark 914, 9052 Gent, Belgium.

[2] Jet Propulsion Laboratory, California Institute of Technology, Pasadena, CA 91109, USA

Abstract

Fouling tendencies of a series of gas condensates were evaluated using comprehensive two-dimensional gas chromatography with flame ionization detection and sulfur chemiluminescence detection. A pixel-based approach was applied in order to identify parts of the chromatograms which were associated with the reactor coil fouling. Particular emphasis is given in this work to evaluate several feature selection methodologies along with various data preprocessing procedures. It was found that both aspects were crucial for studying the fouling tendencies and, as part of the subsequent partial least squares model development, predominantly the feature selection. Based on the flame ionization detector chromatograms and using the RReliefF algorithm for feature selection, a partial least squares regression model with one latent variable resulted in a root mean square error of the cross-validation of 0.65 g deposit/6 h (17%) (see Fig 1.). Based on the sulfur chemiluminescence detector chromatograms, the F-statistics feature selection generated a slightly better partial least squares regression model compared to using RReliefF, thus generating a model using one latent variable with a root mean square error of the cross-validation of 0.81 g deposit/6 h (21%) (see Fig. 2). Heavy aromatic compounds and heavy sulfur containing compounds were negatively associated with the fouling rate. Both were crucial in developing a partial least squares model with good prediction power, however, worked independently as predictors.

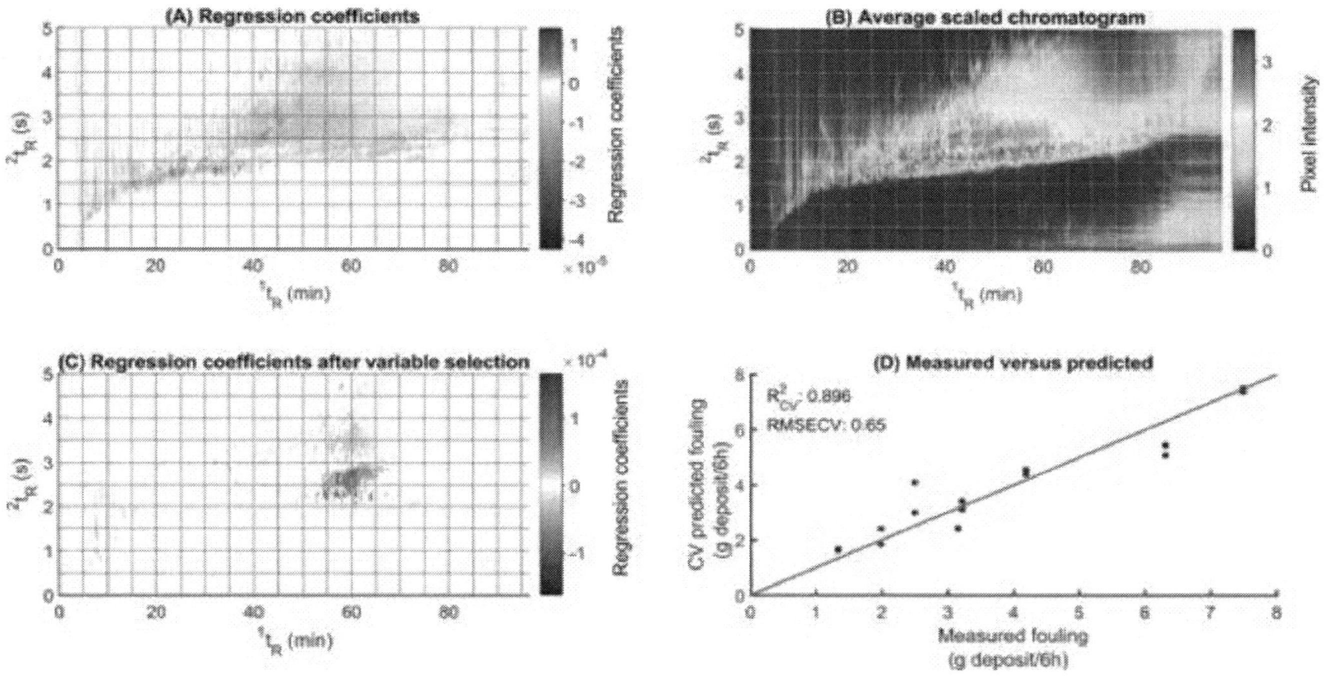

Fig 1: PLSR model based on the scaled FID chromatograms. (A) Presents the regression coefficients prior to feature selection, while Figure C presents the regression coefficients after using RReliefF as feature selection and selecting the optimum cut-off value. (B) Shows an average GC × GC-FID chromatogram based on the arithmetic mean of the scaled and base 10 logarithm transformed chromatograms. The measured versus the cross-validated prediction of the final PLSR model is displayed in (D).

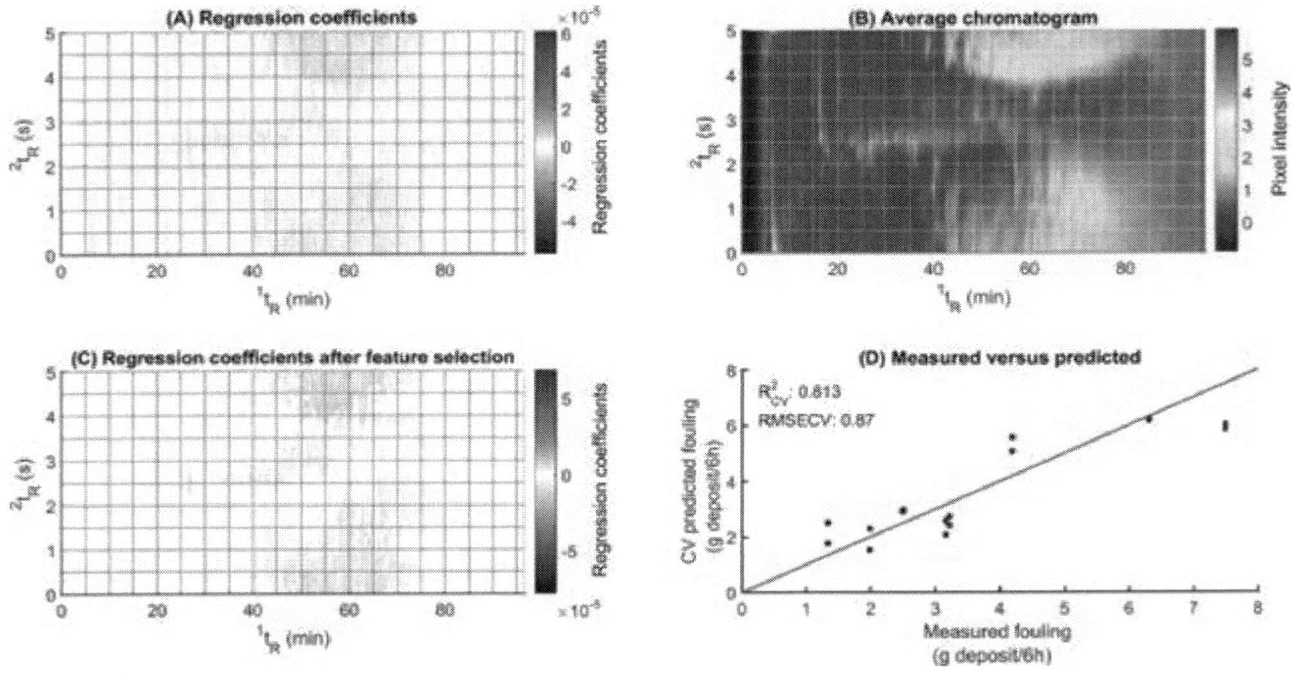

Fig. 2: PLSR model based on the SCD chromatograms. (A) Presents the regression coefficients prior to feature selection, while (C) presents the regression coefficients after using F-statistics as feature selection and selecting the optimum cut-off value. (B) Shows an average GC × GC-FID chromatogram based on the arithmetic mean of the base 10 logarithm transformed chromatograms. The measured versus the cross-validated prediction of the final PLSR model is displayed in (D).

Copyright © American Institute of Chemical Engineers. All rights reserved.

AICHE PAPER NUMBER: 159e

CARBON DISULFIDE AND THIOPHENE: FORMATION AND MITIGATION REVIEW

Dr David Slim

Technology Advisor

Evonik Corporation

Dr Phillip James

Technology Director

Evonik Performance Materials GmbH

Joseph Lally

Technical Service Manager

Evonik Corporation

Prepared for Presentation at the 2020 Spring National Meeting

Houston, TX, March 30 – April 2, 2020

AIChE and EPC shall not be responsible for statements or opinions contained in papers or printed in its publications

CARBON DISULFIDE AND THIOPHENE: FORMATION AND MITIGATION REVIEW

Dr David Slim

Technology Advisor

Evonik Corporation

Dr Phillip James

Technology Director

Evonik Performance Materials GmbH

Joseph Lally

Technical Service Manager

Evonik Corporation

Abstract

Thiophene and carbon disulfide are two compounds that can cause difficulties in the steam cracking process. This paper describes how these species are formed, the levels that are normally experienced, their impact and how they are normally managed. Thiophene and carbon disulfide are found in cracking furnace effluent. Both species are poisons to hydrotreating catalysts. In addition, thiophene boils very closely to benzene and hence is an issue for benzene purification.

Summary

Thiophene (C_4H_4S) and carbon disulfide (CS_2) have been identified as important contaminants to steam crackers. Both have been observed in cracker feed and carbon disulfide can be made in the cracking furnaces. Both compounds cause problems in product quality and catalyst activity. Both compounds have low chemical reactivity and hence are difficult to remove from product streams. However, technology exists to ensure product quality specifications and catalyst activity targets are met. All ethylene plants can manage these contaminants. Such technologies include polyamines and additives for CS_2 and extractive distillation for thiophene

Introduction

The EPC has reviewed various aspects of sulfur species and their impact on the steam cracking process. It is well known that certain sulfur compounds can produce H_2S in the cracking furnaces. This can retard coke formation and enable optimal run lengths and high olefin yields to be obtained. The mechanism and sulfur levels have been described in previous EPC presentations. In summary H_2S forms at the metal interface and undergoes the following inhibition process

- Catalytic coke: H_2S produces [S] which is adsorbed on the surface and poisons the catalytically active metal sites responsible for hydrocarbon decomposition reactions such as dehydrogenation.
- Radical-induced coke: •HS radicals (from H_2S) inhibits by transfer reactions. Continuous sulfur compound addition is required to maintain an optimum amount of adsorbed S needed to inhibit undesirable reactions.

The types of molecules that liberate H_2S include thiols (RSH e.g. C_2H_5SH) sulfides (R_1SR_2 e.g. $C_2H_5SCH_3$) and polysulfides ($R_1(S\text{-}S)_nR_2$ e.g. CH_3SSCH_3. After the cracking furnaces, the lighter S species must be removed or controlled as they can impact downstream process es and product quality. Examples include acid gas treatment for removal of H_2S, thiols and to some extent COS. Heavy sulfur species that concentrate in gas oil cuts are treated outside of the ethylene plant.

The two contaminants that can "slip through the cracks "are CS_2 and thiophene. This paper discusses these components.

Discussion

Properties of thiophene and its comparison to benzene

The general properties of thiophene are described in the sketch and table below

Table 1: properties of thiophene

Chemical Formula	C_4H_4S
Molecular weight	84.14 g/mol
Appearance	colorless liquid
Specific gravity	1.051
Melting point	−38 °C
Boiling point	84 °C

Thiophene is a heterocyclic compound with the formula C_4H_4S. It consists of a planar five membered ring. Thiophene is pseudo-aromatic like benzene. it is resonance stabilized and thus all the bonds in the ring have partial multiple bond character. The resonance stabilization is only slightly less than benzene (120kJ vs 150kJ). This is expected as the S atom is more electronegative than a C atom which draws electrons away from the pi system. Because of its aromaticity, thiophene does not exhibit the properties seen for conventional sulfides. For example, it does not liberate H_2S in cracking furnaces. It is a colorless liquid with a benzene-like odor.

The C–C bonds to the carbons adjacent to the sulfur are about 134 pm, which is comparable to that of benzene, while the other C-C bond is a little weaker (about 141 pm). A normal C-C bond is about 154 pm. Thus, all the C-C bonds in thiophene have multiple bond character. The C–S bond also has multiple bond character (170 pm vs 183 pm as in dimethyl sulfide). These properties show why thiophene is difficult to decompose in a cracking furnace.

Thiophene and especially its derivatives e.g. benzothiophene occur in petroleum sometimes in concentrations up to 1–3%. Thiophene boils close to that of benzene and is found in furnace effluent stream. Benzothiophene and higher thiophenes boil in the gas oil and tar streams and are not issues in steam cracking.

Benzene is a precursor to heavy chemicals, such as ethylbenzene and cumene. As a common contaminant in benzene, thiophene harms the benzene process industry, for example, eroding metallic equipment, making catalyst poisoning, affecting product quality and causing environmental pollution. It is difficult to separate thiophene from benzene by distillation due to their similar boiling points (4°C difference at ambient pressure). Like benzene, thiophene forms an azeotrope with ethanol.

Table 2: Boiling Points of thiophenes and benzene

Compound	Boiling point, °C
Thiophene	84
Benzothiophene	221
Benzene	80

Properties of carbon disulfide

The general properties of carbon disulfide are described in the table below

Table 3: Properties of carbon disulfide

Appearance	Colorless liquid Impure: light-yellow
Density	1.2927 g/cm^3 (0 °C)
Melting point	−111.6 °C
Boiling point	46.2 °C
Water Solubility	2.17 g/L (20 °C)
Solubility in organic solvents	Soluble in alcohol, ether, benzene
Vapor pressure	48.1 kPa (25°C)

Carbon disulfide is a colorless volatile liquid with the formula CS_2. The impure carbon disulfide that is usually used in most industrial processes is a yellowish liquid with an unpleasant odor, like that of rotting radishes. It evaporates at room temperature, and the vapor is more than twice as heavy as air.

CS_2 is a non-polar molecule because it has a linear geometric shape with a symmetric distribution of charge. The CS bonds are double bonds (bond length 155 pm). CS_2 is isoelectronic with carbon dioxide but it is more reactive toward nucleophiles and more easily reduced. These differences in reactivity can be attributed to the weaker π donor-ability of the sulfido centers, which renders the carbon more electrophilic.

The commercial manufacture and uses of CS_2 and thiophene are described in appendix 1

Furnace impact of thiophene and CS_2

The carbon sulfur bonds in both thiophene and carbon disulfide have multiple bond character. This is exemplified in the following table

Table 4: Bond order for CS and thiophene

Molecule	C-S bond length (pm)	C-S bond strength kJ	C-S Bond order
Methane thiol	182	272	1
thiophene	170	~420	1.5
Carbon disulfide	160	573	2

Given the multiple bond character of both thiophene and CS_2, these species and derivatives do not decompose in the cracking furnaces. The C-S bond strengths are significantly stronger than C-C single bonds (~350 Kj). Given this thiophene and CS_2 are termed "non-crackable" and hence will not generate H_2S. Thus CS_2 or thiophene in furnace feed passes through unscathed.

The possibility of CS_2 formation in the radiant section must be considered as well. Up to the 1950s, CS_2 was commercially produced from the elements at about 800°C. However, the partial pressures of sulfur and carbon in the radiant section are mostly regarded as too low and thus formation of CS_2 can only contribute marginally to the overall balance. It has been observed that in many imported feeds the level of CS_2 has recently increased remarkably. This is worth exploring. There have also been reports of low levels of CS_2 being formed as side products from using crackable sulfur compounds.

While mercaptans and organic disulfides are mainly cracked to H_2S in the radiant coils and later washed out in the caustic column, thiophenic, benzothiophenic, and other complex sulfur compounds are reasonably stable and end up in the pyrolysis gasoline fraction. The same happens to carbon disulfide (CS_2), which mostly originates from the cracker feed passing the radiant section without conversion.

There have been reports of a reaction of CS_2 with alkali metal hydroxides (KOH and NaOH). The consensus seems that the rate of dissolution of CS2 is very slow and the kinetics are also unfavorable.

CS_2 reaction with amines

CO_2, COS and CS_2 are iso structural. Since amines e.g. are used to remove high levels of CO_2 from process streams, it is important to compare their effectiveness in treating streams containing COS and CS_2

The reaction of these three components has been established to proceed via a zwitterion formation. Thus, for CO_2

$$CO_2 + RNH_2 = RNH_2^+COO^-$$

$$RNH_2^+COO^- + RNH_2 = RNH_3^+RNHCOO^-$$

For CS_2 the product is termed a dithiocarbamate. The first equation is reversible i.e. the back reaction has to be considered while the second equation is irreversible i.e. the back can be ignored Using conventional kinetics and the measured values of the rate constants, the rate equation can be expressed in the form

$$\text{Rate} \quad = \quad k_3[RNH_2]^2[CO_2]$$

The values of the rate constant k_3 have been measured. The value for CO_2 is approximately 100 times that for COS which in turn is greater than that for CS_2. Note in primary amines, e.g. MEA the thiocarbamate from COS decomposes irreversibly to form a cyclic derivative. It is likely that the dithiocarbamate from CS_2 behaves in a similar manner.

Given the above, conventional amines are relatively ineffective in removing CS_2

However, CS_2 reacts readily with diamines. This forms the basis for CS_2 removal systems. This can be described in the example

$$\underset{\text{Ethylenediamine}}{NH_2CH_2CH_2NH_2} \quad + \quad CS_2 \quad \leftrightarrow \quad \underset{\text{Zwitterion}}{NH_3^+CH_2CH_2NHCS_2^-}$$

Zwitterion $\rightarrow$ 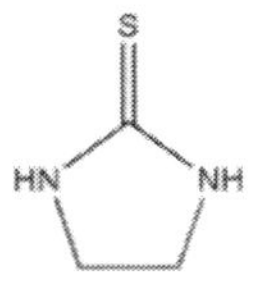 + H_2S

Product is 2-imidazolinethione or ethylene thiourea mp 203°C, bp 347°C

In this case, the dithiocarbamate ion decomposes and liberates a molecule of H_2S to forma cyclic derivative. Presumably the H2S is picked up by the amine. This reaction path forms the basis for CS_2 removal, which is described later

Impacts of thiophene and CS2

CS_2 concentrates in the C5+ stream (pyrolysis gasoline) from the quench water settler and debutanizer bottoms in the steam cracker. For further processing the operator has three main options

- Extract the C5's from the pygas in order to process them in a C5 extraction unit (often a so-called isoprene extraction unit)
- Hydrogenate the C5's for recycle cracking
- Send the C5+ stream to the Pygas hydrotreater.

The extraction and hydrogenation units are the units within the steam cracker process and have reported issues with their performance due to the presence of thiophene and CS_2

<u>PyGas Processing</u>

The purpose of hydrogenation of pyrolysis gasoline is to eliminate highly reactive species deemed to polymerise and potentially cause fouling and plugging, such as alkynes, diolefins, styrene and olefin compounds, as well as to remove sulfur, while minimizing losses of aromatics. This is commonly achieved by hydrogenation in two subsequent stages. The first stage is a trickle phase operation employing palladium or sulfided nickel catalyst to remove alkynes as well as diolefinic and styrenic species. The second stage is a vapour phase operation employing sulfided nickel molybdenum (NiMo) and cobalt molybdenum (CoMo) catalysts to remove mainly olefins and convert sulfur compounds to H_2S

The formation of metal sulfides has an impact on the hydrogenation reactions as it retards all hydrogenation rates. It improves the selectivity of converting di-olefins above the selectivity of converting olefins. Unsulfided Ni or Pd catalyst could hydrogenate aromatics and the presence of sulfur strongly suppresses the hydrogenation of aromatics.

CS_2 acts like a "semi-permanent" poison on both nickel and palladium catalyst applied in the 1st stage. One should not neglect the impact on the 2nd stage pygas reactor as underperformance of the 1st stage reactor cascades to the 2nd stage reactor and a substantial shift in exotherm could occur. The customer example shown in this article shows that catalyst developments and service support (modeling and technical support) can help improve performance by shifting the operation to a more pro-active response.

The deactivation of palladium-based catalyst for the hydrogenation of pyrolysis gasoline has been investigated. the content of CS_2 in the feedstocks was high in the range of 1.8-20.2 μg/g. The CS_2 content in feedstock is recommended to be less than 1 μg/g in well-running hydrogenation units. A deactivation test for the hydrogenation catalyst has been carried out by adding CS_2, dimethyl disulfide and thiophene to feedstock. The effects of the catalyst poisoning on the hydrogenation are in order of CS_2>> dimethyl disulfide > thiophene. Given this while thiophene can be classified as a catalyst poison, it is not regarded as a major poison. Thus, the deactivation of the catalyst can result from CS_2 in the feedstock. The CS_2 content is often linearly related with the C5 content in the feedstock. If the CS_2 content in the feedstock is less than 1 μg/g when the C5 content is less than 0.5%(w), and then the hydrogenation unit can run normally.

Targeting reliable and stable operation, the performance of the catalyst is essential. Knowing the feed composition, operating parameters, and start of run catalyst activity, one can model and predict the catalyst's performance, inclusive of cycle length, for given end of run criteria. This includes adjustments of the model for known feed poisons and inhibitors. If catalyst activity is permanently reduced by a feed contaminant, one classifies such a contaminant as a poison. If catalyst activity is restored after contact with the specific feed contaminant has been stopped, one classifies such a contaminant as a temporary poison or inhibitor.

However, a great advantage is the flexibility to use a broad range of cracker feedstocks. Inherent with the benefit of such flexibility, the composition of the feed and the catalyst poisons and inhibitors in the feed change by nature and quantity.

For stable operation of a pyrolysis gasoline unit, it is of high value to know the effects that various poisons and inhibitors have on the performance of the catalyst as well as to know if and what measures can be taken to counteract and to restore the catalyst's performance.

Other Removal options

Use of polyamines to remove CS_2

Certain polyamines e.g. triethylenetetramine (TETA) are effective in removing CS_2. The structure of TETA is given below

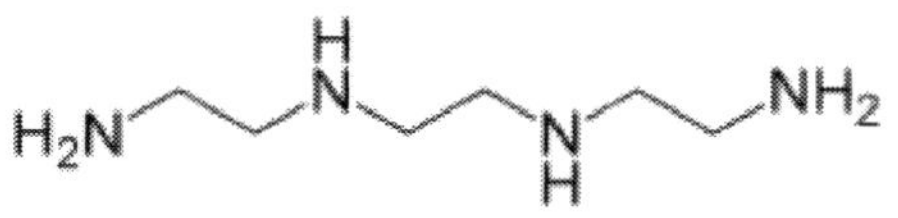

Based on this structure and the chemistry discussed for amines, it is clear that molecules such TETA can be effective in removing CS_2

Separation of thiophene from benzene

There is an urgent need to remove thiophene to ensure the high purity of benzene. Therefore, separation of the benzene-thiophene mixture is of great industrial interest in purification technology. The separation of azeotropic or close boiling binary mixtures is very difficult using ordinary distillation processes. As a result, extractive distillation, a special distillation technique that removes certain compounds from close-boiling mixtures in industrial processes, has received much attention. Such processes are currently being used. Optimization has been achieved recently making the processes very effective. Extractive distillation has effectively replaced the older technology that involved the use of reagents of a group consisting of diuphenylsulfone and dihydroxydiphenylsulfone and distilling the resulting mixture.

Conclusion

Thiophene (C_4H_4S) and carbon disulfide (CS_2) have been identified as important contaminants to steam crackers. They are both found in cracker furnace effluents. Both compounds cause problems in product quality and catalyst activity. Both compounds have low chemical reactivity and hence are difficult to remove. However, technology exists to ensure product quality specifications and catalyst activity targets are met. All ethylene plants can manage these contaminants. Such technologies include polyamines and additives for CS2 and extractive distillation for thiophene

Several references

Sulfur myths – what we know and what we don't know by M Brayden AICHE Proceedings: Volume 23 – 2011 Session: 33

The Influence of Metal Surface and Sulfur Addition on Coke Deposition in the Thermal Cracking of Hydrocarbons by M. Reyniers G.F. Froment AICHE Proceedings: Volume 6 – 1994 Session: 26

Introduction to Sulfur Chemistry as Related to the Olefin Industry by J. Reid AICHE Proceedings: Volume 10 – 1998 Session 23

Carbon Disulfide (CS2) - A New Challenge for Steam Cracker Complexes Pascal Coenen Ernesto Angueira AICHE Proceedings: Volume 28 – 2016 Session: 126

Introduction to Sulfur Compounds and Sulfur Removal as Applied to Olefin Unit Design and Operation by J Reid AICHE Proceedings: Volume 10

Kinetics of removal of carbon disulfide by aqueous monoethanolamine by Sang Cheol Lee, Michael J Snodgrass, Moon Ki Park and Orville C Sandall,; Environmental Science Technology 2001, 35, 2352-2357

Reactions of carbonyl sulfide and methyl mercaptan with ethanolamines by Mahmud A Rahman, Robert N Maddox and C. J. Mains; Industrial Engineering Chemical Research, 1989 28 470-475

Divalent Sulfur Based Functions, by P. J. Ewing Vol 13 Part 2 Treatise on Analytical Chemistry 1966

Method for separating thiophene from mixtures thereof with benzene US patent US40889273 1973-10-23 1973-10-23

APPENDIX 1

COMMERCIAL MANUFACTURE AND USES OF CS_2 AND THIOPHENE

Commercial thiophene synthesis

The first synthesis of thiophene by Meyer, reported the same year that he made his discovery, involves acetylene and elemental sulfur. Thiophenes are classically prepared by the reaction of 1,4-diesters, or dicarboxylates with sulfidizing reagents such as P_4S_{10}. Specialized thiophenes can be synthesized similarly.

Thiophene is produced on a modest scale of around 2,000 metric tons per year worldwide. Production involves the vapor phase reaction of a sulfur source, typically CS_2, and a C4 source, typically butanol These reagents are contacted with an oxide catalyst at 500–550 °C

Thiophene chemistry

General

As a consequence of its aromaticity, thiophene does not exhibit the properties seen for conventional sulfides double bond, giving the thiophene 2,3-epoxide, followed by

Oxidation

Oxidation can occur both at sulfur, giving a thiophene *S*-oxide, as well as at the 2,3-double bond. This gives the thiophene 2-3 epoxide, followed by the subsequent rearrangement. Oxidation of thiophene also demonstrates both reaction pathways. The major pathway forms the *S*-oxide as an intermediate, which undergoes subsequent Diels-Alder type dimerization and further oxidation, forming a mixture of sulphoxide and sulphone products with a combined yield of 83%

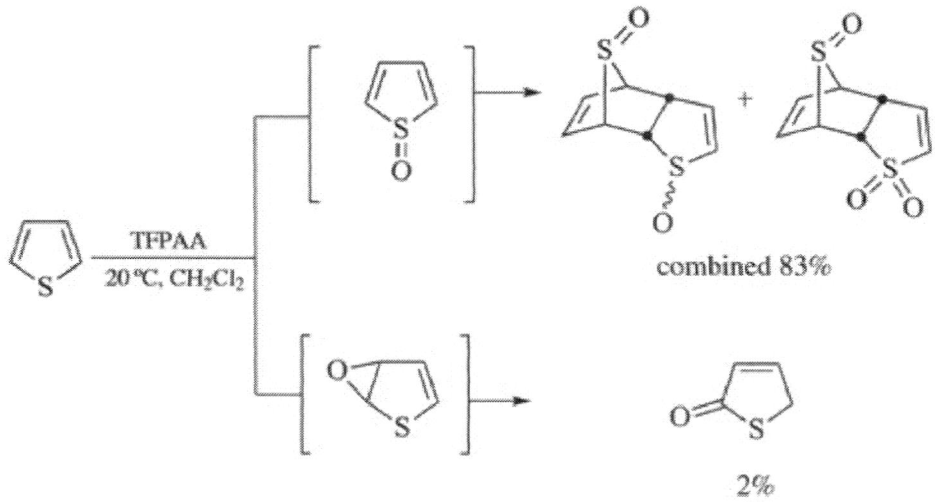

Alkylation

Although the sulfur atom is relatively unreactive, the flanking carbon centers, the 2- and 5-positions, are highly susceptible to attack by electrophiles Halogens give initially 2-halo derivatives followed by 2,5-dihalothiophenes. Thiophene brominates 10^7 times faster than does benzene.

Chloromethylation and chloroethylation occur readily at the 2,5-positions. Reduction of the chloromethyl product gives 2-methylthiophene. Hydrolysis followed by dehydration of the chloroethyl species gives 2-vinylthiophene.

Desulfurization by Raney nickel

Desulfurization of thiophene with Raney nickel affords butane. When coupled with the easy 2,5-difunctionalization of thiophene, desulfurization provides a route to 1,4-disubstituted butanes.

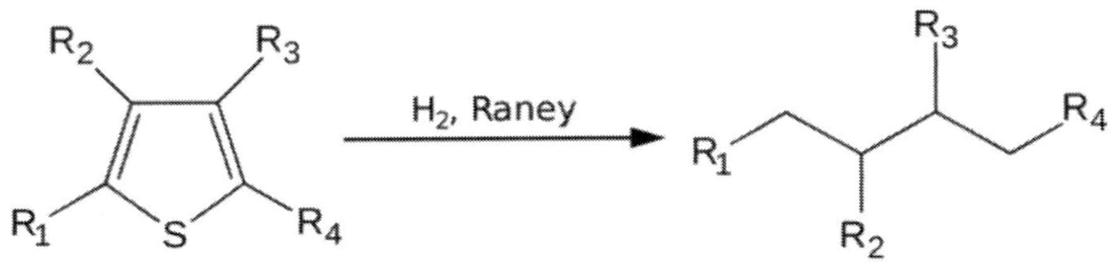

Polymerization

The polymer formed by linking thiophene through its 2,5 positions is called polythiophene Polymerization is conducted by oxidation using electron-transfer reagents. An idealized equation is shown:

$n\ C_4H_4S \rightarrow (C_4H_2S)_n + 2n\ H^+ + 2n\ e^-$

Polythiophene itself has poor processing properties and so is little studied

CS_2 chemistry

At high temperatures CS_2 can be manufactured by combining coke or carbon with sulfur. At lower temperatures, approximately 600°C, in the presence of silica gel or alumina catalysts natural gas can react with sulfur to produce CS_2.

$$2\ CH_4 + S_8 \rightarrow 2\ CS_2 + 4\ H_2S$$

In organic chemistry it is used to manufacture regenerated cellulose, the main ingredient of viscose, rayon and cellophane. It is also used as an industrial and chemical non-polar solvent. Chlorination of CS_2 is the principal route to produce carbon-tetrachloride. From a safety point of view, CS_2 is a toxic and highly flammable component. Its toxicity is close that of H_2S) and as such elevated levels of CS_2 are life threatening as it affects the nervous system. The flash point of CS_2 is -30°C and auto ignition can occur below 100°C

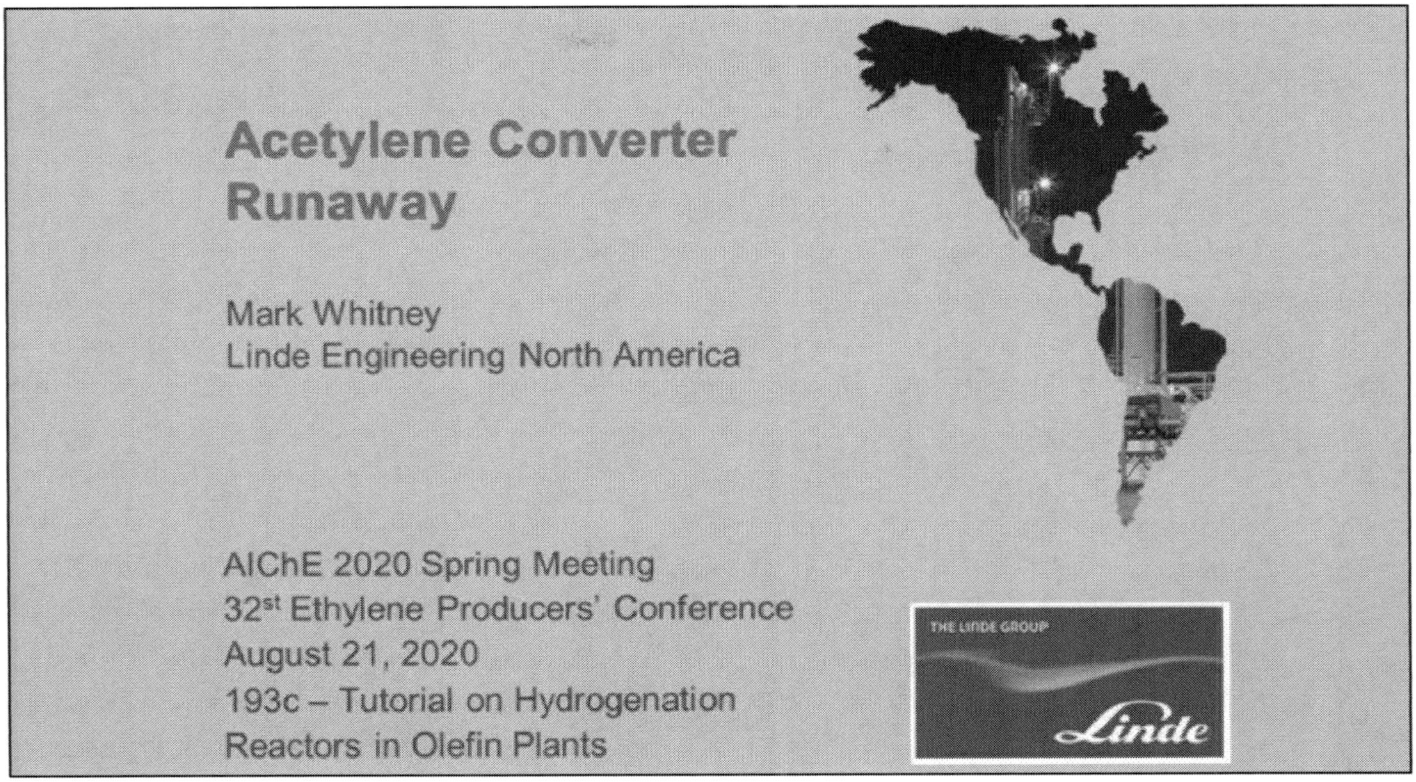

- This is a tutorial to educate engineers new to the ethylene industry.
- This paper is not an in-depth review.
- All the information here is intended to be as generic as possible.
- I'd like to thank several people for helping on this paper including the other members of the EPC's Technology Subcommittee, my co-workers at Linde and other people I have worked with during my career.
- The EPC has produced several papers on hydrogenation reactors. For anyone interested, I have included a list of these papers in the notes that will be available with these slides on the EPC website.

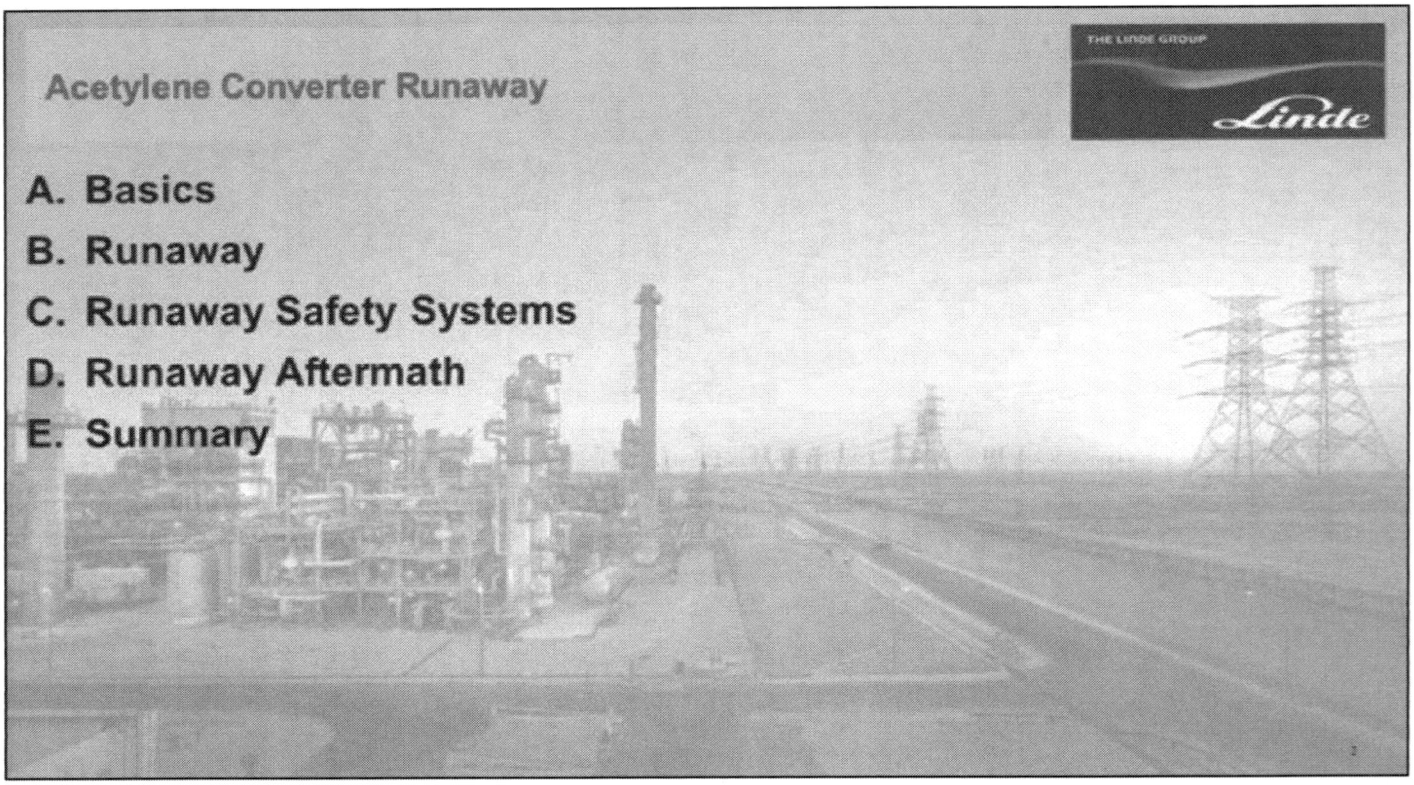

- Despite the best efforts, ethylene furnaces continue to produce some acetylene in the effluent.
- Acetylene is the bane of the olefin plant. It must be removed or destroyed for the ethylene product to be acceptable.
- I'm going to talk in general terms about what happen when the acetylene hydrogenation reaction goes crazy and give you some key points.
- We start with BASICS defining the terms and scope. We'll talk about how acetylene converters work.
- Next, how do RUNAWAYS occur.
- Third, RUNAWAY SAFETY SYSTEMS that detect, predict and remedy a runaway.
- Next, RUNAWAY AFTERMATH. You've survived the runaway, what do you do now?
- Finally, a SUMMARY

- I am going to speed through the basics because there are four other papers in this sessions which will also cover these.
- You have got to know about the reactions which occur in the converter.
- Just as important is how the catalyst adsorption of reactants affects what happens.
- Finally, a brief look at what controls the operators have over the reaction.

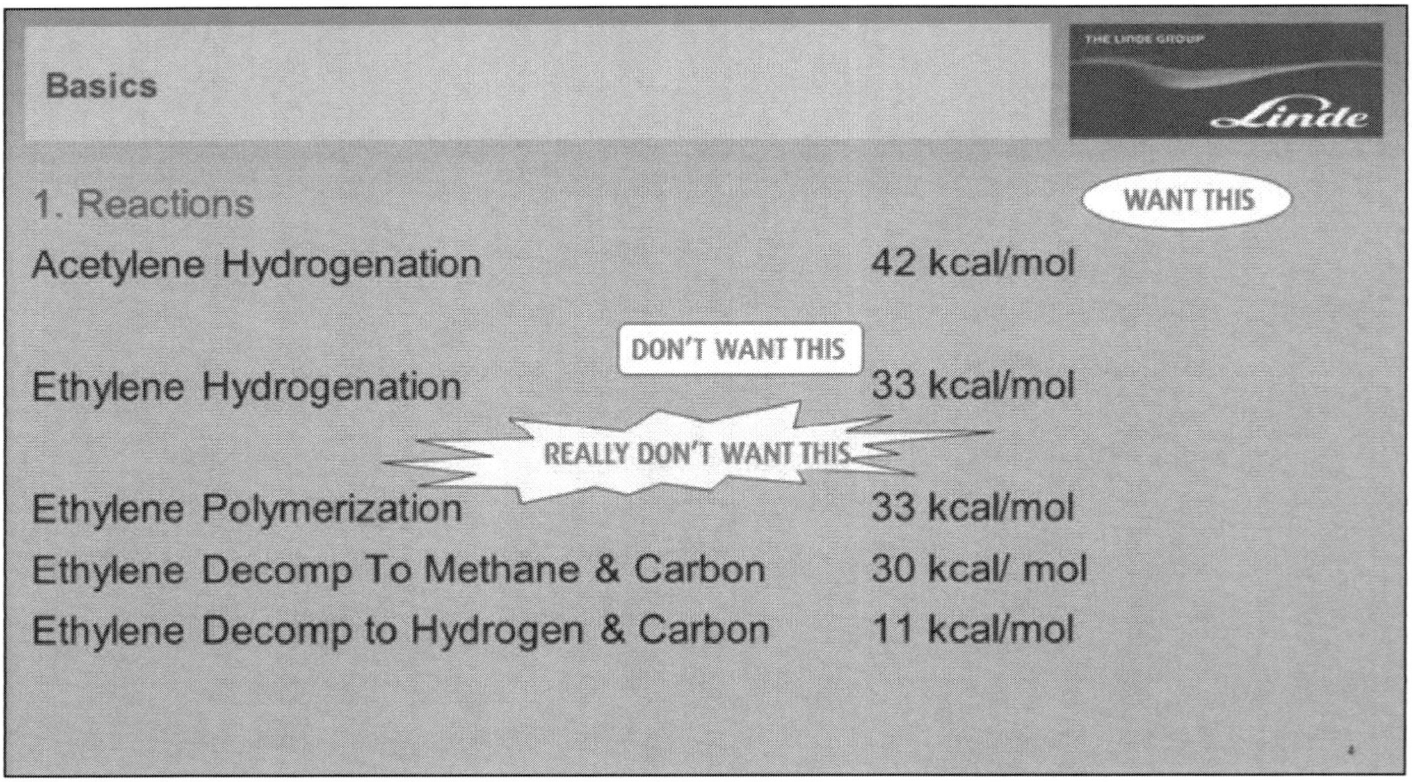

- All of the reactions in the hydrogenation reactor are exothermic.
- Consider this, the converter might see 70 kmol/h of acetylene and 6600 kmol/h ethylene. Converting all the acetylene to ethylene generates just over 3 MW. Converting all of the ethylene to ethane generates over 250 MW.
- If the conditions in the reactor go wrong, other reactions can occur.
    - The ethylene will happily react with itself, making polymers and heat.
    - At over 175°C over catalyst, the ethylene can decompose. It can go to methane and carbon or to hydrogen and carbon.

Basics

THE LINDE GROUP Linde

2. Adsorption

Modern catalysts preferentially absorb in the order

Acetylenes > Diolefins > Carbon Monoxide > Olefins

Material <u>MUST</u> be absorbed on the catalyst, to be hydrogenated.

Note: CO is not hydrogenated.

- A typical reaction would be controlled by temperature and feed concentration.
- Here, the ethylene concentration is about 100 times the acetylene concentration at the converter inlet.
- Further, the rate constant for ethylene hydrogenation is 100 times faster than for acetylene hydrogenation. Logically, that means that ethylene hydrogenation should occur about 10,000 times faster than acetylene hydrogenation.
- Instead the selective adsorption of hydrocarbons on the catalyst controls the reaction. The catalyst adsorbs acetylene first, followed by di-olefins then carbon monoxide and finally olefins. As long as acetylenes, di-olefins and carbon monoxide are present, olefins are kept away from the catalyst and ethylene hydrogenation is minimized.

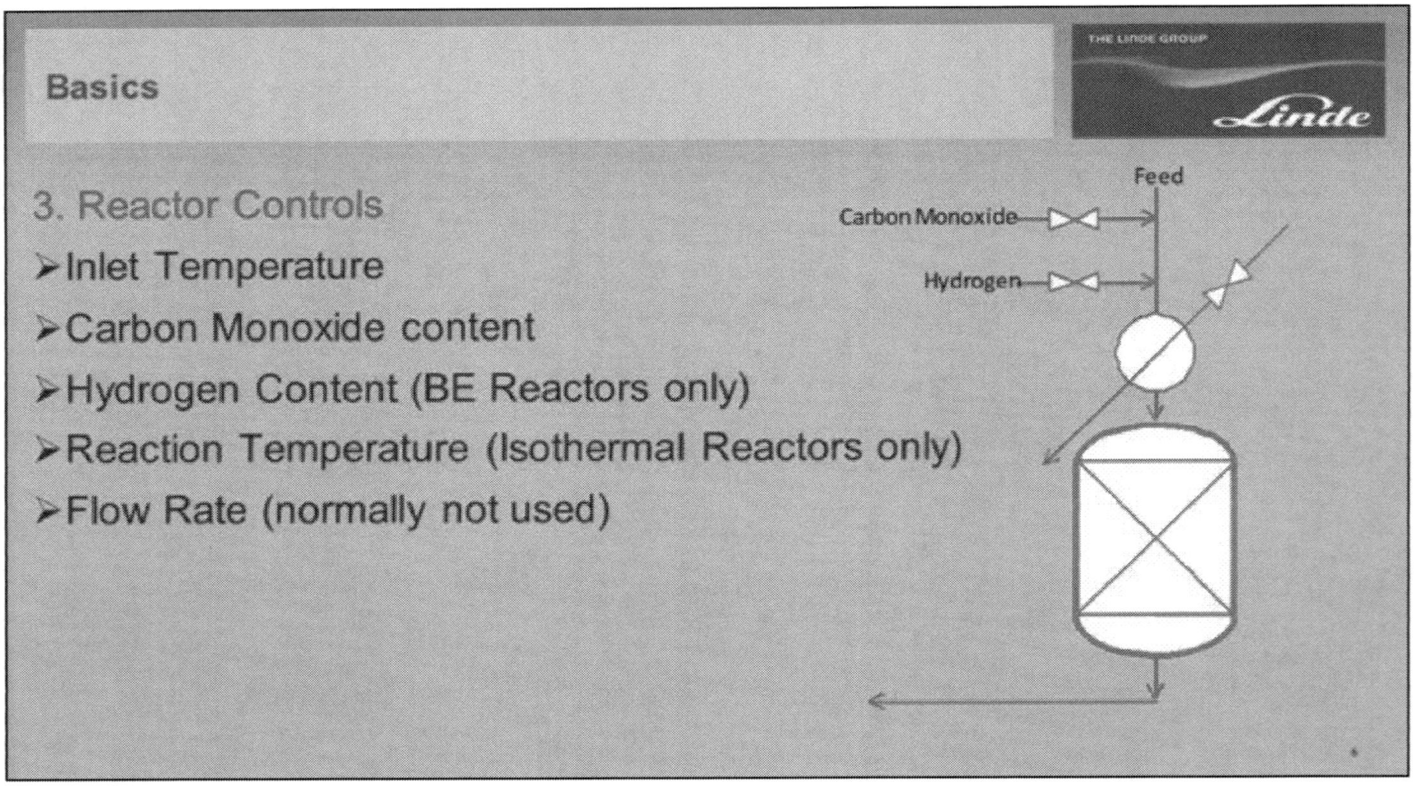

- The operators can control the temperature of the reactor feed.
- The amount of carbon monoxide present can also be adjusted. This might be by direct injection of CO. It might be by adjusting the rate of methanol injection to the furnaces, which is turned into carbon monoxide. It might be by bypassing hydrogen containing carbon monoxide around the hydrogen purification system.
- For back end reactors, hydrogen injection rates can be directly controlled.
- In isothermal reactors, the temperature throughout the reactor can be controlled.
- Finally, in some cases, the feed rate to the converters can be adjusted by using recycles.

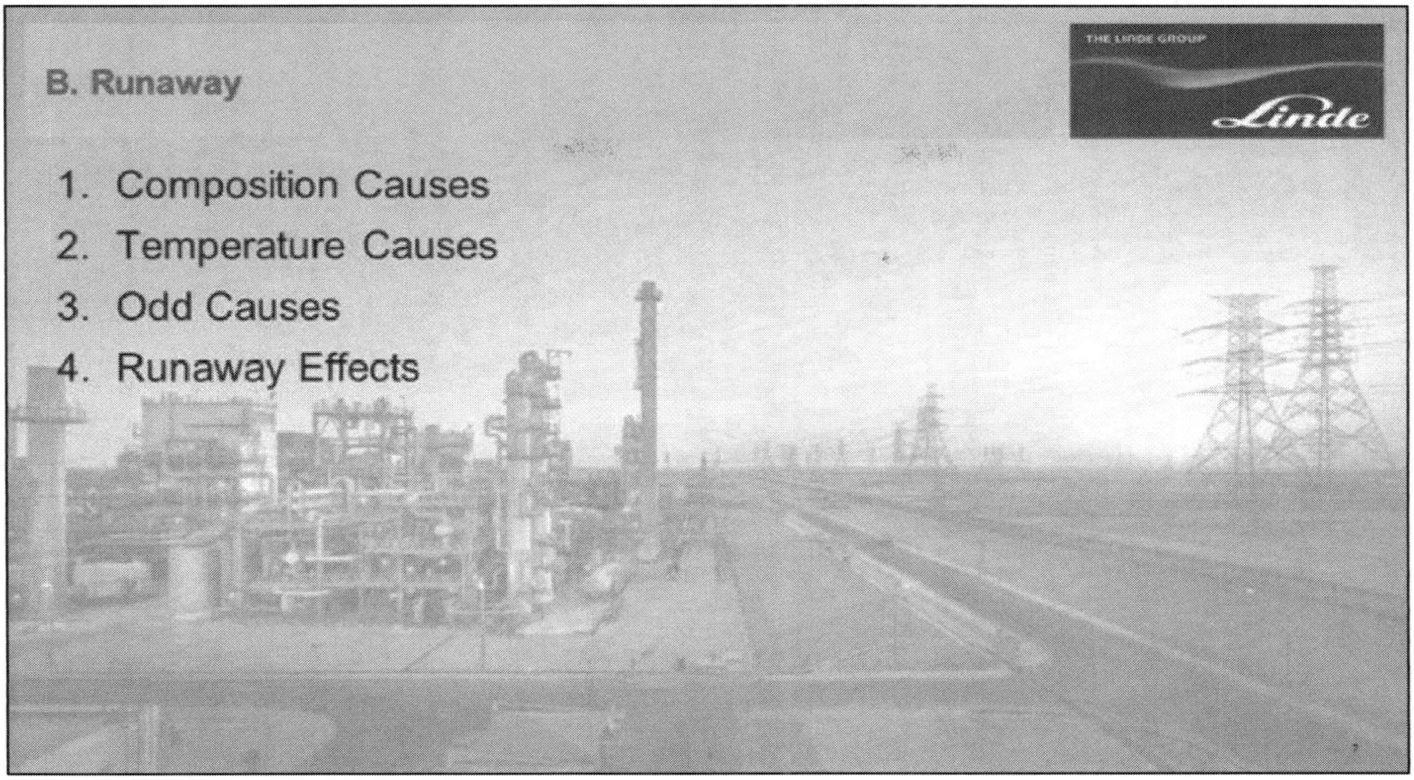

Lets talk about runaways in general.

Changes in composition are frequently the cause of a runaway converter.

Loss of temperature control can also lead to an incident.

There are also the weird causes that pop up. I will mention a few.

Finally, there are the effects of a runaway. What does a runaway look like?

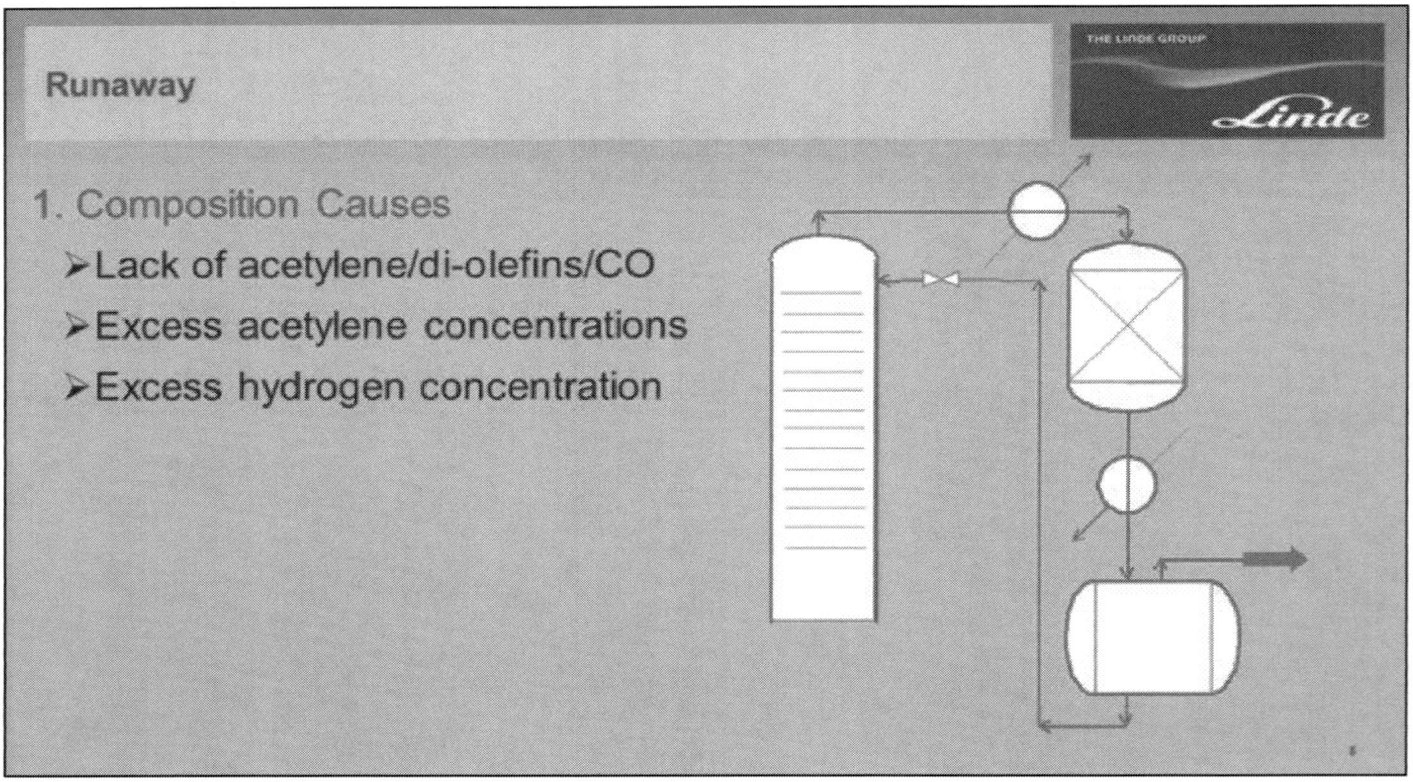

A sudden reduction in acetylene, MAPD or carbon monoxide concentration can cause a runaway. The suddenly available sites on the catalyst are used by ethylene. The ethylene is hydrogenated.

While the reaction is controlled by the adsorption on the catalyst, the addition of highly adsorbing materials such as acetylene will accelerate the reaction. Consider a reactor in the overhead loop of a column. On startup, the reactor feed will not be diluted by the converted acetylene contained in the reflux.

While front-end converters always have lots of hydrogen, back-end converters are always run with minimal excess hydrogen. A sudden surplus of hydrogen can chew up the acetylene that was blocking sites, allowing ethylene access to the catalyst. Or the hydrogen can hydrogenate ethylene already on the catalyst which had been stranded for lack of hydrogen.

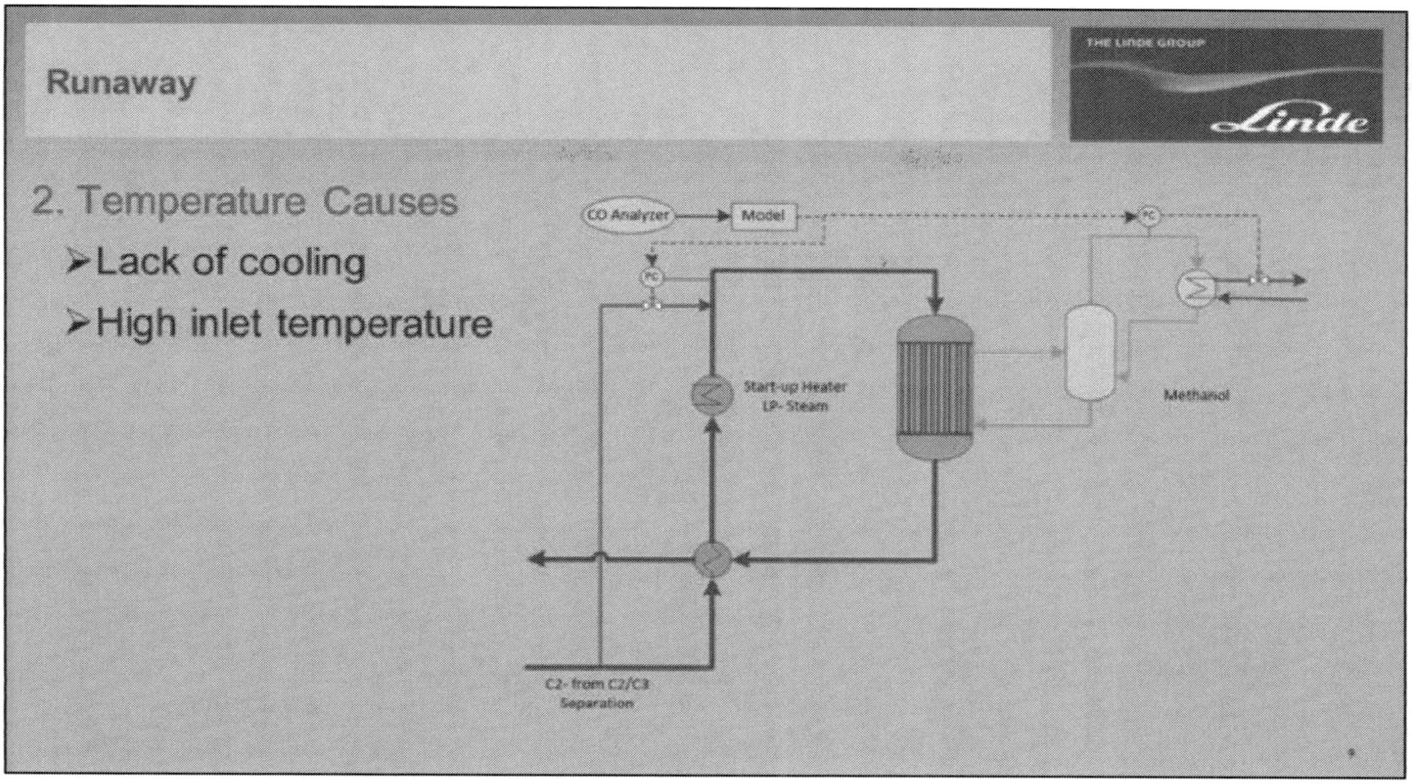

Acetylene converters rely on cooling to control the temperature of the reaction. Cooling failures can have many causes but the result is high reaction rates.

Another, related cause is loss of control of the feed heater. Just a 10° change in the feed temperature can lead to a runaway.

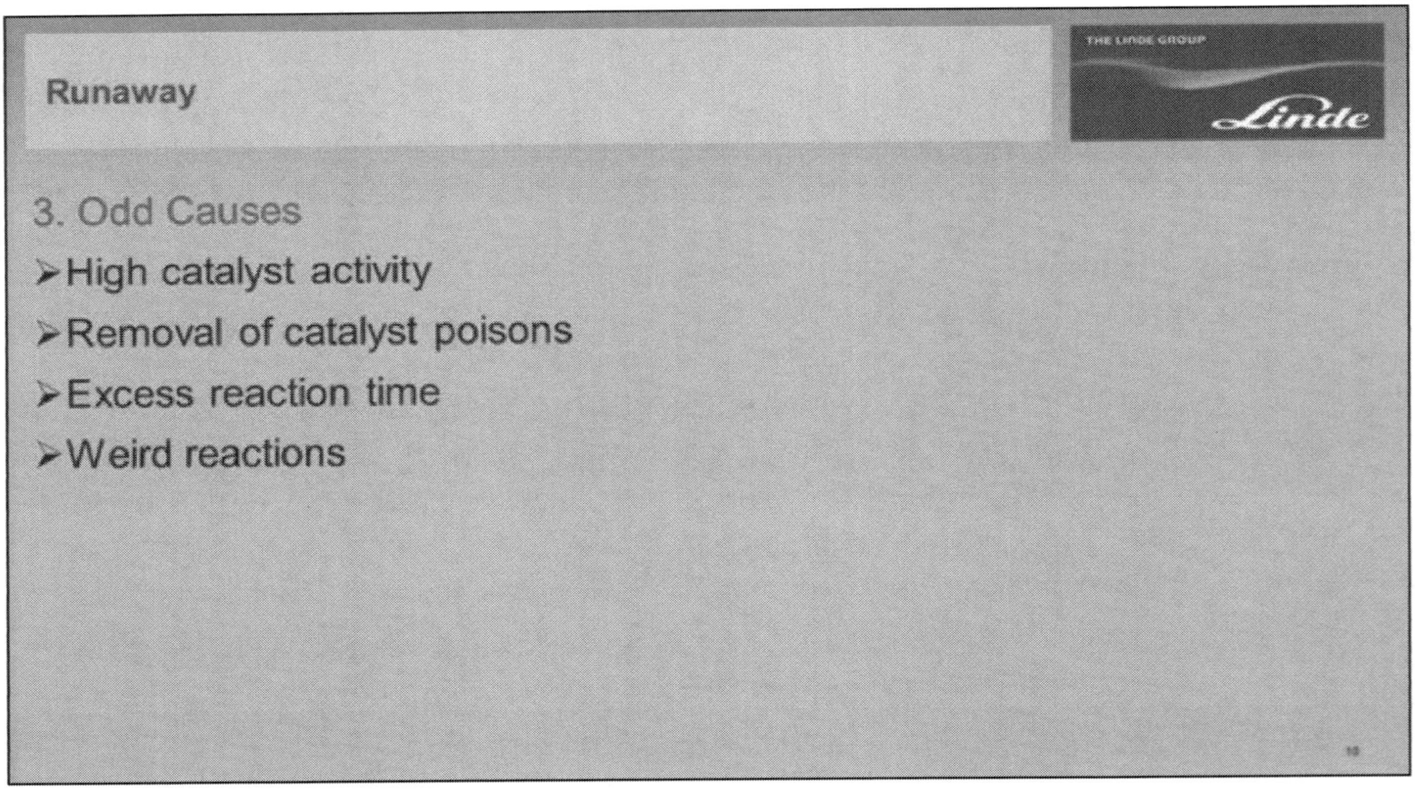

Some of the earliest front end converter catalysts were prone to very high reactivity when first used. Plants were seeing runaway converters at start up despite attempts to minimize acetylene content and lower reactor inlet temperatures. Thankfully, that issue is now behind us.

Certain materials act to inhibit the catalyst activity temporarily. The catalyst activity will increase as that poison comes off the bed. If converter conditions were set based on the presence of the poison and the poison disappears, reaction rates may accelerate to unacceptable limits.

If flow through the convert slows, the reaction will continue at the previous rate. This means that, for adiabatic reactors, the temperatures in the bed will increase, increasing the reaction rate. Further, with the acetylene gone, the ethylene, with its higher reaction rate will now have access to the catalyst sites.

Occasionally unusual material will get into the hydrogenation reactors and weird reactions will occur. I encourage you to look at the EPC papers for examples.

Runaway

4. Runaway Effects

- High temperature
- High acetylene hydrogenation rate
- Olefin hydrogenation occurs
- Polymerization and/or ethylene decomposition may occur

- If the heat generated by the reaction cannot be removed, you have a runaway reaction. This heat translates into high temperatures within the reactor.
- With an uncontrolled reaction, the acetylene will be reacted.
- With the acetylene removed, ethylene will get to the catalyst and get hydrogenated.
- If the temperature gets high enough, the ethylene will begin to polymerize and/or decompose.

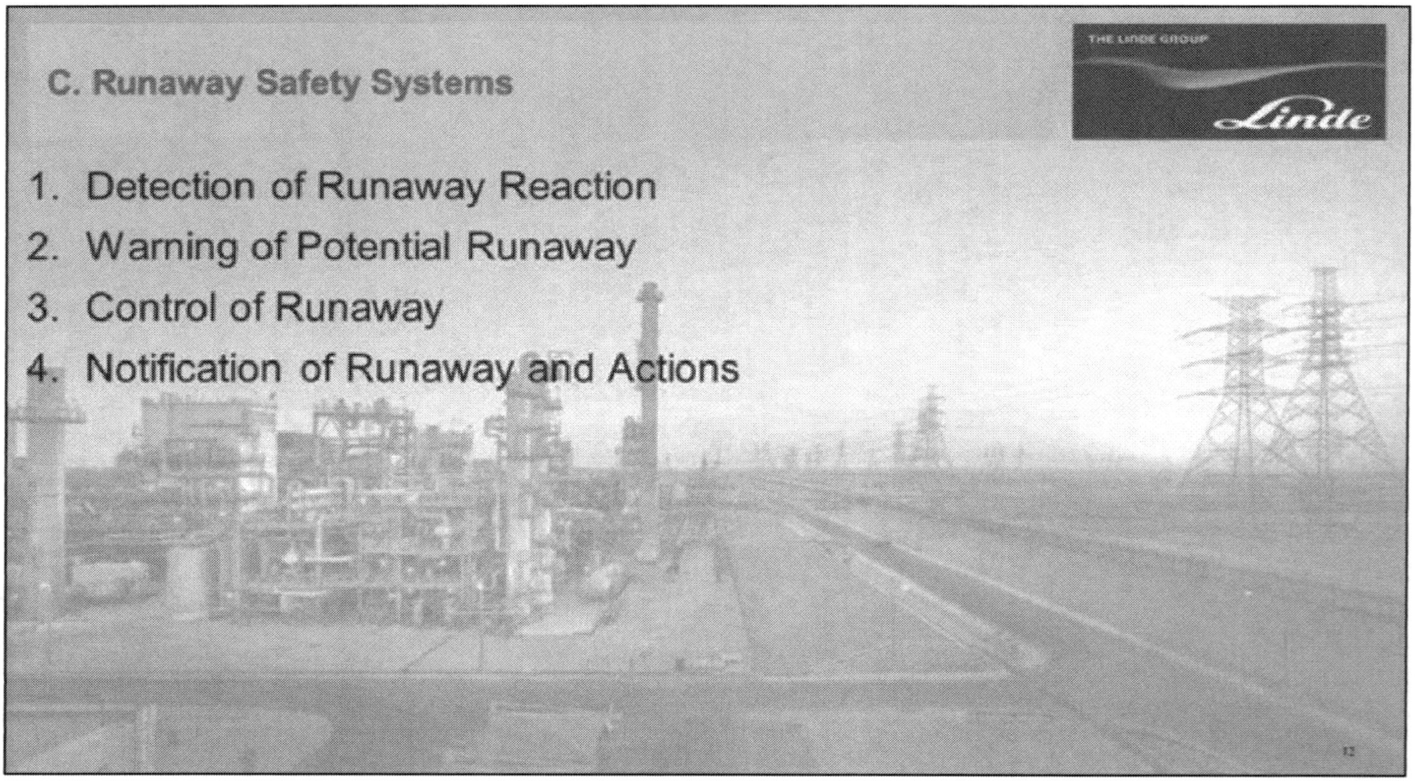

A Converter Safety System needs to

- Detect a runaway reaction. Typically this is done by measuring temperatures in the reactor.
- Detect conditions which may trigger a runaway.
- Control reaction rate. Typically this is done using a stepped approach. First, reactor temperature is minimized. If that fails to control the reaction, more drastic steps are taken.
- Notify operators of the runaway reactor and the actions taken.

Runaway Safety Systems

1. Detection of Runaway Reaction

- Temperature within reactor
- Temperature leaving reactor
- Temperature rise across reactor
- Rate of temperature change
- Operator initiation

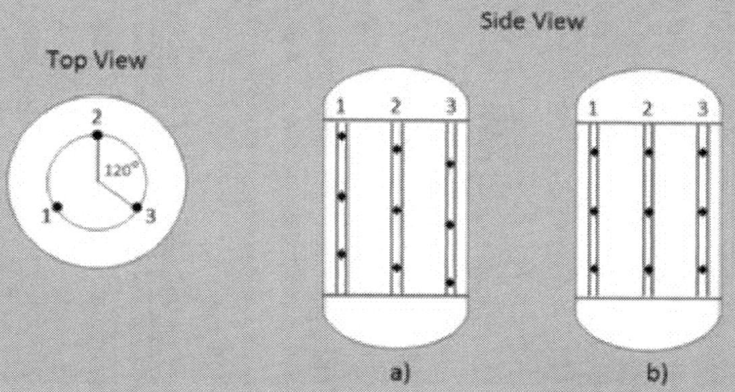

Detecting a runaway acetylene hydrogenation reaction cannot be done directly since there are no reaction rate measuring instruments useful to ethylene plants. Instead the runaway needs to be detected indirectly by measuring temperature.

Because reactor are very large and there are concerns about local "hot spots", temperature measurement within the reactor have evolved to a sophisticated matrix of sensors throughout the bed. These typically have multiple measurements at every location. The reactor effluent temperature is also measured for each bed. On top of this, the temperature rise across a reactor or the rate of temperature change for a measurement point may also be considered.

These system typically require a two out of three voting to trigger the safety system, to prevent a single instrument fault causing a trip while still ensuring that a true event is detected and acted on.

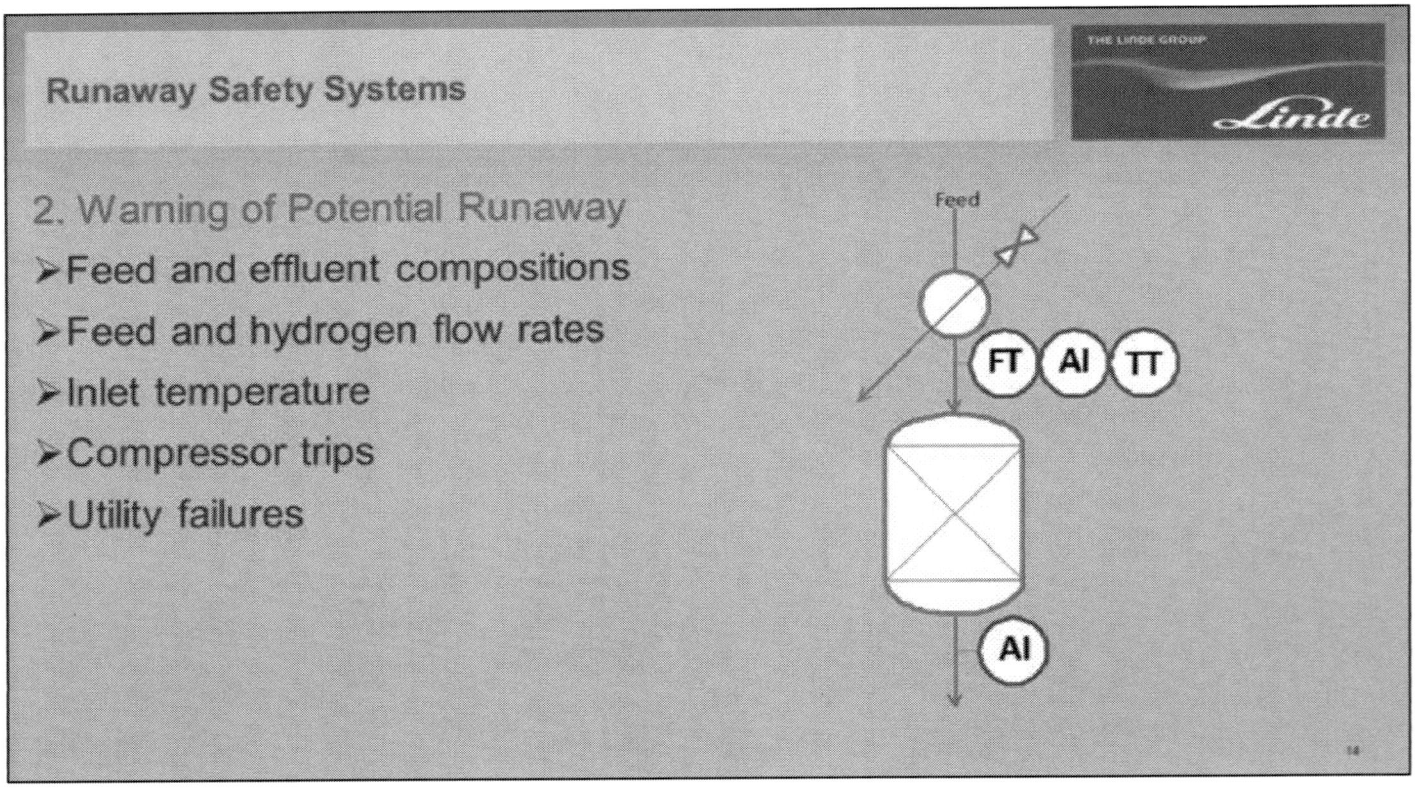

Ideally, a converter safety system should provide warning of an impending runaway. You want to be able to control the reaction without resorting to a plant shutdown.

The feed and effluent compositions can provide warning of runaways. A sudden decrease in carbon monoxide in the feed or lack of any acetylene in the effluent could indicate that a runway is imminent.

Loss of flow through the reactor is certainly a problem and typically causes the reactor safety system to act. Likewise, a high hydrogen flow for back-end reactors may signal an upcoming runaway.

If the reactor inlet temperature is too warm, the reactor safety system needs to intervene.

Loss of the charge gas or refrigerant compressor is frequently used as a trigger for acetylene converter trips.

Likewise, utility failures may also require shutting down the acetylene converter.

Runaway Safety Systems

3. Control of Runaway

- SD-A – Reduce reaction rate while removing acetylene
- SD-B – Reduce reaction rate while avoiding reactor shutdown
- SD-Z – Reduce reactor temperature and pressure to prevent equipment damage

15

Converter safety systems typically have staged responses to a problem. The hope is that by early detection and early action (SD-A), the reactor effluent can stay on-spec for acetylene. Failing that, SD-B allows the converter to continue operating while the off-spec effluent is flared. Finally, the ultimate safety system, SD-Z, will isolate, bypass and depressure the converters.

SD-A will maximize cooling, minimize hydrogen addition and may maximize reactor flow using recycle streams

SD-B will flare reactor effluent before it gets to the ethylene fractionator to prevent contamination of the tower with acetylene. Every effort will be made cool the reactors and minimize reactions.

SD-Z will isolate the converters and bypass the reactor feed. The converters will be quickly vented to flare to minimize the pressure, stopping the reaction and minimizing stress on the equipment.

The speed that a runaway occurs can vary greatly, depending on the reaction rate and the heat capacity of the reactor system. One advantage of an isothermal reactor is that it inherently has a larger heat capacity, which slows runaways.

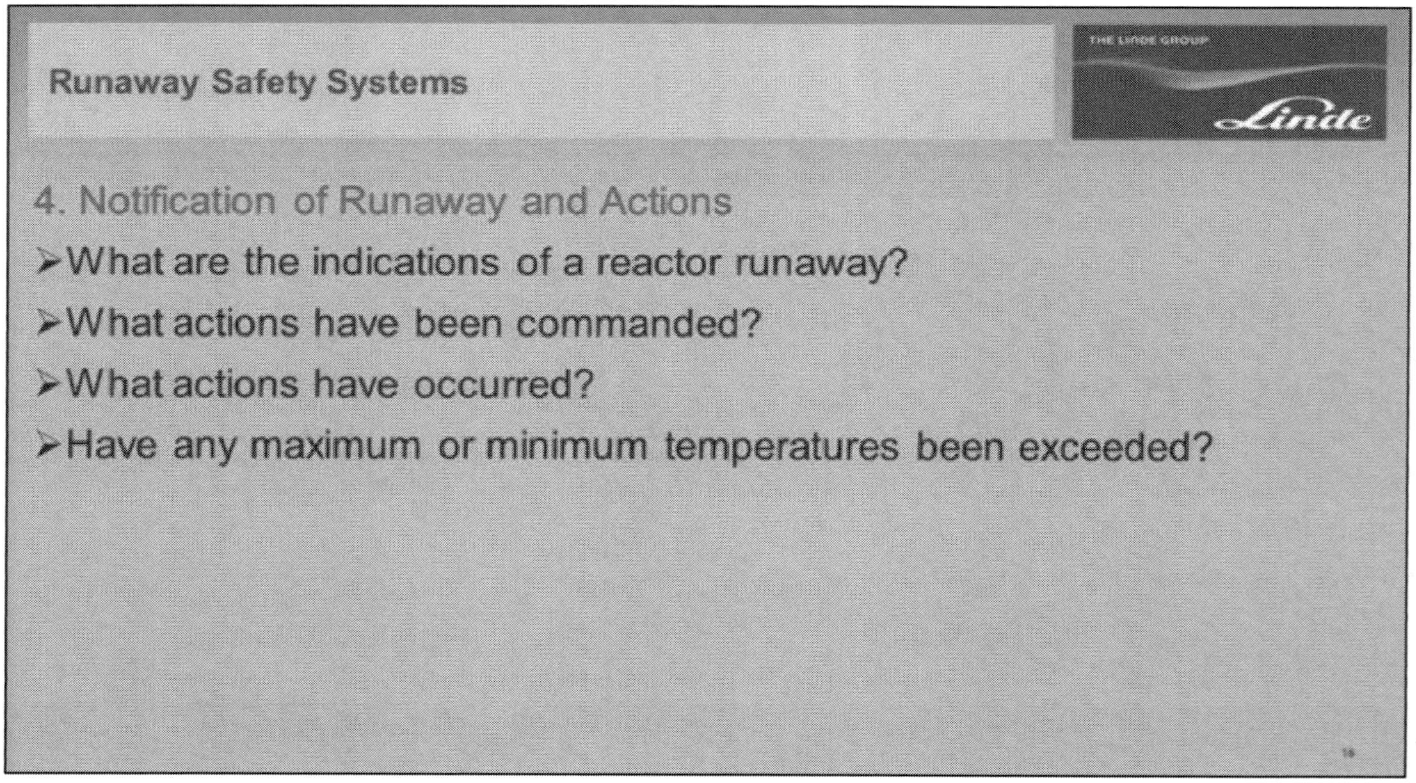

Because a converter runaway can occur quickly and during unusual operating situations, the information regarding the reactor safety system must be presented to the operators as clearly as possible.

- The signal that triggered the safety system must be indicated without ambiguity.
- When a shutdown has been triggered, the actions ordered by the shutdown system should be shown.
- If at all possible, there should be feedback to the operators showing if the ordered actions have actually occurred.
- One of the scary possibilities with an acetylene converter runaway is that local temperatures may get so warm that the steel's strength will become compromised. Also the depressuring of the reactor could cause temperatures below the acceptable minimum temperature. If so, the operators need to be informed so that appropriate actions can be taken.

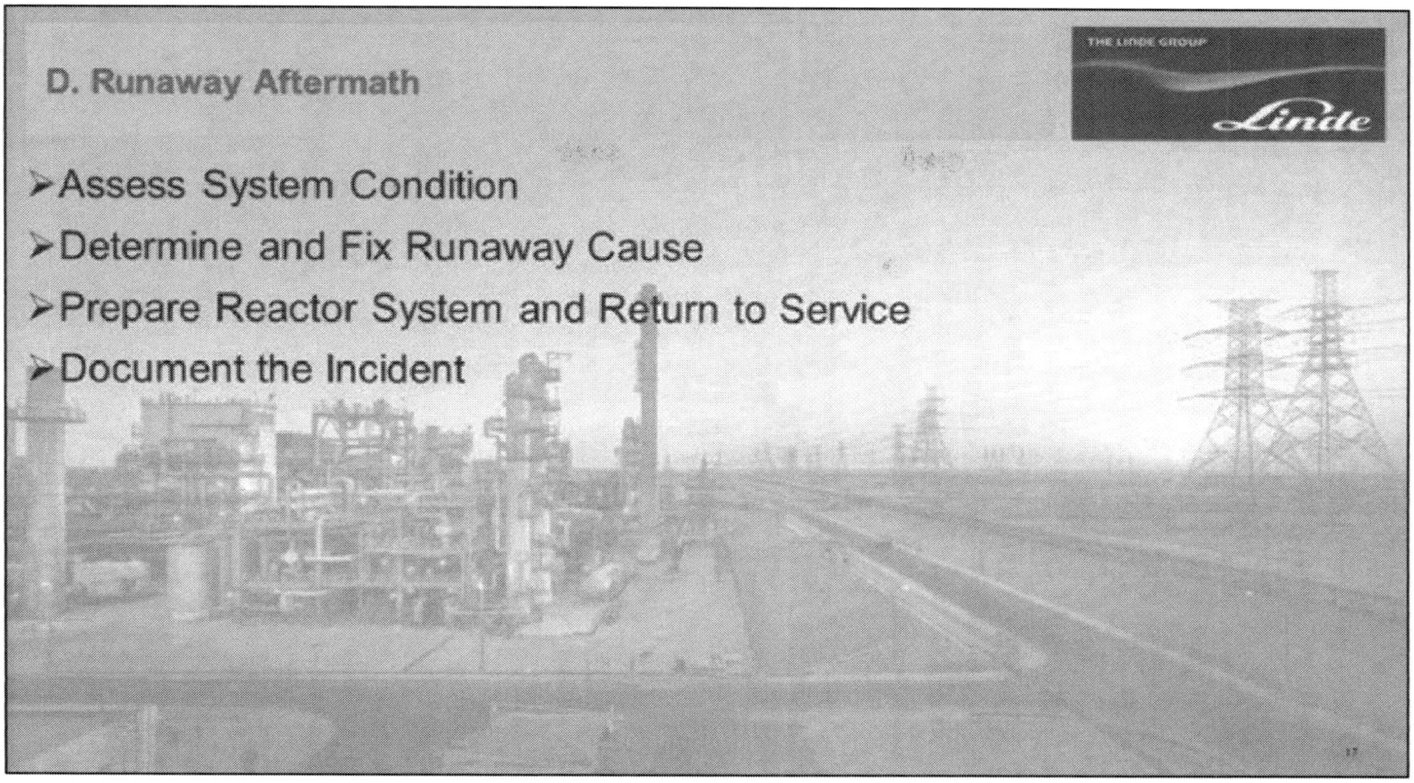

After a runaway and a successful trip of the safety system, the idea is to safely return the plant to operations as quickly as possible.

- ➢ First, the plant must be examined to confirm that no damage has occurred. This includes looking at temperature records of the equipment and determining if vessel internals such as catalyst, mole sieves, demisting pads and trays were displaced or damaged.
- ➢ Second, the cause for the runaway must be determined and remedied.
- ➢ The reactor system must be prepared for service. Reactor may need to be switched. The runaway reactors will need purging with an inert gas and cooling. Reactors and dryers may need to be regenerated.
- ➢ The reactor should be returned to service using normal procedures.
- ➢ Because converter runaways are rare events, every effort should be made to document it so that the information can be used for training.

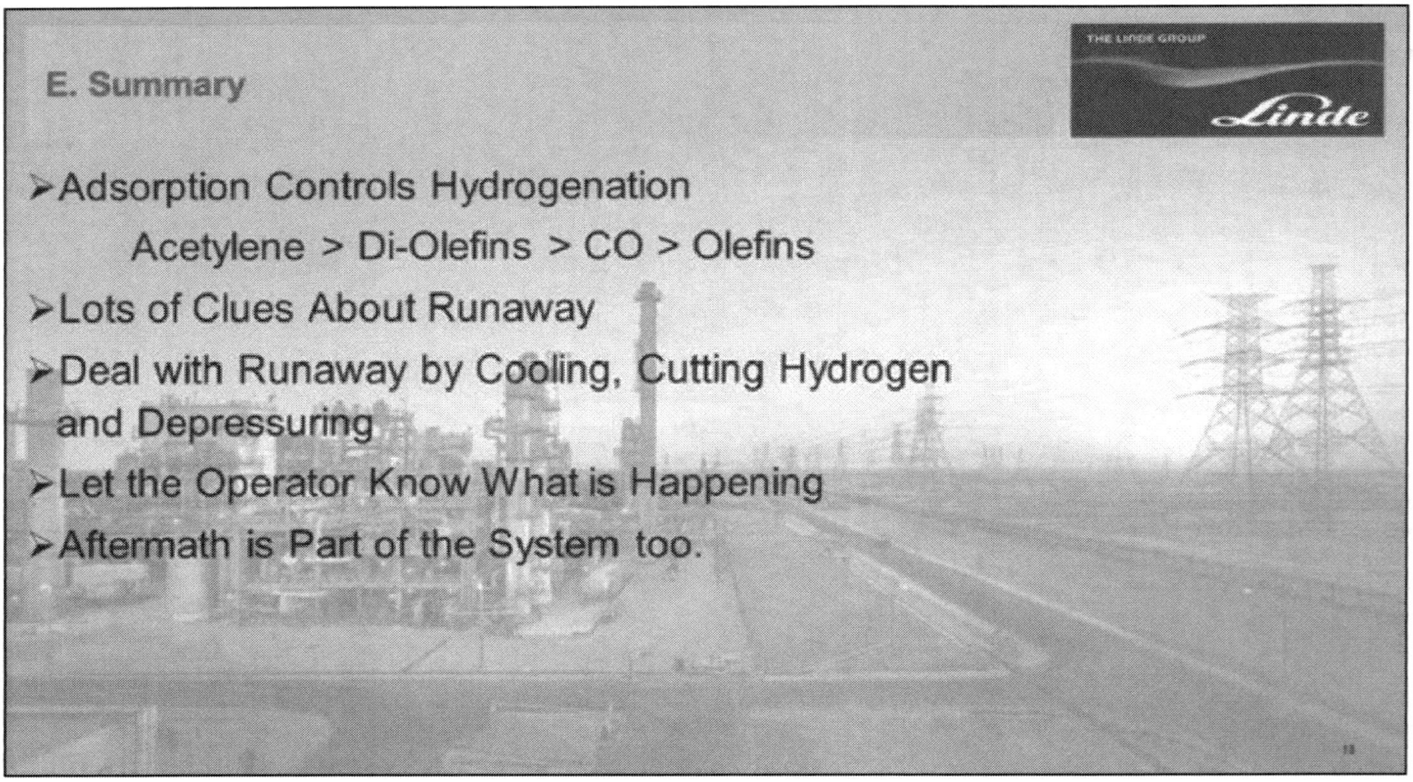
E. Summary
THE LINDE GROUP
Linde
➢Adsorption Controls Hydrogenation
Acetylene > Di-Olefins > CO > Olefins
➢Lots of Clues About Runaway
➢Deal with Runaway by Cooling, Cutting Hydrogen and Depressuring
➢Let the Operator Know What is Happening
➢Aftermath is Part of the System too.

— Remember 2020 is the 100th anniversary of the commercial production of ethylene.

EPC Acetylene Hydrogenation Articles 1992-2019

F.E. Hancock, J.K. Smith, "Design and Performance of Acetylene Hydrogenation Catalysts", 1992_86c

J. Caosyns, L. Quicke, P. Sarrazin, "New Process for Selective Hydrogenation of Acetylene in C_2 Stream", 1992_86d

John A. Krebs, "Large Property Losses in the Hydrocarbon-Chemical Industry", 1992_88f

Ed Sughrue, Marvin Johnson, R. J. Callejas, "Highly Selective Catalysts for Acetylene Hydrogenation", 1993_18d

Mark Kaminsky, "Deactivation Mechanisms for Pd/Al_2O_3 Acetylene Hydrogenation Catalysts", 1994_26b

F.E. Hancock, J.S. Booth, "A Mechanistic Study of the Deactivation by Hydrocarbon Fouling of Front End Acetylene Hydrogenation Catalysts and the Effect of Carbon Monoxide Incorporation", 1994_26e

E.I. Sughrue, R.L. Hair, R.J. Callejas, "The Use of Models in the Operation of Acetylene Hydrogenation Reactors", 1995_21e

R. Voight, J. Merriam, S. Blankenship, "Evolution of Catalyst for Selective Hydrogenation of C_2 Splitters with Interreboilers", 1996_19e

S. Bailey, F. Hancock, J. Booth, "The Application of EPR to the Study of Acetylene Removal and Deactivation Reactions over Pd-based Ethylene Purification Catalysts", 1996_65e

K. Saski, J. Kimble, M Scharre, R.J. Callejas, "Effects of Selectivity on Acetylene Removal Unit Plant Operations", 1997_33b

Mark Eidson, "Runaway Reactor Survey", 1997_35f

Karl Enqvist, "The Operating Errors and the Deficiencies in the Design or in the Safety Systems that Lead to the Acetylene Converter Temperature Runaway", 1999_65d

Jennifer Bancroft, "Temperature Monitoring System for Detecting Hot Spots in Exothermic Reactions", 1999_65e

John Reid, "An Overview of Olefin Unit Accidents and Safety Lessons Learned", 1999_65f

V. Kaiser, J.P. Laugier, R. Di Cintio, and M. Picciotti, "Acetylene Removal from Cracked Gas: A Technology Overview", 1999_68c

Paul B. Himelfarb, Steven M. Moon, William F. Watson, Ray J. Billings, "Reactor Temperature Exotherms in Selective Hydrogenation", 2001_77f

John A. Reid, "An Overview of Hydrogenation Catalyst in Olefins Units", 2001_80a

John A. Reid, "Information Sources for Catalyst Poisons and Contaminate Removal Experiences", 2001_80b

D. Tiedtke, T. Peter Cheung, J. Leger, Stan Zisman, J. Bergmeister, G.A. Delzer, "Chemicals Influencing the Activity of Palladium-Based Catalysts for the Selective Hydrogenation of Acetylene to Ethylene in Acetylene Converters", 2001_80c

Timm E Paxson, Dwight Hines, "Analytical Tools and Methods Employed in Catalyst Analysis", 2001_80d

Sanjeev Kapur, "Contaminant Removal/Treating Technologies", 2001_80e

Tom Kopecky, "Abnormal Operation Caused by Change in Butane Feed Quality", 2001_80f

Gary R. Gildert, “Frequency of Problems in Acetylene Converters”, 2001_80g

Kevin W. Zinke, “Antifoam Impacts in an Olefins Plant: A Case History”, 2001_80h

C. Michael Smith, Naoko Akiya, John Bajer, “Measurement of the Green Oil Formation Kinetics and Catalyst Deactivation During C_2H_2 Hydrogenation”, 2003_53f

Edgar L. Mohundro, “Overview on C2 and C3 Selective Hydrogenation in Ethylene Plants”, 2003_64a

C. Michael Smith, David Cringle, Angelika Heilmann, Steve Kenny, "Acetylene Hydrogenation at Dow Chemical Designing Reactor Operations for Non-Events" , 2003_64b

Kevin McGuire, Darin B. Tiedtke, Gary Delzer, Daryl Bitting, Roxanna Whitfield, Jacques Leger, Peter Cheung, “Tools for Safely Starting Up Pd-Based Acetylene Hydrogenation Catalysts under Extreme Conditions”, 2003_64c

Mario Lesieur, Sanjay Sharma, Ravi Nath, “Advanced Process Control and Optimization of Acetylene Hydrogenation Reactors”, 2003_65f

Frank Mey, Richard Krumpen, Gernfried Belier, Gerard Rosenbrand, “Selective Hydrogenation of Acetylene in Isothermal Reactors”, 2003_66a

Kurt Simon, Darren Adams, “Diagnosis of an Unusual Hydrogenation Reactor Excursion”, 2003_66b

Shawn Cullen, Bala Devakottai, Lucas Hilbrich, Gary Gildert, “Flareless Swapping of C2 and C3 Hydrogenation Reactors”, 2003_66c

David Gent, “Software Upgrade Triggers Events That Lead to Plant Shutdown”, 2004_62d

M.E. Levin, N.O. Gonzales, L.W. Zimmerman, R. Rodriguez, T.E. Paxson, J.E. Fischer, “The Reactivity of Ethylene Containing Minor Amounts of Acetylene”, 2005_38a

Kevin McGuire, “SIS Design Strategy for an Acetylene Hydrogenation Reactor System”, 2005_45b

Allison A. Fielder, Mark B. Schmidt, Raghu K. Menon, “Unintended Consequences of Process Additives on Acetylene Converter Performance in a Gas Feed Steam Cracker”, 2006_31d

Edgar Mohundro, Sam Valleru, Scott Miller, Dimitri Mertens de Wilmars, Bruno Leaustic, Juergen Fuchs, “Upgrade of a Tail-End Acetylene Converter BASF FINA Petrochemicals Limited Partnership Naphtha Cracker”, 2006_135b

Linh Le, J.D. Tate, Mary Beth Seasholtz, Manish Gupta, Doug Baer, Trevor Knittel, Alan Cowie, Jie Zhu, “Rapid Online Analysis of Acetylene for Improved Optimization of Hydrogenation Reactors”, 2006_218e

Bob Strack, “Exothermic Reactor Safety”, 2007_15c

Jeff S. Melching, “Ethylene Decomposition”, 2007_15d

Jennifer L. Romig, Thomas J. Gonzales, Joe Bergmeister, “Operating Experience with Chevron Phillips’ Back End E-Series Acetylene Converter Catalyst”, 2007_91c

Thomas J. Gonzales, Zongxuan Hong, “Unit 24 Operating Experience with Front End E-Series Acetylene Converter Catalyst”, 2008_31e

Edgar L. Mohundro, "Ethylene Plant Operations Acetylene Hydrogenation Tutorial Lecture I: Introduction to Subject", 2009_107a

Richard W. Voight, "Ethylene Plant Operations Acetylene Hydrogenation Tutorial Lecture II: Chemistry & Catalysts", 2009_107b

Rick Callejas, "Ethylene Plant Operations Acetylene Hydrogenation Tutorial Lecture III: Front-End Acetylene Hydrogenation", 2009_107c

Edgar L. Mohundro, "Ethylene Plant Operations Acetylene Hydrogenation Tutorial Lecture V: Catalyst Regeneration", 2009_107e

Edgar L. Mohundro, "Ethylene Plant Operations Acetylene Hydrogenation Tutorial Lecture VI: Safety Considerations", 2009_107f

M. Sun, Raj Rajesh, R. Zoldak, S. Blankenship, M. Urbancic, "Catalyst Technology for the Removal of Acetylenes, NOx and Oxygen from Olefin Streams", 2010_135f

Steven Blankenship, Dr. Mingyong Sun, Dr. Michael Urbancic, Denise B. Cooper, "Maximize Ethylene Gain in Acetylene Removal Units Optimization Strategies and New Generation Catalysts", 2011_96d

Emerentino Quadro, Niara Eckhard, Eliana Vidal, Richard P. Zoldak, Jaildes Britto, "Acetylene Converter Troubleshooting", 2011_119c

Nelson Shen, "Acetylene Converter Safety Systems (A Tutorial)", 2012_150b

Richard D. Swain, Pascal Coenen, Brian Vanderwilp, "Minimizing Risks Associated with First-Time Use of a New C2 Tail-End Catalyst", 2013_52d

Dr. Xiang (Sherry) Liu, Greg Lankford, Adam Chaimowitz, Kuan Ting (Gary) Yeh, Hsu Hung (Steven) Huang, Dr. Wenhai Ji, William Jenko, Dr. Alfred Feitisch, "Assuring Ethylene Purity: Reliable PPM and PPB Measurement of Acetylene with Online TDL Analyzers", 2013_52f

Manzoor Khan, Jennifer L. Nill, Joseph Bergmeister, "Ethylene Plant Operations with Chevron Phillips Chemical Company LP Back End E-Series® Catalyst", 2014_152a

Tyler Conn, Nennette Richard-Jordan, "Root Cause and Resolution of Low Activity in a Tail-End Acetylene Converter", 2015_97c

Ernesto J. Angueira, Pascal Coenen, "C_2 Front-End Hydrogenation Improvements", 2016_192e

Edgar L. Mohundro, "Overview on Acetylene Hydrogenation in Ethylene Plants", 2016_193a

Yu Liu, Lin Luo, Marcos Martinez, Cristina Martinez, Denise B. Cooper, Brian Heasley, Mingyong Sun, "Acetylene Hydrogenation with Olemax 207 for Back End Units", 2016_193b

Geert Vercruysse, Geert Boogaerts, "Lessons Learned from Incidents Case Study: 25 Yrs Operation of a Cracker Plant", 2017_160c

Jane Liang-Weissgerner, Junathan Lunn, "Operating an Unstable Acetylene Hydrogenation Reactor After a Runaway", 2017_81a

Hattachai Aeowjaroenlap, Kritsada Choyiwiriyakun, Nattawat Tiensai, Aljandro Cano, Stephan Spatenka, Sreekumar Maroor, Amit Goda, "Improving Overall Ethylene Plant performance by Optimizing C2 Hydrogenation", 2017_81c

Steven T. Coleman, "Ethylene Decomposition: A Tutorial", 2019_188f

591725 Determining the Frequency and Severity of Vessel Failure Due to Runaway Reactions in Acetylene Reactors

Abstract:

Catastrophic temperature runaways in acetylene reactors that lead to a failure of the reactor are very rare. HAZOP/LOPA teams struggle to determine the frequency and severity of such rare events.

Some HAZOP/LOPAs looked at industry data to assist with determining the frequency of acetylene reactors runaways that resulted in a total failure of the reactor. Each study team reviewed the same data yet came up with different frequencies. A team of SMEs was assembled to resolve this discrepancy. The SMEs did not believe using industry performance was correct as the first assessment requires the initiating even as the response with no consideration given for safeguards. The SMEs looked at metallurgy details, reaction kinetics, and the condition of the reactor contents to come to the correct method to evaluate.

Copyright © American Institute of Chemical Engineers. All rights reserved.

Good afternoon.

Our next presentation offers a framework for understanding the root causes of runaways in Front End Acetylene reactors, along with illustration by several industrial examples.

My name is Darren Adams, head of Clariant's Olefins Purification business.

Our co-authors are Ernie Frank and Yu Liu of Dow Chemicals, and Carlos Vianna, Clariant's Selective Hydrogenation Product Manager. I'd also like to recognize contributions by David Mullenix of Sasol.

Public

Table of contents

1. Front End Runaway Primary Mechanism, Triggers
2. Runaway Root Cause Hazards Categories
3. Catalyst Stability Improvements and Best Practices
4. Industrial Runaway Examples

Since many existing and all new units use Front End configurations, we hope this information will be widely helpful in preparing for Operator training and Process Hazards reviews.

We'll need to move somewhat quickly over the highlights due to time constraints, so we encourage later downloading of the presentation for detailed review.

We will be happy to receive any subsequent questions or discussion requests that we can't get to during today's presentation.

CLARIANT

Runaway Primary Mechanism, Triggers

what is precious to you?

Let's start by discussing the primary mechanism and triggers.

Public

CLARIANT

High concentration H2 + C2H4 in FE sets the stage for a thermally induced, uncontrolled exotherm

- **K1: C_2H_2 is very hydrogenation reactive** (Also 1,3-Butadiene, Methylacetylene)
  - but modest potential Exotherm, limited by low reactant concentrations
- **K2: $C_2H_4 + H_2 \rightarrow C_2H_6$**
  - high mol% = potentially massive heat release, reversing furnace cracking reactions
- **K1, K2 reaction rates** increase with Temperature, = positive Feedback loop
- **Minimizing C_2H_4 hydrogenation is critical**, i.e. maintaining high "Selectivity"
- **Short Residence Time and proper Temperature control** are critical to limit K2 Extent of Reaction

Runaway will result in plant trip and damaged catalyst, at minimum

Runaway risk stems from high concentrations of C2H4 and H2 together over a hydrogenation catalyst. C2H2 is a only a modest contributor, limited by its low concentration.

The general cause of a runaway is "Any combination of factors which appreciably accelerate the ethylene reaction rate". After that, the hydrogenation exotherm provides a positive feed-back loop which can propagate into runaway.

The subsequent trip and outage are less problematic than the unpredictable amount of catalyst damage from polymer laydown. Catalyst performance and life will normally be reduced by even one runaway, with the degree of damage related to the elapsed time and peak temperatures until purged and cooled.

Some plants have experienced greatly impaired production reliability, requiring an unscheduled outage for pre-mature catalyst replacement.

Reliably preventing all causes of ethylene reaction acceleration is a complex challenge.

Public DOW CLARIANT

Understanding the term "Runaway" in different contexts

- **In lab isothermal micro-reactor**, "2 mol% C_2H_6 make" is a commonly used definition
  - Assumption: sufficient exotherm to trigger Runaway if Adiabatic Rx (w/ Gen 3 catalyst, 1990's-2015)
- **Commercially**,
  1. Individual "Hot Spot" TI inflects, accelerating upwards:
     - Localized exotherm initiates → propagation to full reactor
  2. Local Temp → Rxn Rate above convective cooling capability
     - Note: Likely well below typical SD-1 full cooling trip setpoint
  3. Temperature above which polymer laydown (damage) accelerates
     - ~ 150-175°C

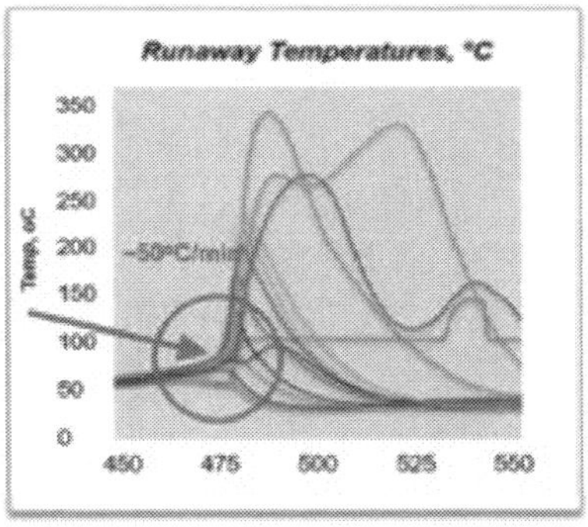

The term "Runaway" has different contextual meanings. In lab isothermal microreactors, 2% Ethane make is used as proxy for the adiabatic exotherm expected to produce an industrial runaway.

Some operators might mis-diagnose quickly rising bed temperatures as <u>NOT</u> a runaway if they have not yet heard a "High Temperature" alarm.

In the example chart, the inlet temperature was reduced to counteract increasing bed TI's. However, somewhere a local temperature exceeded the tipping point where an accelerating exotherm exceeded convective cooling rates.

Note the inflection point. Experienced producers may program TI voting and rate of change analyses into their DCS for early warning, and they are not shy about aggressively cooling the bed to avoid damage.

My operational definition of an <u>impending</u> runaway is when an individual TI starts to rise above others at the same level, with an accelerating deviation or rate of change. With detection limited to discrete TI locations, a runaway is probably already underway.

Runaway initiation temperature depends on the catalyst, its condition, residence time, and gas compositions. Static Alarm and Trip settings, set above anticipated End of Run, are mainly protecting plant and personnel. The catalyst, and the plant's subsequent operability, remain at risk of even a single runaway event.

CLARIANT

Runaway Root Cause Hazard Categories

what is precious to you?

Let's classify potential runaway causes into Categories for discussion

Public

DOW

CLARIANT

What can initiate and accelerate an exothermic runaway?

1. Elevated local catalyst Temperatures
2. Reduction in Flow, while at reaction Temperature
   - Increased Residence Time → higher C_2H_4 hydrogenation exotherm
   - Reduction in cooling Heat Transport
3. Heterogeneous zones of above average catalyst activity, i.e. "Hot Spots"
4. Drop in concentration of activity-suppressing species over active Pd sites
5. Spike in concentration of highly reactive Inlet species

Current catalysts typically offer good temperature separation between complete acetylene removal and acceleration of ethylene reaction. This is aided by activation energies and strong adsorption of C2H2 and CO on the Pd active sites, provided temperatures are not so high as to counteract the adsorptive bond energy.

The first category is thus when local catalyst temperatures are elevated, providing enough system energy to open up sites for ethylene to freely adsorb and react.

The second category is reduced Flow, or increased Residence time. If enough time is available, the exothermic ethylene reaction will eventually take off. Further, low flows mean reduced cooling capacity.

The third category is heterogeneous zones of locally higher catalyst activity.

Finally, a change in the net reactivity of the cracked gas can occur, either by a drop in activity-suppressing species, or an increase in more reactive species.

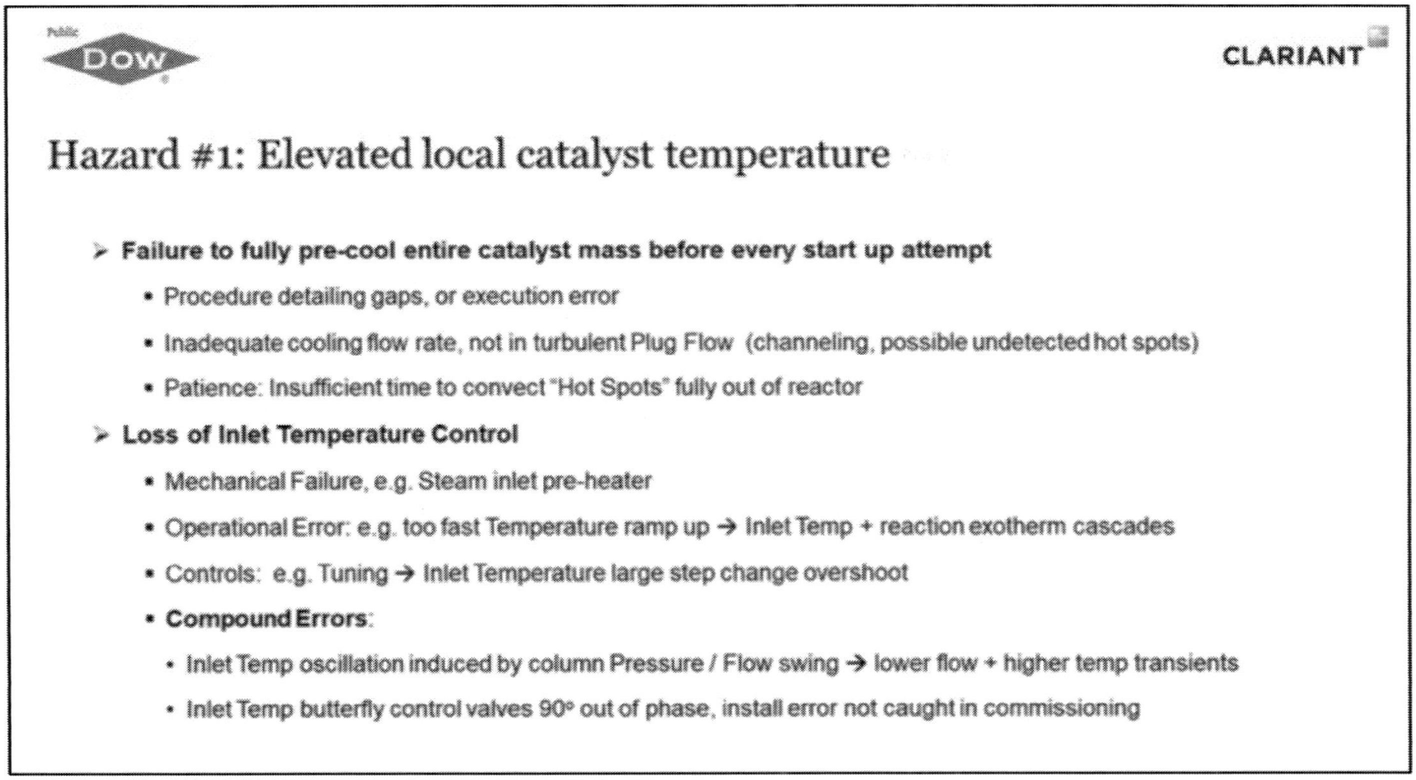

Too high bed temperatures can occur due to loss of positive control of the inlet temperature of the cracked gas, from a variety of reasons.

However, the usual problems are with procedural attention to detail, effectiveness of operator training, and judgement during plant transients. In other words, this should be mostly preventable.

As examples, pre-requisites for a reliable startup include:

- ensuring sufficient cooling flow to enforce even distribution across the reactor cross-section, and
- Ensuring the <u>entire</u> mass is amply cooled to well below reaction initiation temperatures before proceeding.

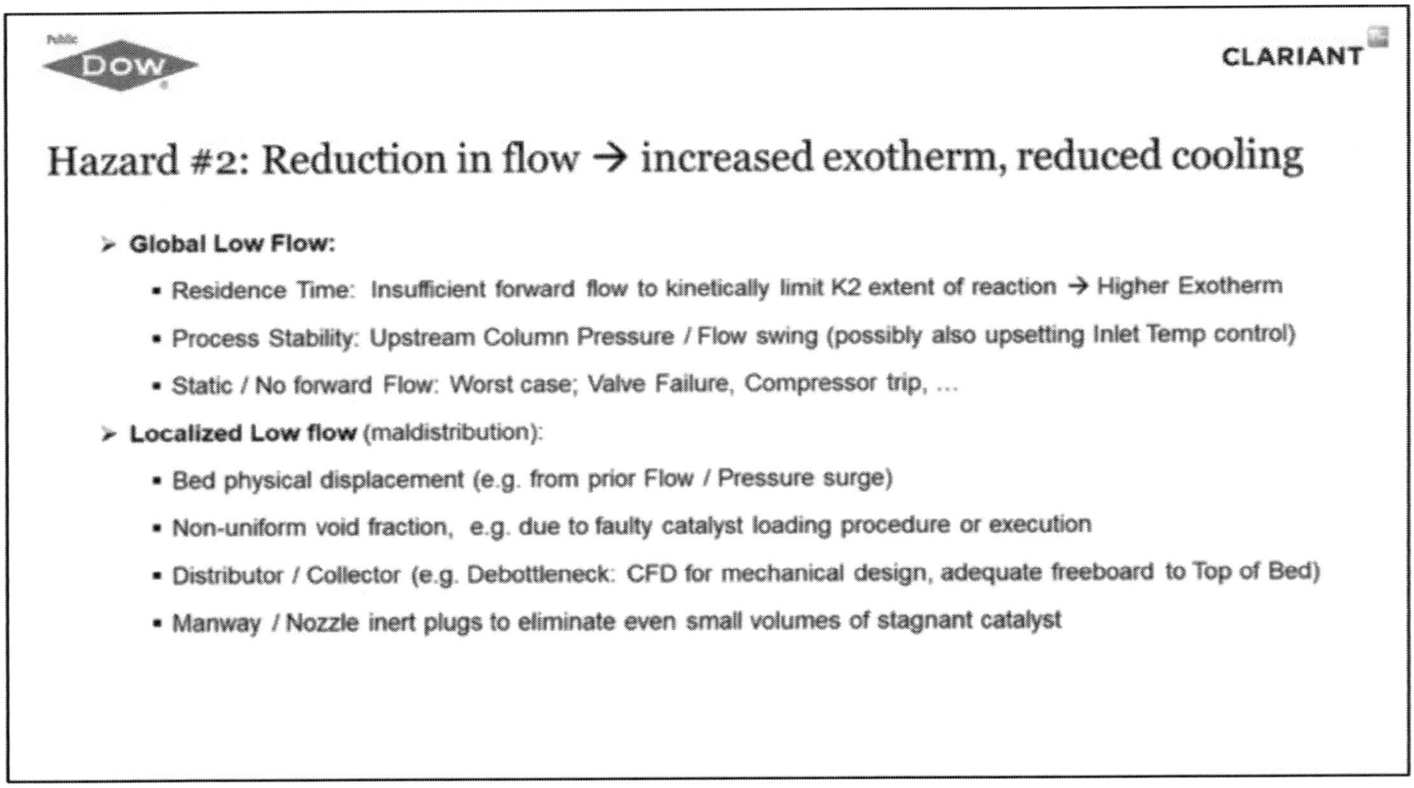

Increased Residence time generally produces control challenges and risk of increasing reaction exotherms.

Licensors specify system minimum flows based on catalyst behavior, compressor anti-surge controls, and other considerations. Given their wider Operating Window, current commercial catalysts are more stable at lower flows than prior generations. Still, since damage is more likely during trips and startups, it's best to maintain a margin above minimum flow trips to reduce this risk.

Operators must also maintain a stable flow. Reacting all the C_2H_2 while limiting ethylene reaction is more difficult when flow and residence time are unstable. Flow surges upset inlet temperature controls, bed temperature profiles, adsorption equilibria, and relative reaction kinetics. Along with GC time lag, proper interpretation of reactor control data is highly compromised in an unstable system.

Localized areas of low flow, for any reason, also result in locally higher conversion and exotherm, and are a likely runaway initiation point. Possible causes include non-uniform loading, Distributor design or damage, or reactor bed disruption due to prior flow or pressure surges.

Public

CLARIANT

Hazard #3: Non-Homogeneous catalyst activity

- Reactor optimization and control are negatively impacted by **non-uniform volumetric activity**
  - **Polymer laydown** from prior Runaway always creates localized activity variation & subsequent challenges
  - **Bed loading procedure**: High Pd fines directly below stationary loading sock → "hot spot" or column
    - Lower void fraction = lower flow + Higher Pd content = high activity, high residence time & conversion
    - Higher local exotherm, probably uninstrumented:
    - **Probable runaway initiation site** during transients

Possible, but less likely:

- Catalyst manufacturing consistency is critical to overall stable control and performance

The predominant cause of non-uniform catalyst activity is prior Runaways and non-uniform deactivation from polymer formation.

Put simply, once you've had one runaway, you are much more likely to have another because of this effect.

Faulty loading procedures have the opposite effect: Loading any amount of dust and fines via a static sock creates a column of lower void fraction, lower flow, and higher volumetric activity from the concentration of Pd-enriched fines.
This "Hot Spot" is a problem waiting to happen.

We understand some plants' safety-related reluctance to have personnel in the reactor continuously distributing the catalyst during loading. However, there is a significant potential for major negative consequences.

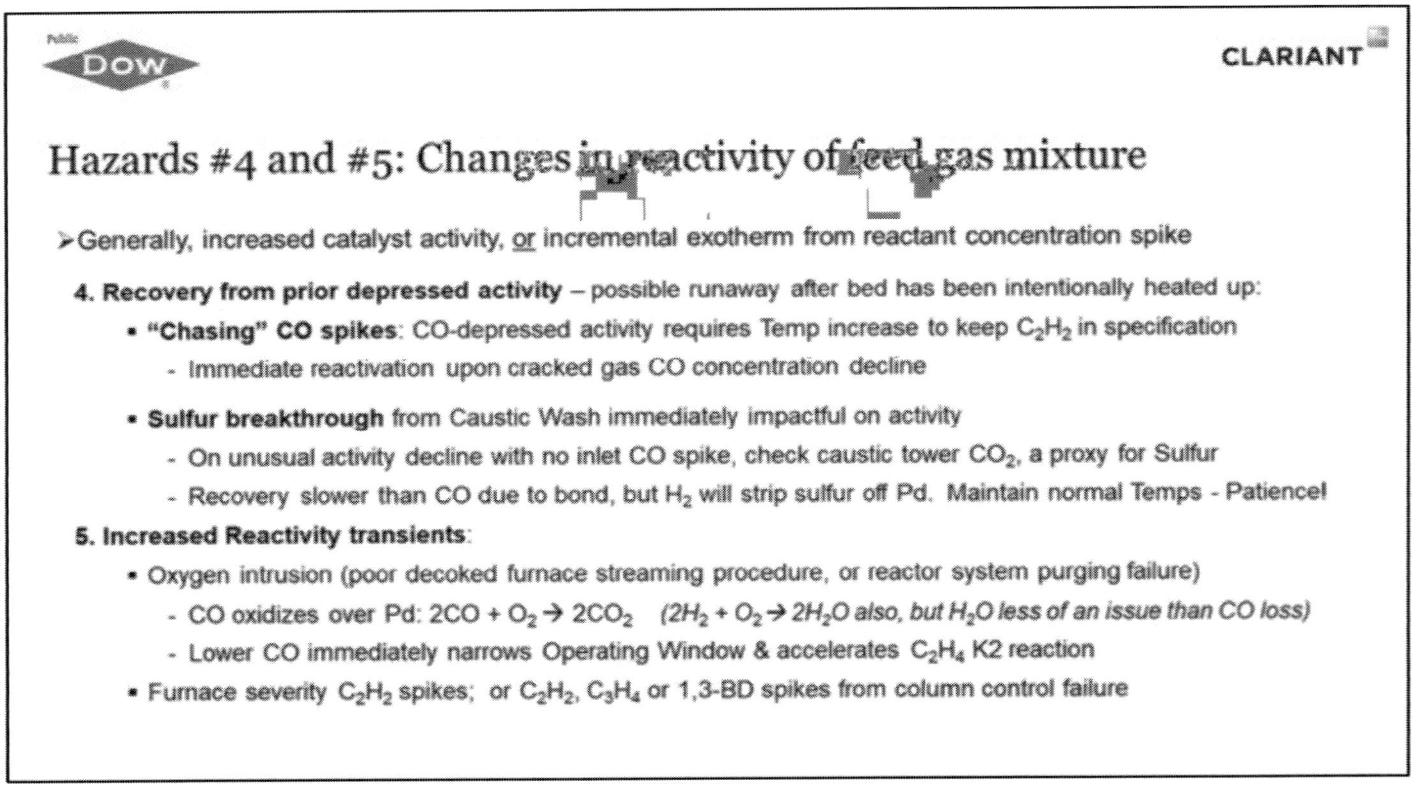

Changes in cracked gas reactivity is usually CO concentration swings.

Newer plants have fewer, very large furnaces, magnifying the effect of individual furnace swings.

Further, enhanced tube metallurgies can produce very low cracked gas CO levels, making reactor operations much more sensitive. We'll expand on this shortly.

Activity loss due to sulfur leakage from the acid gas section is not unprecedented. Sulfur is a reversibly adsorbed poison. High hydrogen levels will help strip it off. Be patient for reactivity to return after getting CO2 levels back in spec, without trying to force reactivity with elevated operating temperatures.

It's not common for increases in other reactants to cause a step change in the exotherm, but we'll mention a few interesting examples in the upcoming case studies.

CLARIANT

Catalyst Stability Improvements, and Best Practice

what is precious to you?

In the next section, we'll touch on the impacts of Catalyst improvements on Operational Stability

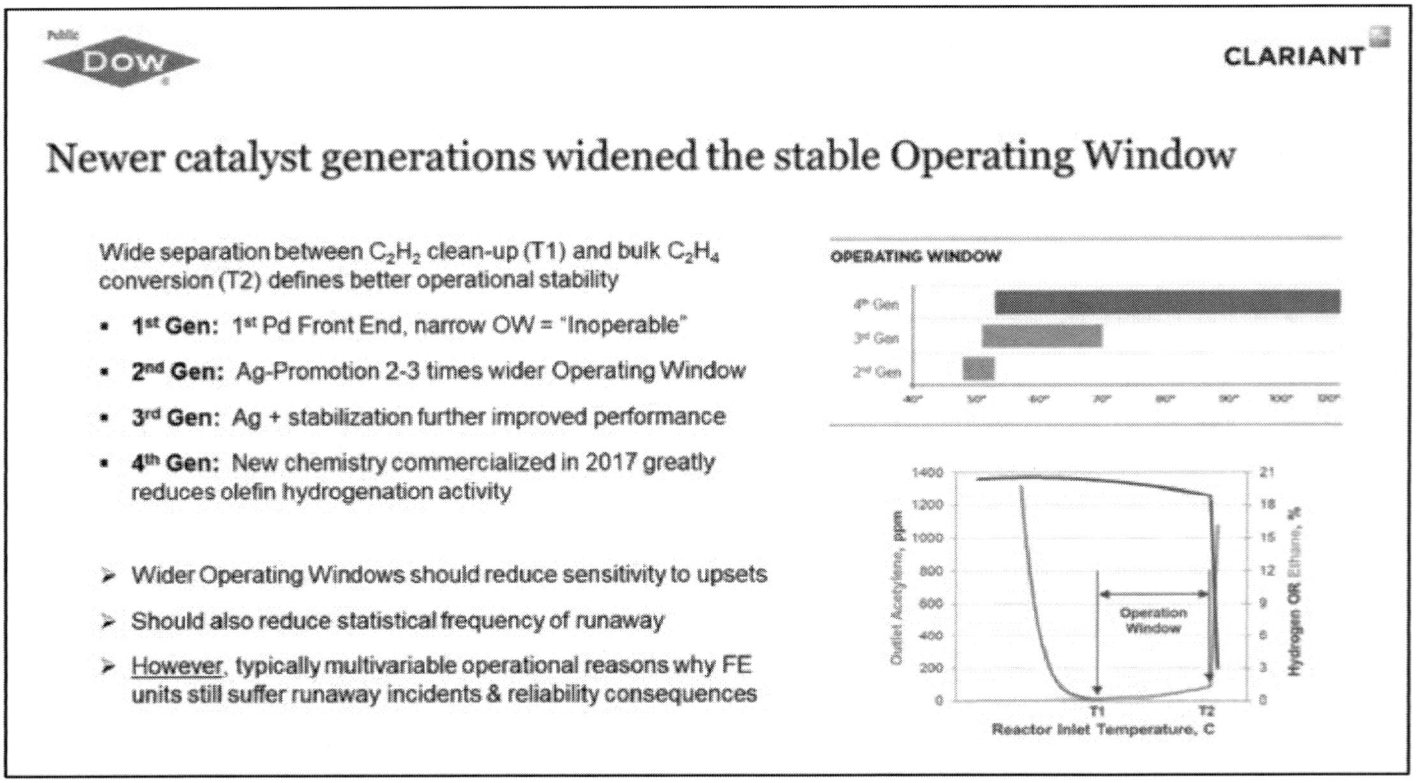

Catalyst performance is often discussed in terms of %Selectivity at C2H2 cleanup. However, the width of the "Operating Window" separating C2H2 cleanup from ethylene runaway is much more important for a high onstream factor and reliable production.

Please note that the lab-measured "Operating Window" is test-specific. Direct comparisons require a single apparatus and protocol.

Of note, the latest catalyst generation offers an extremely wide operating window, adding insurance against runaways in multiple scenarios.

Pd Catalysts have improved greatly over 50 years. They are much easier to operate, and should be statistically more reliable. However plants continue to experience runaways for a variety of mainly operational reasons.

DOW

CLARIANT

Operating Window is a moving target, especially sensitive at lower CO

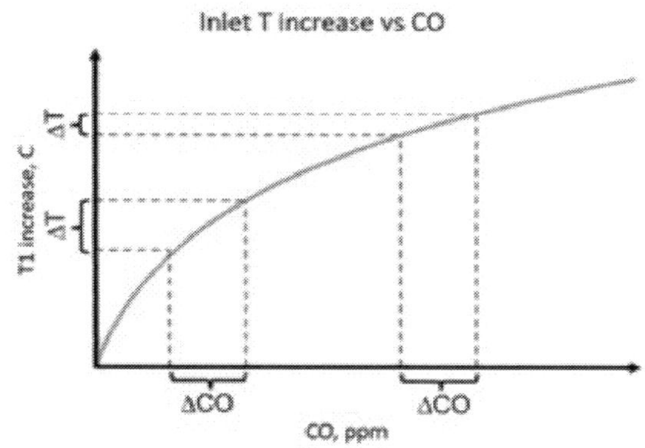

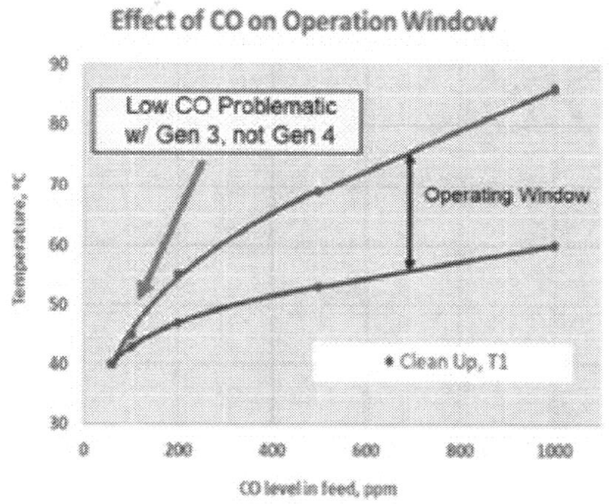

The left chart illustrates the minimum required temperature for acetylene cleanup as a function of absolute CO. Typical catalysts are particularly sensitive at low CO levels. Note the non-linear relationship. The temperature change required for a given change in CO is reduced at higher absolute CO level. Once the active sites are mostly covered, higher CO has a diminishing impact.

The right chart adds the impact of CO adsorption on the active sites and its inhibition of ethylene reaction and runaway. Note that the "Operating Window" between full acetylene cleanup and ethylene runaway gets significantly narrower at lower absolute CO levels.

Taken together, at lower CO levels the operating window is smaller, and moving more.

By extension, a new ethane cracker with few but large furnaces, enhanced tube metallurgy, and low cracked gas CO levels will be much more sensitive to transients than is an older liquids cracker with many small furnaces and typically 3-5 times higher CO levels.

Consequently, selection of an appropriate catalyst is especially critical to success in such units.

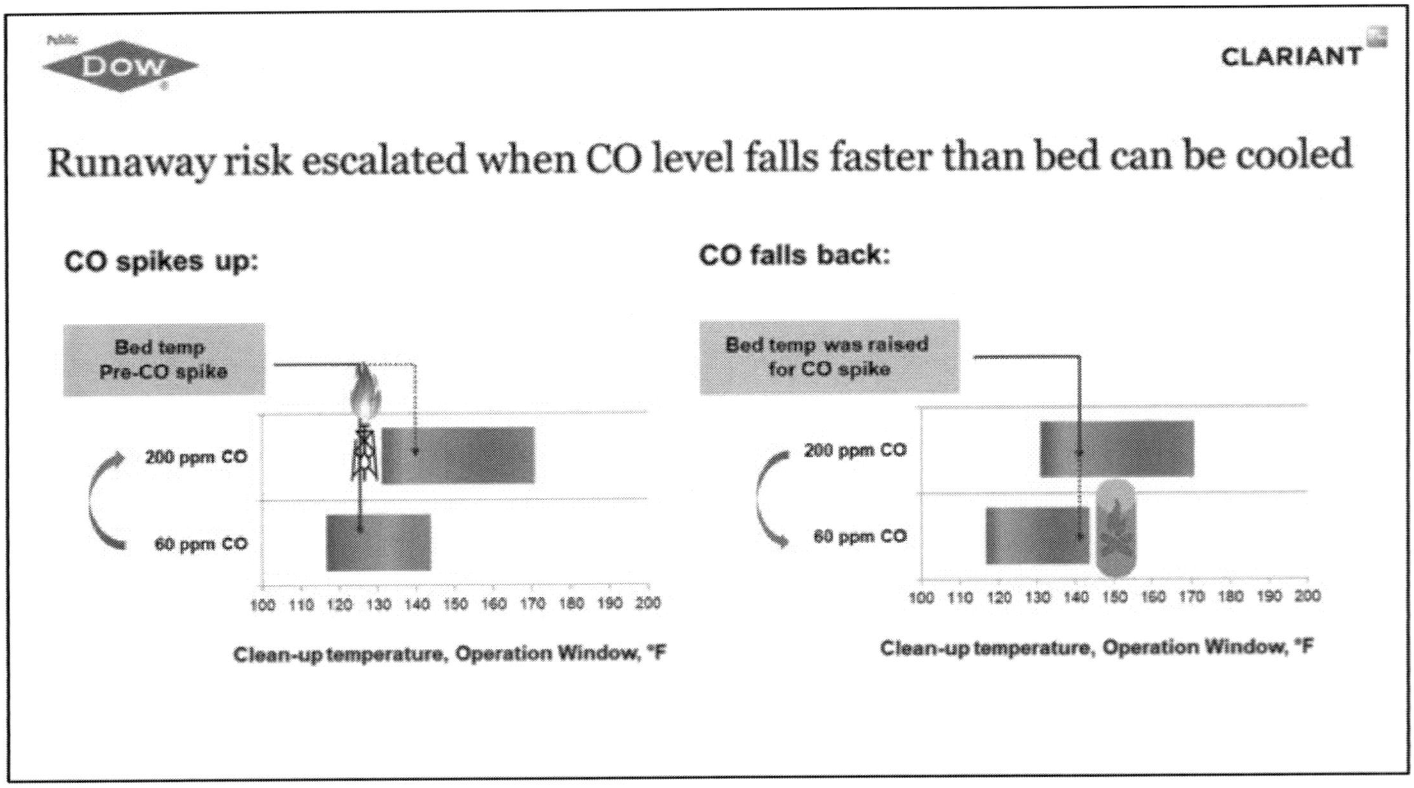

This slide demonstrates the challenge when furnaces are not stable for typical third generation FE catalysts.

In the left chart, the lower bar shows a reactor at a low 60 ppmv CO, with the weighted average bed temperature optimized at 126 oF.

In the upper bar, a CO spike to 200 ppmv widens the “Operating Window”, but also moves it to the right. With activity instantly reduced, a higher temperature is immediately required. C_2H_2 conversion and exotherm are reduced, making the situation worse. Flaring will likely be required until the beds can be heated up to regain specification.

The right chart shows situation reversing, and the more dangerous problem. After specification is regained at a re-optimized 140oF, when CO levels drop back down, typical 3rd gen catalyst will be too hot, overactive, with an increasing exotherm, and on the ragged edge of a runaway. One hopes the CO reversal was slow enough to allow incrementally cooling the catalyst to prevent both an outage and catalyst damage.

It is difficult for the operators to decide what to do, and when to do it, when the plant is unstable. This puts a high premium on effective operator training, and on managerial pre-discussion of operational strategy for CO swings.

Start-up "best practice" – normal procedure, but a high risk transient

- Developed after historical industry "Experience" with early, more susceptible catalyst types
  - Sub-cool entire catalyst mass to well below reaction temperatures (<< 25°C / 80°F)
  - Passivate Catalyst with high dose CO injection (**Caution**: Personnel protection protocols)
  - Stabilize Plant: Cracked Gas flowrate > Minimum Flow, limiting Contact Time
  - Increase Cracked Gas CO (via furnaces: cut DMS or inject Methanol)
  - Pressurize w/ non-reactive gas (C_2H_6, CH_4, N_2) to dilute %H_2 during static Cracked Gas introduction
  - Fast streaming: final pressurization, pop open inlet XV, then exit XV, close bypass
- Not "Fail-Safe" due to a variety of causes

Despite evolutionary catalyst improvements, industry experience with repeated runaways has resulted in drafting increasingly conservative startup procedures.

Consensus Best Practice procedures use all the tools available to ensure no ethylene reaction occurs during startup. Starting with a cold bed, passivating the catalyst with CO, pre-pressuring with a non-reactive gas, and quickly putting a preferably high flow of cracked gas through the reactor set.

While achieving a perfect "all of the above" should not be strictly necessary, the consequences of a single runaway justify following a very conservative procedure.

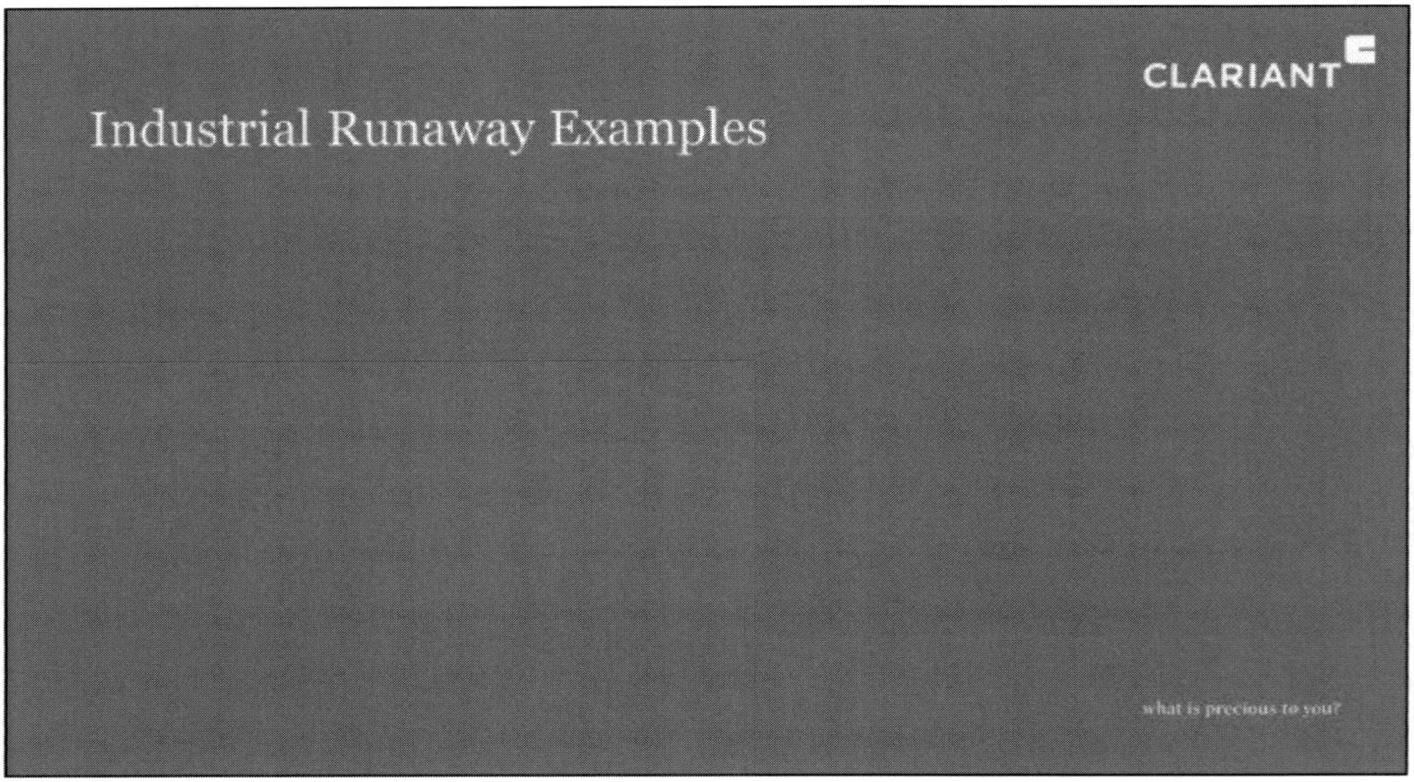

Okay, so we've got good catalyst in a well designed plant with well trained operators. What could possibly go wrong?

The following examples, just a few of many available, are offered as thought leaders toward your hazards and procedures reviews and training programs.

Public

DOW

CLARIANT

Example 1: Multiple faults and errors at new plant 1st startup

500 kta new Ethane cracker at 1st start-up: C2- Feed, 2 adiabatic reactors, Gen 3 catalyst

- Inlet XV failed to open after normal startup procedure including pressurize with Cracked Gas
- Delayed manual reactor trip & de-pressure during failed attempt to quickly locate & correct fault
    - 1st hot TI peaked at 207°C; "Faulty TI Indication?" Later finding DCS readout capped.
    - Runaway delayed >10 minutes by CO pre-dosing, but even inhibited K2 kinetics eventually won
- N_2 purge cooled beds only slowly. Too many N_2 consumers open vs limited supply capacity
- Attempted a quick restart after all Bed TI's dropped down, but immediate 2nd runaway in both beds

A new gas cracker on it's first startup attempt was following a basically good startup procedure. After the reactors were final pressured with cracked gas, the XV failed to open to establish forward flow.

At a decision point, insufficient training allowed an extended and ultimately futile attempt to correct the fault rather than immediately tripping the system as advised.

CO pre-dosing did delay the runaway, but infinite residence time eventually allowed even inhibited ethylene kinetics to proceed.

With the complex in startup mode, too many consumers were pulling on the N2 header, leaving an insufficient gas supply to quickly purge and cool the reactors.

After limited cooling of the reactors before a too-quick second attempt, suspected but unobserved hot pockets triggered a second runaway.

We could stop here for an interesting discussion, but let's press on, shall we?

Public

DOW

CLARIANT

Example 2: Major fluctuations in CO concentration

Older LPG cracker after major revamp turnaround
Wet Raw Gas Feed; Adiabatic reactors with Gen 2 catalyst

- CO cycled from 300 – 1000 ppmv, about once per shift, seemingly not furnace related
- Hot Spot - One TI in Layer 3 of Bed 2 initially running 2-3 oF higher.
  - CO spikes chased up by raising Bed Temps, attempting to stay on Ac specification.
- Multiple runaways initiated upon fast, large CO level drops, typically starting at Hot Spot
  - Rate of Change Alarms programmed into DCS TI's would have helped identify initiation

- Catalyst performance degraded via differential polymer laydown during multiple trips

Once there is one runaway, others become much more likely

At an LPG cracker in its initial startup post revamp, for no apparent reason, cracked gas CO levels were periodically oscillating from 300 to near 1000 ppm, resulting in the previously mentioned operational challenges. Transient feedstock oxygenate spikes were suspected but not confirmed.

From the beginning, one TI was observed running ~3 oF hotter than the others, presumably loading related. The TI positive deviation cyclically increased when CO levels trended down. There were no rate of change alarms programmed.

With the cold side operator busy with the usual startup challenges, he at first did not notice, and then was reluctant to trip the plant, when the hot spot temperature started to ramp up before the analyzer confirmed a CO level drop and before a high temp alarm sounded.

After several runaways originally stemming primarily from chasing CO levels, aggregate heterogenous polymer laydown made it very difficult to control the reactor. The catalyst had to be pre-maturely replaced.

CLARIANT

Example 3: Oxygen intrusion from poor purging of intercooler bypass line between the reactors

Adiabatic reactors in series with Gen 1 catalyst

- After cracked gas flow was established, a runaway occurred in second reactor when the bypass line was opened to warm the second reactor
- Run away caused by the residual Air occluded in the bypass line (red line)
  - CO oxidation reduced CO level, narrowing OW and raising activity
- Root cause: bypass valve was closed during reactor purging step

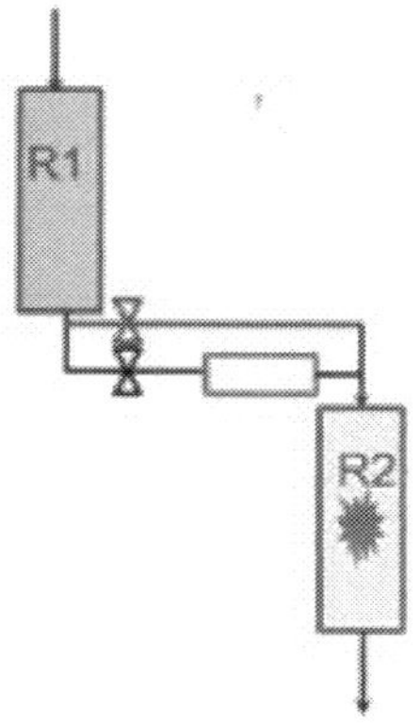

This is an example of a spike in an inlet Reactant, but not one which you might typically expect.

During a routine startup, cracked gas flow was fully established through the 2 reactors and the system was warming up.

The bypass line around the intercooler was cracked open to begin warming the 2nd reactor. Much to everyone's astonishment, a runaway occurred in the second reactor.

Eventually the Root Cause was established: After catalyst loading, the intercooler bypass line was blocked during system purging, leaving a dead leg of air.

Palladium is hydrogenation active, but it is also an excellent oxidation catalyst. And CO is a strong reducing agent, therefore easily oxidized to CO_2.

When the intruding oxygen combusted the CO to CO_2, inhibition of the ethylene reaction was lost due to the instantly lower CO levels, …. And off they went!

Public

DOW

CLARIANT

Example 4: Lesson learned – be specific!

1 MMTa new ethane cracker; 1st start-up; Gen 2 Ag-promoted catalyst

Massive runaway on initial pressurization with Cracked Gas.

- ➢Operating Instructions lacked specificity
  - "Pre-cool to Ambient" not sufficiently enlightening when Ambient = 55°C / 130°F"!!
- ➢New plant - Operator training adequate?
- ➢Start-up Technical Support: Catalyst Supplier onsite for reinforcement of Licensor Start-up training

Incident predated industry experience & adoption of improved start-up procedures

- Sub-cooling, CO pre-dosing, inert pressurization later all standard, Best Practice

Better: " Pre-cool to <25°C / 80 °F, + 4 hours to ensure no hidden hot zones after last TI. "

MMTa = Million Metric Tons per Annum

Now we come to attention to detail in Operating Procedures:

A new ethane cracker experienced a massive runaway during it's first startup attempt upon the start of pressurization with Cracked Gas.

As it turns out, generic guidelines which suggest pre-cooling the reactor to "Ambient" leave quite a lot open to interpretation, depending on whether one is from Canada versus from substantially hotter climates.

Lessons learned?

- Be specific, and require specificity of others
- Install piping to sub-cool the reactors
- Use high dose CO pre-injection to passivate the catalyst, and
- Pressurize to the highest possible pressure with a non-reactive gas before switching to Cracked Gas

Public

DOW

CLARIANT

Example 5: Furnace control is critical to process stability

1.5 MMTa Cracker, de-propanizer overhead w/ Adiabatic Reactors, Gen 3 catalyst;
Cracked Gas C_2H_2 and CO concentrations well below design

- Sudden drop in Inlet Acetylene → loss of reaction exotherm, cooling the catalyst bed
    - **Large transient** input, stemming from Cracking Furnaces control issue
- Inlet temperatures were quickly ramped up to "chase" mid-bed Ac concentration spike (loss of conversion)
- Rebound of inlet Acetylene concentration caused exotherm recovery, added to Inlet Temp addition

Note: low CO level narrows the Operating Window for Gen 1-3 catalysts

– Peak bed temperatures exceeding the Operating Window, T2 at the low prevailing CO level

MMTa = Million Metric Tons per Annum

Furnace instability can really cause issues. A Naphtha cracker was running with well below design CO levels, therefore also quite cool bed temperatures and a narrower operating window.

A sudden drop in acetylene concentration caused a loss of exotherm. C2H2 leakage started to rise because of the cooling beds.

About the time the cold side operators got the catalyst mass heated up to recover acetylene conversion, the furnace operators recovered their severity.

With increased acetylene exotherm added on top of the heated up catalyst at low CO, a runaway initiated.

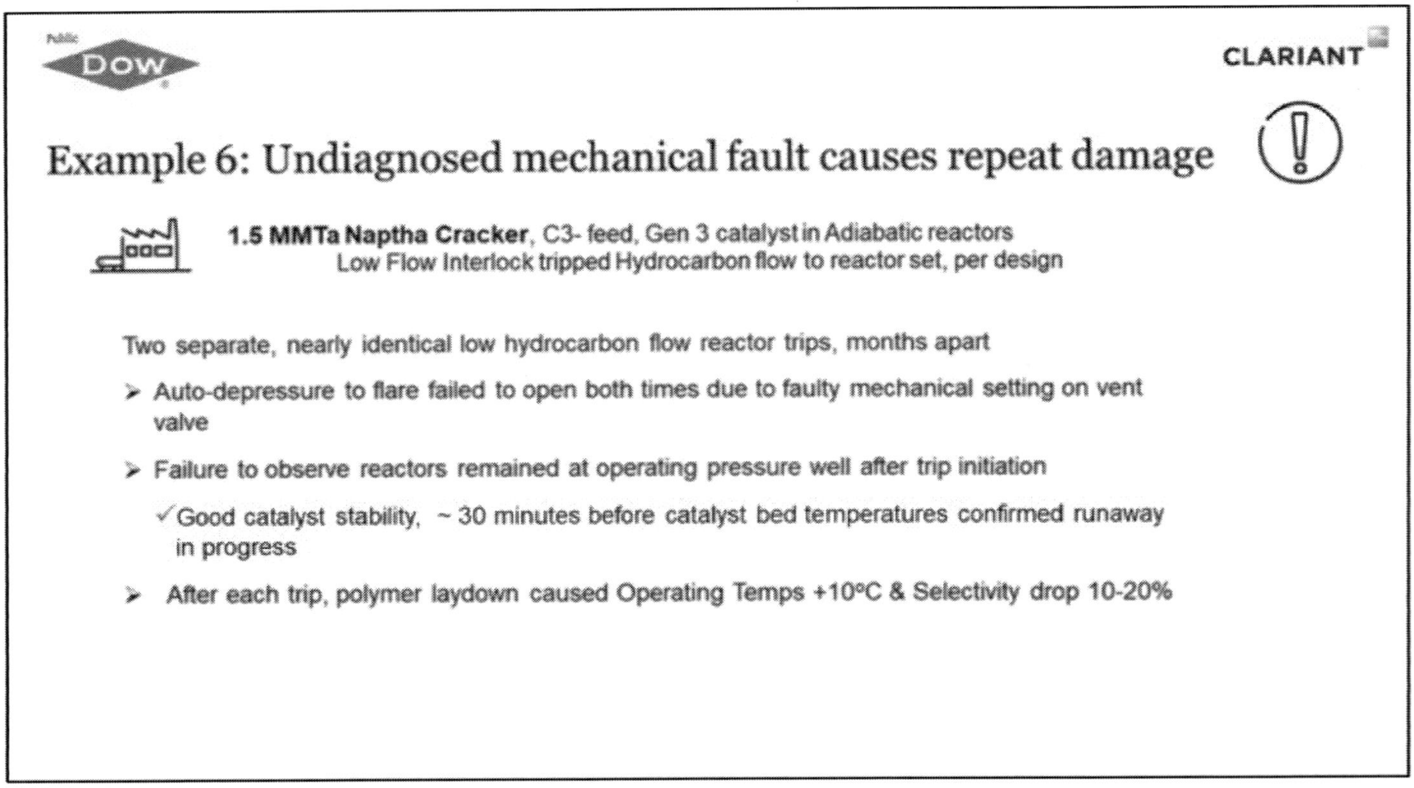

In another naphtha cracker, two separate instances of a low flow interlock tripping the reactors resulted in replicate post-trip runaways.

After the second, it was discovered that a faulty setting on the vent valve was not allowing the reactors to properly de-pressure to flare.

Good catalyst stability resulted in a significant time lag before the runaway commenced. However, prior to the too-late temperature alarms, the busy operators failed to notice that the reactor pressures had not dropped down to allow the vessel to be properly purged.

Worse, the defect was not properly diagnosed after the first incident, setting the plant up for a repeat performance and further catalyst impairment.

Leave aside the question as to why the replicate low flow trips, … Static cracked gas while at operating pressure and temperature will do it every time.

Public
DOW
CLARIANT

Example 7: Egg-shell catalyst dust is very active

Attrition-resistant carriers resist dust generation during transit and loading

- Best Practices: Loading Hopper w/ de-dusting apparatus; Loading Sock to be continually moved, blow down isothermal tubes, and clean dust from Reactor lower head after loading

Multiple experiences of exotherm between Bed Bottom & Exit Line Temperatures

- Generally sufficiently stable <u>during normal operation</u>, but loss of Selectivity

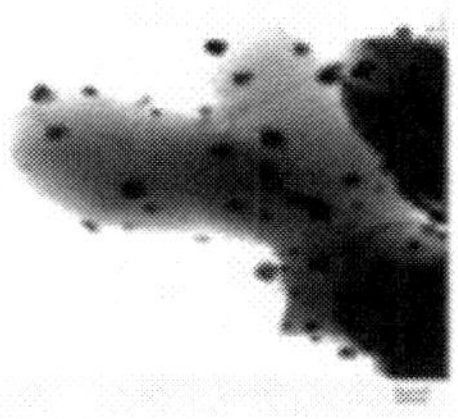

- Runaway can be triggered by catalyst dust in both:
  - Adiabatic reactors during typical challenging transient conditions
  - Isothermal reactor, eg during static pressuring up with Cracked Gas (stagnant, no forward flow)

It's important to note that selective hydrogenation catalysts generally have their Palladium active phase deposited more on the surface than throughout, as shown.

Because any concentration of catalyst dust is therefore highly enriched and very active, it is important to manage this risk.

Since different catalysts have different levels of attrition resistance, in general it's an excellent idea to ask your loading contractors to use a special loading hopper or chute made with screens in the lower sections which can be connected to a vacuum truck for de-dusting during loading.

This would be of particular help in case plant safety regulations prevent the previously mentioned presence of loading personnel in the reactor.

Because of dust generation primarily during unloading, it is best practice to clean the lower head and exit piping prior to buttoning up the reactors.

For isothermals, this should be after blowing each tube to both check pressure drop uniformity and de-dust the tubes given the free-fall that the catalyst is exposed to during loading.

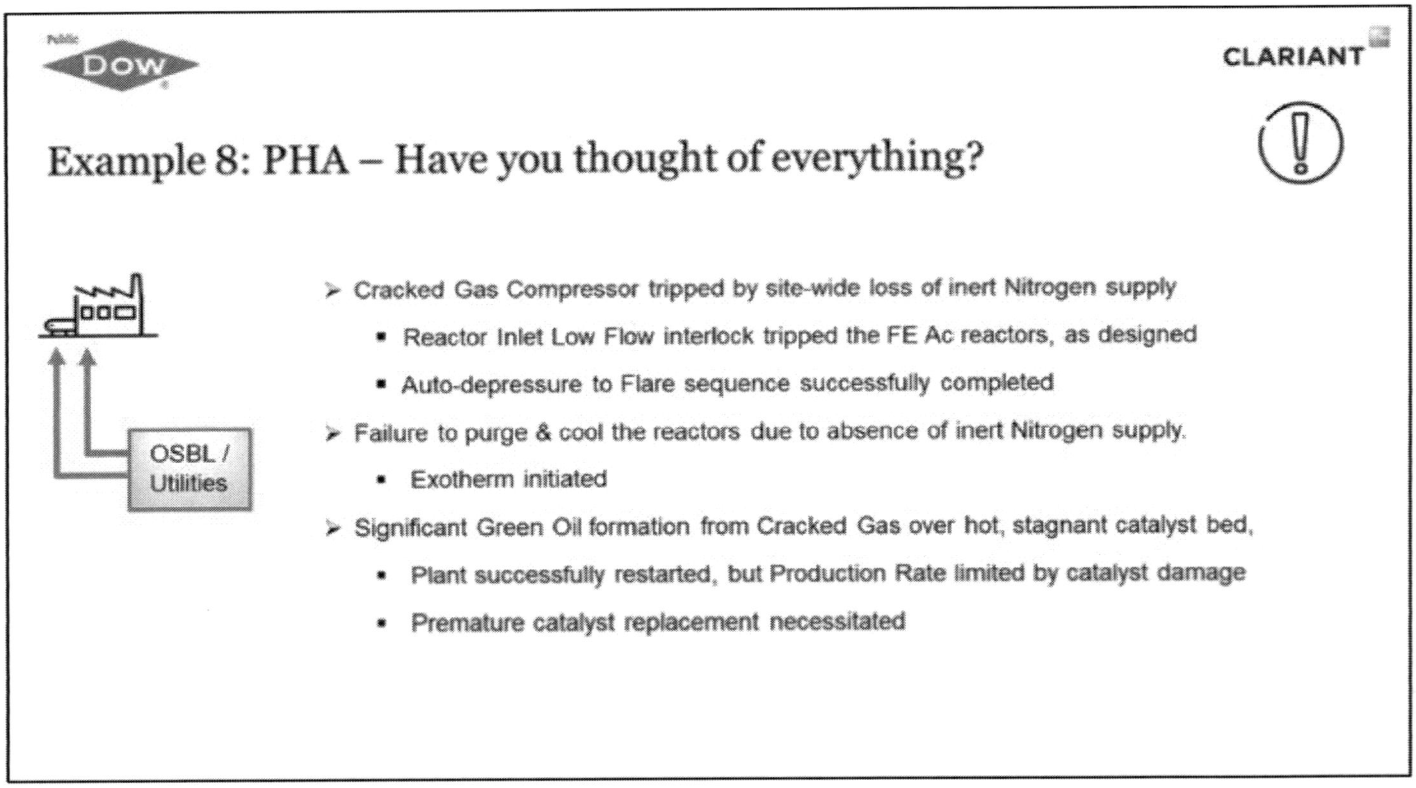

So, … Have you thought of everything in your Process Hazards Assessment?

A large cracker experienced a cracked gas compressor trip due to a site-wide loss of utility Nitrogen.

The reactors tripped and de-pressured as per design, but there was unfortunately no nitrogen supply available to purge and cool the reactors.

An exothermic runaway from the residual cracked gas commenced, generating significant green oil and polymer on the catalyst.

After the plant was re-started, serious catalyst damage prevented the plant from returning to full capacity, and planning for pre-mature catalyst replacement commenced.

Public

DOW

CLARIANT

Conclusions:

- It is amazing how many different ways a Runaway can be operationally triggered, even with very good catalyst
- In the absence of Spared Reactors, even one Runaway can seriously impair plant operability until premature re-fill
- "Avoidable" problems usually start with Decisions made during upsets, or mechanical / instrumental issues
- There are lots of good ideas in our industry on DCS alarm programming.
- Network – learn from as much industry "experience" as possible.
- Hazop your Front End reactor sections rigorously
- Train Operators and Engineers on root causes & especially scenario's with multiple simultaneous faults

- Boosted economic Selectivity is nice, but the real Value of Gen 4 catalyst performance =
  - Inhibited ethylene hydrogenation activity → excellent On-Stream Factor insurance against unpredictable scenarios

In conclusion, despite all the excellent engineering in modern plants and despite today's really very good catalysts, I am continually surprised at how many different ways a runaway can be triggered. These are just a few of the many examples we could have discussed today.

Since the majority of plants don't have spare reactors, and since FE catalyst can clearly be seriously damaged by a runaway, this is an important topic for Producers. Mitigation of the risks requires a multi-pronged approach to address both safety and operating reliability.

If not already done, it's suggested to consult with your Licensor on advanced reactor trip logic and controls. When possible, also network with colleagues to gain as much as possible from industry experience.

Rigorous Hazops are of course critical to plant reliability. Consider also periodic re-training of Operators on catalyst fundamentals toward improving their decision-making during the not always predictable and usually complex plant transients and system failures.

CLARIANT

Thank you

ACKNOWLEDGEMENTS:

CO-AUTHORS: ERNIE FRANK & YU LIU, DOW RESEARCH

CONTRIBUTOR: DAVID MULLENIX, SASOL

QUESTIONS TO:

CARLOS.VIANNA@CLARIANT.COM

DARREN.ADAMS@CLARIANT.COM

what is precious to you?

On behalf of the authors, we hope this presentation provided interesting information which will be useful to Front End producers in their process risk awareness and mitigation efforts.

Carlos and I will be happy to address any questions, now or via email at your convenience.

Thank you!!

AUTHOR INDEX